Dietary Fiber

Chemistry, Physiology, and
Health Effects

Dietary Fiber

Chemistry, Physiology, and Health Effects

Edited by
David Kritchevsky
The Wistar Institute
Philadelphia, Pennsylvania

Charles Bonfield
Astra Associates, Inc.
McLean, Virginia

and
James W. Anderson
Veterans Administration Medical Center
Lexington, Kentucky

Plenum Press ● New York and London

Library of Congress Cataloging-in-Publication Data

George Vahouny Fiber Conference (1988 : Washington, D.C.)
    Dietary fiber : chemistry, physiology, and health effects / edited
by David Kritchevsky, Charles Bonfield, and James W. Anderson.
        p.   cm.
    "Proceedings of the George Vahouny Fiber Conference, held April
19-21, 1988, in Washington, D.C."--T.p. verso.
    Includes bibliographical references.
    ISBN 0-306-43310-9
    1. Fiber in human nutrition--Congresses.   I. Kritchevsky, David,
1920-   . II. Bonfield, Charles.   III. Anderson, James W.
IV. Title.
    [DNLM: 1. Dietary Fiber--congresses.   WB 427 B349d 1988]
TX553.F53G46   1988
613.2'8--dc20
DNLM/DLC
for Library of Congress   90-6710
                                                              CIP

Proceedings of the George Vahouny Fiber Conference,
held April 19-21, 1988, in Washington, D.C.

© 1990 Plenum Press, New York
A Division of Plenum Publishing Corporation
233 Spring Street, New York, N.Y. 10013

Printed in the United States of America

GEORGE VARTKES VAHOUNY
(1932–1986)

George Vahouny's lamentably short but inordinately productive career can be characterized in the words of Albert Szent-Gyorgy: "Research is to see what everybody else has seen and to think what nobody else has thought." He was, above all, an innovator whose originality touched every area of biology in which he worked. The field of dietary fiber was one of his later career interests, but with characteristic skill and devotion he became one of its leading practitioners in a very short time.

Among George Vahouny's admirable attributes was the desire to share and disseminate knowledge. This led him to play a principal role in convening the first of these conferences in 1981. Its resounding success was followed by an even greater one for the second conference, held three years later. The symposium on which this volume is based was being formulated at the time of his death. We, his grieving and admiring colleagues, dedicate this volume to his memory.

Contributors

James W. Anderson • Metabolic Research Group, VA Medical Center, University of Kentucky, Lexington, Kentucky 40511

Elizabeth F. Armstrong • Wolfson Gastrointestinal Laboratories, Department of Medicine, University of Edinburgh, Western General Hospital, Edinburgh EH4 2XU, Scotland

Sheila A. Bingham • University of Cambridge and Medical Research Council, Dunn Clinical Nutrition Centre, Cambridge CB2 1QL, United Kingdom

John E. Blundell • Biopsychology Group, Psychology Department, University of Leeds, Leeds LS2 9JT, United Kingdom

Susan R. Bridges • Metabolic Research Group, VA Medical Center, University of Kentucky, Lexington, Kentucky 40511

Furio Brighenti • Department of Nutritional Sciences, Faculty of Medicine and Division of Endocrinology and Metabolism, St. Michael's Hospital, University of Toronto, Toronto, Ontario M5S 1A8, Canada

N. J. Brown • Sub-Department of Human Gastrointestinal Physiology and Nutrition, Royal Hallamshire Hospital, University of Sheffield, Sheffield S10 2JF, United Kingdom

W. Gordon Brydon • Wolfson Gastrointestinal Laboratories, Department of Medicine, University of Edinburgh, Western General Hospital, Edinburgh EH4 2XU, Scotland

James R. Buchanan • Division of Orthopedic Surgery, Department of Surgery, College of Medicine, The Milton S. Hershey Medical Center, Pennsylvania State University, Hershey, Pennsylvania 17033

Victoria J. Burley • Biopsychology Group, Psychology Department, University of Leeds, Leeds LS2 9JT, United Kingdom

Richard J. Calvert • The U.S. Food and Drug Administration, Washington, D.C. 20204

T. Colin Campbell • Division of Nutritional Sciences, Cornell University, Ithaca, New York 14853

Nicholas C. Carpita • Department of Botany and Plant Pathology, Purdue University, West Lafayette, Indiana 47907

Marie M. Cassidy • Departments of Physiology and Biochemistry, The George Washington University Medical Center, Washington, D.C. 20037

Stephen Cunnane • Department of Nutritional Sciences, Faculty of Medicine and

Division of Endocrinology and Metabolism, St. Michael's Hospital, University of Toronto, Toronto, Ontario M5S 1A8, Canada

Dee A. Deakins • Metabolic Research Group, VA Medical Center, University of Kentucky, Lexington, Kentucky 40511

Martin Eastwood • Wolfson Gastrointestinal Laboratories, Department of Medicine, University of Edinburgh, Western General Hospital, Edinburgh EH4 2XU, Scotland

C. A. Edwards • Sub-Department of Human Gastrointestinal Physiology and Nutrition, Royal Hallamshire Hospital, Sheffield S10 2JF, England; *present address:* Wolfson Gastrointestinal Unit, Western General Hospital, Edinburgh EH4 2XU, Scotland

Hans N. Englyst • MRC Dunn Clinical Nutrition Centre, Cambridge CB2 1QL, United Kingdom

Fresia Fernandez • PHLS-CAMR, Porton Down, Salisbury, Wiltshire SP4 OJG, United Kingdom

S. E. Fleming • Department of Nutritional Sciences, University of California, Berkeley, California 94720

Emily J. Furumoto • Department of Foods and Nutrition, Purdue University, West Lafayette, Indiana 47907

Dennis T. Gordon • Department of Food Science and Nutrition, University of Missouri, Columbia, Missouri 65211

Peter Greenwald • Division of Cancer Prevention and Control, National Cancer Institute, National Institutes of Health, Bethesda, Maryland 20892

Wang Guangya • Institute of Nutrition and Food Hygiene, Chinese Academy of Preventive Medicine, Beijing, China

Kenneth W. Heaton • University Department of Medicine, Bristol Royal Infirmary, Bristol BS2 8HW, United Kingdom

Michael J. Hill • PHLS-CAMR, Porton Down, Salisbury, Wiltshire SP4 OJG, United Kingdom

Ikuo Ikeda • Laboratory of Nutrition Chemistry, Kyushu University School of Agriculture 46-09, Fukuoka 812, Japan

Katsumi Imaizumi • Laboratory of Nutrition Chemistry, Kyushu University School of Agriculture 46-09, Fukuoka 812, Japan

Lucien R. Jacobs • Department of Medicine, University of California, Los Angeles, and Section of Nutrition, Division of Gastroenterology, Cedars-Sinai Medical Center, Los Angeles, California 90048

Alexandra L. Jenkins • Department of Nutritional Sciences, Faculty of Medicine and Division of Endocrinology and Metabolism, St. Michael's Hospital, University of Toronto, Toronto, Ontario M5S 1A8, Canada

David J. A. Jenkins • Department of Nutritional Sciences, Faculty of Medicine and Division of Endocrinology and Metabolism, St. Michael's Hospital, University of Toronto, Toronto, Ontario M5S 1A8, Canada

Robert G. Josse • Department of Nutritional Sciences, Faculty of Medicine and Division of Endocrinology and Metabolism, St. Michael's Hospital, University of Toronto, Toronto, Ontario M5S 1A8, Canada

Chen Junshi • Institute of Nutrition and Food Hygiene, Chinese Academy of Preventive Medicine, Beijing, China

June L. Kelsay • Carbohydrate Nutrition Laboratory, Beltsville Human Nutrition Research Center, Agricultural Research Service, U.S. Department of Agriculture, Beltsville, Maryland 20705

Susan M. Kingman • MRC Dunn Clinical Nutrition Centre, Cambridge CB2 1QL, United Kingdom

David M. Klurfeld • The Wistar Institute of Anatomy and Biology, Philadelphia, Pennsylvania 19104

Scott A. Kripke • Department of Surgery and Otorhinolaryngology, University of Pennsylvania School of Medicine, and Medical Research Service, Veterans Administration Medical Center, Philadelphia, Pennsylvania 19104

David Kritchevsky • The Wistar Institute of Anatomy and Biology, Philadelphia, Pennsylvania 19104

Elaine Lanza • Cancer Prevention Research Program, Division of Cancer Prevention and Control, National Cancer Institute, National Institutes of Health, Bethesda, Maryland 20892

Anthony R. Leeds • King's College, University of London, London WC2B 5RL, England

Geoffrey Livesey • AFRC Institute of Food Research, Norwich Laboratory, Norwich NR4 7UA, United Kingdom

Tom Lloyd • Division of Reproductive Endocrinology, Department of Obstetrics and Gynecology, College of Medicine, The Milton S. Hershey Medical Center, Pennsylvania State University, Hershey, Pennsylvania 17033

Y.-F. Lu • Laboratory of Nutrition Chemistry, Kyushu University School of Agriculture 46-09, Fukuoka 812, Japan

Robert MacLennan • Queensland Institute of Medical Research, Brisbane, Australia

Anthony K. Mallett • The British Industrial Biological Research Association, Carshalton, Surrey SM5 4DS, United Kingdom

Judith A. Marlett • Department of Nutritional Sciences, University of Wisconsin-Madison, Madison, Wisconsin 53706

Banoo Parpia • Division of Nutritional Sciences, Cornell University, Ithaca, New York 14853

John D. Potter • Division of Epidemiology, School of Public Health, University of Minnesota, Minneapolis, Minnesota 55455

A. Venket Rao • Department of Nutritional Sciences, Faculty of Medicine and Division of Endocrinology and Metabolism, St. Michael's Hospital, University of Toronto, Toronto, Ontario M5S 1A8, Canada

N. W. Read • Sub-Department of Human Gastrointestinal Physiology and Nutrition, Royal Hallamshire Hospital, University of Sheffield, Sheffield S10 2JF, United Kingdom

James Robertson • Department of Animal Sciences, Cornell University, Ithaca, New York 14853

John L. Rombeau • Department of Surgery and Otorhinolaryngology, University of Pennsylvania School of Medicine, and Medical Research Service, Veterans Administration Medical Center, Philadelphia, Pennsylvania 19104

Ian R. Rowland • The British Industrial Biological Research Association, Carshalton, Surrey SM5 4DS, United Kingdom

Abigail A. Salyers • Department of Microbiology, University of Illinois, Urbana, Illinois 61801

Subramaniam Satchithanandam • Departments of Physiology and Biochemistry, The George Washington University Medical Center, Washington, D.C. 20037

Barbara Olds Schneeman • Department of Nutrition, University of California, Davis, California 95616

Robert R. Selvendran • AFRC Institute of Food Research, Norwich Laboratory, Colney Lane, Norwich NR4 7UA, United Kingdom

C. P. Sepple • Sub-Department of Human Gastrointestinal Physiology and Nutrition, Royal Hallamshire Hospital, University of Sheffield, Sheffield S10 2JF, United Kingdom

R. Gregg Settle • Department of Surgery and Otorhinolaryngology, University of Pennsylvania School of Medicine, and Medical Research Service, Veterans Administration Medical Center, Philadelphia, Pennsylvania 19104

Jon A. Story • Department of Foods and Nutrition, Purdue University, West Lafayette, Indiana 47907

Michihiro Sugano • Laboratory of Nutrition Chemistry, Kyushu University School of Agriculture 46-09, Fukuoka 812, Japan

Lilian U. Thompson • Department of Nutritional Sciences, Faculty of Medicine and Division of Endocrinology and Metabolism, St. Michael's Hospital, University of Toronto, Toronto, Ontario M5S 1A8, Canada

Hugh Trowell† • Makerere University, Uganda

George V. Vahouny† • Departments of Physiology and Biochemistry, The George Washington University Medical Center, Washington, D.C. 20037

A. Verena F. V. Verne • AFRC Institute of Food Research, Norwich Laboratory, Colney Lane, Norwich NR4 7UA, United Kingdom

Vladimir Vuksan • Department of Nutritional Sciences, Faculty of Medicine and Division of Endocrinology and Metabolism, St. Michael's Hospital, University of Toronto, Toronto, Ontario M5S 1A8, Canada

A. R. P. Walker • Department of Tropical Pathology, School of Pathology, University of the Witwatersrand and the South African Institute for Medical Research, Johannesburg, South Africa 2000

B. F. Walker • Department of Tropical Pathology, School of Pathology, University of the Witwatersrand and the South African Institute for Medical Research, Johannesburg, South Africa 2000

Thomas M. S. Wolever • Department of Nutritional Sciences, Faculty of Medicine, and Division of Endocrinology and Metabolism, St. Michael's Hospital, University of Toronto, Toronto, Ontario M5S 1A8, Canada

Sungshin Yeo • Department of Nutritional Sciences, University of California, Berkeley, California 94720

Chao Zhonglin • Institute of Nutrition and Food Hygiene, Chinese Academy of Preventive Medicine, Beijing, China

†Deceased.

Foreword

Twenty years ago the very idea of an international conference on the fiber contained in plant food would have been totally inconceivable. At that time fiber was generally viewed as an inert component of food of no nutritional value and consequently considered as a contaminant, the removal of which would enhance the purity of a product. It was measured by a now obsolete and almost worthless test introduced in the last century for veterinary rather than human nutrition, and what was measured was referred to as "crude fiber," containing part of the cellulose and lignin but none of the numerous components of fiber now known to play important roles in the maintenance of health.

There were a few lone voices prior to the last two decades who had extolled the laxative properties of the undigested portion of food, assuming that these were related to its irritant action on the bowel mucosa. In retrospect this was a total misconception, and "softage" would have been a more appropriate term than "roughage," since its presence insured soft, not irritating, colon content.

Of the modern workers, those who had entered the field over 20 years ago recognizing that fiber could have specific value include Southgate and Van Soest, who were analyzing its chemical nature; Walker, who became conscious of its relevance to health shortly after World War II; Trowell, who had hinted as its protective action against intestinal and other diseases related to Western culture; Cleave, who perceived the dangers of overrefining plant foods; and Eastwood, who was investigating the influence of fiber on bile acid metabolism in the late 1960s.

The 1981 and 1984 Washington International Conferences on Dietary Fiber were organized by David Kritchevsky and George Vahouny. The 1988 conference was convened under the name of the George Vahouny Fiber Conference in honor of the man who had contributed so much to the field and unfortunately died before the conference took place. Contributions to these conferences together with other meetings on dietary fiber have followed the path of progress that is common to much scientific endeavor. Initial clinical and epidemiologic observations provide the data on which tentative hypotheses of disease causation can be formulated. These can then be tested and modified, discarded or enforced, in the light of the progressive results of both clinical and experimental studies. The investigations then move to specific effects of particular components of fiber on individual disease processes. Only thus can hypotheses become theories and theories move into the realm of confirmed fact.

The 1984 Washington Fiber Conference was largely devoted to epidemiologic and clinical observations. The 1988 conference retained some clinical and epidemiologic content, but by that time experimental studies had made considerable progress. I could only be an interested spectator rather than an active participant at this stage. One valuable lesson that I learned, however, was that hard data accumulated through painstaking observation and careful recording usually remains secure, although its interpretation may repeatedly have to be revised or modified.

An analogy that may be used to illustrate the growth and development that has taken place in regard to the so-called "fiber hypothesis," and which is reflected in the progressive change in nature of these successive conferences, is the manner in which a pathologist examines a specimen microscopically. He begins with the overall view obtained through a low-power lens. By this means he selects particular areas for more detailed study through a high-power lens. The final study of particular characteristics is performed through an oil-immersion lens or more recently through an electron microscope.

Perhaps a more vivid and yet appropriate picture is to visualize methods of examining a wood. It can be observed from a balloon, and its situation with regard to the surrounding landscape can be determined. The next step might be to look at the trees and see how they relate to one another to constitute the wood. Going deeper into the study, each tree would be recognized as consisting of branches, and then concentrating on details of the pattern of the leaves from which each branch is made up could be an intriguing study. Finally, each leaf would prove a fountain of information in itself.

So it has been and continues to be with the story of dietary fiber. Some decades ago little more than woods were discerned; scientists have now got down to branches and have in some instances started on leaves.

The importance of the role that dietary fiber is now deemed to play in adequate nutrition is emphasized by a statement recently made by Donald McLaren (1988). He suggested that there had been three revolutions in the concept of adequate nutrition in recent history:

1. That some diseases could result from a deficiency of vitamins in the diet.
2. That good nutrition is not just prevention of deficiency but involves avoiding excessive intake of certain dietary components such as saturated fat, salt, and sugar.
3. That adequate dietary fiber, which is not an essential nutrient, could have a profound effect on the maintenance of health.

These sentiments put dietary fiber in a place of real importance in clinical nutrition.

The general acceptance of the role of fiber in the maintenance of health is perhaps best illustrated by the fact that a review of 65 highly authoritative recommendations on diet produced in 11 countries between 1965 and 1986 (Cannon, 1988) reveals near unanimity in recommending that diets in more economically developed countries should contain more fiber and starch and less fat, salt, and sugar. In all that made recommendations on fiber, there was unanimous agreement that it should be increased.

Future research seems likely to be focused on the special varieties of dietary fiber likely to be most protective against particular diseases, their recommended doses, and the best means of insuring these changes in the national diet. It could then be confidently anticipated that not only could the health of Western populations be significantly improved, but also that Third World populations could be warned against their greatest health hazard, the copying of Western lifestyles and their eating habits in particular.

Denis Burkitt

REFERENCES

Cannon, G., 1988, Diet and the food industry, *RSA J.* **May:**399–412.
McLaren, D., 1988, Fibre, fibre burning bright, *Med. Dig. (Asia)* **6:**1–158.

Denis Burkitt, Alec Walker, and Hugh Trowell

Preface

In the planning for the symposium on which this volume is based, we thought it necessary to begin to explore the field of dietary fiber beyond the available rich descriptive background. We have a fairly good idea of *what* dietary fiber can do, but the *where* and *how* require elucidation. Thus the opening session was devoted to the structure of plant cell wall material as well as to its analysis. The influence of fiber on nutrient absorption, including minerals and vitamins, was discussed in the next session, which was followed by expositions of the physiological effects of fiber as well as its effects on intestinal microflora. Satiety and obesity were the subjects of the fourth discussion. The last two sessions centered on conditions such as diabetes, gallstones, and hyperlipidemia, and on two fiber-related areas which are certain to grow in importance, namely, resistant starch and short-chain fatty acids. The last day of the meeting comprised a symposium on future directions of research in cancer—covering epidemiology, treatment, and experimentally induced tumors—sponsored by the National Cancer Institute.

We were also fortunate during this meeting to have the opportunity to bring together three of the pioneers in fiber work—Denis Burkitt, Alec Walker, and Hugh Trowell—and to offer a modest token of recognition for their trailblazing achievements.

David Kritchevsky

Contents

The Chemistry and Properties of Plant Cell Walls and Dietary Fiber

ROBERT R. SELVENDRAN and A. VERENA F. V. VERNE

1. INTRODUCTION

The role of dietary fiber (DF) in human nutrition and health has become increasingly topical in recent years as its physiological effects have been more widely appreciated and as its chemistry has been better understood. In our institute we have studied the composition, structure, properties, and organization of cell walls (CW) derived from edible plant organs and from fiber preparations used in clinical feeding trials. Our objectives were to define the physiological role of DF, to improve methods of CW and DF analysis, and to understand the factors that determine texture in fresh, cooked, and processed vegetables and in fruits during ripening.

Trowell *et al.* (1976) defined DF as all the polysaccharides and lignin in the diet that are not digested by the endogenous secretions of the human digestive tract. The definition covers both plant CW constituents and the 2% of polysaccharide food additives that are included in our diet. The additives are present as plant gums, algal polysaccharides, pectins, modified celluloses, and modified starches. They have similar structures to CW polysaccharides and have, accordingly, been used as model compounds to study the action of DF, although CW themselves provide the major source of fiber in the diet. For analytical purposes, DF refers mainly to the nonstarch polysaccharides and lignin we consume (Southgate, 1976).

Most of our DF intake comes from CW of fruits, vegetables, cereal products, and other seeds. The CW composition depends on the maturity of the plant organ on harvesting and also, in part, on its postharvest storage conditions. The principal DF components are complex polysaccharides, some of which are associated with polyphenolics (including lignin), and proteins. Noncarbohydrate components make up a

ROBERT R. SELVENDRAN and A. VERENA F. V. VERNE • AFRC Institute of Food Research, Norwich Laboratory, Colney Lane, Norwich NR4 7UA, United Kingdom.

small proportion of the CW (~5–10%), but nevertheless, some have a significant effect on the properties and physiological actions of DF. Examples are lignin and phenolic esters in wheat bran, cutin and waxes in leafy vegetables, and suberin in roots and tubers.

Parenchymatous tissues are the most important source of vegetable fiber because the vascular bundles and parchment layers of cabbage leaves, runner bean pods, asparagus stems, carrot roots, etc. are still relatively immature and only slightly lignified when the vegetables are harvested and eaten. The seeds of soft fruits such as strawberries form only a small proportion of the total DF complement of the fruit. Lignified tissues are of greater importance in cereal sources such as wheat bran and bran-based and dehulled oat products. Wheat bran contains the lignified outer layers of the grain and the aleurone layer. It makes a significant contribution to cereal fiber intake.

The main components of CW from different tissues of food plants and the structural features of some major CW polysaccharides are summarized in Tables 1 and 2 of Selvendran (1984). For an extended discussion on CW polymers of edible plants, see Selvendran (1983, 1985); for accounts of the chemistry and properties of DF see Southgate (1976), Southgate and Englyst (1985), Selvendran (1984), and Selvendran *et al.* (1987). For reviews of methods for analyzing CW see Selvendran and O'Neill (1987), and for analyzing DF see Selvendran *et al.* (1988). The above give references to a number of other major reviews and to publications in the CW and DF areas.

2. THE MAIN COMPONENTS OF CW AND DF

The proportions of the various types of polymer found in the CW vary according to the plant organ and its maturity. The percentage occurrence of the different polymer types is given in Table I for parenchymatous and lignified tissues of both vegetables and cereals. The table was compiled from the amounts of polymers solubilized during sequential extraction studies on CW of a range of edible plant organs. The traditional extraction sequence of hot water, oxalate, chlorite/acetic acid (the delignification may

TABLE I. Polymer Composition of the Cell Walls of Parenchymatous and Lignified Tissues of Edible Plant Organs[a]

	Fruits and vegetables		Cereal products	
	Parenchymatous	Lignified[b]	Parenchymatous	Lignified
Pectic polysaccharides	35–40	5	<0.5	<0.5
Cellulose	35	40	3–5	30–35
Hemicelluloses	10	25–30	80–85	45–50
Proteoglycans	5–10	} 5	} 10	5–10
Glycoproteins	5–10			—
Polyphenolics[c]	5	20–25	5	15

[a]The values are on a percentage dry weight basis and are approximate.
[b]The cell walls of lignified tissues of dicotyledons make only a small contribution to the DF content of plant foods.
[c]Polyphenolics include lignin and phenolic esters.

be omitted for parenchymatous tissues of cereals), and alkali of increasing concentrations was used to derive total amounts of pectic polysaccharides, hemicelluloses, cellulose, proteoglycans and glycoproteins, and polyphenolics. The structural features of the polysaccharides are discussed in Selvendran (1983).

The types of polymer present in the CW can be inferred from the overall sugar analysis of a destarched alcohol-insoluble residue or CW preparation. Hydrolyses should be performed using 1 M H_2SO_4 and 72% H_2SO_4 followed by dilution to 1 M acid strength (Saeman hydrolysis conditions), and the sugar analysis should include an estimation of uronic acid. Significant levels of uronic acid, galactose, and arabinose indicate pectic substances; xylose and glucose (obtained by 1 M H_2SO_4 hydrolysis) indicate hemicellulosic xyloglucan. Glucose released by Saeman hydrolysis derives from cellulose. It should be remembered that 15–20% of the pectic polysaccharides are usually associated with the α-cellulose residue. Detailed fractionation using mild, nondegradative conditions allowed three types of pectic substances to be identified and estimated: pectins bound within the wall matrix by bridging calcium ions only, pectins bound by both Ca^{2+} and ester cross links, and pectins associated with the α-cellulose. This type of detailed study may be useful in correlating DF composition with its effects in the alimentary tract. Tables II and III compare the results obtainable for potato cell wall material (CWM) using the nondegradative and traditional extraction procedures. Table IV gives the results of extracting beeswing bran CWM using the refined method only. The results are discussed in Section 2.3. Overall methylation analysis of the CW gives information about the nature of constituent polymers in terms of glycosidic link-

TABLE II. Sequential Extraction of Cell Wall Material of Potatoes, Refined Method[a]

Fraction	Percentage (w/w) recovered	Sugars in decreasing order of abundance and the major polysaccharides
CDTA, 0.05 M, 20°C, 2 × 2 hr	22	GalpA >> Ara, Gal, Glc, Rha *Slightly branched pectic polysaccharides of middle lamellae, cross linked by Ca^{2+}*
Na_2CO_3, 0.05 M, 1°C, 20 hr	14	GalpA, Ara, Gal, Glc >> Rha *Highly branched pectic polysaccharides of primary walls, cross linked by Ca^{2+} and ester bonds*
Na_2CO_3, 0.05 M, 20°C, 2 hr	6	GalpA, Ara, Gal, Glc >> Rha *Very highly branched pectic polysaccharides, cross linked by Ca^{2+} and ester bonds*
KOH, 1 M, 20°C, 4 hr	6	Glc, Ara, Xyl, Gal, GalpA >> Rha *Xyloglucans and small amounts of highly branched pectic polysaccharides*
KOH, 4 M, 20°C, 2 hr	3	Glc, Ara, Xyl, Gal, GalpA >> Rha *Mainly xyloglucans*
α-Cellulose residue	40	Glc, Gal, GalpA, Ara, Rha *Mainly cellulose and cross-linked, highly branched pectic polysaccharides*

[a]P. Ryden and R. R. Selvendran, unpublished results.

TABLE III. Sequential Extraction of Cell Wall Material of Potatoes,
Traditional Method[a]

Fraction	Yield (% w/w)	Sugars in decreasing order of abundance and the major polysaccharides
Hot water, 80°C, 2 × 2 hr	15	Gal, GalpA, Ara, Rha *Degraded pectic-galactans and -arabinogalactans*
(NH₄)₂C₂O₄, 0.5 M, 80°C, 2 hr	32	GalpA, Gal, Ara, Rha *Degraded pectic arabinogalactans*
KOH, 1 M, 20°C, 2 hr	10	Gal, GalpA, Ara, Glc, Rha *Degraded pectic galactans and some proteoglycans*
KOH, 4 M, 20°C, 2 hr	11	Glc, Gal, Ara, GalpA, Rha *Xyloglucans and small amounts of degraded pectic polysaccharides*
α-Cellulose residue	32	Glc, GalpA, Gal, Man, Ara, Rha *Cellulose and cross-linked pectic polysaccharides*

[a]Results taken from Ring and Selvendran (1981).

ages and degrees of branching. From this information it may be possible to predict the degradability of a fiber source by bacteria (Stevens and Selvendran, 1988).

We describe the CW of different plant tissues/organs in the following order: parenchymatous tissues of fruits and vegetables (dicotyledonous plants), parenchymatous tissues of cereal grains, and lignified tissues of wheat. Since lignified tissues make such a small contribution to the DF content of fruits and vegetables, their chemistry is not discussed here.

TABLE IV. Carbohydrate Composition of the Fractions Obtained by
Sequential Extraction of Beeswing Wheat Bran CWM.

Fraction	Yield (% w/w)	Major sugars (% w/w)[a]					
		Ara	Xyl	Man	Gal	Glc	Total
CWM		31	30	0.5	2.4	34	98
NaOH, 0.05 M, 2°C, 4 hr	2	1.3	1.2	—	—	1.4	3.9
KOH, 1 M, 2°C, 4 hr	10	36	43	3	2.3	5.7	90
KOH, 1 M, 2°C, 4 hr	8.6	Similar to above fraction					
KOH, 1 M, 20°C, 4 hr	6.7						
KOH, 4 M, 20°C, 4 hr	7.9	34	35	1.7	3.8	10	85
0.6% Chlorite/HOAc	2.6	Mainly Ara and Xyl					
KOH, 1 M, 20°C, 2 hr	5.4						
KOH, 4 M, 20°C, 2 hr	2.4	42	41	0.6	3.4	4.4	91
α-Cellulose	21	15	14	0.5	1.1	69	100

[a]All fractions contain ~5% GlcpA. Results taken from DuPont and Selvendran (1987).

2.1. Parenchymatous Tissues of Dicotyledons

Pectic polysaccharides are the main type of polymer found in the parenchyma of dicotyledonous and certain monocotyledonous plants (such as leeks and onions). They are present to ~35–40% of the walls. The main hemicellulosic polymers of dicotyledon parenchyma walls are xyloglucans, and these make up ~7–10% by weight. Xyloglucans are closely associated with cellulose (indeed, they enable its dispersion throughout the primary wall matrix) and are only released from the walls by strong alkali. The primary CW matrix also contains small but significant amounts of proteoglycans—complexes containing polysaccharide, protein, and some polyphenolic moieties—which appear to serve as cross-linking agents. Glycoproteins are also present in small amounts and likewise seem to cross link the wall polysaccharides. The hydroxyproline contents of both types of protein complex vary. Cell walls of suspension-cultured tissues such as potato and beetroot generally contain significantly high levels of hydroxyproline-rich glycoproteins, whereas parenchymatous tissues of the same vegetables have relatively small amounts. Interestingly, the CW of legume parenchyma, such as runner bean pods, also contain fairly high levels of hydroxyproline-rich glycoproteins.

The wall polymers are further cross linked by small amounts of polyphenolics and, in some instances such as sugarbeet fiber, by esters of ferulic acid, which are covalently bound to the pectic polysaccharides. The ferulic acid and other phenolic ester cross-linking agents cause autofluorescence in beetroot and sugarbeet parenchyma.

The bulk of the pectic polysaccharides of sugarbeet primary CW are cross linked to cellulose and therefore much less soluble in hot water than those of runner beans or onions. Cellulose is a major DF polymer and is composed of $\beta(1 \rightarrow 4)$-linked glucose residues. Strong hydrogen bonds exist between the C6 hydroxyl groups and glycosidic oxygen of adjacent polymer chains (see Fig. 1 of Selvendran 1983). These render cellulose insoluble in water. Cellulose is soluble in 72% sulfuric acid because sulfonation takes place at the C6 hydroxyl group, breaking the hydrogen bonds. Substitution of the hydroxyl group at C6 with xylose (or short oligosaccharides containing xylose) as in xyloglucans renders cellulose soluble in alkali and water, again because the hydrogen bonds are broken. In primary walls the cellulose microfibrils are "coated" with a layer of xyloglucans (bound by hydrogen bonds), and this enables the insoluble cellulose to be dispersed within the wall matrix. A proportion of the C6 hydroxyl groups can be chemically replaced with carboxyl groups to give carboxymethyl cellulose (CMC), which is water soluble. The water-holding capacity of cellulose is ~5 g water/g cellulose, whereas that of CMC is 50–60 g water/g CMC. This shows how the properties of DF can be altered.

Cross linking of various kinds and hydrogen bonding influence the extractability of CW polymers from CW. We shall show how polymers from CW of fresh potatoes can be extracted using two sets of aqueous inorganic solvents. Table II shows the extractability of CW polysaccharides by sodium cyclohexanediaminetetraacetate ($CDTA^-Na^+$) at ambient temperature followed by alkali of increasing strength until only the insoluble α-cellulose residue remains. The CDTA is a powerful chelating agent that can abstract calcium ions from the pectic polysaccharides. The calcium ions form ionic "bridges" between pectic polymers, and once Ca^{2+} is removed, those polysaccharides held by the

ionic bonds only are solubilized. Most of these pectic polysaccharides are highly meth-
ylesterified and slightly branched and originate in the middle lamellae region (see
Selvendran, 1985). The ratio of rhamnose to galacturonic acid is ~1 : 30–40 for this type
of pectin. Further extractions with very dilute Na_2CO_3 in the cold solubilized a large
proportion of highly branched pectic polysaccharides. This treatment ensured maximum
deesterification with minimum transeliminative degradation. The rhamnose : galacturonic acid ratio of these branched polysaccharides is ~1 : 10. Most of the pectins are
probably derived from the primary CW and appear to be solubilized only after "bridging" Ca^{2+} ions are removed and ester cross links between galacturonic and neutral
sugar hydroxyl groups (elsewhere in the wall matrix) are hydrolyzed. The two groups of
pectic polysaccharides differ structurally, having different degrees of branching of the
rhamnogalacturonan backbone and different lengths of arabinogalactan side chain (Redgwell and Selvendran, 1986; P. Ryden and R. R. Selvendran, unpublished results).

Strong alkali treatment solubilized significant amounts of xyloglucans, some small
amounts of highly branched pectic polysaccharides, and some proteoglycans. These
polymers are hydrogen bonded to cellulose but are released during strong alkali treatment. The alkali abstracts protons from the C6 hydroxyl groups of the $(1 \rightarrow 4)$-linked
glucose residues of cellulose. The existing hydrogen bonds are broken, the negatively
charged oxygen atoms are readily hydrated, and the microfibrils swell. The alkali-
insoluble α-cellulose residue still contains a significant amount of highly cross-linked
pectic polysaccharides and a proportion of the hydroxyproline-rich glycoproteins (Stevens and Selvendran, 1980; O'Neill and Selvendran, 1983; P. Ryden and R. R. Selvendran, unpublished results).

Table III shows the traditional sequential extraction of CW (Ring and Selvendran,
1981). Hot water and oxalate extractions simulate cooking conditions (with and without
salt). Hot water solubilized pectic polysaccharides with a very high galactose content,
indicating that β-elimination had taken place. It is likely that some of the pectic galactans associated with the α-cellulose residue (Table I) were degraded and solubilized. Hot
oxalate abstracted the "bridging" Ca^{2+} and solubilized additional degraded pectic
polysaccharides from the middle lamellae and primary CW. Molar KOH then solubilized
a significant amount of the degraded ester-cross-linked pectic galactans. Then 4 M KOH
solubilized the xyloglucans. The α-cellulose residue contained cellulose and cross-
linked, degraded, pectic polysaccharides. The two extraction sequences show that CW
polysaccharides have different solubility characteristics and are degradable to varying
degrees depending on the conditions used. This is important in the DF context, as we
shall show with reference to raw and cooked apples.

Fresh apples undergo little cell sloughing on ingestion and mastication, and the
hydrochloric acid in the stomach only solubilizes a very small proportion of the pectins.
Little hydrolysis of methyl-esterified pectins or ester-cross-linked pectins occurs in the
small intestine, and so the pectins of raw apples contribute little to the viscous properties
of the digesta. Cooking encourages cell sloughing (Selvendran, 1985), and a significant
proportion of the middle lamellae pectic polysaccharides are solubilized. These make
the digesta more viscous, and this may have therapeutic effects in controlling hyperglycemia (in diabetics) and hypercholesterolemia by minimizing movement of sugars
across the villi in the small intestine. The HCl in the stomach increases the solubility of

the degraded pectic polysaccharides by abstracting Ca^{2+} ions. The pH in the ileum is ~7–8 and unlikely to release the ester-cross-linked pectic polysaccharides of the primary CW. This is an oversimplification but serves to illustrate the importance of the form in which food is consumed in determining its efficacy as a dietary fiber source.

2.2. Cell Walls of Endospermous Tissues of Cereals

Cereal endosperm CW contain ~90% hemicelluloses—arabinoxylans and β-D-glucans—some of which may be associated with phenolics. They contain ~2–3% cellulose in close association with glucomannans (or mannans?). Pectic polysaccharides appear to be either absent or present in very small amounts. A third of the endosperm hemicelluloses are soluble in hot water, and significant amounts are soluble even in cold water. The absence of middle lamellae pectins accounts for these solubility characteristics. Most of the hot-water-insoluble hemicelluloses are soluble in molar KOH—the requirement for alkali shows that the hemicelluloses are bound within the walls by phenolic ester cross links.

The predominant hemicelluloses of wheat endosperm are highly branched arabinoxylans, which have small amounts of phenolics such as ferulic acid associated with them. In contrast, barley and oat endosperm CW are very rich in mixed-linkage β-glucans and relatively poor in arabinoxylans. The hot-water-soluble β-glucans of oat endosperm walls exhibit the viscosity-enhancing properties of pectins and certain food gums such as guar gum. The β-glucans may thus show the therapeutic effects of certain other DF sources. Viscous solutions of guar gum apparently slow glucose absorption in the ileum by interacting with intestinal mucosa.

2.3. Cell Walls of Beeswing Wheat Bran

Commercial wheat bran contains a range of tissues, including the pericarp with attached testa, aleurone layer, and some endosperm. We have studied the CW polymers of the beeswing bran layer, the major part of the pericarp, in considerable detail. Beeswing wheat bran is comprised of cuticle, epidermis, and hypodermis (DuPont and Selvendran, 1987). Detailed fractionation studies on the CWM revealed some interesting characteristics of the constituent polymers. These are discussed briefly. Table IV shows the carbohydrate composition of the fractions obtained by sequential extraction of beeswing wheat bran CWM. Mild alkali extraction (0.05 M NaOH, 20°C, 4 hr) solubilized very small amounts of the acidic arabinoxylans. Comparable conditions extracted primary CW pectic polysaccharides in significant amounts from onion and potato CW previously depectinated with CDTA (twice) at 20°C. The phenolic ester cross links of beeswing wheat bran are thus much stronger than the pectin ester cross links. Subsequent extraction of the beeswing bran CWM with molar KOH at 2°C and then 20°C solubilized significant amounts of arabinoxylans and small amounts of ferulic and p-coumaric acids. A large proportion of the solubilized acidic arabinoxylans are probably cross linked by phenolic esters, although a small proportion may be linked to lignin.

A proportion of the polymers that remained insoluble after 4 M KOH treatment were rendered alkali soluble after delignification. The acidic arabinoxylans released by alkali after delignification were probably ester cross linked to lignin via glucuronic acid or 4-O-Me-glucuronic acid.

Graded alcohol precipitation of the polymers solubilized by the first alkali treatment (1 M, 2°C, 4hr) followed by ion-exchange chromatography and methylation analysis showed three main types of hemicellulosic polymer. These are, in decreasing order of abundance, highly branched acidic arabinoxylans, slightly branched acidic arabinoxylans, and acidic arabinoxylan–xyloglucan complexes. The study showed that the heterogeneity found within the three classes of polymer is native to the walls and not a result of degradation during extraction and purification.

Cross linking of polymers within the wall matrix and their degrees of branching determine the rate of degradation of fiber by colonic bacteria.

2.4. Cell Walls of the Wheat Aleurone Layer

The walls of the wheat aleurone cells are ~2 μm thick and composed mainly of arabinoxylans, β-D-glucans, and small amounts of cellulose and glucomannan, which are highly cross linked by phenolic (e.g., ferulic) esters. The walls contain no lignin. A proportion of the arabinoxylans are soluble in hot water, and the rest require alkaline conditions. The arabinoxylans appear to be neutral unlike those of the beeswing bran layer. Glucuronic acid and/or 4-O-Me-glucuronic acid was found in very small amounts and in covalent association with xylose in the partial acid hydrolysates (R. R. Selvendran and S. E. King, unpublished results). The arabinoxylans of the aleurone layer are much less branched than the majority of the arabinoxylans found in the beeswing bran layer. The aleurone layer arabinoxylans have comparable structures to the slightly branched beeswing bran arabinoxylans, but the absence of large amounts of glucuronic acid indicates that they are not linked to lignin. The slightness of branching also contributes to their relative ease of degradation by fecal bacteria *in vitro*. The outer lignified layers are much less degradable.

3. FACTORS THAT INFLUENCE DEGRADATION OF DF BY FECAL BACTERIA

Although DF is not digested by the enzymes of the alimentary tract, it is degraded by colonic bacteria, which use the component monosaccharides as energy sources. For general references see Salyers (1985); Salyers *et al.* (1977); Cummings (1984, 1985); and Stephen and Cummings (1980).

Over 95% of DF is derived from CW of various tissues of edible plant organs (Selvendran, 1984). The extent to which colonic bacteria degrade the CW depends on the polymers present within the wall matrix, their general solubility, and the ways in which they are associated together. It is well known that DF increases fecal bulk (Kelsay, 1978; Eastwood and Brydon, 1985). The increased mass results from undegraded fiber, increased bacterial mass, and water retained by both fiber and bacteria. Stephen and Cummings (1980) concluded that bacteria account for 55% of fecal solids. This proportion is increased when the diet is supplemented with a readily fermentable fiber source such as cabbage or apple. A wheat bran supplement increases the proportions of undegraded fiber and water. Cummings *et al.* (1978) measured increases in fecal weight and decreases in transit time when fiber from single sources (20 g/day) was added to a standard British diet (22 g fiber/day). Bran increased fecal weight by 127% and cabbage, apple, and guar gum caused increases of 69, 40, and 20%, respectively. Bran also produced the greatest change in mean transit time. Our studies (Selvendran, 1984; Selvendran *et al.*, 1987) linked these figures with the types of polysaccharides present in the fiber sources. We then studied degradation by fecal bacteria of fiber preparations derived from apple, wheat bran, and alkali-treated wheat bran (Stevens *et al.* 1988; Stevens and Selvendran, 1988). The fecal incubate was centrifuged, and the pellet boiled in absolute ethanol for 10 min to give the insoluble substrate fraction in which the undegraded fiber remained. The supernatant was spun again at higher velocity to separate the bacteria (with some extracellular polysaccharide material) from the soluble substrate fraction.

Cell wall material was preferred for the experiments rather than the alcohol-insoluble residue (AIR) because the latter contains intracellular proteins and starch, which might provide a substrate on which the bacteria could multiply very quickly. In practice apple AIR and CWM, which contain 8% and 5% protein, gave similar results. In wheat bran, the level of intracellular proteins and starch is much higher in the AIR (15%), and the wall polymers are very highly cross linked and therefore inaccessible to bacteria. In this case AIR and CWM would show very different degradation patterns, and so only CWM was used.

Table V shows degradation of apple and wheat bran CWM with time. In the case of apple, degradation began slowly. There was little carbohydrate loss in the first 6 hr. Only the pectic arabinan began to be degraded, but the level of protein rose, showing that the bacteria had multiplied. After 24 hr most of the CW had been degraded. The proportion of protein rose from 7% to 64% with a fall in carbohydrate from 78% to 3%. There was little further degradation after 72 hr. Our studies showed that parenchymatous tissues were "completely" degraded by colonic bacteria regardless of the types of polymers present in the CW. We suggest that the bacteria degrade the pectins present in the α-cellulose and thus gain access to degrade the (insoluble) cellulose itself. In lignified tissues, the bacteria cannot degrade the cellulose, but where the tissue has been treated with alkali, the microfibrils are hydrated and swollen, the bacteria find access, and the cellulose is degraded.

The bran showed a rapid decrease in levels of arabinose, xylose, and glucose over the first 24 hr as a result of degradation of the arabinoxylans and β-glucans of the aleurone layer. On further incubation the bacteria began to degrade the cellulose of the pericarp layer slowly. Bran CWM was treated with 1 M KOH at 1°C for 2 hr, dialyzed, and freeze dried. After 40 hr incubation with the fecal inoculum, the aleurone layer was completely degraded as before, and, in addition, most of the arabinoxylans and cellulose of the pericarp layer were also degraded. The alkali-soluble acidic arabinoxylans were readily degraded by bacteria. The electron micrographs in Figs. 1 and 2 show portions of CW of untreated and alkali-treated pericarp layer of wheat bran after incubation with fecal inoculum. In the untreated sample the bacteria enter the cell, but the CW remain intact; however, in the alkali-treated sample, the walls have been (considerably) de-

TABLE V. Composition and Extent of Degradation of Dietary Fiber Substrates of Apple and Wheat Bran by Fecal Bacteria

Substrate	Incubation time (hr)	Insoluble substrate recovered[a] (g)	Protein[b] (g kg^{-1})	Total sugars recovered (mg g^{-1})[a]							
				Deoxyhexose	Ara	Xyl	Man	Gal	Glc	Uronic acid	Total
Apple CWM	0[c]	0.854	77	27.7	119.2	62.3	25.1	60.7	303.3	189	787.3
Apple AIR	6[d]	1.080	172	25.3	96.1	63.1	26.2	67.2	322.9	121	722.8
Apple CWM	24	0.117	640	3.0	1.3	1.9	2.3	2.4	18.5	5	34.4
Apple CWM	72	0.077	688	2.2	1.4	1.2	1.0	1.9	4.6	3	15.3
Bran CWM	0[c]	0.885	60	4.2	175.5	286.3	7.5	12.2	195.3	63	744.0
Bran CWM	24	0.703	176	5.2	122.6	125.3	3.9	9.8	137.0	47	450.8
Bran CWM	72	0.653	150	7.4	116.3	105.9	3.7	9.7	126.7	48	417.7
1 M KOH-treated bran CWM	40	0.298	526	4.9	17.1	11.3	1.2	3.4	26.4	6	65.4
1 M KOH-soluble bran CWM	40	0.142	624	2.9	2.7	0.6	0.5	1.7	8.0	2	18.4

[a]Values relate to 1 g of original substrate. Results taken from Stevens et al. (1988) and Stevens and Selvendran (1988).
[b]N × 6.25. The bulk of the protein in the bacterially degraded fractions was from adherent bacteria.
[c]Substrate and medium only.
[d]Subsequent experiment. Sugar values relate to 1 g of original substrate.

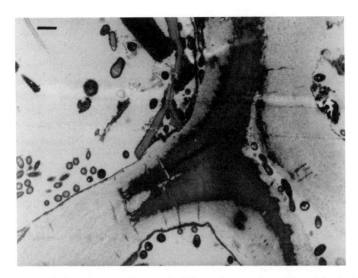

FIGURE 1. Transmission electron micrograph of the pericarp region of a particle of wheat bran after incubation with a fecal slurry for 72 hr at 37°C. Bar marker = 1 μm. (From Stevens et al., 1988.)

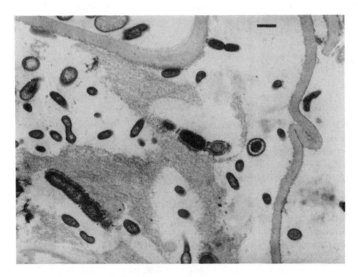

FIGURE 2. Transmission electron micrograph of the pericarp region of a particle of wheat bran after extraction with 1 M KOH for 2 hr at 2°C and subsequent incubation with a fecal slurry for 72 hr at 37°C. Some bacteria are closely associated with the degraded cell wall. Bar marker = 1μm. (From Selvendran et al., 1987.)

graded by the bacteria. After 40 hr incubation, the glucuronic acid level of the alkali-treated bran had dropped from 47% to 6%, showing that the highly branched arabinoxylans had been completely degraded following saponification of the ester linkages with lignin. Molar alkali treatment rendered the bran (including cellulose) completely degradable by bacteria. From this we inferred it is the degree of cross linking between polymers that prevents bacterial degradation and not the degree of branching of individual polysaccharides. The cross-linked pectins of sugarbeet fiber (results not shown) were much less readily degraded than the apple pectins and showed a degradation pattern comparable with the breakdown of arabinoxylans in the aleurone layer of wheat bran.

4. CONCLUSIONS

Vegetable and cereal fiber sources contain different types of polysaccharides. The solubility characteristics of the polymers determine their effects in the small and large intestines and influence their degradability by bacteria. Those vegetable fiber sources that are low in ferulic-acid-ester cross links are almost completely degraded by fecal bacteria. Highly cross-linked polymers (especially acidic arabinoxylans) and the cellulose in lignified tissues are resistant to degradation. On exposure to fecal bacteria, the polysaccharides of the aleurone layer of wheat bran are preferentially degraded, but prior alkali treatment of the CWM hydrolyzes ester cross links, and this increases degradation of polymers from both aleurone and pericarp layers. The cellulose is also rendered more susceptible to bacterial attack. There is a case for further clinical trials into different fiber sources. The information gained could be used to tailor, chemically or enzymically, fiber sources with physiologically beneficial properties.

REFERENCES

Cummings, J. H., 1984, Microbial digestion of complex carbohydrates in man, *Proc. Nutr. Soc.* **43**:35–44.

Cummings, J. H., 1985, Polysaccharide breakdown in the human colon, in: *Proceedings of the XII International Congress of Nutrition* (T. G. Taylor and N. K. Jenkins, eds.), John Libbey, London, Paris, pp. 185–187.

Cummings, J. H., Southgate, D. A. T., Branch, W., Houston, H., Jenkins, D. J. A., and James, W. P. T., 1978, Colonic response to dietary fibre from carrot, cabbage, apple, bran and guar gum, *Lancet* **1**:5–9.

DuPont, M. S., and Selvendran, R. R., 1987, Hemicellulosic polymers from the cell walls of beeswing wheat bran: Part 1, polymers solubilised by alkali at 2°C, *Carbohdr. Res.* **163**:99–113.

Eastwood, M. A., and Brydon, W. G., 1985, Physiological effects of dietary fibre on the alimentary tract, in: *Dietary Fibre, Fibre-Depleted Foods and Disease* (M. Trowell and D. Burkitt, eds.), Academic Press, London, pp. 105–131.

Kelsay, J. L., 1978, A review of research on effects of fiber intake on men, *Am. J. Clin. Nutr.* **31**:142–159.

O'Neill, M. A., and Selvendran, R. R., 1983, Isolation and partial characterisation of a xyloglucan from the cell walls of *Phaseolus coccineus, Carbohydr. Res.* **111**:239–255.

Redgwell, R. J., and Selvendran, R. R., 1986, Structural features of cell wall polysaccharides of onion *Allium cepa*, *Carbohydr. Res.* **157**:183–199.

Ring, S. G., and Selvendran, R. R., 1981, An arabinogalactoxyloglucan from the cell wall of *Solanum tuberosum*, *Phytochemistry* **20**:2511–2519.

Salyers, A. A., 1985, Catabolism of dietary fibre by human colonic bacteria, in: *Proceedings of the XII International Congress of Nutrition* (T. G. Taylor and N. K. Jenkins, eds.), John Libbey, London, Paris, pp. 195–198.

Salyers, A. A., West, S. E. H., Vercellotti, J. R., and Wilkins, T. D., 1977, Fermentation of mucins and plant polysaccharides by anaerobic bacteria from the human colon, *Appl. Environ. Microbiol.* **34**: 529–533.

Selvendran, R. R., 1983, The chemistry of plant cell walls, in: *Dietary Fibre* (G. G. Birch and K. J. Parker, eds.), Applied Science Publishers, London, pp. 95–147.

Selvendran, R. R., 1984, The plant cell wall as a source of dietary fiber: Chemistry and structure, *Am. J. Clin. Nutr.* **39**:320–337.

Selvendran, R. R., 1985, Developments in the chemistry and biochemistry of pectic and hemicellulosic polymers, *J. Cell Sci. [Suppl.]* **2**:51–88.

Selvendran, R. R., and O'Neill, M. A., 1987, Isolation and analysis of cell walls from plant material, in: *Methods of Biochemical Analysis,* Vol. 32 (D. Glick, ed.), John Wiley & Sons, New York, pp. 25–153.

Selvendran, R. R., Stevens, B. J. H., and DuPont, M. S., 1987, Dietary fiber: Chemistry, analysis, and properties, in: *Advances in Food Research,* Vol. 31 (C. O. Chichester, E. M. Mrak, and B. S. Schweigert, eds.), Academic Press, London, pp. 117–209.

Selvendran, R. R., Verne, A. V. F. V., and Faulks, R. M., 1989, Methods for analysis of dietary fibre, in: *Modern Methods of Plant Analysis,* New Series, Vol. 10: *Plant Fibers* (H. F. Linskens and J. F. Jackson, eds.), Springer-Verlag, Berlin, Heidelberg, pp. 234–259.

Southgate, D. A. T., 1976, The chemistry of dietary fiber, in: *Fiber in Human Nutrition* (G. A. Spiller and R. J. Amen, eds.), Plenum Press, New York, pp. 31–72.

Southgate, D. A. T., and Englyst, H., 1985, Dietary fibre: Chemistry, physical properties and analysis, in: *Dietary Fibre, Fibre-Depleted Foods and Disease* (H. Trowell, D. Burkitt, and K. Heaton, eds.), Academic Press, London, pp. 31–55.

Stephen, A. M., and Cummings, J. H., 1980, Mechanism of action of dietary fibre in the human colon, *Nature* **284**:283–284.

Stevens, B. J. H., and Selvendran, R. R., 1980, The isolation and analysis of cell wall material from the alcohol-insoluble residue of cabbage (*Brassica oleracea* var. *capitata*), *J. Sci. Food Agr.* **31**:1257–1267.

Stevens, B. J. H., and Selvendran, R. R., 1988, Changes in composition and structure of wheat bran resulting from the action of human faecal bacteria *in vitro, Carbohydr. Res.* **183**:311–319.

Stevens, B. J. H., Selvendran, R. R., Bayliss, C. E., and Turner, R., 1988, Degradation of cell wall material of apple and wheat bran by human faecal bacteria *in vitro, J. Sci. Food Agr.* **44**:151–166.

Trowell, H., Southgate, D. A. T., Wolever, T. M. S., Leeds, A. R., Gassull, M. A., and Jenkins, D. J. A., 1976, Dietary fiber redefined, *Lancet* **1**:967.

2

The Chemical Structure of the Cell Walls of Higher Plants

NICHOLAS C. CARPITA

1. INTRODUCTION

Without question, the fundamental chemistry of dietary fiber resides in the chemical structure of plant cell walls. These complicated networks of polysaccharide, structural protein, and phenolic substances are the basis of cell shape and integrity in plants and the slowly digested components that make up now-recognized important dietary supplements. Although the original definition of dietary fiber has been modified to include "soluble" as well as "insoluble" fibers, the recent advances in analysis of polysaccharide structure and survey of a wider range of higher plants have demonstrated that their chemistry is more complex than once imagined. Flowering plants alone comprise a broad range of orders that are grouped into two major evolutionary classes, the Dicotyledonae and Monocotyledonae (Fig. 1). In the Monocotyledonae, the order Graminales contains the cereal grasses that constitute the major foodstuffs of the world. In grasses, the primary cell walls are vastly different from all of the Dicotyledonae and other Monocotyledonae studied.

The chemistry known for cell walls of the vast, diverse orders represented in Fig. 1 is indeed small, and to catalogue variations in wall structure will require an enormous effort for the numerous commercial plants and some of the less exploited species. Wall structure not only differs markedly among species but also during the normal development within one species or even a single cell. The cotton fiber, for example, is a single cell that elongates about 3000 times its original diameter to about 1.5 inches long. During this elongation, the chemical structure of the primary cell wall is not unlike that of many other dicot species (Meinert and Delmer, 1977). When elongation is complete,

Abbreviations: AGP, arabinogalactan-protein; HS-GAX, highly substituted glucuronoarabinoxylan; GAX, glucuronoarabinoxylan; PGA, polygalacturonic acid; RG, rhamnogalacturonan.

NICHOLAS C. CARPITA • Department of Botany and Plant Pathology, Purdue University, West Lafayette, Indiana 47907.

FIGURE 1. The major orders of flowering plants. Dicotyledonae are all the orders to the left of the speculated progenitors, and the Monocotyledonae are to the right. The cell walls of Graminales and Palmales are fundamentally different from all dicots and many other monocots, including the Liliales. (After Benson, 1979.)

however, there is a massive synthesis of a single type of polysaccharide, cellulose, which forms a secondary wall that nearly fills the interior of the cell and constitutes at maturity over 98% of the dry mass of the fiber (Table I). Likewise, fiber, tracheid, and vessel cells of woody Gymnosperms and Angiosperms have primary walls that are very similar, but secondary wall formation produces xylans and cellulose that comprise most of the polysaccharide mass. In addition, lignin, a complex network of aromatic compounds fused together through ether and diphenyl linkages, can make up 20 to 30% of the mass of the mature cells (Table I). This cellulose–lignin matrix imparts the great rigidity of woody tissues.

It is now recognized that dietary fiber constitutes an array of polymers assembled into a complex matrix imparting different properties on different kinds of fibers. This review generalizes the chemical structure of primary cell walls of dicots and graminaceous monocots that may constitute the basis of at least some of these different properties. One must also bear in mind that the cell walls are made by living organisms, and as the cell walls form, they are constantly altered to conform to a specific developmental pattern that results in the unique shape of any given species. Only a few of these morphogenic alterations have been documented or studied in any detail. This review is also from a biological rather than a strict chemical point of view of the various polysaccharides, and the dynamic changes in polymer structure are related to development of the plant cell.

In all eukaryotic cells, an internal membrane packaging system called the Golgi apparatus is central to the processing, sorting, and secretion of glycoproteins. Although the Golgi apparatus certainly performs these functions in plants, it is obviously different from that of animals because it also participates in the synthesis and secretion of the vast majority of the polysaccharides of the cell wall. With respect to the sheer mass of material synthesized, processing of polysaccharides is the major function of the plant Golgi apparatus.

TABLE I. Distribution of Cell Wall Components in Various Cell Types

Cell type	Percentage of cell wall (by weight)			
	Pectin (+protein)	Hemicellulose	Cellulose	Lignin
Cell cultures				
Sycamore maple[a]	53	24	23	—
Proso millet[b]	10	65	25	—
Tomato fruit[c]	55	15	30	—
Wheat straw[d]	—	40	30	20
Conifer wood[e]	—	30	40	30
Cotton fibers[f]	—	2	98	—

[a]Data from Keegstra et al., 1973.
[b]Data from Carpita et al., 1985.
[c]Data from Gross, 1984.
[d]Data from Higuchi et al., 1967.
[e]Data from Grisebach, 1981.
[f]Data from Meinert and Delmer, 1977.

In growing cells, at least 80% of the cell wall comprises noncellulosic polysaccharides, and these polymers are synthesized in association with the Golgi apparatus (Delmer and Stone, 1988; Fincher and Stone, 1981). These polymers, membrane bound in vesicles, are secreted to the plasma membrane surface, where coordinated assembly with the growing cellulose crystals forms a rigid but dynamic matrix. Cellulose and callose, which is a $(1\rightarrow3)\beta$-D-glucan* produced either by wounding or synthesized transiently in specialized walls of pollen tubes or cotton fibers (Bacic *et al.*, 1988; Maltby *et al.*, 1979), are the only known polysaccharides made at the plasma membrane (Anderson and Ray, 1978; Delmer and Stone, 1988; Fincher and Stone, 1981; Giddings *et al.*, 1980). Because cell expansion is controlled, at least in part, by interaction of cellulose and noncellulosic polymers, considerable effort has been devoted to determining their chemical structure and how this structure imparts dynamic properties to the expanding wall. Documentation of this chemistry has also sparked interest in how these polymers are synthesized, how they are organized for secretion to the exterior compartment, how they are assembled into a functional matrix, and how polymer structure is then modified after assembly. Unfortunately, there is an enormous lack of information on the chemical linkage structure of these polymers, the chemical factors that govern their solubility and assembly within secretory vesicles, and the chemical modifications of these polymers after export and incorporation into the existing wall. Some ideas are also presented in this report on how solubility of newly synthesized pectins and hemicelluloses might be controlled by chemical modification until they arrive at the sites of assembly in the cell wall. These properties may give additional insight to subtle diversity in plant fibers inherent in any developing tissue as well as a few ideas on how polysaccharides can be modified chemically to provide new, beneficial properties.

2. PRIMARY CELL WALLS OF DICOTYLEDONAE

Chemical structures of cell walls of many fruits and vegetables have been studied, primarily with respect to the structure of pectic polysaccharides. The primary cell wall, the wall capable of stretching to define ultimate cell shape, has been studied more extensively in cells in liquid culture, such as sycamore maple (*Acer pseudoplatanus*), carrot (*Daucus carota*), and tobacco (*Nicotiana tabacum*). From these studies, a generalized wall structure has emerged (Fig. 2A). Cellulose microfibrils, which are condensed crystals of about 30 to 50 linear $(1\rightarrow4)\beta$-D-glucan chains each about 1000 to 5000 units long, form the foundation matrix (Delmer, 1987). These microfibrils are coated with a monolayer of more complex "hemicellulosic" polymers held tightly by

---

*Nomenclature is based on the kinds of linkages that monosaccharides can make. For example, glucose is a hexose in which carbons 1 and 5 form a heterocyclic ring with oxygen. The oxygens (O) of carbons 2, 3, 4, and 6 are each available for linkage. The carbon 1 is the anomeric carbon on which the oxygen can form either an equatorial (β) or axial (α) configuration. The anomeric carbon is always in the linkage; hence, a $(1\rightarrow3)$ β-D-glucan is a polymer composed of glucose units in which all the anomeric carbons are attached to the O-3 of the next glucose. A $(1\rightarrow3)$ β-D-glucosyl unit is a member of a polymer attached through its anomeric carbon to the O-3 of any sugar, whereas a "3-linked" glucosyl unit is a member of a polymer that has any sugar attached to its O-3 position.

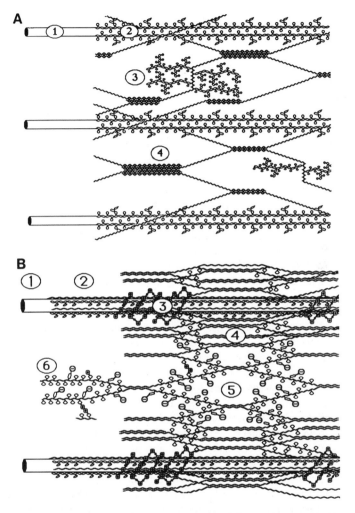

FIGURE 2. Comparison of models of primary cell walls of dicots and grasses. (From Carpita *et al.*, 1987.) A: Dicot cell wall. Cellulose microfibrils [1] are coated with a monolayer of xyloglucan [2]. Not shown is additional xyloglucan and arabinoxylan that may span the microfibrils (Hayashi *et al.*, 1987). The cellulose–xyloglucan framework is embedded in a gel matrix of polygalactu-ronic acids cross linked in part by Ca^{2+} [4]. Additional polymers containing mostly neutral sugars constitute the major side groups and are attached to the rhamnosyl units of rham-nogalacturonan [3]. Not shown is the hydroxyproline-rich extensin that can cross-stitch the cellulose fibrillar network (Cooper *et al.*, 1987; Lamport, 1986). B: Grass cell wall. Cellulose microfibrils [1] are coated primarily with glucuronoarabinoxylans (GAX) and some xyloglucan [2], a portion of which are immobilized by phenolic ether cross stitches [3]. Other acidic arabinox-ylans are hydrogen bonded to each other and may span the matrix [4]; additional GAX may define the pores of the matrix [5], replacing the function of pectic substances in these species (Carpita, 1984). The highly substituted GAX (Carpita and Whittern, 1986) may be the newly synthesized GAX cross linked loosely by diferulic acid [6]. Not shown is the developmental stage-specific β-D-glucan that is synthesized when cell expansion begins.

hydrogen bonds, and this framework is embedded in a gel of pectic polysaccharides (Bacic et al., 1988; Carpita, 1987).

2.1. Pectic Substances

Besides polygalacturonic acid (PGA), a simple polymer composed of $(1\rightarrow4)\alpha$-D-galactosyluronic acid units, there are more complex rhamnogalacturonans (RG I), which are contorted rodlike polymers owing to repeating -$(1\rightarrow2)\alpha$-L-rhamnosyl-$(1\rightarrow4)\alpha$-D-galactosyluronic acid units (Bacic et al., 1988; Jarvis, 1984). About one-half of the rhamnosyl units are substituted on the O-4 with side groups, and they comprise helical $(1\rightarrow5)\alpha$-L-arabinosyl chains and more highly branched chains with both 2,5- and 3,5-linked arabinosyl branched residues. Other significant side chains include $(1\rightarrow4)\alpha$-D-galactans and two kinds of arabinogalactans with mostly single nonreducing terminal arabinofuranosyl units attached to the O-3 of 4-linked galactan chains (type I) or the O-3 and O-6 of 3- and 6-linked galactan chains (type II) (Bacic et al., 1988; Fincher et al., 1983). Rhamnogalacturonan II also comprises a small portion of the pectin gel (Lau et al., 1985). This remarkably complex polymer is too low in amount to play a major structural role in the dicot cell wall, and the physiological role of this molecule is not yet known.

Pectic polysaccharides define the porosity of the cell walls of dicots (Baron-Epel et al., 1988), and this porosity may be governed by several factors. Principal cross-linking is provided by contortion of the helical $(1\rightarrow4)\alpha$-D-galactosyluronic groups from adjacent polysaccharides and condensation with Ca^{2+}, linking once soluble polymers into rigid, "egg-box" structures (Rees, 1977). These Ca^{2+}-cross-linked gels are dissolved by Ca^{2+}-chelating agents such as EDTA, EGTA, and ammonium oxalate. Actually, "pectic substances" are classically defined as material solubilized by such treatment regardless of its chemical composition. In addition to PGA, RG I is also solubilized by chelating agents, indicating that the repeating rhm-galA units do not constitute the entire polymer but that stretches of contiguous galA units are connected to the rhamnose-rich regions to provide regions in which Ca^{2+} cross bridging may occur (Jarvis, 1984). After exhaustive extraction of pectic substances with the chelators, additional uronic-acid-rich material is still attached to the matrix, which must be removed by dilute alkali such as Na_2CO_3, $NaBH_4$-Na_2BO_4, or 0.1 M KOH solutions. This second group of polymers also contains PGA and RG I but is probably held by alkali-labile ester linkages in addition to forming the Ca^{2+} cross bridges (Fry, 1986a). Hydroxycinnamic acids, such as ferulic and p-coumaric acid, may participate in cross bridging through formation of ester bonds with neutral sugar side chains of the RG I, and subsequent biphenyl or ether bond formation of these aromatic linkages may "wire" the pectins onto the hemicellulosic matrix to provide two functional domains of pectin in the primary wall. The extent of the Ca^{2+} cross bridging, of esterification through aromatic linkages, and even of branching and size of neutral sugar side chains can each influence gel flexibility, porosity, and interaction with hemicellulosic polymers (Fig. 2A).

The individual components of the pectic matrix are actually quite soluble polymers in the absence of Ca^{2+}, so control of solubility of newly synthesized polymers in secretory vesicles is not difficult. Formation of a gel in the cell wall could depend on control of Ca^{2+} flux or levels in the wall. Nakajima et al. (1981) have shown that the ratio of Ca^{2+} to Mg^{2+} in cell walls of pea increases almost sixfold between the elongating and mature cells of the stem of the seedling. However, fine control of cross linking may lie within the polymer. A large proportion of the carboxyl groups of the galactosyluronic acids may be methyl esterified during synthesis (Kauss and Hassid, 1967; Moustacas et al., 1986), and these relatively uncharged polymers may be the form that is secreted. Cleavage of the methyl esters by pectin methylesterase localized in the cell wall would render carboxyl groups available for cross linking with Ca^{2+} (Moustacas et al., 1986).

2.2. Hemicelluloses

Once the pectin polysaccharides are extracted or digested, the cellulose—hemicellulose framework remains. This framework provides part of the true rigidity of the cell wall. The principal hemicellulose of dicots is xyloglucan, a linear $(1\rightarrow4)\beta$-D-glucan chain substituted regularly at the O-6 with xylosyl units, some of which are substituted further to form galactosyl-$(1\rightarrow2)\beta$-D-xylosyl or fucosyl-$(1\rightarrow2)\alpha$-D-galactosyl-$(1\rightarrow2)\beta$-D-xylosyl units (Bacic et al., 1988). A repeating unit structure composed of hepta- and nonasaccharides has been proposed for dicot xyloglucan based on methylation (linkage) analysis and use of "restriction endoglycosidases" (Fig. 3A). These are enzymes that not only hydrolyze specific linkages but also require additional structural information from the polymer. Such enzymes sometimes can cleave a huge polysaccharide into smaller and smaller oligomers representative of the repeating unit structure that are amenable to sequence analysis (Bacic et al., 1988; Carpita, 1987). The substituted xylosyl unit of the nonasaccharide constitutes a "molecular hook" of yet unknown significance (Fig. 3A), although two reports indicate that it constitutes part of a feedback control system for hormone-induced elongation (McDougall and Fry, 1988; York et al., 1984). A smaller proportion of the hemicellulose is a glucuronoarabinoxylan, a linear $(1\rightarrow4)\beta$-D-xylan chain substituted on the O-2 and O-3 with terminal α-L-arabinofuranosyl and α-D-glucosyluronic acid units (Darvill et al., 1980).

Xyloglucans bind tightly to cellulose, and it is suspected that this conjugation occurs quickly at the plasma membrane surface and is coordinated with the formation of the cellulose microfibrils (Hayashi et al., 1987). Xyloglucans are fragmented enzymically during growth, and these fragments are soluble and often found in substantial amounts along with AGPs in the growth medium of cells in liquid culture (Aspinall et al., 1969; Stevenson et al., 1986). The larger xyloglucans extracted from the wall by alkali exist in both soluble and insoluble forms after neutralization of the alkali extract and subsequent dialysis, so considerable hydrogen bonding may occur between the large xyloglucan polymers as well. Xyloglucans are probably acetylated (Fry, 1986b), a modification that may prevent condensation of the xyloglucans in secretory vesicles until they reach sites of assembly.

2.3. Structural Protein

Protein is also a significant part of the walls of growing cells (Cooper et al., 1987; Lamport, 1986). Extensin, a structural hydroxyproline-rich glycoprotein, is the major constitutent in dicots. The "polyproline-II-like" tight helix is reinforced by an unusual

A

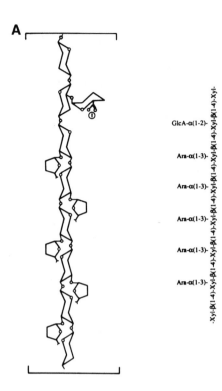

GlcA-α(1-2)-

Ara-α(1-3)-

Ara-α(1-3)-

Ara-α(1-3)-

Ara-α(1-3)-

Ara-α(1-3)-

-Xyl-β(1-4)-Xyl-β(1-4)-Xyl-β(1-4)-Xyl-β(1-4)-Xyl-β(1-4)-Xyl-β(1-4)-Xyl-

B

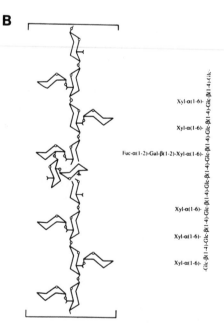

Xyl-α(1-6)-

Xyl-α(1-6)-

Fuc-α(1-2)-Gal-β(1-2)-Xyl-α(1-6)-

Xyl-α(1-6)-

Xyl-α(1-6)-

Xyl-α(1-6)-

-Glc-β(1-4)-Glc-β(1-4)-Glc-β(1-4)-Glc-

-Glc-β(1-4)-Glc-β(1-4)-Glc-β(1-4)-Glc-β(1-4)-Glc-β(1-4)-Glc-β(1-4)-Glc-

FIGURE 3. Principal hemicelluloses of cell walls of flowering plants. (From Carpita, 1987.) A: Molecular model of the hepta- and nonasaccharide repeating unit structure of dicot xyloglucan. B: Molecular model of the highly substituted glucuronoarabinoxylan of grasses.

amino acid, isodityrosine, which is formed by intramolecular fusion of two tyrosine residues of the glycoprotein (Fry, 1982). The isodityrosine is presumably formed by an extracellular peroxidase, and the linkage reinforces the protein into a rodlike molecule long enough to span the wall anticlinally, that is, at right angles to the cellulose microfibrils (Stafstrom and Staehelin, 1986). This "warp–weft" model literally forms a molecular cloth that gives the wall its tensile strength. Long stretches of the protein contain four contiguous hydroxyproline residues, each substituted with arabinose tetrasaccharides that coil around the rodlike protein and provide additional stability to the stiff protein (van Holst and Varner, 1984). Intermolecular isodityrosine linkages or covalent protein–protein or protein–polysaccharide linkages may contribute to formation of a rigid, inextensible wall (Cooper et al., 1987). Formation of this protein–polysaccharide network signals the end of cell expansion and may be the first stage of differentiation (Sadava and Chrispeels, 1973). The network is also quite calcitrant to extraction by both alkali and enzymic digestion (Shea et al., 1989).

3. PRIMARY CELL WALLS OF GRAMINEAE

Cell walls of the graminaceous monocots, the cereal grasses, are completely different from walls of dicots (Fig. 2B). A comparison of the relative abundance of specific kinds of polymers found in pectic and hemicellulosic substances is given in Table II. Chelator-soluble pectic polysaccharides constitute only 10% of the wall, and only about half of this material is PGA and RG I similar to dicot pectins (Carpita, 1988; Shibuya

TABLE II. Composition of the Primary Cell Walls of Dicots and Graminaceous Monocots

Component	Acer pseudoplatanus cell suspension culture[a] (% cell wall)		Zea mays coleoptile[b] (% cell wall)
Pectin	34		10
(1→4)α-D-Galacturonic acid		+++	+
Rhamnogalacturonan I		++	+
Rhamnogalacturonan II		+	?
Arabinan, galactan, AGP		++	+
Hemicellulose	24		65
Xyloglucan		+++	+
Glucuronoarabinoxylan		+	+++
Mixed-linkage β-D-glucan		?	++
Protein	19		10
Extensin		++	+/−
AGP		++	++
Cellulose	23		15

[a]Data from Keegstra et al., 1973.
[b]Data from Carpita, 1983; Carpita and Kanabus, 1988.

and Nakane, 1984). The other half is glucuronoarabinoxylan (GAX) similar to the hemicellulosic polymers extracted with more concentrated alkali. About 50 to 70% of the cell wall is hemicellulose, and a majority of these polysaccharides are GAX consisting of linear $(1\rightarrow4)\beta$-D-xylan backbones substituted primarily at the O-3 with terminal α-L-arabinofuranosyl units and at the O-2 with a smaller number of terminal glucosyluronic and 4-O-methyl glucosyluronic acid units (Bacic *et al.*, 1988; Carpita, 1983a; Carpita and Whittern, 1986) (Fig. 3B). Although dicots also have GAX, it is a slightly different polymer, for the dicot GAX has most of the arabinosyl residues on O-2 rather than O-3 of the xylosyl units (Darvill *et al.*, 1980). Small amounts of xyloglucan similar to that of dicots are also found; the α-D-xylosyl units are less frequent and more random along the glucan backbone such that no heptasaccharide repeating unit structure and no further elaboration of xylosyl units are observed (Wada and Ray, 1978; Bacic *et al.*, 1988).

3.1. Glucuronoarabinoxylans

The GAXs represent a heterogeneous group of polymers that vary in their resistance to alkali extraction and in the proportion of substitution of the arabinofuranosyl units (Carpita, 1983a,b, 1984, 1987). In growing cells of the maize seedling, a discrete fraction of GAX is extracted with weak alkali, and this GAX is highly substituted (HS-GAX). Six of every seven xylosyl units is a branch point, primarily for terminal α-L-arabinosyl units (Carpita, 1983b; Carpita and Whittern, 1986), and more concentrated alkali extracts additional GAX with just one-third the proportion of substitution (Carpita, 1983a,b, 1984). Periodate oxidation (Smith degradation) demonstrated that a substantial proportion of the HS-GAX has six contiguously substituted xylosyl units (Carpita and Whittern, 1986).

When similar fractions from walls of coleoptile cells are examined throughout elongation, the proportion of substitution falls markedly as cells fully elongate (Carpita, 1983b, 1984). Walls of mature leaves have little HS-GAX and increased amounts of GAX that is recalcitrant to alkali extraction and with less than one in six xylosyl units substituted with arabinosyl or glucosyluronic acid units (Carpita, 1983a). These data indicate that the HS-GAX is the soluble precursor to other GAX and that enzymic cleavage of arabinosyl units renders contiguously unsubstituted xylosyl units long enough to permit hydrogen bonding to cellulose or to other GAX (Carpita, 1983a,b, 1987) (Fig. 2B). In support of this hypothesis, a portion of radioactive HS-GAX prepared from coleoptile sections pulse-labeled with [^{14}C]glucose turns over to the recalcitrant GAX during subsequent chase with unlabeled glucose (Carpita, 1983b). Further, once extracted from the cell wall, HS-GAX can no longer bind to cellulose *in vitro*, whereas GAX and HS-GAX in which the arabinofuranosyl units were removed by mild acid hydrolysis each bind to cellulose (Carpita, 1983b). Hence, HS-GAX may be a soluble polysaccharide held to other wall polymers by ester linkages of diferulic acid, and once these esters are broken the polymer remains soluble.

3.2. β-D-Glucans

A polymer truly unique to the grasses is the mixed-linkage β-D-glucan, a large unbranched polymer composed of $(1\rightarrow3)\beta$-D- and $(1\rightarrow4)\beta$-D-glucosyl units in a ratio of about 1:3 (Bacic *et al.*, 1988; Carpita, 1987). A basic repeating unit structure has been proposed based on enzymic hydrolysis of the macromolecule with a restriction endo-glycosidase purified from *Bacillus subtilis* culture filtrates. This enzyme has an unusual property in that it is specific for (1→4)β-D-glucosyl linkages but requires a penultimate (1→3)β-D-glucosyl unit for full activity. Digestion of the glucan gives nearly equal amounts of cellotriosyl- and cellobiosyl-(1→3)β-D-glucose (Kato and Nevins, 1986; Staudte *et al.*, 1983), suggesting a repeating unit of cellotriose and cellotetraose con-nected by a (1→3)β-D-glucosyl linkage (Fig. 4A). This enzyme is quite useful in quantifying the amounts of β-D-glucan in mixtures of other polymers or intact cell walls (Carpita and Kanabus, 1988).

There are several enzymes of the cell walls of cereals that can also hydrolyze the β-D-glucan. Two exo-β-D-glucanases cleave the terminal glucosyl units (Hatfield and Nevins, 1986), and germinating grains have an endoglucanase similar in activity to that of *B. subtilis*, cleaving the polymer into the tri- and tetraoligomers (Mundy *et al.*, 1985). Further, Huber and Nevins (1981) isolated another endoglucanase from developing maize seedlings that exhibits more limited activity toward the β-D-glucan. With exo-glucanase activity suppressed with Hg^{2+}, the endoglucanase cleaves the macro-molecule into homogeneous fragments about 50 residues long that have the cellotriose-cellotetraose repeating unit structure (Huber and Nevins, 1981). The enzyme has been partially purified, and at least four contiguous (1→4)β-D-glucosyl units are necessary for full activity (Hatfield and Nevins, 1987), indicating that these special regions are engineered in the polymer about every 50 units. A small portion of contiguous (1→3)β-D-glucosyl oligomers in extensive digests with the endoglucanase mixtures in β-D-glucan from maize seedlings and barley and oat endosperm (Kato and Nevins, 1984; Staudte *et al.*, 1983).

From molecular models, Kato and Nevins (1984) suggested that three contiguous (1→3)β-D-glucosyl linkages would cause the molecule to fold back on itself, producing a "molecular sheet" (Fig. 4B). Further, this molecular sheet could interlace the cellulose microfibrils, which are spun in a shallow helix around an elongating cell (Fig. 4C), and this reinforcement may be the principal tensile component in the longitudinal plane. The β-D-glucans are probably quite heterogeneous, however, and the model proposed in Fig. 4C is only one of many possibilities.

A sizable portion of the β-D-glucans from soft grains are soluble in hot water (Woodward *et al.*, 1983), whereas those from developing maize seedlings require 2 M KOH to be removed from the wall (Carpita, 1983b). This behavior may reflect either differences in molecular structure of the β-D-glucan that alter its organization in the wall or differences in the strength of its interaction with cellulose microfibrils. Computer models demonstrate that randomly dispersed (1→3)β-D-glucosyl units among the cel-lotriosyl and cellotetraosyl units confer increased flexibility to the molecular form (Buliga *et al.*, 1986). The position and contiguity of the (1→4)β-D- and (1→3)β-D-units at specific intervals within the macromolecule must be examined more rigorously with a wider range of β-D-glucans. Determination of the fine structure of this group of polymers is truly necessary to obtain a clear picture of the heterogeneity of molecular form and to determine how these structures interact directly with the cellulose microfibrils.

The β-D-glucan is also unusual in that it is developmental stage-specific. In re-cently divided cells of the embryonal coleoptile or cells of proso millet in liquid culture,

A

-Glc-β(1-4)-Glc-β(1-4)-Glc-β(1-4)-Glc-β(1-3)-Glc-β(1-4)-Glc-β(1-4)-Glc-

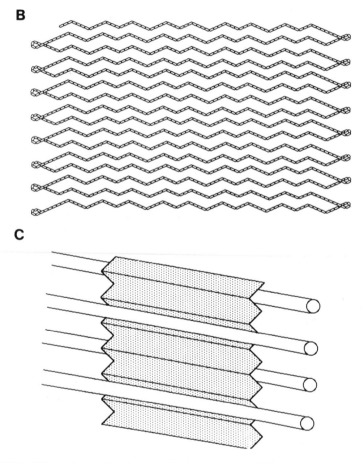

B

C

FIGURE 4. A: Molecular model of the repeating unit structure of the β-D-glucan of grasses. (From Carpita, 1987.) Structure was deduced by a combination of controlled enzyme digestion with an enzyme purified from *Bacillus subtilis* and methylation analysis of the oligomer products (Staudte *et al.*, 1983; Kato and Nevins, 1986). B: Proposed molecular "sheet" produced by regularly spaced contiguous (1→3) β-D-glucosyl units (Kato and Nevins, 1984). C: Proposed integration of the β-D-glucan sheet into the cellulose fibrillar network (Carpita, 1987).

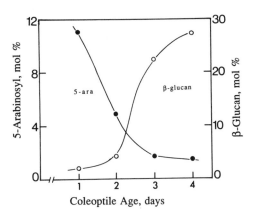

FIGURE 5. Comparison of the mole fractions of 5-arabinosyl units and β-D-glucan in the hemicellulosic fractions of cell walls during development of the maize coleoptile. Cell number is fixed between day 1 and 2 after germination, and the coleoptile elongates from 0.5 cm to over 4 cm between days 2 and 4.

the β-D-glucan is virtually absent (Carpita, 1984; Carpita *et al.*, 1985). Synthesis of β-D-glucan begins when cells expand, and amounts, determined either by linkage analysis or by hydrolysis with the *B. subtilis* endoglucanase, reach a level of nearly 30% of the alkali-extractable hemicellulosic fraction (Carpita, 1984; Luttenegger and Nevins, 1985 (Fig. 5). Further, β-D-glucan decreases markedly when the coleoptile senesces (Luttenegger and Nevins, 1985). Expanded cells of the endosperm of cereal grains are remarkably enriched in β-D-glucan, constituting over 70% of the entire wall (Stuart *et al.*, 1986).

3.3. Cross-Linking Molecules

Unlike dicots, the hydroxyproline-rich extensin is virtually absent from the walls of grasses. There are similar threonine- and proline-rich proteins found in maize pericarp and cells in culture (Hood *et al.*, 1988; Kieliszewski and Lamport, 1987) and soluble AGPs (the type II polymers containing the 3- and 6-linked galactan substituted with terminal arabinosyl units) (Fincher *et al.*, 1983), but the cross-linking function has mostly been replaced by esterified and etherified aromatic compounds (Carpita, 1986; Scalbert *et al.*, 1985). The diphenyl diferulic acid may cross link a small portion of GAX (Markwalder and Neukom, 1976), but most of the GAX is anchored by stronger covalent linkages than provided by simple esters (Carpita, 1983b, 1986). Like the dicot cellulose–extensin matrix, hemicelluloses of the cellulose–aromatic (grass lignin) matrix are more resistant to both extraction to alkali and enzymic digestion (Carpita, 1986), and, hence, their chemical activity may be altered in unpredictable ways.

4. SUMMARY

Substantial progress has been made in identifying the major polysaccharide, protein, and phenolic substances that make up the primary cell walls of typical dicot and grass species, but we actually know very little about how they are assembled into a functional matrix. New technologies, such as nuclear magnetic resonance, x-ray crystallography, and differential scanning calorimetry, have been useful to study the behavior of pectic and hemicellulosic polysaccharides in solution. We also know something about how cells chemically modify the polysaccharides to impart special properties needed for moving polymers from their site of synthesis to deposition at the growing wall. Because these modifications are performed by enzymes expressed at specific stages of plant cell development, there is promise that these modifications can be controlled to produce different kinds of polymers that may be more beneficial dietary fibers.

ACKNOWLEDGMENTS. Much of the work reviewed was the result of support from grant DCM-8415027 from the National Science Foundation and grant DE-FG02-88ER-13903 from the Department of Energy. Journal paper No. 11845 of the Purdue University Agricultural Experiment Station.

REFERENCES

Anderson, R. L., and Ray, P. M., 1978, Labeling of the plasma membrane of pea cells by a surface localized glucan synthetase, *Plant Physiol.* **61:**723–730.

Aspinall, G. O., Molloy, J. A., and Craig, J. W. T., 1969, Extracellular polysaccharides from suspension-cultured sycamore cells, *Can. J. Biochem.* **47:**1063–1070.

Bacic, A., Harris, P. J., and Stone, B. A., 1988, Structure and function of plant cell walls, in: *The Biochemistry of Plants,* Vol. 14 (J. Preiss, ed.), Academic Press, New York, pp. 297–371.

Baron-Epel, O., Gharyl, P. K., and Schindler, M., 1988, Pectins as mediators of wall porosity in soybean cells, *Planta* **175:**389–395.

Benson, L., 1979, *Plant Classification,* 2nd ed., D. C. Heath, Lexington, MA.

Buliga, G. S., Brandt, D. A., and Fincher, G. B., 1986, Sequence statistics and solution conformation of a barley (1→3),(1→4)-β-D-glucan, *Carbohydr. Res.* **157:**139–159.

Carpita, N. C., 1983a, Fractionation of hemicelluloses from maize cell walls with increasing concentrations of alkali, *Phytochemistry* **23:**1089–1093.

Carpita, N. C., 1983b, Hemicellulosic polymers of cell walls of *Zea* coleoptiles, *Plant Physiol.* **72:**515–521.

Carpita, N. C., 1984, Cell wall development in maize coleoptiles, *Plant Physiol.* **76:**205–212.

Carpita, N. C., 1986, Incorporation of proline and aromatic amino acids into cell walls of maize coleoptiles, *Plant Physiol.* **80:**660–666.

Carpita, N. C., 1987, The biochemistry of "growing" cell walls, in: *Physiology of Cell Expansion during Plant Growth* (D. J. Cosgrove and D. P. Knievel, eds.), American Society of Plant Physiology, Rockville, MD., pp. 28–45.

Carpita, N. C., 1988, Pectic polysaccharides of maize coleoptiles and proso millet cells in liquid culture, *Phytochemistry* **28:**121–125.

Carpita, N. C., and Kanabus, J., 1989, Chemical structure of the cell walls of dwarf maize and changes mediated by gibberellin, *Plant Physiol.* **88:**671–678.

Carpita, N. C., and Whittern, D., 1986, A highly-substituted glucuronoarabinoxylan from developing maize coleoptiles, *Carbohydr. Res.* **146:**129–140.

Carpita, N. C., Mulligan, J. A., and Heyser, J. W., 1985, Hemicelluloses of a proso millet cell suspension culture, *Plant Physiol.* **79:**480–484.

Cooper, J. B., Chen, J. A., van Holst, G.-J., and Varner, J. E., 1987, Hydroxyproline-rich glycoproteins of plant cell walls, *Trends Biochem. Sci.* **12:**24–27.

Darvill, J. E., McNeil, M., Darvill, A. G., and Albersheim, P., 1980, The structure of plant cell walls.

XI. Glucuronoarabinoxylan, a second hemicellulose in the primary cell walls of suspension-cultured sycamore cells, *Plant Physiol.* **66:**1135–1139.

Delmer, D. P., 1987, Cellulose biosynthesis, *Annu. Rev. Plant Physiol.* **38:**259–290.

Delmer, D. P., and Stone, B. A., 1988, Biosynthesis of plant cell walls, in: *The Biochemistry of Plants* (J. Preiss, ed.), Academic Press, Orlando, FL, pp. 373–420.

Fincher, G. B., and Stone, B. A., 1981, Metabolism of noncellulosic polysaccharides, in: *Plant Carbohydrates II. Encyclopedia of Plant Physiology, New Series,* Vol. 13B (W. Tanner and F. A. Loewus, eds.), Springer-Verlag, Berlin, pp. 68–132.

Fincher, G. B., Stone, B. A., and Clarke, A. E., 1983, Arabinogalactan-proteins: Structure, biosynthesis, and function, *Annu. Rev. Plant Physiol.* **34:**47–70.

Fry, S. C., 1982, Isodityrosine, a new cross-linking amino acid from plant cell-wall glycoprotein, *Biochem. J.* **204:**449–455.

Fry, S. C., 1986a, Cross-linking of matrix polymers in the growing cell walls of angiosperms, *Annu. Rev. Plant Physiol.* **37:**165–186.

Fry, S. C., 1986b, *In-vivo* formation of xyloglucan nonasaccharide: A possible biologically active cell-wall fragment, *Planta* **169:**443–453.

Giddings, T. H., Jr., Brower, D. L., and Staehelin, L. A., 1980, Visualization of particle complexes in the plasma membrane of *Micrasterias denticulata* associated with the formation of cellulose fibrils in primary and secondary walls, *J. Cell Biol.* **84:**327–339.

Grisebach, H., 1981, Lignins, in: *The Biochemistry of Plants,* Vol. 7 (P. K. Stumpf and E. E. Conn, eds.), Academic Press, New York, pp. 457–478.

Gross, K. C., 1984, Fractionation and partial characterization of cell walls from normal and non-ripening mutant tomato fruit, *Physiol. Plant.* **62:**25–32.

Hatfield, R., and Nevins, D. J., 1986, Purification and properties of an endoglucanase isolated from the cell walls of *Zea mays* seedlings, Carbohydr. Res. **148:**265–278.

Hatfield, R., and Nevins, D. J., 1987, Hydrolytic activity and substrate specificity of an endoglucanase from *Zea mays* seedling cell walls, *Plant Physiol.* **83:**203–207.

Hayashi, T., Marsden, M. P. F., and Delmer, D. P., 1987, Pea xyloglucan and cellulose. V. Xyloglucan–cellulose interactions *in vitro* and *in vivo, Plant Physiol.* **83:**384–389.

Higuchi, T., Ito, Y., and Kawamura, I., 1967, *p*-Hydroxyphenylpropane component of grass lignin and role of tyrosine–ammonia lyase in its formation, *Phytochemistry* **6:**875–881.

Hood, E. E., Shen, Q. X., and Varner, J. E., 1988, A developmentally regulated hydroxyproline-rich glycoprotein in maize pericarp cell walls, *Plant Physiol.* **87:**138–142.

Huber, D. J., and Nevins, D. J., 1981, Partial purification of endo- and exo-β-D-glucanase enzymes from *Zea mays* L. seedlings and their involvement in cell wall autohydrolysis, *Planta* **151:**206–214.

Jarvis, M. C., 1984, Structure and properties of pectin gels in plant cell walls, *Plant Cell Environ.* **7:** 153–164.

Kato, Y., and Nevins, D. J., 1984, Enzymic dissociation of *Zea* shoot cell wall polysaccharides. II. Dissociation of (1→3),(1→4)-β-D-glucan by purified (1→3),(1→4)-β-D-glucan 4-glucanohydrolase from *Bacillus subtilis, Plant Physiol.* **75:**745–752.

Kato, Y., and Nevins, D. J., 1986, Fine structure of (1→3),(1→4)-β-D-glucan from *Zea* shoot walls, *Carbohydr. Res.* **147:**69–85.

Kauss, H., and Hassid, W. Z., 1967, Enzymic introduction of the methyl ester groups of pectin, *J. Biol. Chem.* **242:**3449–3453.

Keegstra, K., Talmadge, K. W., Bauer, W. D., and Albersheim, P., 1973, The structure of plant cell walls. III. A model of the walls of suspension-cultured sycamore cells based on the interactions of the macromolecular components, *Plant Physiol.* **51:**188–196.

Kieliszewski, M., and Lamport, D. T. A., 1987, Purification and partial characterization of a hydroxyproline-rich glycoprotein in a graminaceous monocot, *Zea mays, Plant Physiol.* **85:**823.

Lamport, D. T. A., 1986, The primary cell wall: A new model, in: *Cellulose: Structure, Modification and Hydrolysis* (R. A. Yound and R. M. Rowell, eds.), John Wiley & Sons, New York, pp. 77–90.

Lau, J. M., McNeil, M., Darvill, A. G., and Albersheim, P., 1985, Structure of the backbone of rhamnogalacturonan I, a pectic polysaccharide in the primary cell walls of plants, *Carbohydr. Res.* **137:**111–125.

Luttenegger, D. G., and Nevins, D. J., 1985, Transient nature of a $(1\rightarrow3),(1\rightarrow4)$-β-D-glucan in *Zea mays* coleoptile walls, *Plant Physiol.* **77:**175–178.

Maltby, D., Carpita, N. C., Montezinos, D., Kulow, C., and Delmer, D. P., 1979, β-1,3-Glucan in developing cotton fibers. Structure, localization, and relationship of synthesis to that of secondary wall cellulose, *Plant Physiol.* **63:**1158–1164.

Markwalder, H.-V., and Neukom, H., 1976, Diferulic acid as a possible crosslink in hemicelluloses of wheat germ, *Phytochemistry* **15:**836–837.

McDougall, G. J., and Fry, S. C., 1988, Inhibition of auxin-stimulated growth of pea stem segments by a specific nonasaccharide of xyloglucan, *Planta* **175:**412–416.

Meinert, M. C., and Delmer, D. P., 1977, Changes in biochemical composition of the cell wall of the cotton fiber during development, *Plant Physiol.* **59:**1088–1097.

Moustacas, A.-M., Nari, J., Diamantidis, G., Noat, G., Crasnier, M., Borel, M., and Ricard, J., 1986, Electrostatic effects and the dynamics of enzyme reactions at the surface of plant cells. 2. The role of pectin methylesterase in the modulation of electrostatic effects in soybean cell walls, *Eur. J. Biochem.* **155:**191–197.

Mundy, J., Brandt, A., and Fincher, G. B., 1985, Messenger RNAs from the scutellum and aleurone of germinating barley encode $(1\rightarrow3,1\rightarrow4)$β-glucanase, α-amylase, and carboxypeptidase, *Plant Physiol.* **79:**867–871.

Nakajima, N., Morikawa, H., Igarashi, S., and Senda, M., 1981, Differential effect of calcium and magnesium on mechanical properties of pea stem cell walls, *Plant Cell Physiol.* **22:**1305–1315.

Rees, D. A., 1977, *Polysaccharide Shapes,* Chapman and Hall, London.

Sadava, D., and Chrispeels, M. J., 1973, Hydroxyproline-rich cell wall protein extensin role in the cessation of elongation in excised pea epicotyls, *Dev. Biol.* **30:**49–55.

Scalbert, A., Monties, B., Lallemand, J.-Y., Guttet, E., and Rolando, C., 1985, Ether linkage between phenolic acids and lignin fractions from wheat straw, *Phytochemistry* **24:**1359–1362.

Shea, E. M., Gibeaut, D. M., and Carpita, N. C., 1989, Structural analysis of the cell walls regenerated by carrot protoplasts, *Planta* **179:** 293–308.

Shibuya, N., and Nakane, R., 1984, Pectic polysaccharides of rice endosperm walls, *Phytochemistry* **23:** 1425–1429.

Stafstrom, J. P., and Staehelin, L. A., 1986, Cross-linking patterns in salt-extractable extensin from carrot cell walls, *Plant Physiol.* **81:**234–241.

Staudte, R. G., Woodward, J. R., Fincher, G. B., and Stone, B. A., 1983, Water-soluble $(1\rightarrow3),$- $(1\rightarrow4)$-β-D-glucans from barley (*Hordeum vulgare*) endosperm. III. Distribution of cellotriosyl and cellotetraosyl residues, *Carbohydr. Polym.* **3:**299–312.

Stevenson, T. T., McNeil, M., Darvill, A. G., and Albersheim, P., 1986, Structure of plant cell walls. XVIII. An analysis of the extracellular polysaccharides of suspension-cultured sycamore cells, *Plant Physiol.* **80:**1012–1019.

Stuart, I. M., Loi, L., and Fincher, G. B., 1986, Development of $(1\rightarrow3,1\rightarrow4)$-β-D-glucan endo-hydrolase isoenzymes in isolated scutella and aleurone layers of barley (*Hordeum vulgare*), *Plant Physiol.* **80:**310–314.

van Holst, G.-J., and Varner, J. E., 1984, Reinforced polyproline II conformation in a hydroxyproline-rich cell wall glycoprotein from carrot root, *Plant Physiol.* **74:**247–251.

Wada, S., and Ray, P. M., 1978, Matrix polysaccharides of oat coleoptile cell walls, *Phytochemistry* **17:** 923–931.

Woodward, J. R., Phillips, D. R., and Fincher, G. B., 1983, Water-soluble $(1\rightarrow3),(1\rightarrow4)$-β-D-glucans from barley (*Hordeum vulgare*) endosperm. I. Physicochemical properties, *Carbohydr. Polym.* **3:** 143–156.

York, W. S., Darvill, A. G., and Albersheim, P., 1984, Inhibition of 2,4-dichlorophenoxyacetic acid-stimulated elongation of pea stem segments by a xyloglucan oligosaccharide, *Plant Physiol.* **75:** 295–297.

3

Analysis of Dietary Fiber in Human Foods

JUDITH A. MARLETT

1. INTRODUCTION

The heterogeneity of the components of dietary fiber and the complexity and differences among the plant matrices from which fiber is extracted have made it difficult to develop methods of analysis that can be applied to all classes of fiber-containing foods, i.e., fruits, vegetables, legumes, and grain products. Different philosophies as to what constitutes dietary fiber have also led to confusion. Our laboratory defines dietary fiber as lignin plus the polysaccharides that cannot be digested by monogastric endogenous enzymes. In this chapter, emphasis is given to methods of fiber analysis developed or used for a variety of human foods. This chapter is not intended to be an exhaustive review of the many methods of fiber analysis that have been developed for specific or general analytical use. Analytical differences that may account for different results among the methods are illustrated. However, because of the paucity of comparative analyses, the review of the possible causes of variable fiber values should be viewed as incomplete as well.

Divergent needs and interests for the dietary fiber composition of foods and forages have led to a proliferation of methods for its analysis (James and Theander, 1981; Robertson and Van Soest, 1981; Southgate, 1981; Asp *et al.*, 1983; Prosky *et al.*, 1985; Theander and Westerlund, 1986; Englyst and Hudson, 1987; Anderson and Bridges, 1988). Most methods of dietary fiber analysis can be generally divided into two approaches: gravimetric and chemical analysis (Pilch, 1987). In a gravimetric procedure, enzymatic and chemical steps are used to extract nonfiber components. The weight of the remaining residue, after correction for incompletely removed materials, is dietary fiber. In those methods that approach dietary fiber analysis chemically, the sum of neutral and acidic sugar constituents, and usually Klason lignin, is used as the measure of dietary fiber.

JUDITH A. MARLETT • Department of Nutritional Sciences, University of Wisconsin-Madison, Madison, Wisconsin 53706.

Nearly all analyses for dietary fiber begin with a dry, ground sample (Fig. 1). Drying or grinding procedures that involve heat will damage the fiber carbohydrates and complicate subsequent analyses. Homogenization in a blender with adequate water followed by lyophilization does not cause heat damage and will help to insure a uniform sample and a representative aliquot (Neilson and Marlett, 1983).

Most of the gravimetric procedures and some of the chemical approaches do not utilize 80% ethanol to extract soluble sugars such as the endogenous monosaccharides in fruits or sucrose added to cereals (Fig. 1). Because most fiber-containing foods do not contain significant amounts of lipid, inclusion of a lipid extraction step is not considered to be essential (Asp *et al.*, 1983). None of those methods that do include such a step extract lipid quantitatively. Those procedures that do not include a lipid solvent will usually recommend a modification to extract lipid early in the analysis from foods containing more than 2–5% lipid.

The gravimetric methods contain a proteolytic step, although nitrogen extraction is incomplete and a correction must be applied to the final fiber residue weight for the contribution of crude protein determined as Kjeldahl nitrogen × 6.25 (Fig. 1). Neutral detergent solution, particularly when the fiber residue is subsequently incubated with an amylase, appears to be the most effective at removing nitrogen, although it also extracts some fiber (Neilson and Marlett, 1983; Marlett and Chesters, 1985). None of the chemical approaches include a step to remove nitrogenous materials. All methods of dietary fiber analysis incorporate one or more enzymatic steps to extract starch (Fig. 1). Inadequately removed starch will be weighed as part of the fiber residue in gravimetric

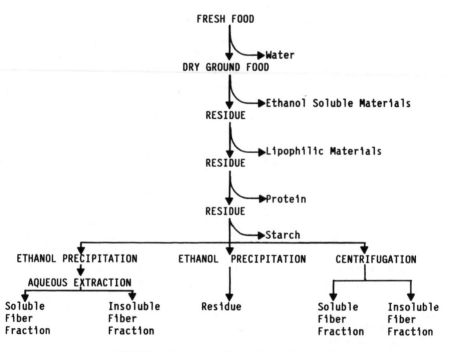

FIGURE 1. Measuring fiber in human foodstuffs.

analyses or detected as glucose that is indistinguishable from cellulose-derived glucose in a chemical analysis.

Ethanol (80%) is used to precipitate fiber polysaccharides in most gravimetric and chemical procedures that measure total dietary fiber (Fig. 1). The subsequent residue is dried, weighed, and the weight corrected for crude protein and in some instances ash in the gravimetric approaches, whereas the residue is acid hydrolyzed for subsequent analysis in the chemical approaches. In some chemical analyses, the polysaccharides representing total dietary fiber are also precipitated with ethanol, and the soluble fiber fraction extracted from the ethanol precipitate with water (Englyst et al., 1982; Anderson and Bridges, 1988). In other chemical analyses the soluble fraction is obtained by centrifugation of the sample after starch hydrolysis, and the soluble fiber polysaccharides are concentrated by dialysis and lyophilization (Theander and Westerlund, 1986) (Fig. 1). Whether total dietary fiber or soluble and insoluble fiber values are the aim, the chemical approaches will determine monomeric constituents of fiber following acid solubilization or hydrolysis of the polysaccharides. The neutral sugars released by acid hydrolysis can be measured indirectly by colorimetric techniques or directly with high-performance liquid chromatography (HPLC) or with gas chromatography (GC) following derivatization. Uronic acids are measured colorimetrically or as carbon dioxide released during decarboxylation of the acid moiety. Lignin is usually determined gravimetrically as the residue insoluble in 72% sulfuric acid, that is, Klason lignin.

Most methods have continued to evolve as experience was gained with different foods. Although this continual process has generally improved the methods, it has made it difficult to interpret the utility of some of the dietary fiber data obtained by different modifications of a method when the modification produced a significant change in the result. Incomplete evaluation of a proposed method of analysis prior to publication also has led to confusion over the utility of the method and the fiber values it generates. Difficulties frequently encountered by those who are incorporating dietary fiber analysis or the use of fiber values into their research programs can be minimized by a thorough, careful review of the relevant literature.

2. GRAVIMETRIC DIETARY FIBER ANALYSIS

The limitations of the early gravimetric approaches to fiber analysis, i.e., crude fiber and acid detergent fiber, have been recently summarized (Pilch, 1987). Four other gravimetric approaches have been used for the fiber analysis of human foods (Robertson and Van Soest, 1981; Asp et al., 1983; Prosky et al., 1985; Mongeau and Brassard, 1986). The neutral detergent fiber (NDF) method, originally developed for animal forage analysis, recovers lignin, cellulose, and variable amounts of the hemicelluloses from human foodstuffs (Van Soest and Wine, 1967; Robertson and Van Soest, 1981). When the method includes a step to hydrolyze starch, the NDF residue is usually not contaminated with significant starch or nitrogen. Several approaches to starch extraction have been proposed (Schaller, 1976; Marlett and Lee, 1980; Robertson and Van Soest, 1981). However, the efficacy of the starch extraction was not always verified by analyzing the NDF for residual starch, and it is apparent that some procedures do not remove all of the starch from the NDF (Marlett and Lee, 1980; Marlett and Chesters, 1985).

Although the number of comparisons was limited, our previous analyses of wheat bran, peas, apples, and a food composite representing a typical daily food intake suggested that the NDF is similar in amount and composition to the insoluble fraction recovered by the Theander (Theander and Aman, 1979) method of chemical analysis (Neilson and Marlett, 1983) (Table I). In a separate study NDF was also found to be similar in fiber content and composition to the insoluble fraction of five foods recovered by the Southgate (1981) method (Marlett and Chesters, 1985). Cummings and Englyst (1987) reported that NDF values "corresponded well" to those of the insoluble non-starch polysaccharides.

More recent data from our laboratory suggest that this relationship may not hold for other foods (Table I). For example, the NDF content of white wheat bread was 2.0% (Johnson and Marlett, 1986), whereas the insoluble fiber content determined by a direct chemical analysis was 2.8% (Table I). Thus, the NDF method failed to measure 40% of the insoluble fiber fraction in this refined grain product. Similarly, the NDF content (percentage dry wt.) of macaroni was 1.9% (Johnson and Marlett, 1986), and the insoluble fiber fraction 3.2% (Table I). The NDF procedure failed to measure even greater amounts of the insoluble fiber from some vegetables (Table I). In contrast, the NDF and insoluble fiber fraction of rice were the same 1.7% (Table I). The NDF procedure is simple and rapid compared to most other methods. Its main limitation is the loss of some fiber components and thus incomplete measurement of the total dietary fiber. For refined grain products, which have very little pectic material, measurement of uronic acids in the material not recovered in the NDF residue would not provide a measure of the fiber lost.

To meet the limitations of the early gravimetric methods, Asp, DeVries, Furda, and Schweizer developed a gravimetric procedure to measure total dietary fiber that, following two interlaboratory studies (Prosky et al., 1984, 1985), has received first action by the Association of Official Analytical Chemists (AOAC). The second interlaboratory

TABLE I. Comparison of Neutral Detergent Fiber
and Insoluble Fraction of Dietary Fiber

	Percentage dry weight	
	NDF	Insoluble
Wheat bran	36.3	37.3
Peas, canned	15.6	17.9
Apples, with peel	7.2	9.2
Food composite	4.2	5.2
Rice, cooked	1.7	1.5
Macaroni, cooked	1.9	3.2
White wheat bread	2.0	2.8
White wheat flour	1.0	2.1
Green beans, canned	16.3	22.3
Carrots, raw	8.3	28.3
Corn, whole kernel, frozen	4.7	8.2
Orange, fresh	6.2	7.1

study reported coefficients of variation of >10% for the two samples, rice and soy isolate, that contained small amounts (<1.5% dry wt.) of fiber (Prosky et al., 1985). The number of samples with which the method was tested was small, and commercial laboratories have experienced unsatisfactory variability with this method (AACC Technical Committee, 1987). We have found that this method generated dietary fiber values that were within 10% of chemically determined values for canned peas, kidney beans, rice, and wheat bran, providing careful analytical technique was used (Table II).

The AOAC method overestimated the fiber content of apple by 14% and that of a food composite by 18% (Table II). Results of the chemical analysis of the AOAC residues of apples and food composite suggested that some of the simple sugars in these samples (88% of the dry wt. of apple and 30% of the food composite) are coprecipitated with the fiber polysaccharides in the 80% ethanol precipitate and thus are measured as fiber (Marlett and Navis, 1988). This contention is supported by P. Theander (personal communication, 1987). In contrast to the chemical approach we used for these comparative analyses, there is no initial 80% ethanol extraction step to remove these soluble simple sugars in the AOAC procedure. Recently, the AOAC method has been modified to determine soluble and insoluble fractions of fiber, although it was concluded from the first interlaboratory study that further study was needed because of the variability among the soluble dietary fiber results (Prosky et al., 1988). Asp and co-workers (1983) have developed a gravimetric–enzymatic method that extends the AOAC method. A different approach, consisting of pepsin and pancreatin steps, is used for protein digestion, and the fiber is separated into soluble and insoluble fractions.

The fourth gravimetric method developmental effort is under way in Canada (Mongeau and Brassard, 1986). This procedure defines soluble fiber as those polysaccharides solubilized at 100°C and subsequently precipitated with 100% ethanol after enzymatic starch hydrolysis. The insoluble fiber is measured as NDF, which is determined by a separate analysis. The fact that the two fiber fractions are determined on separate samples assumes that there is neither overlap nor loss of fiber components in measuring the two fractions of fiber in this manner. The first interlaboratory study of this method has been completed (AACC Technical Committee, 1987). Significantly different fiber values for some foods were obtained by this method compared to those obtained with the AOAC or chemical approach of Englyst (Mongeau and Brassard, 1986).

TABLE II. Gravimetric versus Chemical Analysis of Total Dietary Fiber

	Percentage dry weight	
	Theander chemical	AOAC gravimetric
Peas, canned	24.0	25.1
Food composite	7.7	9.1
Kidney beans, canned	21.5	19.6
Rice, cooked	1.5	1.5
Apple, with peel	12.2	13.9
Wheat bran	45.9	47.7

3. CHEMICAL DIETARY FIBER ANALYSIS

The predominant chemical approaches to dietary fiber analysis are those of Theander of Uppsala, Sweden (Theander and Aman, 1979, 1981; Theander, 1983; Theander and Westerlund, 1986), which is used by our laboratory (Neilson and Marlett, 1983; Shinnick *et al.*, 1988; Marlett, 1989; Marlett *et al.*, 1989) and is the official method in Australia (P. Theander, personal communication, 1987), and the procedure by Englyst (Englyst *et al.*, 1982), a modification of Southgate's method that is being adopted in Britain. Recently, Anderson and Bridges (1988) developed a modification of the Southgate method that incorporates aspects of the procedure developed by Englyst and co-workers to measure fiber in foods and diets. The original Southgate procedure (1969) was used to generate most of the dietary fiber data that are in the British Food Composition Tables (Paul and Southgate, 1978). The measurement of the fiber-derived sugars colorimetrically, which is relatively nonspecific, and incomplete starch hydrolysis are the likely causes of the high fiber values produced by this early method (Southgate and White, 1981). More recent modifications by Southgate have addressed these sources of error (Southgate, 1981).

Englyst's modification of Southgate's pioneering work has undergone several additional modifications. Englyst's first major modification involved the determination of fiber components by GC analysis for neutral sugars and colorimetry for uronic acids (Englyst *et al.*, 1982). In 1982 Englyst and co-workers proposed three procedures. Procedure A involved enzymatic hydrolysis of the starch with a mixture of α-amylase and pullulanase followed by precipitation of all fiber polysaccharides with ethanol. The GC measurement of neutral sugars following a two-step Saeman acid hydrolysis of the precipitate provides a measure of total dietary fiber. In procedure B a similar residue is prepared using a separate aliquot of sample; it is treated with dilute acid to hydrolyze noncellulosic polysaccharides. The difference in glucose content of samples analyzed by procedures A and B is taken as a measure of cellulose. Water-insoluble nonstarch polysaccharides are measured in procedure C, which is applied to a third aliquot. The essential feature of procedure C is that an aqueous extraction step is incorporated after starch hydrolysis to remove soluble fiber components; the insoluble fiber fraction is the residue obtained by centrifugation.

Lignin is not measured in Englyst's procedures. Several small changes in the conditions of the various steps and the addition of DMSO to solubilize starch were published in 1984 (Englyst and Cummings, 1984). The evolution of this method, along with a description of its most recent form and the protocol for using colorimetric techniques instead of GC to quantitate neutral sugars (Englyst and Hudson, 1987), has been recently reviewed (Cummings and Englyst, 1987). If a full set of data is desired, the procedure is as time consuming as the other chemical approaches that use direct sugar analysis and begin with a single sample for analysis. One negative feature of his method is the use of dimethyl sulfoxide (DMSO), which, because of its lipophilic nature, can be a hazardous compound for routine laboratory use.

Theander's analytical approach consists of several analyses of a single sample that provide total, soluble, and insoluble dietary fiber content, soluble and insoluble hemicellulosic and pectic components, cellulose, and Klason lignin. Since nearly all of the glucose in an acid hydrolysate of an insoluble fiber fraction is from cellulose, it is taken as a measure of cellulose (Theander and Aman, 1981). A major strength of this method is that recoveries were performed during development and modification of the method to assess any possible loss of fiber components. As with the other chemical methods of fiber analysis, this method has undergone several modifications (Theander, 1983; Theander and Aman, 1979, 1981; Theander and Westerlund, 1986). Three approaches were described in 1986: one that measures total dietary fiber with starch, a second for total dietary fiber after starch extraction, and a modification of the original detailed procedure that measures soluble and insoluble dietary fiber components. Early modifications of the steps to prepare the extractive-free residue (EFR) included replacement of sequential refluxing with 80% ethanol and chloroform with sequential extraction with 80% ethanol and light petroleum (b.p. 60–70°C) in a sonicating waterbath (Neilson and Marlett, 1983; Salomonsson et al., 1984). It also became evident in early experiments that the thermostable enzyme Termamyl did not hydrolyze starch to monomers but rather to short polymers that were recovered in the soluble fiber fraction (Theander and Aman, 1979; Neilson and Marlett, 1983). The current approach is to use a two-step starch hydrolysis procedure: a short incubation with Termamyl followed by an overnight incubation with an amyloglucosidase (Theander and Westerlund, 1986). Experiments were conducted to make certain that no fiber components were lost in the preparation of the EFR, during starch hydrolysis, and during dialysis of the soluble fraction (Theander and Westerlund, 1986). Because we apply this method to a wide variety of foods, all essentially "unknowns," we continue to determine recoveries of the soluble and insoluble fiber fractions by measuring starch, crude protein, and moisture in the fractions as well as the fiber components (Neilson and Marlett, 1983; Marlett and Navis, 1988; Shinnick et al., 1988; Marlett, 1989; Marlett et al., 1989). The fiber content of the fractions is also corrected for any contaminating starch.

Since these two methods appear to be emerging as the major chemical approaches to dietary fiber analysis, the ensuing comparison of the two approaches is provided not only to highlight subtle but possibly significant differences but also to illustrate the complexity of the issues of dietary fiber analysis. Theander provides an analytical method that is applied to a single sample, whereas more than one aliquot must be analyzed by the Englyst approach to obtain the same information. Use of more than one analysis on separate samples assumes that fiber components are neither lost nor are measured more than once. The use of the three separate Englyst procedures on three separate samples, however, does provide a calculated measure of cellulose. This approach to separating cellulose and noncellulosic polysaccharides operationally defines these two classes of polysaccharides on the basis of their susceptibility to acid hydrolysis. In Theander's method the glucose in the insoluble fiber fraction is taken as a measure of cellulose. A lipid extraction is in the Theander but not the Englyst method; this is not likely to have a significant effect on the outcome of dietary fiber analysis.

The use of 80% ethanol to extract soluble endogenous sugars in one (Theander and Westerlund, 1986) but not the other (Englyst et al., 1982) may lead to different results from the two methods. These sugars may coprecipitate with the fiber polysaccharides in amounts sufficient to be measured as "fiber-derived" neutral sugars at the later step in which 80% ethanol is used to precipitate these fiber polysaccharides. We have evidence that this type of coprecipitation may occur during the recovery of fiber polysaccharides by 80% ethanol precipitation (Marlett and Navis, 1988). Theander's method minimizes this possibility by incorporating two extractions with ethanol to remove simple sugars, an approach that was determined by measuring the efficacy of four sequential ethanol extractions to remove low-molecular-weight sugars from carrots (Theander and Westerlund, 1986). We have found comparable results with four ethanol extractions of oranges and peas (J. A. Marlett, unpublished data).

Each laboratory approaches starch hydrolysis differently. Englyst uses DMSO (1 hr, 100°C) to disperse starch, as proposed earlier by Leach and Schoch (1962), prior to incubation (16 hr, 42°C) with a mixture of pancreatin as a source of pancreatic amylase and pullulunase. The DMSO disperses resistant starch (Englyst and Cummings, 1984). Resistant starch also can be dispersed with potassium hydroxide (Englyst et al., 1982; Cummings and Englyst, 1987; Westerland et al., 1989). Theander uses a two-step enzymatic approach consisting of a short incubation (0.5 hr, 96°C) with a thermostable α-amylase and a longer incubation with an amyloglucosidase (16 hr, $\leq$60°C). Whether Theander's method includes resistant starch as defined by Englyst is unknown. When starch was measured enzymatically, we have found 0.1–0.7% starch (percentage dry wt.) remaining in rice and macaroni fiber fractions analyzed by the Theander method, 1–3% starch in pea and kidney bean fiber, and none in oat bran fiber fractions (Marlett et al., 1989). Theander reported ~1% in samples of extrusion-cooked wheat flour that could be hydrolyzed if the termamyl step was 96°C (Theander and Westerlund, 1986). The relationship of starch remaining undigested in an in vitro fiber analysis to starch escaping digestion in vivo is unknown, as none of the fiber analysis procedures include conditions that mimic gastric digestion of the plant matrix, i.e., pepsin at an acidic pH.

A third major difference between the two laboratories is the manner in which the soluble dietary fiber fraction is recovered and analyzed. In the Theander method the sample is centrifuged after enzymatic starch hydrolysis, and the soluble fiber is recovered as the supernatant and concentrated by dialysis and lyophilization (Theander and Westerlund, 1986). The fraction measured as noncellulosic polysaccharides by dilute acid hydrolysis in procedure B of the Englyst method is the soluble fraction (Englyst et al., 1982). The use of ethanol to precipitate fiber polysaccharides in Englyst's procedures may lead to coprecipitation of the products of starch hydrolysis that subsequently will be measured as fiber. In addition, it has been reported that some polysaccharides are incompletely precipitated by ethanol (Larm et al., 1975). Whether these are significant sources of error is not known. The heterogeneity of the fiber complex from different foods makes it difficult to state that these two possible sources of error will not occur during an analysis of any food.

Theander's group has developed a separate method to measure β-glucans (Aman and Graham, 1987), as have we (Shinnick et al., 1988), whereas Englyst has not. The product of cellulose, starch, or β-glucan acid hydrolysis is the same, glucose. Because of its probable role in the hypocholesterolemic effects of some fibers, a prudent analytical course is to analyze specifically for, rather than to calculate a value for, β-glucans in those foods rich in β-glucans, i.e., oats and barley products.

Another major difference between the Theander and Englyst approaches to fiber analysis is in the measurement of Klason lignin; Theander includes it, whereas Englyst excludes lignin from the dietary fiber complex.

4. ACID HYDROLYSIS OF FIBER FRACTIONS

A variety of conditions have been used to acid hydrolyze fiber fractions. Most of them utilize sulfuric acid, and each laboratory has developed its own set of conditions for the two basic types, dilute acid hydrolysis and concentrated or Saeman hydrolysis. The Saeman hydrolysis involves two steps, a primary hydrolysis and a secondary hydrolysis, and it was originally developed by Saeman (Saeman *et al.*, 1954) to hydrolyze wood. The basic approach of the primary hydrolysis, solubilizing cellulose in concentrated 72% sulfuric acid at ambient temperature, has been retained in most modifications, although the duration of the incubation varies from approximately 1 to 3 hr and the ratio of sample to concentrated acid varies. The sample is diluted with water for the secondary hydrolysis, the step in which hydrolysis of the polymers actually occurs. Again, considerable variation in the acid concentration (~ 0.8 to 2.0 N) and time (2 to 6 hr) exists among approaches taken by fiber analysis laboratories. Further, the temperature of the secondary step has been 100°C for ≥ 2 hr or higher (125°C, 15 psi) in an autoclave for shorter time periods, e.g., 1 hr. Theander recommends conditions for the secondary hydrolysis step of 0.8 N sulfuric acid, 1 hr, 125°C in a preheated autoclave, Englyst 2 hr at 100°C, 2 N sulfuric acid. A Saeman hydrolysis is appropriate for any sample containing cellulose, which would be either the insoluble or the total dietary fiber fraction.

Dilute acid hydrolysis, usually 2 N sulfuric acid at 100°C for 2–3 hr, is used to hydrolyze the soluble fiber fractions, although Theander recently recommended trifluoroacetic acid (Theander and Westerlund, 1986).

Completeness of acid hydrolysis is critical to accurate quantitation of fiber polysaccharides as their monomeric constituents, neutral sugars. Highly lignified tissue may require more than a 1-hr primary hydrolysis to assure the solubilization of cellulose. One way to evaluate the efficacy of acid hydrolysis of a total dietary fiber residue or of an insoluble fiber fraction is to rehydrolyze any residue that remains after treatment. We have also used cellobiose, which we detect with the HPLC column we use for analysis of fiber-derived monosaccharides, as an indication of the efficacy of the hydrolysis. Cellobiose serves as a marker of incomplete hydrolysis and reflects di- and oligosaccharides that are soluble but, because of their size, would not be detected either with GC or with HPLC. The GC analysis used by most laboratories to measure fiber neutral sugars would not detect any di- or oligosaccharides, including cellobiose.

Hydrolysis of both soluble and insoluble fractions can be incomplete. The aim is to hydrolyze fiber polysaccharides to their neutral monosaccharides under conditions that will minimize the destruction of neutral sugars after their release. These hydrolysis losses, and derivitization losses in the case of GC quantitation, should be determined and applied to the individual sugars prior to their expression as polymers. Based on analysis of onion, macaroni, pea, and orange fiber fractions in 1987, we recommended that dilute acid hydrolysis was adequate for the soluble fraction and that a 0.5-hr secondary step in the Saeman hydrolysis was adequate for insoluble fiber fraction (Marlett, 1989). More recent analytical work indicates that neither set of conditions completely hydrolyzes fiber polysaccharides from some fruits and vegetables (J. Marlett, unpublished data, 1989). Acid hydrolysis conditions that we now apply to both soluble and insoluble fiber fractions consist of 12 M H_2SO_4 (1 ml/150 mg of sample), 1 hr, ambient temperature with frequent mixing followed by 0.4 M H_2SO_4, 1 hr, 121°C, 15 psi.

5. CRUDE LIGNIN

Inclusion by Theander versus exclusion by Englyst of Klason lignin has been proposed as the primary basis for lower fiber values obtained by Englyst. Using data from our own laboratory, Table III illustrates the effect of Klason lignin on total dietary fiber values. Exclusion of Klason lignin produced substantively lower fiber values for only four of the 12 foods: wheat bran, rice, peas, and apple. For one of these foods, wheat bran, the Klason lignin probably does reflect real lignin (Neilson and Marlett, 1983). Lignin is a heterogeneous mixture of phenolic compounds (Theander and James, 1979). When measured as the residue remaining after concentrated acid hydrolysis, i.e., Klason lignin, it may contain Maillard products, tannins, cutin, silica, and precipitated protein as well (Theander and James, 1979).

Inclusion of lignin in a dietary fiber complex is a point of considerable discussion, with some investigators arguing that it should be excluded (Cummings and Englyst, 1987), while others argue for its inclusion (Robertson and Van Soest, 1981; Theander and Aman, 1981; Bingham, 1987). Most human foods do not have any significant amounts of silica or cutin. Contribution of protein or Maillard products to the "artifact" fraction of lignin can be indirectly assessed by analyzing for Klason lignin in samples from which a significant proportion of the crude protein has been removed, such as insoluble fiber fractions prepared from samples treated with protease or pepsin (Marlett *et al.*, 1989). The NDF also contains very little nitrogen (Neilson and Marlett, 1983).

TABLE III. Effect of Klason Lignin (KL) on Total Dietary Fiber (TDF) and Distribution of Fiber between Soluble and Insoluble Fractions

	TDF (% dry wt.)		Soluble fiber (% TDF)	
	With KL	Without KL	With KL	Without KL
Wheat bran	39.4	33.3	4	5
White wheat bread	3.8	3.4	26	29
Macaroni, cooked	4.2	4.1	24	24
Rice, cooked	1.7	1.0	11	22
Kidney beans, canned	21.7	20.1	24	26
Peas, canned	21.2	17.1	13	16
Onions	14.0	14.0	3	3
Green beans, canned	27.9	26.7	4	4
Corn, whole kernel, frozen	9.2	8.1	4	5
Carrots, raw	29.7	28.8	5	5
Apple, with peel	12.6	10.7	23	27
Orange, fresh	10.0	9.5	27	28

We have found that inclusion of a protease or pepsin step in the fiber analysis procedure significantly decreased the Klason lignin of only one, macaroni, of five foods; the Klason lignin content of rice, kidney bean, peas, or oat bran was not changed (Marlett *et al.*, 1989). These results suggest that removal of half or more of the crude protein from a fiber fraction would not result in a better estimate of lignin. In contrast, results were significantly different with hydrolysis of NDF residues, which contain much less nitrogen. The NDF of only one of four foods, wheat bran, contained measurable amounts of Klason lignin, whereas the insoluble fiber fraction of apple, peas, or food composite collected by the Theander method contained 0.9–3.5% (dry wt.) Klason lignin (Neilson and Marlett, 1983). Determination of Klason lignin content of foods by analysis of the NDF or of the acid detergent fiber or by potassium permanganate oxidation as proposed by Van Soest (Robertson and Van Soest, 1981) are approaches that should be evaluated as more accurate estimates of lignin. We argue that lignin should be included in the dietary fiber complex for two reasons. First, Klason lignin measured in unrefined grains probably largely reflects lignin (Neilson and Marlett, 1983; Shinnick *et al.*, 1988; Marlett, 1989). Second, phenolic compounds are becoming increasingly linked to carcinogenic and anticarcinogenic activities, and it is possible that they are responsible for some of the effects attributed to dietary fiber (Newmark, 1987).

6. SOLUBLE VERSUS INSOLUBLE DIETARY FIBER

Because of its modulation of upper gastrointestinal events, considerable interest in the soluble dietary fiber fraction has been expressed by the scientific, industrial, and lay communities. In our laboratory the soluble fiber fraction is usually ≤25% of the total fiber in typical foods when Klason lignin is included as part of fiber (Table III). Concentrated or processed fiber sources are exceptions; wheat bran fiber is only about 5% soluble, whereas the soluble fiber fraction in oat products is 33–43% of the total fiber in unprocessed and 49–55% of the total fiber in processed oat products (Shinnick *et al.*, 1988). Analyses from Theander's laboratory agree (Theander and Aman, 1979, 1981); those from Englyst's laboratory are almost but not always higher (Englyst, 1981; Englyst *et al.*, 1982, 1983; Englyst and Cummings, 1984). Recently, Anderson and Bridges (1988), using a method of analysis that combined features of Southgate's and Englyst's methods, reported soluble fiber contents (percentage of total dietary fiber) of 32% for cereals, 32% for vegetables, 25% for dried beans, and 38% for fruits.

We recently evaluated two possible causes for these differences in the proportion of total fiber recovered in the soluble fraction: (1) the inclusion of Klason lignin in the insoluble fiber fraction and (2) enzymatic and chemical variations among the methods. Excluding lignin from the total fiber does not appear to have a substantial effect on the proportion that is recovered in the soluble fiber fraction of foods except for rice (Table III). However, the method of fiber analysis does significantly influence the amount of total dietary fiber extracted into the soluble fraction, although we found that there was considerable difference among foods (Marlett *et al.*, 1989). Incorporation of a protease step into the analysis had only a modest effect on the recovery of soluble fiber from peas, kidney beans, and oat bran; the soluble fraction of rice, however, increased from 6% to 16% of the total, and that of macaroni from 15% to 26% of the total fiber. Incorporation of a pepsin digestion step into the dietary fiber analysis solubilized a greater proportion of the total fiber from all of the foods except oat bran (Marlett *et al.*, 1989). The soluble fiber content of peas and kidney beans increased by ~50% compared to the soluble fiber obtained by the standard Theander method, whereas the soluble fiber fraction of rice increased from 6% to 21% of the total fiber and of macaroni from 15% to 33%. Since neither of these protein digestion steps is used by Englyst or Anderson, they would not explain the differences between their data and those analyzed by the Theander method. Two modifications that are used by Englyst and Anderson, the use of DMSO to solubilize starch and of an elevated temperature (100°C) to extract the soluble fraction, had no significant effect on the recovery of either total dietary fiber or the soluble fraction (Marlett *et al.*, 1989). It is possible that other analytical differences, e.g., acid hydrolysis conditions or recovery of fiber fractions with 80% ethanol, may account for the variations in soluble fiber analysis among the three laboratories.

7. WHY MEASURE DIETARY FIBER?

There are many reasons for measuring dietary fiber yield and its composition in human foods; among them are (1) the need for a data base for determining intakes for total dietary fiber and its various components, (2) the ability to relate dietary fiber composition to a physiological effect or function, and (3) the determination of the mechanisms of action of different dietary fibers.

7.1. Dietary Fiber Data Bases

Both the many epidemiologic correlations between dietary fiber intake and health maintenance and apparent disease prevention and the emerging research that is identifying possible roles for dietary fiber as a component of a treatment regimen for various diseases highlight the need for accurate values of both fiber content and composition of human foods (Pilch, 1987). The two relatively complete data bases currently available are those for crude fiber (Watt and Merrill, 1963) and those for total dietary fiber (Paul and Southgate, 1978), generated largely with the early method of Southgate. Both have limitations, although those of the Southgate data base are less significant, and it has been used to generate most of the available fiber intake data (Pilch, 1987). Because of the simplicity of NDF procedure, many foods have been analyzed by this method (Neilson and Marlett, 1983; Johnson and Marlett, 1986; Patrow and Marlett, 1986). Our lack of knowledge about what fiber is not measured by this method when it is applied to a wide variety of foods, however, limits the utility of these values. The data in Table IV illustrate differences among nonstarch polysaccharides, i.e., the sum of hemicelluloses, cellulose, and pectins but not lignin, reported by three laboratories using detailed methods of analysis. The differences among the data from the different laboratories indicate to us that there are unanswered analytical issues that need to be addressed. Undoubtedly, some differences result from the analysis of different samples. However, for other foods that are standardized, e.g., white flour and bread in the United States and commercial cereals, this cannot be the basis for the differences. We can identify three possible sources for these differences: (1) loss of some fiber component; (2) incomplete acid hydrolysis of fiber fractions; and (3) recovery with the fiber of some nonfiber component

TABLE IV. Differences among Fiber-Derived Polysaccharide Values from
Three Laboratories

	Laboratory		
	A[a]	B[b]	C[c]
Apple	10.5	14.7	9.9
Beans, green	31.4	32.2	27.8
Bread, white wheat	3.1	2.6	3.3
Carrot	22.8	23.4	28.9
Corn, whole kernal, canned	8.9	—	6.7
Cornflakes	1.1	0.7	2.4
Flour, white wheat	3.7	3.2	3.0
Oats, rolled	9.5	7.1	8.5
Peas, canned	20.4	—	22.3
Rice Krispies™	0.8	0.9	0.8
Special K™	2.2	—	2.7

[a] From Anderson and Bridges (1988).
[b] From Englyst (1981); Englyst and Cummings (1984); Englyst et al. (1982, 1983).
[c] From J. A. Marlet (unpublished data).

such as simple sugars endogenous to the sample or starch hydrolysis products. Combining values from a variety of analyses would provide a data base suitable for making comparisons of relative dietary fiber intakes among individuals and groups, but this approach would not provide accurate fiber intakes.

7.2. Relating Composition to Function

Relating the composition of dietary fiber to its physiological effects is a step toward determining its mechanisms of action, and this is an active area of research for many laboratories. Detailed chemical characterization of the neutral and acidic sugars in dietary fiber, however, is only one means of assessing how dietary fiber and its metabolic products affect gastrointestinal function. Largely ignored are a variety of other properties of the plant matrix and fiber sources that undoubtedly play an important role in the physiological effects we attribute to dietary fiber (Table V). Loss of the physiolog-

TABLE V. Differences among
Dietary Fiber Sources

Structure
Chemical composition
Degree of polymerization
Charge
Water-holding capacity
Particle size
Source
Extraction process

TABLE VI. Composition[a] of Two Lots
of AACC[b] Soft White Wheat Bran

	Purchase date	
	1983	1985
Dietary fiber		
Neutral sugars	31.4	39.1
Klason lignin	5.9	5.4
Uronic acids	1.6	1.4
Total	38.9	45.9
NDF	36.6	34.5
Phytate	3.6	3.8
Crude protein	22.7	23.4

[a] Percentage dry weight.
[b] American Association of Cereal Chemists, St. Paul, MN.

ical response on purification of a dietary fiber is evidence that the ultrastructure of the polysaccharides and their relationship to other constitutents in the plant are important aspects of dietary fiber research.

In some instances "standard" fiber sources have been used without analysis on the assumption that the chemical composition does not change. In our laboratory we have found that the compositions do differ. For example, different lots of a standard wheat bran and different psyllium seed products had different compositions (Tables VI and VII). The physiological responses we observed with these products varied as well (J. A. Marlett, unpublished data).

7.3. Determining Mechanisms of Action of Dietary Fiber

The presence of dietary fiber in the lumen of the gastrointestinal tract significantly influences gastrointestinal function. Some of the mechanisms of action of dietary fiber are a consequence of its presence in the gut; others are undoubtedly a result of its metabolism. The colon is the major site of fiber metabolism, although fiber affects and is affected by gut activities proximal to the colon. The net result is that dietary fiber

TABLE VII. Composition of Psyllium Seed Products[a]

	Husk: Searle[b]	Whole seed	
		Meer	Hathaway
Neutral sugars	68.7	63.1	57.4
Uronic acids	4.7	1.8	1.8
Klason lignin	11.1	5.0	6.5
Crude protein	4.4	19.0	19.8
Starch	Neg.	Neg.	1.2

[a] Percentage of EFR, which is the residue remaining after treatment with 80% ethanol and light petroleum.
[b] Manufacturer.

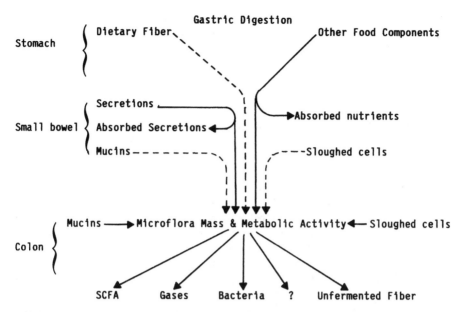

FIGURE 2. Interrelationships among metabolism of dietary fiber and its physiological effects on gastrointestinal function.

influences the environment in which it is metabolized not only by its presence in the lumen but also by altering the nutrient supply available to the microflora. These interrelations are diagrammed in Fig. 2. Characterizing the changes that dietary fiber undergoes during movement through the lumen of the gut requires additional analytical techniques. Most methods of food fiber analysis cannot be used to measure dietary fiber in the gut lumenal contents or excreta unless they are modified to exclude other sources of polysaccharides, e.g., bacterial cell walls and exopolysaccharides, endogenous mucin, and carbohydrate-containing mucosal cell debris.

8. CONCLUSIONS

Significant progress has been made during this decade toward our understanding of the analysis of dietary fiber. But, as is often the situation, progress has identified many new issues, so that there are nearly as many questions now as there were when Southgate developed the first detailed method of analysis in the 1960s. As with his pioneering work, all methods have continued to evolve, and this evolution should be reviewed before any method is used.

The proportion of total dietary fiber extracted as soluble fiber may vary by >100% and is determined by the method of analysis. Other significant differences among the analytical data obtained by different methods, however, remain unexplained. These differences cannot always be explained by the inclusion or exclusion of Klason lignin or by the use of different starting samples. More comparisons among these methods in one laboratory using the same food sample to minimize variation are needed. Consideration should be given to those steps in an analytical procedure where dietary fiber may be lost or not detected, such as completeness of sample hydrolysis or completeness of sample recovery with ethanol precipitation. Application of a method to a wide range of food types is also a prerequisite to determining the range of its use. Crude lignin remains a controversial issue. Because of its potential role in carcinogenesis, consideration should be given to other methods of estimating lignin in human foods, particularly grain products.

The selection of the appropriate method for use should be based on the intended use and whether the method was developed considering that use. For example, a method developed primarily by analyzing cereals may not be appropriate for analyzing fresh fruits and vegetables. Each method offers a unique set of advantages and disadvantages that should be thoroughly evaluated before a decision is made. Gravimetric procedures obviously require substantially less time but not necessarily less technical expertise; however, as they are modified to increase the amount of data obtained, e.g., insoluble and soluble fiber fractions instead of just total dietary fiber, the time difference between gravimetric and detailed approaches to analysis will become less significant. The source and method of analysis by which existing data were obtained for the dietary fiber content of human foods should be explored before the values are incorporated into a data base.

Despite the significant progress that has been made toward understanding the measurement of dietary fiber content and composition, it should be recognized that such information describes only one property of dietary fiber. Other physiochemical properties and the relationship of the fiber components to other constitutents in the plant matrix will undoubtedly explain some of fiber's actions. The relationships of our *in vitro* analyses and characterizations to *in vivo* composition and effects are unknown. Data emerging in the next few years should expand substantially our understanding of the analysis of dietary fiber and its constituents.

ACKNOWLEDGMENTS. Research reported in this paper was funded in part by the National Institutes of Health Grant DK 21712, Contract NO1 CN 45181, and the College of Agricultural and Life Sciences, University of Wisconsin-Madison.

REFERENCES

AACC Technical Committee on Dietary Fiber Analysis, 1987, in: American Association of Cereal Chemists 72nd Annual Meeting, November 1–5, Nashville, TN.

Aman, P., and Graham, H., 1987, Analysis of total and insoluble mixed linked (1,3),(1,4)-β-D-glucans in barley and oats, *J. Agric. Food Chem.* **35**:704–709.

Anderson, J. W., and Bridges, S. R., 1988, Dietary fiber content of selected foods, *Am. J. Clin. Nutr.* **47**:440–447.

Asp, N. G., Johansson, C. G., Hallmer, H., and Siljestrom, M., 1983, Rapid enzymatic assay of insoluble and soluble dietary fiber, *J. Agr. Food Chem.* **31**:476–482.

Bingham, S., 1987, Definitions and intakes of dietary fiber, *Am. J. Clin. Nutr.* **45**:1226–1231.

Cummings, J. H., and Englyst, H. N., 1987, The development of methods for the measurement of "dietary fibre" in food, in: *Cereals in a European Context. 1st European Congress of Food Science and Technology* (I. D. Morton, ed.), VCH Publishers, New York, pp. 188–220.

Englyst, H. N., 1981, Determination of carbohydrates and its composition in plant materials, in: *The Analysis of Dietary Fiber in Food* (W. P. T. James and O. Theander, eds.), Marcel Dekker, New York, pp. 71–93.

Englyst, H. N., and Cummings, J. H., 1984, Simplified method for the measurement of total non-starch polysaccharides by gas–liquid chromatography of constituent sugars as alditol acetates, *Analyst* **109:**937–942.

Englyst, H. N., and Hudson, G. J., 1987, Colorimetric method for routine measurement of dietary fibre as non-starch polysaccharides. A comparison with gas–liquid chromatography, *Food Chem.* **24:**63–76.

Englyst, H. N., Wiggins, H. S., and Cummings, J. H., 1982, Determination of non-starch polysaccharides in plant foods by gas–liquid chromatography of constituent sugars as alditol acetates, *Analyst* **107:**307–318.

Englyst, H. N., Anderson, V., and Cummings, J. H., 1983, Starch and non-starch polysaccharides in some cereal foods, *J. Sci. Food Agr.* **34:**1434–1440.

James, W. P. T., and Theander, O. (eds.), 1981, *The Analysis of Dietary Fiber in Food,* Marcel Dekker, New York.

Johnson, E. J., and Marlett, J. A., 1986, A simple method to estimate neutral detergent fiber (NDF) content in typical daily menus, *Am. J. Clin. Nutr.* **44:**127–134.

Larm, D., Theander, O., and Aman, P., 1975, Structural studies on a water-soluble arabinan isolated from rapeseed (*Brassica napus*), *Acta Chem. Scand.* **B29:**1011–1014.

Leach, H. W., and Schoch, T. J., 1962, Structure of the starch granule. III. Solubilities of granular starches in dimethyl sulfoxide, *Cereal Chem.* **39:**318–324.

Marlett, J. A., 1989, Analysis of dietary fiber, *Anim. Feed Sci. Tech.* **23:**1–13.

Marlett, J. A., and Chesters, J. G., 1985, Measuring dietary fiber in human foods, *J. Food Sci.* **50:**410–414, 423.

Marlett, J. A., and Lee, S. C., 1980, Dietary fiber, lignocellulose and hemicellulose contents of selected foods determined by modified and unmodified Van Soest procedures, *J. Food Sci.* **45:**1688–1693.

Marlett, J. A., and Navis, D., 1988, Comparison of gravimetric and chemical analyses of total dietary fiber in human foods, *J. Agr. Food Chem.* **36:**311–315.

Marlett, J. A., Chesters, J. G., Longacre, M. J., and Bogdanske, J. J., 1989, Recovery of soluble dietary fiber is dependent on the method of analysis, *Am. J. Clin. Nutr.* **50:**479–485.

Mongeau, R., and Brassard, R., 1986, A rapid method for the determination of soluble and insoluble dietary fiber: Comparison with AOAC total dietary fiber procedure and Englyst's method, *J. Food Sci.* **51:**1333–1336.

Neilson, M. J., and Marlett, J. A., 1983, A comparison between detergent and nondetergent analyses of dietary fiber in human foodstuffs, using high-performance liquid chromatography to measure neutral sugar composition, *J. Agr. Food Chem.* **31:**1342–1347.

Newmark, H. L., 1987, Plant phenolics as inhibitors of mutational and precarcinogenic events, *Can. J. Physiol. Pharmacol.* **65:**461–466.

Patrow, C. J., and Marlett, J. A., 1986, Variability in the dietary fiber content of wheat and mixed grain commercial breads, *J. Am. Diet Assoc.* **86:**794–796.

Paul, A. A., and Southgate, D. A. T., 1978, *McCance and Widdowson's The Composition of Foods,* 4th ed.), Elsevier/North-Holland Biomedical Press, Amsterdam, p. 418.

Pilch, S. M. (ed.), 1987, *Physiological Effects and Health Consequences of Dietary Fiber,* FASEB, Bethesda, MD.

Prosky, L., Asp, N. G., Furda, I., DeVries, J. W., Schweizer, T. F., and Harland, B. F., 1984, Determination of total dietary fiber in foods, food products and total diets: Interlaboratory study, *J. Assoc. Off. Anal. Chem.* **67:**1044–1052.

Prosky, L., Asp, N. G., Furda, I., DeVries, J. W., Schweizer, T. F., and Harland, B. F., 1985, Determination of total dietary fiber in foods and food products: Collaborative study, *J. Assoc. Off. Anal. Chem.* **68:**677–679.

Prosky, L., Asp, N. G., Schweizer, I. F., DeVries, J. W., and Furda, I., 1988, Determination of insoluble, soluble, and total dietary fiber in foods and food products: Interlaboratory study, *J. Assoc. Off. Anal. Chem.* **71**:1017–1023.

Robertson, J. B., and Van Soest, P. J., 1981, The detergent system of analysis and its application to human foods, in: *The Analysis of Dietary Fiber in Food* (W. P. T. James and O. Theander, eds.), Marcel Dekker, New York, pp. 123–158.

Saeman, J. F., Moore, W. E., Mitchell, R. L., and Millett, M. A., 1954, Techniques for the determination of pulp constituents by quantitative paper chromatography, *Tappi* **37**:336–343.

Salomonsson, A. C., Theander, O., and Westerlund, E., 1984, Chemical characterization of some Swedish cereal whole meal and bran fractions, *Swed. J. Agr. Res.* **14**:111–117.

Schaller, D., 1976, Analysis of cereal products and ingredients. Paper presented at the 61st annual meeting of the American Association of Cereal Chemists, October 6, 1976, New Orleans.

Shinnick, F. L., Longacre, M. J., Ink, S. L., and Marlett, J. A., 1988, Oat fiber: Composition vs. physiological function, *J. Nutr.* **118**:144–151.

Southgate, D. A. T., 1969, Determination of carbohydrates in foods. II. Unavailable carbohydrates, *J. Sci. Food Agr.* **20**:332–335.

Southgate, D. A. T., 1981, Use of the Southgate method for unavailable carbohydrates in the measurement of dietary fiber, in: *The Analysis of Dietary Fiber in Food* (W. P. T. James and O. Theander, eds.), Marcel Dekker, New York, pp. 1–19.

Southgate, D. A. T., and White, M. A., 1981, Commentary on results obtained by the different laboratories using the Southgate method, in: *The Analysis of Dietary Fiber in Foods* (W. P. T. James and O. Theander, eds.), Marcel Dekker, New York, pp. 37–50.

Theander, O., 1983, Advances in the chemical characterization and analytical determination of dietary fibre components, in: *Dietary Fibre* (G. G. Birch and K. J. Parker, eds.), Applied Science Publishers, London, pp. 77–93.

Theander, O., and Aman, P., 1979, Studies on dietary fibres. 1. Analysis and chemical characterization of water-soluble and water-insoluble dietary fibres, *Swed. J. Agr. Res.* **9**:97–106.

Theander, O., and Aman, P., 1981, Analysis of dietary fibers and their main constituents, in: *The Analysis of Dietary Fiber in Food* (W. P. T. James and O. Theander, eds.), Marcel Dekker, New York, pp. 51–70.

Theander, O., and James, P., 1979, European efforts in dietary fiber characterization, in: *Dietary Fibers: Chemistry and Nutrition*, (G. E. Inglett and I. Falkehag, eds.), Academic Press, New York, pp. 245–249.

Theander, O., and Westerlund, E., 1986, Studies on dietary fiber. 3. Improved procedures for analysis of dietary fiber, *J. Agr. Food Chem.* **34**:330–336.

Van Soest, P. J., and Wine, R. H., 1967, Use of detergents in the analysis of fibrous foods. IV. Determination of plant cell-wall constituents, *J. Assoc. Off. Anal. Chem.* **50**:50–55.

Watt, B. K., and Merrill, A. L., 1963, *Composition of Foods, Agriculture Handbook No. 8*, United States Department of Agriculture, Washington, p. 190.

Westerlund, E., Theander, O., Andersson, R., and Aman, P., 1989, Breadmaking. 2. Effects of baking on polysaccharides in white bread fractions, *J. Cereal Sci.* **10**: 149–156.

4

Dietary Fiber and Resistant Starch

A Nutritional Classification of Plant Polysaccharides

HANS N. ENGLYST and SUSAN M. KINGMAN

1. INTRODUCTION

Plant polysaccharides may be separated into two broad categories. Starch, a ubiquitous storage polysaccharide, is an α-linked glucan and is the major carbohydrate of dietary staples such as cereal grains and potatoes. The nonstarch polysaccharides (NSP) of plants, such as cellulose, pectin, and hemicellulose, are non-α-glucan polysaccharides. The NSP tend to have a structural function and are the principal components of the plant cell wall. Recently NSP have become the objective in the measurement of dietary fiber.

Because of its α-glucosidic linkages, starch is susceptible to hydrolysis by pancreatic α-amylase. In contrast, NSP completely resist digestion by the enzymes secreted into the small intestine of man.

Until recently, the physiological separation of the two polysaccharide categories was thought to be as complete as the chemical, starch being entirely digested and NSP entirely resisting digestion in the small intestine. However, it is now clear from a number of studies in man (Englyst and Cummings, 1985, 1986, 1987) and animals (Millard and Chesson, 1984; Graham *et al.*, 1985) that this is not the case and that a significant proportion of dietary starch may pass intact into the colon, where it becomes available for fermentation by gut bacteria. In addition, for some animals a considerable amount of NSP (and starch) may be fermented in the small intestine. The metabolic consequences of starch resisting digestion are not yet clear, but it is assumed that starch is fermented with the production of volatile fatty acids (VFA), which are then absorbed.

HANS N. ENGLYST and SUSAN M. KINGMAN • MRC Dunn Clinical Nutrition Centre, Cambridge CB2 1QL, United Kingdom.

This chapter describes the development of methods for the classification and measurement of starch and nonstarch polysaccharides (dietary fiber). A novel classification of starch based on its *in vitro* digestibility is proposed as a framework for nutritional investigation of starch resisting digestion in the small intestine of man.

2. DEFINITION OF DIETARY FIBER

In 1976, Trowell *et al.* defined dietary fiber as the "plant polysaccharides and lignin which are resistant to hydrolysis by the digestive enzymes of man." This definition was intended to encompass all polysaccharides other than starch, the only plant polysaccharide known to be digested by human enzymes (Trowell *et al.*, 1985). However, the definition was also open to interpretation as "plant polysaccharides and lignin resisting or escaping hydrolysis by the digestive enzymes of man." This interpretation was inevitably unsuitable for analytical purposes, constituting an ill-defined analytical goal. The discovery that significant amounts of starch may also escape digestion in the small intestine made it critical that dietary fiber be redefined in unequivocal chemical terms. Thus, at the workshop on dietary fiber in Cambridge in 1978, sponsored by the EEC Committee on Medical Research, it was proposed by Cummings and Englyst (James and Theander, 1981) that for analytical purposes dietary fiber should be defined as the sum of the nonstarch polysaccharides in food.

Since lignin is not a polysaccharide, it is not included in the measurement. This exclusion is defensible on the grounds that lignin is a very minor component in the human diet, has physiological effects (mainly shown in animals) very different from those of NSP, and is difficult to determine accurately. Quantitatively lignin is insignificant, but its inaccurate measurement may obscure the true significance of individual NSP measurements. Should lignin in the future be shown to be an important food component, a case must be made for its separate measurement. Defining dietary fiber for analytical purposes solely as NSP gives the best index of cell wall polysaccharides, is in keeping with the original concept of dietary fiber, and provides a clear objective for analysts (Englyst *et al.*, 1987a).

3. MEASUREMENT OF DIETARY FIBER AS NONSTARCH POLYSACCHARIDES

The measurement of dietary fiber as NSP requires a method that is specific and precise. The enzymatic–gravimetric techniques for dietary fiber analysis, such as the AOAC method (Prosky *et al.*, 1984), are not chemically specific but isolate a fraction of the diet that *in vitro* resists hydrolysis by bacterial amylase and protease. The isolated fraction is assessed gravimetrically and may contain some starch, lignin, Maillard browning products, and other components of plant or animal origin not digested by the *in vitro* technique. In contrast, the Englyst method for dietary fiber analysis measures an enzymatically isolated polysaccharide fraction (NSP) in terms of its constituent mono-saccharides. It is therefore a highly specific and informative method measuring chemically well-defined components.

The Englyst method has been described in detail elsewhere (Englyst et al., 1982, 1987b,c; Englyst and Cummings, 1984, 1988; Englyst and Hudson, 1987), and a kit is available from Novo Biolabs. The method is outlined in Fig. 1, and a summary is given below.

3.1. Measurement of Total NSP

Starch in a milled sample is dispersed with dimethylsulfoxide and subsequently hydrolyzed by incubation with amylase and pullulanase. After precipitation of NSP with ethanol, the starch-free residue is hydrolyzed with sulfuric acid. The neutral sugars released on hydrolysis are measured as alditol acetates by gas–liquid chromatography (GLC), and the uronic acids by colorimetry. Total NSP is then calculated as the sum of the individual neutral sugars and uronic acids.

Alternatively a single value for NSP may be obtained using a rapid colorimetric technique allowing dietary fiber to be measured within an 8-hr working day. After acid hydrolysis of NSP (see Fig. 1), the sample is neutralized with sodium hydroxide and heated for 15 min with a dinitrosalicylate reagent. The absorbance for test samples and standards is read at 530 nm, allowing the calculation of NSP as reducing sugars in the hydrolysate.

3.2. Measurement of Soluble and Insoluble NSP

An estimate of the insoluble NSP present in a sample may be obtained by replacing the precipitation of soluble NSP with ethanol (see Fig. 1) by extraction with phosphate buffer, pH 7, at 100°C. In all other respects the procedure for the measurement of insoluble NSP is identical to that for total NSP, and the result may be obtained by GLC or colorimetry. The values obtained represent insoluble NSP; soluble NSP is given by the difference between total NSP and insoluble NSP.

3.3. Measurement of Cellulose and Noncellulosic Polysaccharides

In the procedure for total NSP, cellulose is dispersed with 12 M sulfuric acid prior to hydrolysis. If this dispersion is omitted from the procedure and replaced by direct hydrolysis with 2 M sulfuric acid, cellulose is not measured, and a value for non-cellulosic polysaccharides (NCP) is obtained. A value for cellulose may be calculated as the difference between the glucose content of the total NSP and that of the NCP obtained either by GLC of the individual sugars or by a specific colorimetric method using glucose oxidase.

Table I gives values for NSP, measured by the Englyst procedure, in a selection of plant foodstuffs.

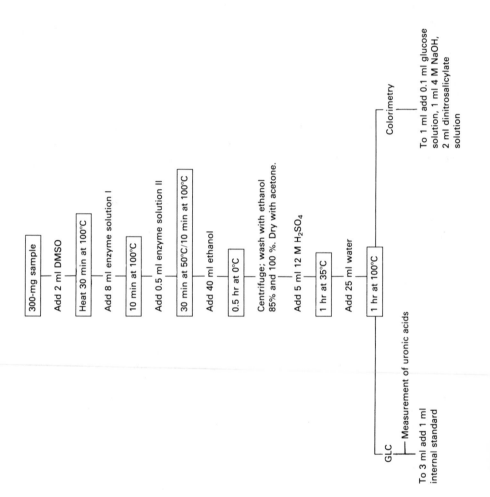

300-mg sample

Add 2 ml DMSO

Heat 30 min at 100°C

Add 8 ml enzyme solution I

10 min at 100°C

Add 0.5 ml enzyme solution II

30 min at 50°C/10 min at 100°C

Add 40 ml ethanol

0.5 hr at 0°C

Centrifuge; wash with ethanol 85% and 100 %. Dry with acetone.

Add 5 ml 12 M H_2SO_4

1 hr at 35°C

Add 25 ml water

1 hr at 100°C

GLC — Measurement of uronic acids

To 3 ml add 1 ml internal standard

Colorimetry

To 1 ml add 0.1 ml glucose solution, 1 ml 4 M NaOH, 2 ml dinitrosalicylate solution

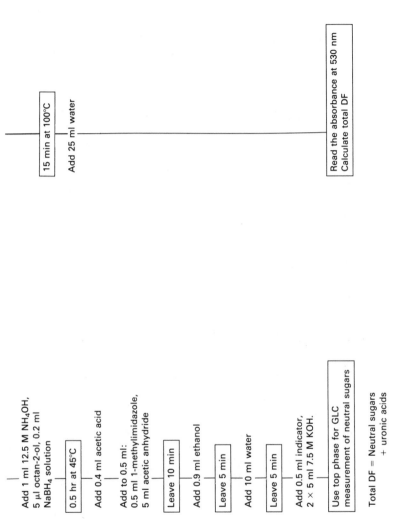

FIGURE 1. A flow chart of the Englyst procedures for measurement of total DF by gas–liquid chromatography (GLC) or by colorimetry. For measurement of insoluble DF, replace the 40 ml of ethanol with 40 ml of buffer, pH 7, and extract for 30 min at 100°C. Soluble DF = Total DF – Insoluble DF.

TABLE I. Nonstarch Polysaccharides in Plant Products

		Total (g/100 g fresh weight)	Total (g/100 g dry weight)	Composition (g/100 g dry weight)									
				Cellulose	Noncellulosic polysaccharides[a]								
					Rha	Fuc	Ara	Xyl	Man	Gal	Glu	U.Ac.
Whole wheat flour, Allinsons													
	Soluble NSP	2.5	2.8	—	t	t	0.8	1.3	0.1	0.2	0.3	0.1
	Insoluble NSP	7.2	8.1	1.6	t	t	2.3	3.3	t	0.1	0.6	0.2
DM = 0.89	Total NSP	9.7	10.9	1.6	t	t	3.1	4.6	0.1	0.3	0.9	0.3
Rice, brown													
	Soluble NSP	t	t	—	t	t	t	t	t	t	t	t
	Insoluble NSP	1.9	2.2	0.7	t	t	0.4	0.5	t	0.1	0.1	0.1
DM = 0.86	Total NSP	1.9	2.2	0.7	t	t	0.4	0.5	t	0.1	0.1	0.1
White bread, Sunblest													
	Soluble NSP	0.9	1.7	—	t	t	0.5	0.8	t	0.1	0.3	t
	Insoluble NSP	0.6	1.1	0.2	t	t	0.3	0.4	0.1	t	0.1	t
DM = 0.54	Total NSP	1.5	2.8	0.2	t	t	0.8	1.2	0.1	0.1	0.4	t
Ryvita, original													
	Soluble NSP	3.9	4.1	—	t	t	1.2	1.7	0.1	0.1	1.0	t
	Insoluble NSP	7.8	8.2	1.0	t	t	2.2	3.4	0.1	0.2	1.2	0.1
DM = 0.95	Total NSP	11.7	12.3	1.0	t	t	3.4	5.1	0.2	0.3	2.2	0.1
Digestive, McVities													
	Soluble NSP	1.1	1.1	—	t	t	0.3	0.6	t	0.1	0.1	t
	Insoluble NSP	1.1	1.1	0.2	t	t	0.3	0.4	0.1	t	0.1	t
DM = 0.99	Total NSP	2.2	2.2	0.2	t	t	0.6	1.0	0.1	0.1	0.2	t

		NSP (as eaten)	NSP (dry matter)	Cellulose	Rha	Fuc	Ara	Xyl	Man	Gal	Glu	U.Ac
Corn flakes, Kelloggs	Soluble NSP	0.4	0.4	—	t	t	t	0.2	t	t	0.1	0.1
	Insoluble NSP	0.5	0.5	0.3	t	t	0.1	0.1	t	t	t	t
DM = 0.96	Total NSP	0.9	0.9	0.3	t	t	0.1	0.3	t	t	0.1	0.1
Whole wheat spaghetti, cooked	Soluble NSP	0.8	2.0	—	t	t	0.5	0.7	0.1	0.2	0.4	0.1
	Insoluble NSP	2.7	6.8	1.7	t	t	1.9	2.5	0.1	0.1	0.3	0.2
DM = 0.39	Total NSP	3.5	8.8	1.7	t	t	2.4	3.2	0.2	0.3	0.7	0.3
Beans, red kidney, dried, cooked	Soluble NSP	3.2	9.2	—	0.2	t	4.3	1.1	0.2	1.2	0.4	1.8
	Insoluble NSP	3.5	9.9	4.8	0.1	t	2.7	0.9	0.1	0.3	t	1.0
DM = 0.35	Total NSP	6.7	19.1	4.8	0.3	t	7.0	2.0	0.3	1.5	0.4	2.8
Sweet corn kernels, cooked	Soluble NSP	0.1	0.5	—	t	t	0.1	0.1	t	0.1	0.1	0.1
	Insoluble NSP	1.3	5.5	1.7	t	t	1.1	1.6	t	0.3	0.5	0.3
DM = 0.24	Total NSP	1.4	6.0	1.7	t	t	1.2	1.7	t	0.4	0.6	0.4
Dates, dried	Soluble NSP	1.2	1.4	—	t	t	0.1	t	t	0.2	t	1.1
	Insoluble NSP	2.8	3.1	1.5	t	t	0.2	1.0	t	0.2	t	0.2
DM = 0.89	Total NSP	4.0	4.5	1.5	t	t	0.3	1.0	t	0.4	t	1.3
Oranges, raw	Soluble NSP	1.4	9.8	—	0.3	t	1.9	0.1	0.1	1.4	0.1	5.9
	Insoluble NSP	0.7	5.2	3.4	t	t	0.3	0.5	0.3	0.4	t	0.3
DM = 0.14	Total NSP	2.1	15.0	3.4	0.3	t	2.2	0.6	0.4	1.8	0.1	6.2
Peanuts, fresh	Soluble NSP	1.9	2.0	—	0.1	t	0.6	0.2	t	0.1	t	1.0
	Insoluble NSP	4.3	4.6	2.1	t	t	1.3	0.4	t	0.2	0.1	0.5
DM = 0.94	Total NSP	6.2	6.6	2.1	0.1	t	1.9	0.6	t	0.3	0.1	1.5

[a]Rha, rhamnose; Fuc, fucose; Ara, arabinose; Xyl, xylose; Man, mannose; Gal, galactose; Glu, glucose; U.Ac, uronic acid; t, trace.

4. INTERFERENCE OF STARCH IN THE MEASUREMENT OF NSP

Glucose molecules from hydrolyzed starch are not distinguished from those released by the hydrolysis of NSP. Thus, any starch remaining in a sample following enzymatic hydrolysis will be measured as NSP and artificially elevate the dietary fiber content of the sample. Accurate measurement of dietary fiber is therefore critically dependent on the complete removal of starch from the sample.

During the development of the Englyst procedure (Englyst *et al.*, 1982) it was observed that certain processed foods such as white bread and cooked potato exhibited higher apparent values for NSP than their raw counterparts. This increase in NSP was surmised to be artifactual, since cell wall polysaccharides cannot be increased by processing, and it was subsequently demonstrated that the increment was entirely accounted for by an increase in glucose. The glucan involved was found to be hydrolyzed by pancreatin or amyloglucosidase following dispersion using KOH or dimethylsulfoxide and was thus identified as a form of starch. This fraction was termed resistant starch (RS) and defined as starch resistant to dispersion in boiling water and hydrolysis with pancreatic amylase and pullulanase.

5. DIGESTIBILITY OF NSP AND STARCH

The discovery that a small proportion of starch in certain foods was resistant to α-amylase activity *in vitro* prompted a series of experiments designed to investigate the digestibility of different starch forms *in vivo*. The study of digestion within the small intestine was made feasible by the use of ileostomy subjects—people who have undergone surgical removal of their large bowel. In these people the terminal ileum forms a fistula on the abdominal wall from which small intestinal effluent empties into a bag for collection or disposal. This effluent represents the digesta that, in normal subjects, enters the colon and acts as a substrate for the colonic flora.

In a series of experiments using ileostomy subjects, the digestibilities of NSP and starch from white bread, corn flakes, oats, banana, and potato were assessed. Following a 24-hr period on a plant-polysaccharide-free diet, the subjects were given a test meal at 9 a.m. Every 2 hr thereafter the resulting effluent was collected from the ileostomy bag. The rate of passage of the test meal through the small intestine was monitored using the amount of NSP-xylose recovered in the effluent.

Using this technique it was shown that NSP are not digested in the human small intestine. The recoveries of NSP from the ileostomy effluent were virtually complete for each of the foods tested, as shown in Table II. Thus, it may be assumed that the amount of NSP entering the colon in normal subjects is equivalent to that ingested in the diet.

If the lack of digestibility of NSP in the small intestine was expected, the poor digestibility of certain starches was not. Table III shows that the test foods exhibited a broad range of starch digestibility that was not necessarily related to the percentage of resistant starch (as originally defined) present in the food. Although the starch from cereal products and freshly cooked potato was well digested, a large proportion (up to 89%) of the starch from banana escaped digestion in the small intestine. Cooled cooked

TABLE II. Recovery of NSP in Ileostomy Effluent

	NSP (g)	
	Fed	Recovered
White bread	2.3	2.3
Corn flakes	0.6	0.7
Oats	6.6	6.3
Bananas	2.1	2.0
Potato	3.3	3.4

potato was less well digested than freshly cooked potato. Although starch digestion was partly ameliorated by reheating the cooled potato, it was still less digestible than its freshly cooked counterpart.

It is clear from these studies that the amount of starch that actually escapes digestion in the small intestine may be greatly in excess of that fraction of starch originally defined as RS. Thus, the analytical procedure previously developed for the measurement of RS is now considered on its own to be insufficient for application to physiological studies on the digestibility of starch. It is also inappropriate that the fraction originally identified as RS (mainly retrograded amylose) should retain this title exclusively, since it represents only a small proportion of the starch that resists digestion. All starch entering the large intestine is suitable as a substrate for bacterial fermentation and, on the basis of this physiological property, should be grouped together as resistant starch.

6. FACTORS INFLUENCING STARCH DIGESTION

Most of the starch that reaches the colon is not totally resistant to pancreatic amylase, but for one reason or another its hydrolysis is retarded so that it is not

TABLE III. Digestion of Starch in the Small Intestine of Man

	Starch fed		Starch recovered	
	Total (g)	RS[a] (%)	(g)	(%)
White bread	62	2	1.6	3
Oats	58	0	1.2	2
Cornflakes	74	4	3.7	5
Banana	19	0	17.2	89
Potato				
Freshly cooked	45	2	1.5	3
Cooled	47	3	5.8	12
Reheated	47	3	3.6	8

[a] As originally defined (i.e., mainly retrograded amylose).

completely digested during its passage through the small intestine. The reasons for this incomplete digestion may be separated into intrinsic factors (i.e., properties of the starchy food itself) and extrinsic factors.

6.1. Intrinsic Factors

6.1.1. Physical Inaccessibility

This occurs when starch is contained within undisrupted plant structures such as whole or partly milled grains and seeds. Here, the cell walls may entrap starch and prevent its complete swelling and dispersion (Wursch et al., 1986), thus delaying or preventing its hydrolysis with pancreatic amylase in the small intestine. In humans, whole or coarsely milled grains of wheat, maize, and oats have been shown to elicit smaller plasma insulin responses than finely ground flours (Heaton et al., 1988). It has also been observed (Englyst, 1985) that after a meal of sweet corn, peas, and beans, up to 20% of fecal solids may be starch contained in recognizable, undigested food. Other examples of foods whose gross physical structure may retard hydrolysis and result in incomplete digestion of starch in the small intestine are parboiled white rice (Wolever et al., 1986) and spaghetti (Hermansen et al., 1986).

6.1.2. Resistant Starch Granules

In the plant, starch is contained within granular structures where it exists in a closely packed, partially crystalline form. The crystal structure of starch within the granule can be of x-ray diffraction pattern A, B, or C as distinguished by Katz (1937). The x-ray pattern may be explained by envisaging a double-helical conformation that packs into a hexagonal array. In the A-type starch, the center of the array is occupied by a further helix, whereas in the B type this position is occupied by water (Wu and Sarko, 1978a,b; Gidley, 1987).

The actual crystalline structure of the starch granule is suggested to depend on the chain length of amylopectin. The A, B, and C types of starches are reported to contain amylopectin molecules with average chain lengths of 23–29, 30–44, and 26–29, respectively (Hizukuri, 1985). A is the normal pattern for cereal starch granules, B is typical of potato, amylomaize, and retrograded starch, and C (a combination of A and B patterns) is characteristic of certain pea and bean starches. In general starch granules showing x-ray diffraction patterns B or C tend to be the most resistant to pancreatic amylase, although the degree of resistance is dependent on the plant source (Fuwa et al., 1980). Cooking disrupts the granules and facilitates the hydrolysis of the starch contained within them. Thus, the resistant crystal structure of potato starch granules is of no nutritional importance when potatoes are cooked before consumption. The highly resistant nature of banana starch granules is of more nutritional significance since bananas are normally eaten raw.

6.1.3. Retrograded Starch

Controlled drying of a heated starch gel can produce any of the A, B, or C x-ray diffraction patterns depending on the temperature (Katz, 1937).

Crystallization at higher temperature and lower water content will favor the A, and lower temperature and high water content the B pattern. Therefore, on cooling, gelatinized starchy foods will retrograde and develop x-ray diffraction pattern B (Katz, 1934, 1937; Wu and Sarko, 1978a). During retrogradation, solubility of the starch molecule decreases, and so does its susceptibility to hydrolysis by acid and enzymes.

Chain length and linearity are important factors affecting retrogradation. The longer the starch chains, the greater the number of interchain hydrogen bonds formed. The extent of crystalline bonding in amylopectin is limited by the branch length. Therefore, amylopectin retrogrades to a lesser extent than amylose, and retrograded amylopectin is not so firmly bound as retrograded amylose (Sterling, 1978).

Pure amylose crystallized with the B pattern can only be solubilized by autoclaving (Wu and Sarko, 1978a). The resistance to dissolution arises from the extensive network of intra- and interhelical hydrogen bonds that stabilize the double-helical structure of crystalline amylose. In mixtures, amylopectin has an inhibitory effect, and the properties described for pure B-amylose are enhanced in proportion to the content of amylose (Wu and Sarko, 1978a). In a separate study (Ring *et al.*, 1987) it was shown that the dissolution of the amylose and amylopectin gels required temperatures of 153 and 59°C, respectively.

Retrograded starch may be separated into that redispersed at 100°C (mainly retrograded amylopectin) and that resistant to dispersion in boiling water (mainly retrograded amylose). Studies in man suggest that the mainly-retrograded-amylose fraction virtually completely resists digestion in the small intestine of man (Englyst and Cummings, 1985, 1987).

6.1.4. Other Factors

Other factors that are intrinsic to starchy foods have been shown to affect α-amylase activity *in vitro*. These include specific amylose–lipid complexes (Holm *et al.*, 1983), native α-amylase inhibitors (Shainkin and Birk, 1970), and various nonstarch polysaccharides that may have a direct effect on enzyme activity (Dunaif and Schneeman, 1981). However, it is not clear to what extent these factors affect the digestibility of starch *in vivo*.

6.2. Extrinsic Factors

The digestibility of starch within the small intestine is also influenced by a number of highly variable factors that may be independent of crystallinity and physical form of the starchy food itself. These include the extent of chewing, the transit time of the food along the small intestine, the concentration of amylase available for breakdown of the starch, the amount of starch, and the presence of other food components that might retard enzymatic hydrolysis. It is not possible to predict with any certainty the extent to which such extrinsic factors may influence the digestion of a starchy food. Thus, as is the case for dietary fiber, a physiological definition of resistant starch can never be completely compatible with an *in vitro* analytical technique.

However, despite their limitations in terms of describing physiologically resistant starch, *in vitro* analytical techniques may be used to obtain a reproducible measure of the potential of various starchy foods to resist digestion in the small intestine.

7. *IN VITRO* NUTRITIONAL CLASSIFICATION OF STARCH

Total starch (TS) is measured as the glucose released from a milled or homogenized sample, gelatinized at 100°C, dispersed with KOH, and then incubated with pancreatin and amyloglucosidase.

The fractionation of starch into the subfractions shown in Table IV is based on the incubation of starchy foods with pancreatin and amyloglucosidase under specified conditions.

Rapidly digestible starch (RDS) is measured as the glucose released from a sample after 20 min. It mainly consists of amorphous and dispersed starch.

Slowly digestible starch (SDS) is measured as the glucose released from a sample after 120 min minus RDS. It mainly consists of starch with x-ray diffraction patterns A and C, but it also includes the proportion of B-type starch (granular or retrograded) hydrolyzed within 120 min.

Resistant starch (RS) is determined as TS − (RDS + SDS). RS_1 may be measured as the increase in glucose released within 120 min obtained by milling or homogenizing the food sample. RS_2 is measured as TS − RS_3 and the glucose released after 120 min for a milled or blended sample. RS_3 is measured as the starch in a milled sample that resists dispersion in boiling water and hydrolysis with pancreatic amylase and pullulanase as described previously (Englyst *et al.*, 1982). It mainly consists of retrograded amylose.

The resistant starch fractions described here are determined solely by properties inherent in the food sample and are not the result of variable physiological factors. The fractions therefore represent reproducible measurements by which starchy foods can be compared and validated within a nutritional framework.

8. THE DIGESTIBILITY OF STARCH IN PROCESSED FOODS

A critical factor in the digestibility of starch is its crystalline state. Amorphous and dispersed starch is readily accessible to hydrolytic enzymes, but as a crystal structure forms, resistance to digestion increases to a degree dependent on the crystal pattern. As

TABLE IV. *In Vitro* Nutritional Classification of Starch

Type of starch	Example of occurrence	Probable digestion in small intestine
Rapidly digestible starch	Freshly cooked starchy food	Rapid
Slowly digestible starch	Most raw cereals	Slow but complete
Resistant starch		
1. Physically inaccessible starch	Partly milled grain and seeds	Resistant
2. Resistant starch granules	Raw potato and banana	Resistant
3. Retrograded starch	Cooled cooked potato, bread, and corn flakes	Resistant

discussed earlier, temperature and moisture are important determinants of crystal structure.

In a series of experiments in this laboratory, using different types of flour, it has been demonstrated that the availability of water during the processing of starchy foods is critical to starch digestibility. Each of four flours was made into biscuits prepared with (1) butter, (2) oil, or (3) oil or butter plus water, and the biscuits were analyzed using the *in vitro* scheme described in this chapter. The results are shown in Table V. The baking of biscuits with butter (low water) or oil did not fully gelatinize the starch granules. For wheat and maize with the A-type crystal structure, this resulted in a high proportion of SDS. For the B-type starch in banana and potato, the incomplete gelatinization resulted in a high proportion of RS in the form of resistant starch granules (RS_2). In each case the inclusion of water in the biscuit dough increased the proportion of RDS at the expense of SDS and RS_2.

Breads made with white or whole-meal flours were highly digestible with 95% and 94% of the starch measured as RDS, respectively (Table VI). During baking 2% RS_3 was formed in the breads.

For freshly cooked beans 45% of the starch was measured as SDS and 11% as RS_1 (Table VII). When cooled, RS_1 and RS_3 increased to a total of 31%. This was especially at the expense of RDS. Lentils were more digestible than the beans, with 54% of the starch as RDS. For peas and sweet corn a very high proportion was measured as physically inaccessible starch, RS_1.

TABLE V. *In Vitro* Digestibility of Biscuits Made from Different Flours

Preparation	Rapidly digestible starch (RDS) (%)	Slowly digestible starch (SDS) (%)	Resistant starch (%)		
			RS_1	RS_2	RS_3
Maize starch					
Flour + butter	42	45	—	12	1
Flour + oil	38	53	—	8	1
Flour + oil + water	89	2	—	7	2
Banana Flour					
Flour + butter	15	30	—	55	t
Flour + oil	9	13	—	78	—
Flour + oil + water	70	17	—	13	t
Potato flour					
Flour + butter	19	27	—	53	t
Flour + butter + water	72	21	—	6	t
Flour + oil	15	20	—	65	—
Wheat flour					
Flour + butter	38	59	—	3	t
Flour + butter + water	99	t	—	1	t
Flour + oil	52	48	—	t	t
Digestive biscuits	50	48	—	1	t

TABLE VI. *In Vitro* Digestibility of Wheat Flour and Breads

	Rapidly digestible starch (RDS) (%)	Slowly digestible starch (SDS) (%)	Resistant starch (%)		
			RS$_1$	RS$_2$	RS$_3$
White flour	49	48	—	3	t
White breads	95	3	—	—	2
Whole wheat breads	94	4	—	—	2

9. DISCUSSION AND CONCLUSIONS

The definition of dietary fiber as nonstarch polysaccharides allows the analyst to identify and measure a chemically well-defined component of the diet. The Englyst method, presented here, measures dietary fiber as the sum of monosaccharides released from NSP by acid hydrolysis and gives the option of measuring subfractions of dietary fiber (soluble and insoluble, cellulose and noncellulosic polysaccharides) using simple modifications. With the GLC procedure it is also possible to obtain detailed information about the constituent sugars of the NSP being measured. This is valuable for interpreting the results of physiological and epidemiologic studies where disease may be related to the type of dietary fiber. When detailed information is not required, values for total, soluble, and insoluble dietary fiber can be obtained using the colorimetric option of the method.

Existing enzymatic–gravimetric techniques for the determination of dietary fiber include in the measurement a small amount of starch that is resistant to pancreatic amylase and pullulanase *in vitro*. It has been suggested that this starch consisting of RS$_3$ or part of this fraction should be included with NSP under the heading of dietary fiber (Asp and Johansson, 1984; Berry, 1984). Like NSP, RS$_3$ escapes digestion in the small intestine of man, but it is now clear that this particular fraction is not representative of the starch that resists digestion by human small intestinal enzymes *in vivo*. Other types

TABLE VII. *In Vitro* Digestibility of Legumes and Sweetcorn

	Rapidly digestible starch (RDS) (%)	Slowly digestible starch (SDS) (%)	Resistant starch (%)		
			RS$_1$	RS$_2$	RS$_3$
Beans, freshly cooked	37	45	11	t	6
Beans, cooled	29	39	20	—	11
Lentils, cooled	54	21	14	1	9
Peas, freshly cooked	29	36	27	—	7
Sweet corn, freshly cooked	32	43	25	—	t

of starch also pass into the colon (Englyst and Cummings, 1985, 1986, 1987). In addition to NSP and various types of starch, other substances such as protein, lactose, and oligosaccharides may escape digestion in the small intestine. This characteristic is therefore not unique to NSP and RS_3 and is not a rational basis for the inclusion of RS_3 with dietary fiber.

Types of starch that resist digestion or are only slowly hydrolyzed in the small intestine may be important food components with benefits to health. Since the fraction of starch measured as RS_3 *in vitro* represents only a small proportion of that which resists digestion *in vivo*, it is inappropriate to measure this fraction alone, whether included as dietary fiber or not. It seems most logical that the types of starch that resist digestion in the small intestine of man should be treated as a separate entity from dietary fiber and measured as such. For this purpose an *in vitro* technique for the classification and measurement of starch for nutritional purposes has been developed.

The inclusion of RS_3 in dietary fiber as by the AOAC procedure poses a separate problem for the analyst because the quantity of RS_3 in a food is critically dependent on processing conditions. RS_3 may be introduced into a food as a result of heating, cooling, freezing, or drying. Thus, by the AOAC procedure (Prosky *et al.*, 1984) different values for dietary fiber in a single food will be reported if it is hot, cold, frozen, or dried. The apparent dietary fiber content will also change if test samples before analysis are heated, cooled, frozen, or dried. The inclusion of an RS fraction in dietary fiber serves to increase variability and has the potential to obscure the true meaning of NSP and starch values. If a single food yields a range of dietary fiber values depending on the treatment it has received prior to analysis, the construction of food tables and the calculation of the fiber content of mixed dishes is made difficult if not impossible. Dietary fiber measured as NSP is not affected by food processing, so the amount of fiber in processed foods and mixed diets can be calculated simply from knowing the amount in the raw product.

REFERENCES

Asp, N.-G., and Johansson, C.-G., 1984, Dietary fibre analysis, *Nutr. Abstr. Rev.* **54:**735–752.

Berry, C. S., 1984, Resistant starch: Analytical nuisance or man-made dietary fibre? *FMBRA Bull.* **6:** 236–248.

Dunaif, G., and Schneeman, B. O., 1981, The effect of dietary fiber on human pancreatic activity *in vitro, Am. J. Clin. Nutr.* **34:**1034–1035.

Englyst, H. J. N., 1985, *Dietary Polysaccharide Breakdown in the Gut of Man,* Ph.D. Thesis, University of Cambridge, U.K.

Englyst, H. N., and Cummings, J. H., 1984, Simplified method for the measurement of total non-starch polysaccharides by gas–liquid chromatography of constituent sugars as alditol acetates, *Analyst* **109:**937–942.

Englyst, H. N., and Cummings, J. H., 1985, Digestion of the polysaccharides of some cereal foods in the human small intestine, *Am. J. Clin. Nutr.* **42:**778–787.

Englyst, H. N., and Cummings, J. H., 1986, Digestion of the carbohydrates of banana (*Musa paradisiaca sapientum*) in the human small intestine, *Am. J. Clin. Nutr.* **44:**42–50.

Englyst, H. N., and Cummings, J. H., 1987, Digestion of polysaccharides of potato in the small intestine of man, *Am. J. Clin. Nutr.* **45:**423–431.

Englyst, H. N., and Cummings, J. H., 1988, Improved method for measurement of dietary fiber as the non-starch polysaccharides in plant foods, *J. Assoc. Off. Anal. Chem.* **71:**808–814.

Englyst, H. N., and Hudson, G. J., 1987, Colorimetric method for routine measurement of dietary fibre as non-starch polysaccharides. A comparison with gas–liquid chromatography, *Food Chem.* **24**:63–76.

Englyst, H., Wiggins, H. S., and Cummings, J. H., 1982, Determination of the non-starch polysaccharides in plant foods by gas–liquid chromatography of constituent sugars as alditol acetates, *Analyst* **107**:307–318.

Englyst, H. N., Trowell, H. W., Southgate, D. A. T., and Cummings, J. H., 1987a, Dietary fiber and resistant starch, *Am. J. Clin. Nutr.* **46**:873–874.

Englyst, H. N., Cummings, J., and Wood, R., 1987b, Determination of dietary fibre in cereals and cereal products—collaborative trials. II. Study of a modified Englyst procedure, *J. Assoc. Publ. Analysts* **25**:59–71.

Englyst, H. N., Cummings, J. H., and Wood, R., 1987c, Determination of dietary fibre in cereals and cereal products—collaborative trials. III. Study of further simplified procedures, *J. Assoc. Publ. Analysts* **25**:73–110.

Fuwa, H., Takaya, T., and Sugimoto, Y., 1980, Degradation of various starch granules by amylases, in: *Mechanisms of Saccharide Polymerization and Depolymerization* (J. J. Marshall, ed.), Academic Press, New York, pp. 73–100.

Gidley, M. J., 1987, Factors affecting the crystalline type (A–C) of native starches and model compounds: A rationalisation of observed effects in terms of polymorphic structures, *Carbohydr. Res.* **161**:301–304.

Graham, H., Hesselman, K., and Aman, P., 1985, The effect of wheat bran, whole crop peas, and beet pulp on the digestibility of dietary components in a cereal-based pig feed, in: *Proceedings of the 3rd International Seminar on Digestive Physiology in the Pig, Copenhagen. Report from the National Institute of Animal Science, Denmark* (A. Just, H. Jorgensen, and J. A. Fernandez, eds.), National Institute of Animal Science, pp. 195–198.

Heaton, K. W., Marcus, S. N., Emmett, P. M., and Bilton, C. H., 1988, Particle size of wheat, maize and oat test meals: Effects on plasma glucose and insulin responses and on the rate of starch digestion *in vitro*, *Am. J. Clin. Nutr.* **47**:675–682.

Hermansen, K., Rasmussen, O., Arnfred, J., Winther, E., and Schmitz, O., 1986, Differential glycaemic effects of potato, rice and spaghetti in type 1 (insulin dependent) diabetic patients at constant insulinaemia, *Diabetalogia* **29**:358–361.

Hizukuri, S., 1985, Relationship between the distribution of the chain length of amylopectin and the crystalline structure of starch granules, *Carbohydr. Res.* **141**:295–306.

Holm, J., Björck, I., Ostrowska, S., Eliasson, A.-C., Asp, N. G., Larsson, K., and Lundquist, I., 1983, Digestibility of amylose–lipid complexes *in-vitro* and *in-vivo*, *Starch/Stärke* **35**:294–297.

James, W. P. T., and Theander, O. (eds.), 1981, *The Analysis of Dietary Fiber in Food*, Marcel Dekker, New York, p. 259.

Katz, J. R., X-ray investigation of gelatinization and retrogradation of starch and its importance for bread research, *Bakers Weekly* **81**:34–37.

Katz, J. R., 1937, The amorphous part of starch in fresh bread, and in fresh pastes and solutions of starch, *Recl. Trav. Chim. Pays Bas* **18**:55–59.

Millard, P., and Chesson, A., 1984, Modification to swede (*Brassica napus* L.) anterior to the terminal ileum of pigs: Some implications for the analysis of dietary fibre, *Br. J. Nutr.* **52**:583–594.

Prosky, L., Asp, N.-G., Furda, I., Devries, J. W., Schweizer, T. F., and Harland, B. F., 1984, Determination of total dietary fiber in foods, food products and total diets: Interlaboratory study, *J. Assoc. Off. Anal. Chem.* **67**:1044–1052.

Ring, S. G., Colonna, P., I'Anson, K. J., Kalichevsky, M. T., Miles, M. J., Morris, V. J., and Orford, P. D., 1987, The gelation and crystallisation of amylopectin, *Carbohydr. Res.* **162**:277-293.

Shainkin, R., and Birk, Y., 1970, α-Amylase inhibitors from wheat. Isolation and characterization, *Biochim. Biophys. Acta* **221**:502–513.

Sterling, C., 1978, Textural qualities and molecular structure of starch products, *J. Texture Stud.* **9**:225–255.

Trowell, H., Southgate, D. A. T., Wolever, T. M. S., Leeds, A. R., Gussell, M. A., and Jenkins, D. J. A., 1976, Dietary fiber redefined, *Lancet* **1**:967.

Trowell, H., Burkitt, D., and Heaton, K., eds., 1985, *Dietary Fibre, Fibre-Depleted Foods and Disease,* Academic Press, London.

Wolever, T. M. S., Jenkins, D. J. A., Kalmusky, J., Jenkins, A., Giordano, C., Giudici, S., Josse, R. G., and Wong, G. S., 1986, Comparison of regular and parboiled rices: Explanation of discrepancies between reported glycemic responses to rice, *Nutr. Res.* **6**:349–357.

Wu, H.-C., and Sarko, A., 1978a, The double-helical molecular structure of crystalline B-amylose, *Carbohydr. Res.* **61**:7–25.

Wu, H.-C., and Sarko, A., 1978b, The double-helical molecular structure of crystalline A-amylose, *Carbohydr. Res.* **61**:27–40.

Wursch, P., Del Vedovo, S., and Koellreutter, B., 1986, Cell structure and starch nature as key determinants of the digestion rate of starch in legume, *Am. J. Clin. Nutr.* **43**:25–29.

5

Quantitative and Qualitative Adaptations in Gastrointestinal Mucin with Dietary Fiber Feeding

MARIE M. CASSIDY,
SUBRAMANIAM SATCHITHANANDAM,
RICHARD J. CALVERT, GEORGE V. VAHOUNY,†
and ANTHONY R. LEEDS

1. INTRODUCTION

The viscous multicomponent mucin gel that coats the epithelial lining of the gastrointestinal tract is a heretofore neglected component of this organ system. The unique structural and functional characteristics of this biological material enable it to act as a protective physiological barrier to potentially deleterious agents. In addition, it may function to limit access to the intestinal surface and consequently limit absorption of nutrients.

Gastrointestinal mucus is a complex substance, primarily consisting of glycoproteins, water molecules, macromolecules of blood or tissue origin, microorganisms, and dead cells, which are continually shed into the lumen of the gastrointestinal tract. The highly specialized cells that synthesize and secrete mucin glycoproteins are a discrete population of enterocytes. They may be found predominantly on the surface, as in the gastric region, or interspersed at regular intervals with the absorptive epithelial cells, as in the villi of the small intestine.

Nutrients such as sugars, amino acids, and small peptides, in addition to the major

†Deceased August 1, 1986.

MARIE M. CASSIDY, SUBRAMANIAM SATCHITHANANDAM, and GEORGE V. VAHOUNY† • Departments of Physiology and Biochemistry, The George Washington University Medical Center, Washington, D.C. 20037.   RICHARD J. CALVERT • The U.S. Food and Drug Administration, Washington, D.C. 20204.   ANTHONY R. LEEDS • King's College, University of London, London WC2B 5RL, England.

electrolytes, must traverse the "unstirred water layer" before digestive and absorptive processes can ensue. A substantial portion of this layer is contributed by a continuous mucus gel barrier of variable thickness. Since mucin appears to play a significant role in protection of the mucosal surface against harmful materials in the lumen (Cassidy et al., 1981), dietary factors that influence the synthesis, composition, secretion, and degradation of mucin are of growing importance. Certain types of dietary fibers have been shown to have significant effects in reducing the carcinogenic potential of malignancies of the large bowel in animal models (Pilch, 1987). In general, experimental evidence suggests that depletion of mucin, possibly accompanied by alterations in mucin type, may be implicated in the development of colon cancer.

The search for a reliable early diagnostic marker of the progression from polyp formation to frank carcinoma has prompted the recent growth of interest in alterations in mucin associated with malignant transformation. In previous studies, we have demonstrated that 10% wheat bran fed to rats stimulated intestinal cell turnover and villus transit time (Vahouny et al., 1985). Simultaneously, the incorporation of radioactively labeled metabolic precursors of mucin was enhanced. Morphologically, goblet cell secretory activity was also elevated. In the studies reported here, we have developed a new polyclonal antibody to purified rat mucin that enables the assay of mucin mass in tissue or luminal compartments with relative ease and reliability.

There are indications that the presence of fiber constituents or their derivatives in the luminal contents may lead to organic chemical interactions with the secreted mucin, thus altering the nature of the mucosal mucus barrier. As more knowledge is developed concerning the structural and functional properties of both fiber and mucin that promote or impede hydration and gel formation, there may be many additional hypothetical compositional interactions deserving the attention of chemists interested in this area.

Finally, there are two major types of mucin, which may be defined chemically as either sialomucins or sulfomucins. The fetal/early neonatal pattern consists predominantly of less sulfated mucins (Colony, 1983). In the adult gut, the proportion of the sulfated component increases from proximal to distal colon. Although there are unresolved quantitative discrepancies between results obtained with biochemical versus histochemical methods, this trend is similar. Of major putative significance is the fact that the expression of epithelial cell dysplasia and adenomatosis in colon cancer is temporally associated with a reversion from the adult pattern to the neonatal one. In this time period, the expression of carcinoembryonic antigen reactivity is also observed (Greaves et al., 1984). Clearly, the possible beneficial effects of some types of dietary fiber with respect to this disease entity may reside in the capacity of certain fiber components to stimulate cell production and the provision of a healthy mucin profile. In this regard, we have examined the sialo/sulfomucin cytochemical staining characteristics of the small and large intestine in the rat model and the impact of a variety of soluble and insoluble fiber feeding regimens on this parameter.

2. INTERACTION OF DIETARY FIBER WITH THE INTESTINAL MUCOSA OF RATS

Certain dietary fibers, most particularly those that are soluble and form viscous gels, modify the absorption of nutrients such as glucose or lipids. This acute response to fiber ingestion appears to be a function of the viscosity of the material present in the lumen of the gastrointestinal tract, thereby affecting the resistance to the absorption of solutes from the small intestine. A spectrum of both *in vivo* and *in vitro* studies also suggests that dietary fiber components may indirectly and transiently interact with the intestinal mucosa. Thorough rinsing of isolated intestinal preparations, prior to transport studies, abolishes the inhibitory effects of fiber administration (Sigleo *et al.*, 1984). In the following series of experiments, we investigated the possibility that dietary fiber may directly affect the mucosal surface of the intestine. Fiber materials were labeled covalently with the dye Remazol brilliant blue R (C.I. reactive blue 19). Viscosity and water-holding capacity measurements were performed with the dye-labeled fiber and with unlabeled control preparations. No differences were observed in these parameters between control and tagged fibers.

A linear relationship between dye released from the conjugated fibers by alkaline hydrolysis and fiber quantity is shown in Fig. 1. Thus, liberation of the dye indicates the presence of fiber. The labeled fibers were used to examine fiber–mucosal interactions in both isolated rat small intestinal segments and in live intact rats given the labeled fibers by gastric intubation. Experimental methods are described in detail elsewhere (Vahouny *et al.*, 1986). Briefly, after a 2-hr period of incubation with the labeled fiber preparations, the small intestinal mucosa was scraped off, sonicated in 10 ml of buffer, and centrifuged. The resulting pellet was assayed for Remazol content to determine tissue-associated labeled fiber. Viscosity was measured in the supernatant, after which glycoproteins were precipitated with cetyltrimethylammonium bromide (CTAB). This precipitate was analyzed for Remazol content to estimate mucin-associated labeled fiber.

Table I summarizes the results of these studies in terms of the quantities of labeled cellulose or guar gum per average content of protein (Lowry *et al.*, 1951) in the rat intestinal homogenates. Following the 2-hr incubation of the intestinal segments, 0.72 ± 0.01 mg of dye-labeled cellulose was still associated with the tissue preparation. The precipitated glycoprotein fraction contained a threefold higher concentration of fiber

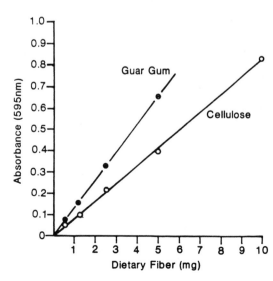

FIGURE 1. Relationship between solubilized Remazol blue dye and original dietary fiber concentrations following alkaline hydrolysis.

TABLE I. Association of Remazol-Labeled Cellulose
and Guar Gum with Rat Intestinal Mucosa[a]

	Remazol-labeled	
Measurement conditions	Cellulose	Guar gum
In vitro incubation Fiber concentrations (mg/mg protein)		
Total mucosa	0.72 ± 0.01	0.94 ± 0.03
CTAB precipitate	1.22 + 0.19	1.26 ± 0.35
Soluble	0.41 ± 0.05	0.80 ± 0.22
In vivo incubation Fiber concentrations (mg/mg protein)		
Total mucosa	1.01	0.57

[a] These values represent the means ± S.E.M. of seven to nine experiments in each condition.

than the remaining soluble fraction, suggesting a preferential association between fiber and glycoprotein. *In vitro* incubations with guar gum yielded a concentration of labeled guar gum in the mucosal homogenate of 0.94 ± 0.03 mg/mg tissue protein. The highest concentration of labeled fiber was again found in the glycoprotein precipitate. Results of viscosity measurements of the small intestinal mucosal homogenate after fiber treatment were remarkably similar in the *in vitro* and *in vivo* incubation studies (Fig. 2).

These studies, although preliminary in nature, do offer some direct evidence that certain semipurified dietary fiber components possess the capacity for interaction and association with the mucosal barrier. A substantial proportion of these fibers coprecipitate with mucosal glycoproteins. Our tentative conclusion is that a considerable propor-

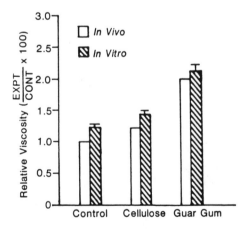

FIGURE 2. Relative viscosities of homogenates of small intestinal mucosa from rats administered cellulose or guar gum *in vivo* or from intestine incubated for 2 hr *in vitro* with the labeled dietary fibers.

tion of the derivatized polysaccharides possess an affinity for the cellular surface or its adherent mucin coat.

3. QUANTITATION OF GASTROINTESTINAL MUCIN USING AN ENZYME-LINKED IMMUNOSORBENT ASSAY (ELISA) METHOD

Although mucin is a functionally important substance in the gastrointestinal tract, measurement of this complex mixture of glycoproteins has proved difficult. Histochemical methods, such as that of Spicer (1965), can be used to stain differentially for sulfo- versus sialomucins but do not lend themselves to quantitative assessment of mucin production. Methods employing radiolabeled mucin precursors such as [^{35}S]sulfate or [^{3}H]monosaccharides as tracers to monitor mucin synthesis are not mucin-specific and may also reflect tracer incorporation into nonmucin components (Bennett et al., 1974). Chemical methods, such as the orcinol method (Weimer and Moshin, 1953), may be subject to interference from free proteins or nucleic acids, whereas precipitation of mucin by CTAB (Ofosu et al., 1978) probably recovers only 70–80% of the mucin actually present.

In order to improve quantitation of gastrointestinal mucin, we developed a poly- clonal antibody to purified rat intestinal mucin. Such an immunologic method can potentially determine small quantities of mucin accurately and specifically. Because mucin is a complex mixture of substances, a polyclonal antibody may possess advan- tages over monoclonal antibodies in that a variety of mucin structural variants may be determined, measuring "total" mucin rather than only one mucin subtype. With rat small intestinal mucin purified by column chromatography used as antigen, a polyclonal antibody was developed in rabbits by repeated administration of the purified mucin by intradermal injection. After production of the antibody, it was carefully evaluated for specificity for rat mucin and for responsiveness to pharmacological agents known to enhance or diminish mucin production.

3.1. Preparation of Purified Rat Intestinal Mucin

Small intestinal mucin for purification was collected from three male Wistar rats 2 hr after intravenous administration of [^{14}C]glucose. After the small intestinal lumen was flushed with ice-cold normal saline, the intestine was slit open longitudinally and pinned flat to a cork board. The mucus layer was removed by vacuum aspiration from the mucosal surface and collected in a flask containing a 1% solution of ice-cold sodium azide. This material was homogenized for 30 sec with a Teflon rotary homogenizing pestle in a glass tube and centrifuged at $6000 \times g$ for 30 min at 4°C. The supernatant was removed and dialyzed against 40 volumes of 0.02% sodium azide solution overnight.

To purify this crude extract, the dialyzed material was concentrated to one-fourth its original volume using an Amicon XM50 filter. The concentrate was assayed for protein (Lowry et al., 1951) and glycoprotein (Mantle and Allen, 1978) and applied to a Sephadex G200 column (2.5 × 50 cm) equilibrated with 0.1 M phosphate buffer (pH

7.1) containing 0.02% sodium azide. Column fractions were monitored for protein and glycoprotein content as above and for ^{14}C radioactivity by scintillation counting. Fractions containing maximal levels of both glycoprotein and ^{14}C radioactivity were pooled and concentrated as described above. This concentrate was reapplied to a Sepharose 4B column equilibrated in the same manner as noted previously. Protein, glycoprotein, and ^{14}C radioactivity of the column fractions were assayed as before. Coincident peaks of radioactivity and glycoprotein content (column fractions 10 to 15) were pooled. This peak was located in close proximity to that for dextran blue (molecular weight 2×10^6).

Prior to purification by column chromatography, the recovered mucus contained nearly equal amounts of protein and glycoprotein, whereas after the final step of purification (elution on the Sephadex 4B column), the relative proportion of glycoprotein to protein had increased to approximately 10. Major constituent sugars of the purified mucin were hexosamine, fucose, and sialic acid, present at 58.0, 12.0, and 6.5 mg/ml, respectively.

3.2. Production of Specific Mucin Antibodies

Production of specific mucin antiserum followed the general procedure described in detail by Vaitukaitis et al. (1971). A 1.0-ml aliquot of purified mucin (100 μg mucin protein/ml) was first combined with 1.0 ml of Freund's complete adjuvant and injected intradermally into 2- to 3-kg New Zealand rabbits. Four subsequent injections were given at weekly intervals, substituting incomplete adjuvant for the complete adjuvant.

One week following the final injection, 50 ml of blood was collected by cannulation of an ear vein. Serum was recovered by centrifugation at $1000 \times g$ for 20 min. The recovered immune serum was frozen at $-20°C$ in 1-ml aliquots for subsequent use as specific mucin antiserum in further mucin studies.

3.3. Characterization of the Mucin Antiserum

The mucin antiserum prepared above was confirmed to react appropriately with the original mucin antigen when tested via Ouchterlony (1958) plates. As another check of the specificity of the antiserum, mucin glycoproteins were separated using 5% SDS-polyacrylamide gel electrophoresis (Irwin et al., 1984). The gel pattern was transferred to nitrocellulose paper using the Western blot technique (Towbin et al., 1979). Treatment of the paper with the mucin antiserum was followed by washing with an antirabbit IgG goat serum. This complex was then treated with antigoat IgG-peroxidase followed by incubation with diaminobenzidinetetrahydrochloride (DAB) to produce staining. Staining corresponded well to the high-molecular-weight glycoprotein bands isolated by gel electrophoresis.

3.4. Development of an ELISA Method for Mucin Using the Mucin Antiserum

This method was developed as a modification of the enzyme-linked immunosorbent assay (ELISA) method of Teerlink et al. (1984) to allow quantitation of mucin from gastrointestinal specimens. The specimens were diluted in 0.06 M carbonate buffer (pH 9.8) to contain 25–50 ng protein per milliliter. A 200-μl aliquot of the diluted specimens was added to each well of a microtiter plate. The test solutions were incubated overnight at 37°C. Following incubation, each well was washed with phosphate-buffered saline (PBS)/Tween (0.01 M PBS, pH 7.2, containing 0.05% Tween 20). This wash was removed by inverting the plate, and 200 μl of PBS/gelatin/goat serum (0.01 M PBS containing 0.3% gelatin and 1% goat serum) was added to each well. The plate was incubated for 30 min at 37°C. The PBS/gelatin/goat serum was removed by inverting the plate, and 200 μl of antimucin rabbit serum (diluted 1000-fold) was added to one set of wells, while 200 μl of nonimmune rabbit serum was added to a second set of duplicate wells. Again the plate was incubated for 30 min at 37°C. Immune and nonimmune sera were removed by inversion, and the wells were washed with PBS/Tween/

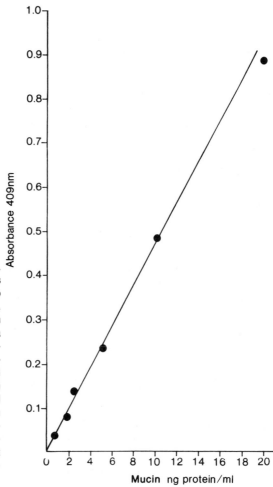

FIGURE 3. Mucin standard curve. Purified rat intestinal mucin was diluted with carbonate buffer to contain 0–20 ng mucin protein/ml. Aliquots (200 μl) of each dilution were placed in separate wells of a microtiter plate. The plate was incubated with mucin antiserum followed by biotin-conjugated IgG and, finally, the peroxidase–avidin conjugate. Samples were washed between incubations and then treated with the peroxidase substrate ABTS. After color development, absorbance was read at 409 nm in a microtiter plate reader.

gelatin (0.01 M PBS, pH 7.2, containing 0.05% Tween 20 and 0.1% gelatin). A 200-μl aliquot of biotin-conjugated goat antirabbit IgG, diluted 1:800, was added to each well and incubated for 30 min as described previously. After removal of biotin-conjugated goat antirabbit IgG and washing as above, 200 μl of peroxidase–avidin conjugate, diluted 1:800, was added to each of the wells. After removal of this conjugate and washing with PBS/Tween/gelatin, 200 μl of the substrate for the peroxidase enzyme, 2,2′-azino-bis(3-ethylbenz-thiazolinesulfonic acid) (ABTS), was added to each well. The plate was placed in the dark at room temperature for 10 to 20 min to allow color development. The color-producing reaction was stopped by adding 50 μl of 0.037 M sodium cyanide to each well. Absorbance at 409 nm was read using a microtiter plate reader.

Purified mucin ranging in concentration from 0 to 20 ng protein/ml was used to produce an ELISA standard curve. Mucin immunoreactivity of the test specimens was then calculated from this curve. A typical mucin assay standard curve is shown in Fig. 3.

3.5. Examination of Mucin Immunoreactivity in Specimens from Different Tissues

Mucin specimens were obtained from swine, human stomach, and human colon as well as rat stomach, small intestine, and colon. Mucin immunoreactivity in each of these tissues is illustrated in Fig. 4. Only specimens from the rat had appreciable immunoreactivity, suggesting that the mucin antiserum is specific for rat mucin.

4. IN VIVO STUDIES USING THE ELISA METHOD TO EXAMINE DIET- OR DRUG-INDUCED CHANGES IN MUCIN IMMUNOREACTIVITY

Since we had developed an ELISA method and established its specificity toward rat mucin, and chemically characterized the purified mucin antigen, we proceeded to examine the effects of various *in vivo* treatments on mucin immunoreactivity. In the first study, the acute effects of acetylsalicylic acid (ASA) and several prostaglandins on rat gastrointestinal mucin were compared using radioactive tracer incorporation as well as the newly developed ELISA technique. This was followed by a chronic feeding study to examine the effect of two types of dietary fiber on rat mucin immunoreactivity as measured by the ELISA technique.

4.1. The Acetylsalicylic Acid and Prostaglandin Study

In this study, groups of anesthetized male Wistar rats were treated with either ASA, 100 mg/kg body weight; prostaglandin E_2 (PGE$_2$), 100 μg/kg body weight; dimethylprostaglandin E_2 (dmPGE$_2$), 100 μg/kg body weight; a mixture of 100 mg/kg ASA and 100 μm/kg PGE$_2$ or a mixture of 100 mg/kg ASA and 100 μg/kg dmPGE$_2$. These materials were administered intragastrically 2 hr prior to necropsy. Control ani-

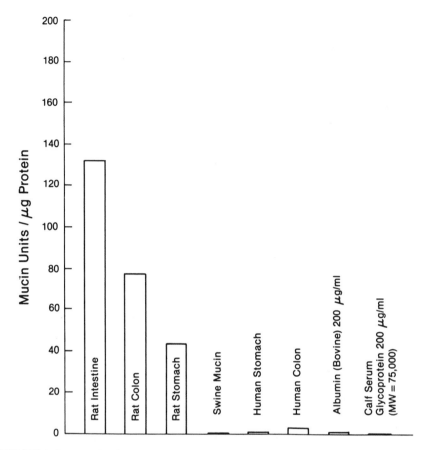

FIGURE 4. Comparative immunoreactivities of mucins from various sources. These materials were incubated with the mucin antisera, and quantitative results were obtained with the ELISA assay as described in the text.

mals received only the PBS carrier solution. One hour prior to necropsy the animals were given 10 μCi of [³H]glucose and sodium [³⁵S]sulfate via the jugular vein. Just prior to necropsy, 3 ml of blood was collected by cardiac puncture, following which the rats were killed by decapitation.

The entire small intestine was removed and transferred to ice-cold PBS, rinsed, slit open longitudinally, and pinned to a cork board mucosal sosal surface upward. The luminal mucus from each rat intestine was drawn by gentle suction into a flask containing ice-cold PBS with 0.02% sodium azide. The mucosa was then scraped off with a glass microscope slide and transferred to a measured volume of cold PBS. Both the luminal and scraped samples were homogenized for 30 sec and centrifuged at 6000 × g for 30 min at 4°C. The supernatant was removed and precipitated with CTAB (Ofosu et al., 1978) and incubated overnight at room temperature. This material was centrifuged at $6000 \times g$ for 30 min, and the pellet dissolved in 1.0 ml of PBS. A 100-μl aliquot was then counted in a liquid scintillation counter. Protein was determined by the method of Lowry *et al.* (1951). Aliquots were also prepared for the previously described ELISA assay by diluting them with carbonate buffer to contain 25 to 50 ng protein/ml.

To allow for isotopic decay during the *in vivo* portion of the experiment, isotopic

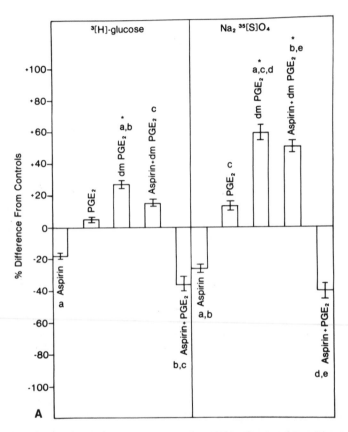

FIGURE 5. A: Incorporation of [³H]glucose and sodium [³⁵S]sulfate into CTAB-precipitated luminal mucin from rat small intestine. Vacuumed luminal mucin obtained from rats treated with various pharmacological agents was assayed for radioactive tracer incorporation. Results are expressed as percentage differences from control values. PGE₂, prostaglandin E₂; dmPGE₂, dimethylprostaglandin E₂. Values for [³H]glucose incorporation that share a common letter differ significantly from one another ($P < 0.05$ or better). Values for sodium [³⁵S]sulfate incorporation are marked similarly. Asterisk indicates these values differ significantly from control ($P < 0.05$ or better). B: Incorporation of [³H]glucose and sodium [³⁵S]sulfate into CTAB-precipitated tissue mucin from rat small intestine. After removal of luminal mucin by vacuuming, tissue mucin was obtained by scraping the small intestine of rats treated with various pharmacological agents. Radioactive tracer incorporation was assayed by liquid scintillation counting. Results are expressed as percentage differences from control values.

incorporation of the radiolabeled precursors was expressed as a labeling index as noted below:

Labeling index = (counts/min per mg protein)/(counts/min per μl serum)

Results of radiolabeled tracer incorporation into CTAB-precipitable mucin are shown in Fig. 5. The ASA treatment reduced tracer incorporation into both luminal (vacuumed) and tissue (scraped) mucin, although these effects were statistically significant only for the luminal mucin samples. Prostaglandin treatments tended generally to increase tracer incorporation, although once again these effects were statistically significant only in the luminal mucin samples. Dimethylprostaglandin appeared to be able to counteract the effect of ASA alone, resulting in enhanced tracer incorporation, although the reverse was true for PGE_2 and ASA given simultaneously.

Examination of the results obtained with the ELISA method show a remarkably similar pattern of responses to the experimental treatments (Fig. 6). Prostaglandin treatments again increased detectable mucin, and ASA resulted in a decrease. The largest

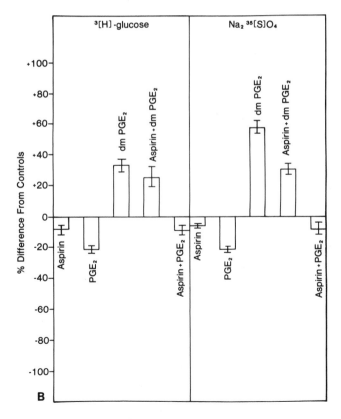

FIGURE 5. (Continued)

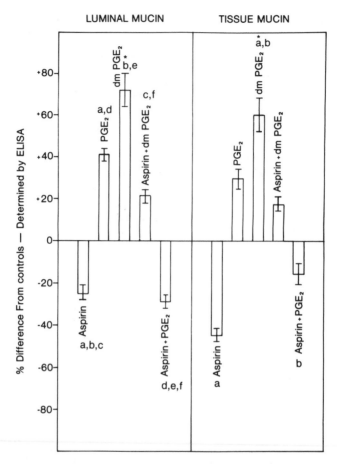

FIGURE 6. Mucin immunoreactivity of CTAB-precipitated luminal and tissue mucus from rat small intestine as determined by the ELISA technique. Luminal or tissue mucins were obtained by vacuuming or scraping the small intestine of rats that received various pharmacological agents. The results are expressed as percentage differences from control. PGE_2, prostaglandin E_2; $dmPGE_2$, dimethylprostaglandin E_2. Luminal mucin values that share a common letter differ significantly from each other ($P < 0.05$ or better). Tissue mucin values are marked similarly. Asterisk denotes values that differ significantly from control ($P < 0.05$ or better).

number of statistically significant differences were seen in the luminal mucin samples, but several of the alterations observed in tissue mucin samples were also statistically significant. Results of the ASA–prostaglandin combination treatments were also similar to those obtained with the radioactive tracer methodology.

Overall agreement between the results of the two methods was very good. Furthermore, effects of the pharmacological agents on mucin synthesis were consonant with those reported elsewhere in the literature (Keress *et al.*, 1982; Lukie and Forstner, 1972).

These results suggest that the ELISA technique is a useful method to quantitatively assess alterations in gastrointestinal mucin.

4.2. Alterations of Gastrointestinal Mucin Induced by Chronic Feeding of Various Dietary Fibers

A number of lines of evidence suggest that consumption of some types of dietary fibers may result in quantitative alterations in gastrointestinal mucin. Increases in the relative number of mucin-secreting goblet cells have been noted after feeding fiber-containing diets (Schneeman, 1982). Acutely altered absorption of nutrients such as simple carbohydrates has been reported in the presence of dietary fiber (Johnson and Gee, 1981), and slowed rates of lipid absorption have been observed in animals prefed various dietary fibers for 4 weeks (Vahouny et al., 1988). Mucin, an important constituent of the intestinal "unstirred water layer," may be a rate-limiting factor in nutrient absorption, and therefore quantitative alterations in mucin levels may be responsible for fiber-induced alterations in nutrient absorption. Previous work in this laboratory using isotopic tracer methodologies suggested that consumption of wheat bran or cellulose enhanced tracer incorporation into CTAB-precipitable mucin (Cassidy et al., 1981). Further examination of the effects of chronic fiber feeding on gastrointestinal mucin levels using the newly devised ELISA technique provided a opportunity to explore these effects in a more quantitative manner than was possible with previous methods.

In this study, male Wistar rats (300–350 g) consumed one of three semipurified diets ad libitum for 4 weeks. The basal diet contained (g/100 g) dextrose 55, casein 25, corn oil 14, salt mix 5, vitamin mix 1, as described in detail previously (Sigleo et al., 1984). This diet was consumed by the control group, while the other groups consumed modifications of this diet in which either guar gum or citrus fiber was added at 5% levels at the expense of dextrose. The citrus fiber was a commercial preparation containing cellulose plus lignin, hemicellulose, and pectin in an approximate ratio of 1:1:1.5.

At the conclusion of the feeding period, the rats were killed by decapitation. The stomach, small intestine, and colon were removed and transferred to ice-cold normal saline. Each organ was slit open, rinsed with cold saline, and pinned flat on a cork board. Luminal mucin was removed by gentle vacuum aspiration, and tissue mucin recovered by scraping off the mucosa with a glass microscope slide. These were processed and analyzed for mucin immunoreactivity by the ELISA method as described in the proceeding experiment.

Animal body weight gains and food consumptions were satisfactory regardless of dietary treatment. The effects of the dietary treatments on gastric, small intestinal, and colonic mucin are shown in Figs. 7, 8, and 9, respectively. Data in the figures are shown as luminal mucin and total mucin (luminal plus tissue mucin) for each organ and dietary group. Feeding of 5% citrus fiber enhanced gastric luminal mucin levels compared to either the fiber-free control group or the guar-gum-fed group. In the small intestine, both guar gum and citrus fiber feeding resulted in a significant elevation in total mucin compared to the control values. Luminal mucin was elevated in the citrus-fiber group as well. Citrus fiber also tended to increase levels of colonic mucin, but this change was not statistically significant.

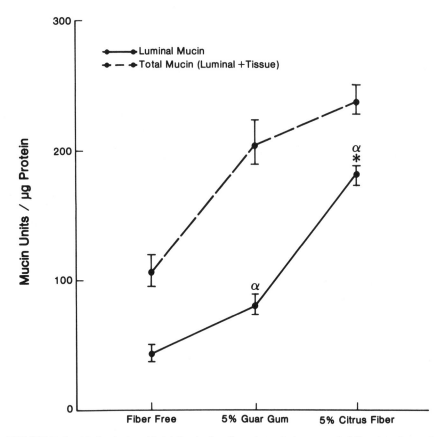

FIGURE 7. Gastric luminal and total (luminal + tissue) mucin from rats fed fiber-free (control), 5% guar gum, or 5% citrus fiber diets for 4 weeks. Values marked with a common letter differ significantly from each other ($P < 0.05$ or better). Values marked with asterisk differ significantly from the control group ($P < 0.05$ or better).

These results provide quantitative evidence of significant alterations in gastric and small intestinal mucin resulting from chronic consumption of two types of dietary fibers. The effect was more pronounced with the citrus fiber, which is a mixture of fiber types consisting mainly of cellulose, hemicellulose, and pectin, than with guar gum feeding. The functional significance of the observed fiber-induced alterations in gastrointestinal mucin may be inferred but has not as yet been fully established. Certainly mucin has been suggested to function as a barrier to nutrient absorption. Cholesterol has been shown to be effectively adsorbed to mucin (Mayer et al., 1985), possibly limiting its rate of absorption. Prefeeding animals for 4 weeks with guar gum, pectin, or Metamucil has been shown to slow lymphatic lipid absorption in response to an oleic acid test meal (Vahouny et al., 1988). It has been proposed that mucin acts to protect the gastric

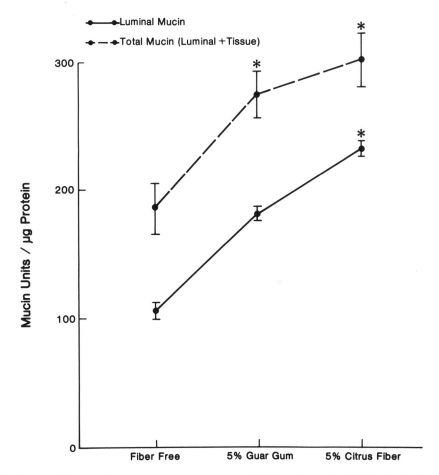

FIGURE 8. Small intestinal luminal and total (luminal + tissue) mucin from rats fed fiber-free (control), 5% guar gum, or 5% citrus fiber diets for 4 weeks. Values marked with asterisk differ significantly from the control group ($P < 0.05$ or better).

mucosa from hydrogen ions (Sarosiek *et al.*, 1984) or that it may function in protecting the colon from the effects of carcinogens or promoters (Calvert *et al.*, 1987). Mucin may also act as a gastrointestinal lubricant, promoting more rapid passage of the intestinal contents. This type of effect may be important in producing the more rapid transit time observed with feeding of some types of dietary fibers such as wheat bran (Calvert *et al.*, 1987). With development of an effective quantitative assay for mucin and the existence of phamacological agents that inhibit mucin synthesis, it should be possible to design controlled experiments to test the validity of each of these hypothetical roles for gastrointestinal mucin.

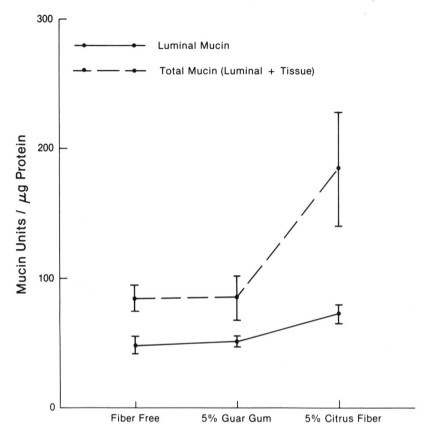

FIGURE 9. Colonic luminal and total (luminal + tissue) mucin from rats fed fiber-free (control), 5% guar gum, or 5% citrus fiber diets for 4 weeks.

5. CYTOCHEMISTRY OF MUCIN IN THE SMALL AND LARGE INTESTINE: THE EFFECTS OF CHRONIC FIBER FEEDING

We also have been interested in whether feeding the spectrum of fiber sources we have previously employed in a variety of studies might affect the qualitative mucin composition of the constantly renewed intestinal goblet cells. We employed the methodology of Spicer (1965) to perform an assessment of the sulfated and nonsulfated sialomucins in the rat model, examining tissues from the terminal ileum and the proximal, middle, and distal colon. Preliminary studies using the KOH/AB 1.0/PAPS technique (Reid *et al.*, 1984a) yielded similar results, but the high-iron diamine/Alcian blue technique of Spicer (1965) yielded sharper clarity and superior ease of quantitation in our laboratory. This technique uses high-iron diamine/Alcian blue to produce black staining of sulfomucins and aquamarine (pale blue) staining of sialomucins. A small population of the goblet cells demonstrated both types of reaction, and we have designated these as "mixed."

Table II shows the results of staining formalin-fixed, paraffin-embedded sections of intestinal tissues of animals fed a fiber-free diet for a period of 4 weeks. The table depicts the mean percentage counts ± statistical variation for a minimum of 200 villi or colonic crypts from three to five animals. There is clearly an incremental rise in the number of cells exhibiting the sulfomucin pattern, with longitudinal progression along the gastrointestinal tract from proximal to distal, which has been previously observed by other investigators (Filipe and Branfoot, 1976; Reid et al., 1984b).

Figure 10 shows the modification of the sulfomucin goblet cell population with the ingestion of 5% levels of the gelling fibers pectin, guar gum, Metamucil® (Procter and Gamble, Cincinnati, OH), or the bile acid sequestrant cholestyramine. Compared to the fiber-free control group, larger numbers of sulfomucin-staining cells were present, most particularly in the terminal ileum and the mid- and distal colon. The effect of various insoluble bulking fiber diets on this same parameter is depicted in Fig. 11. These regimens contained 10% levels of either alfalfa, wheat bran, cellulose, or a mixed-fiber diet containing barley bran and pectin (3 : 1). Cellulose had no effect in the distal ileal region of the small intestine and little impact in the proximal colon. Each of the fiber types, however, tended to produce increases in the percentage sulfated goblet cells in the mid- and distal colon.

Histochemically, a loss of sufated mucin and an elevation of sialomucin has been reported in a variety of gastrointestinal disorders that range from inflammatory bowel disease or premalignant dysplasia to frank adenocarcinoma (Reid et al., 1984b). Considerable controversy exists as to whether this phenomenon is a nonspecific transitional mucosal alteration or whether it is an important criterion for differential diagnosis and prognosis. One significant caveat concerning this technique is that staining intensity does not correlate well with actual biochemical measurements of mucin sulfate levels. Table III, which sumarizes some of the known characteristics of gastrointestinal mucus, demonstrates that there is only a slight to moderate increase in the average sulfate content of colonic versus small intestinal mucus sources. Although this discrepancy remains to be addressed and resolved (possibly by the joint application of immunologic

TABLE II. Percentage of Various Goblet Cell Types at Several Anatomic Sites in the Intestine of Rats Fed a Fiber-Free Diet for 4 Weeks[a]

| Site | Percentage of goblet cell type | | |
	Sulfated	Nonsulfated	Mixed
Terminal ileum	28.3 ± 6.1	22.8 ± 5.9	48.9 ± 4.3
Proximal colon	25.1 ± 7.3	50.8 ± 8.2	24.0 ± 3.4
Midcolon	26.6 ± 9.9*	45.1 ± 14.9*	28.3 ± 6.1
Distal colon	71.6 ± 11.8*	1.53 ± 0.9*	26.9 ± 2.6

[a]These values represent the means ± S.E.M. of the differentially staining subtypes of goblet cells following 4 weeks of fiber-free diet consumption. At least three to five animals and 200 villi or colonic folds were sampled in each location. Values marked with an asterisk within a given column differ significantly from each other.

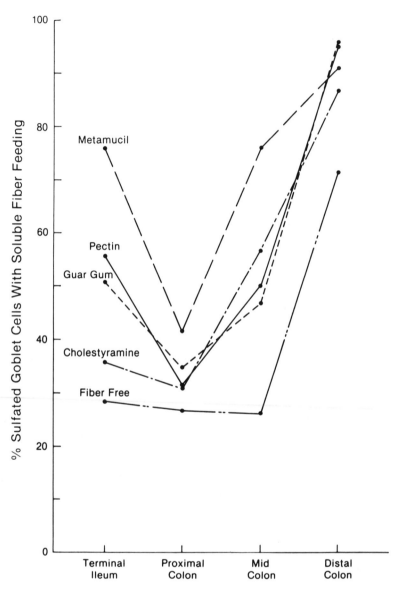

FIGURE 10. Effect of feeding various soluble dietary fibers on the percentage of ileal and colonic goblet cells staining positively for sulfomucins. Groups of rats were fed the various fibers at a 5% level in the diet for 4 weeks.

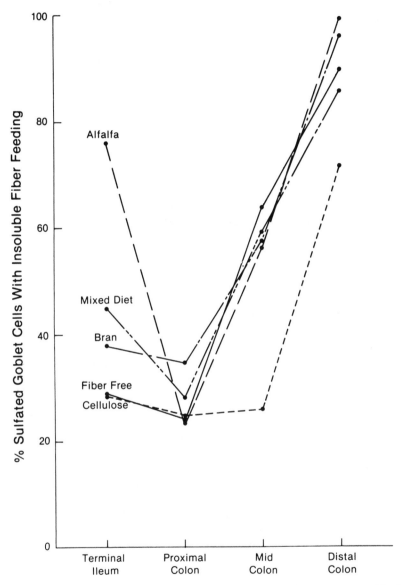

FIGURE 11. Effect of feeding various insoluble dietary fibers on the percentage of ileal and colonic goblet cells staining positively for sulfomucins. Groups of rats were fed the various fibers at a 10% level in the diet for 4 weeks.

TABLE III. Characteristics of Gastrointestinal Mucin[a]

	Gastric	Intestinal	Colonic
Size of polymer (mol. wt. $\times 10^{-6}$)	2–44	2	1–15
Size of subunit (mol. wt. $\times 10^{-6}$	0.5–2.6	0.25–0.5	0.5–6.0
Protein (% weight)	13–20	13–16	13–48
Sulfomucin staining	±	++	++++
Sialomucin staining	++	+++	+
Sulfate of purified mucin (% by weight)	1–3	2–3	3–5
Sialic acid of purified mucin (mol %)	<2	8–16	10–20

[a] Adapted from Neutra and Forstner (1987).

and immunocytochemical approaches), the findings presented here are worthy of more detailed pursuit. Since secreted mucin plays a central role in mucosal protection, fiber-containing diets, by altering both the quantity and compositional characteristics of the synthesized mucins, may lessen the undesirable effects of endogenous, dietary, or bacterially produced carcinogens.

5. SUMMARY AND CONCLUSIONS

Results of both qualitative and quantitative studies of gastrointestinal mucin suggest that fiber consumption alters mucin production. These results include the following.

1. Feeding of 10% cellulose or 10% wheat bran to rats induced increased crypt-cell turnover and accelerated cell movement up the small intestinal villus. Morphological evidence of enhanced mucin secretion was qualitatively apparent. With [^{3}H]glucose and sodium $^{35}SO_4$ used as kinetic tracers, the cellulose- or wheat-bran-fed animals demonstrated greater incorporation of both isotopes into intestinal glycoproteins (Vahouny *et al.*, 1985).

2. With Remazol dye-labeled cellulose and guar gum, a significant increase in mucosal viscosity was observed with the soluble fiber guar gum but not with the insoluble cellulose. A large fraction of both of the "bound" mucosal fiber preparations coprecipitated with the intestinal mucins. Thus, certain dietary fiber components, both soluble and insoluble, exhibit extensive surface interaction with mucins.

3. A quantitative immunologic assay was developed to purified rat mucin using specific mucin antisera. Comparison of this new method with tracer-incorporation methodologies showed excellent agreement between the results obtained with the two methods. The antiserum was specific for rat mucin, showing virtually no reactivity to human or porcine mucin.

4. After feeding with 5% citrus fiber, gastric and intestinal luminal mucin (as determined by the ELISA assay) was significantly elevated compared to values in the fiber-free control group. Feeding with 5% guar gum resulted in a significant increase in gastric luminal mucin.

5. By use of a differential cytological staining technique for sialo- and sulfomucin-containing goblet cells, it was found that the population of sulfomucin-staining cells was

significantly increased in the terminal ileum, midcolon, and distal colon following feeding with a wide variety of dietary fiber sources and derivatives.

6. Alterations in gastrointestinal mucin may have a number of functional implications. Such alterations may increase the diffusional barrier to nutrient absorption. Protection of the colonic mucosa against carcinogens or promoters may be enhanced. Alteration of the colonic goblet cell population from predominantly sialo- to sulfomucin occurs during fetal/neonatal development or in response to dietary fiber feeding. The reverse process is known to occur in the rat colon during the induction of neoplastic growth. Alterations in mucin type or quantity may therefore be one mechanism by which some fiber types may inhibit experimental colon carcinogenesis.

ACKNOWLEDGMENTS. The authors would like to thank the following individuals for their important contributions to the work described in this chapter, Dr. I. Adamson, F. A. Asskaryar, Dr. R. Chanderbhan, L. Grau, Dr. A Kharroubi, Dr. T. Lee, and F. Lightfoot. We would also like to recognize the imagination and enterprise of the late Dr. George V. Vahouny in helping to bring this work to fruition, and express our deep regret at the passing of a fine colleague and friend.

REFERENCES

Bennett, G., Leblond, C. P., and Haddad, A., 1974, Migration of glycoprotein from the Golgi apparatus to the surface of various cell types as shown by autoradiography after labeled fucose injection into rats, *J. Cell Biol.* **60**:258–284.

Calvert, R. J., Klurfeld, D. M., Subramaniam, S., Vahouny, G. V., and Kritchevsky, D., 1987, Reduction of colonic carcinogenesis by wheat bran independent of fecal bile acid concentration, *J. Natl. Cancer Inst.* **79**:875–880.

Cassidy, M. M., Lightfoot, F. G., and Vahouny, G. V., 1981, Structural–functional modulation of mucin secretory patterns in the gastrointestinal tract, in: *Membrane Biophysics. Structure and Function in Epithelia* (M. Dinno and A. Callahan, eds.), Alan R. Liss, New York, pp. 97–127.

Colony, P. C., 1983, Qualitative changes in glycoprotein staining characteristics in the rat colonic mucosa during development, *Gastroenterology* **84**:1129.

Filipe, M. I., and Branfoot, A. C., 1976, Mucin histochemistry of the colon, *Curr. Top. Pathol.* **63**: 143–178.

Greaves, P., Filipe, M. I., Abbas, S., and Ormerod, M., 1984, Sialomucins and carcinoembryonic antigen in the evolution of colorectal cancer, *Histopathology* **8**:825–834.

Irwin, D., O'Looney, P. A., Quinet, E., and Vahouny, G. V., 1984, Application of SDS gradient polyacrilamide slab gel electrophoresis to analysis of apolipoprotein mass and radioactivity of rat lipoproteins, *Atherosclerosis* **53**:163–172.

Johnson, I. T., and Gee, J. M., 1981, Effect of gel-forming gums on the intestinal unstirred water layer and sugar transport *in vitro*, *Gut* **22**:398–403.

Keress, S., Allen, A., and Garner, A., 1982, A simple method for measuring thickness of mucus gel layer adherent to rat, frog, and human gastric mucosa: Influence of feeding prostaglandins, N-acetylcysteine and other agents, *Clin. Sci.* **63**:187–195.

Lowry, O. H., Rosebrough, N. J., Farr, A. L., and Randall, R., 1951, Protein measurement with the Folin phenol reagent, *J. Biol. Chem.* **193**:265–275.

Lukie, B. E., and Forstner, G. G., 1972, Synthesis of intestinal glycoproteins: Inhibition of [^{14}C]glucosamine incorporation by sodium salicylate *in vitro*, *Biochim. Biophys. Acta* **833**:34–43.

Mantle, M., and Allen, A., 1978, A colorimetric assay for glycoprotein based on the periodic acid/Schiff's stain, *Biochem. Soc. Trans.* **6**:607–609.

Mayer, R. M., Treadwell, C. R., Gallo, L. L., and Vahouny, G. V., 1985, Intestinal mucins and cholesterol uptake in vitro, Biochim. Biophys. Acta 833:34–43.

Neutra, M. R., and Forstner, J. F., 1987, Gastrointestinal mucus: Synthesis, secretion and function, in: Physiology of the Gastrointestinal Tract, 2nd ed. (L. R. Johnson, ed.), Raven Press, New York, pp. 975–1010.

Ofosu, F., Forstner, J., and Forstner, G., 1978, Mucin degradation in the intestine, Biochim. Biophys. Acta 543:476–483.

Ouchterlony, O., 1958, Diffusion-in-gel methods for immunological analysis, Prog. Allergy 5:1–78.

Pilch, S. M. (ed.), 1987, Physiological Effects and Health Consequences of Dietary Fiber, Life Sciences Research Office, Federation of American Societies for Experimental Biology, Bethesda, MD, pp. 129–134.

Reid, P. E., Dunn, W. L., Ramey, C. W., Coret, E., Trueman, L., and Clay, M. G., 1984a, Histochemical studies of the mechanism of the periodic acid–phenylhydrazine–Schiff (PAPS) procedure, Histochem. J. 16:641–649.

Reid, P. E., Culling, C. F. A., Dunn, W. L., and Clay, M. G., 1984b, Chemical and histochemical studies of normal and diseased gastrintestinal tract I. A comparison between histologically normal colon, colonic tumors, ulcerative colitis, and diverticular disease of the colon, Histochem. J. 16: 235–251.

Sarosiek, J., Slomiany, A., Takagi, A., and Slomiany, B. L., 1984, Hydrogen ion diffusion in dog gastric mucus glycoprotein: Effect of associated lipids and covalently bound fatty acids, Biochem. Biophys. Res. Commun. 118:523–531.

Schneeman, B. O., 1982, Pancreatic and digestive function, in: Dietary Fiber in Health and Disease (G. V. Vahouny and D. Kritchevsky, eds.), Plenum Press, New York, pp. 73–83.

Sigleo, S., Jackson, M. J., and Vahouny, G. V., 1984, Effects of dietary fiber constituents on intestinal morphology and nutrient transport, Am. J. Physiol. 246:G34–G39.

Spicer, S. S., 1965, Diamine methods for differentiating mucosubstances histochemically, J. Histochem. Cytochem. 13:211–234.

Teerlink, T., Van der Krift, T. P., Van Heusden, P. H., and Wirtz, K. W., 1984, Determination of nonspecific lipid transfer protein in rat tissues and Morris hepatomas by enzyme immunoassay, Biochim. Biophys. Acta 793:251–259.

Towbin, H., Staehelin, T., and Gordon, J., 1979, Electrophoretic transfer of proteins from polyacrilamide gel to nitrocellulose sheets. Procedure and some applications, Proc. Natl. Acad. Sci. U.S.A. 76:4350–4354.

Vahouny, G. V., Le, T., Ifrim, I., Satchithanandam, S., and Cassidy, M. M., 1985, Stimulation of intestinal cytokinetics and mucin turnover in rats fed wheat bran or cellulose, Am. J. Clin. Nutr. 41: 895–900.

Vahouny, G. V., Adamson, I., Asskaryar, F. A., Chanderbhan, R., Satchithanandam, S., and Cassidy, M. M., 1986, Interaction of dietary fiber with the intestinal mucosa in rats, Nutr. Rep. Int. 34:985–993.

Vahouny, G. V., Satchithanandam, S., Chen, I., Tepper, S. A., Kritchevsky, D., Lightfoot, F. G., and Cassidy, M. M., 1988, Dietary fiber and intestinal adaptation: Effects on lipid absorption and lymphatic transport in the rat, Am. J. Clin. Nutr. 47:201–206.

Vaitukaitis, J., Robbins, J. B., Nieschlag, E., and Ross, G. T., 1971, A method for producing specific antisera with small doses of immunogen, J. Clin. Endocrinol. 33:988–991.

Weimer, H. E., and Moshin, J. R., 1953, Serum glycoprotein concentrations in experimental tuberculosis in guinea pigs, Am. Rev. Tuberc. 68:594–602.

Premenopausal Osteoporosis
Contributions of Exercise
and Dietary Practices

TOM LLOYD and JAMES R. BUCHANAN[†]

1. INTRODUCTION

Osteoporosis is the reduction in amount of bone per unit volume, and it leads to increased susceptibility to fracture with little or no trauma. The disease is insidious because osteoporosis remains largely asymptomatic until sufficient bone matrix has been lost for fractures to occur. The most common fractures occur in the spine, followed by fractures of the hip and forearm. Vertebral fractures of the spine are usually unrecognized; they occur in more than 33% of Caucasian women over the age of 65 and can lead to a reduction in height of several inches between the ages of 50 and 70. Hip fractures now occur in 33% of all women in the United States over the age of 75 (Cummings *et al.*, 1985; Riggs and Melton, 1986; Price *et al.*, 1987). The complications associated with hip fractures result in the death of 25% of these patients and in long-term nursing care for 50% of those who survive. In the United States alone, the estimated annual cost for *acute* care of fractures resulting from osteoporosis rose from $1 billion in 1984 to approximately $10 billion in 1987!

A rapid escalation of osteoporosis has been seen in the developed Western countries. This has generally been attributed to the rapid growth of older populations in these countries. In conflict with this explanation, however, is the fact that life expectancy in underdeveloped non-Caucasian countries has increased without an attendant rise in osteoporosis. Moreover, for the past 100 years, reliable medical records have been kept in developed countries, and throughout this time, a significant number of the population

† Deceased.

TOM LLOYD • Division of Reproductive Endocrinology, Department of Obstetrics and Gynecology, College of Medicine, The Milton S. Hershey Medical Center, Pennsylvania State University, Hershey, Pennsylvania 17033.   JAMES R. BUCHANAN • Division of Orthopedic Surgery, Department of Surgery, College of Medicine, The Milton S. Hershey Medical Center, Pennsylvania State University, Hershey, Pennsylvania 17033.

have lived beyond 65 years of age. Among this elderly population, the *frequency* of osteoporosis-related problems has increased during the last 50 years (Melton *et al.*, 1982; Frandsen and Kruse, 1983; Johnell *et al.*, 1984; Elabdien *et al.*, 1984; Boyce and Vessey, 1985). These observations bring us to the working hypothesis that in the past several decades dietary and/or life-style factors have developed that play pivitol roles in the increase of osteoporosis. By analogy, coronary heart disease was a minor cause of death at the turn of the century (Levy and Moskowitz, 1972). Changes in life style and dietary practices were the major contributors to the escalation of coronary heart disease during the next 50 years. After recognition of these factors, the subsequent public medical education has been credited for much of the decrease in coronary heart disease seen in recent years (Keys, 1970; Lipid Research Clinics, 1984).

1.1. Bone Biology

Bone is constantly being renewed, although at a slower rate than most other tissues. All bone is made up of both cortical bone, the outer dense portion, and trabecular bone, which is less dense and is found in the interior of bone. Different bone structures have different proportions of cortical and trabecular bone: the middle of the radius is 95–99% cortical; in contrast, the spinal column is about 75% trabecular bone and 25% cortical (Nottestad *et al.*, 1987). These proportions are important since trabecular bone is more responsive to mechanical, hormonal, and metabolic changes and can be lost at a rate eight times faster than cortical bone. The loss in height in older women occurs largely as a result of loss of trabecular bone density in the spinal column.

The two cell classes responsible for the constant remodeling of bone are osteoblasts, which lay down new bone matrix, and osteoclasts, which are responsible for bone resorption. Clearly, the regulation of the activities of these two cell types is a major factor, if not the major factor, in development of osteoporosis. Both cell types are known to be reactive to hormonal influences, but our understanding of overall regulation of the cell types is imperfect (Coarpron, 1981; Klearekoper *et al.*, 1981). The endogenous factors that have been most thoroughly studied to date as regulators of bone remodeling include sex steroids, calcitonin, vitamin D, parathyroid hormone, and calcium.

Estradiol is the major ovarian hormone in premenopausal women. Although its level varies during the menstrual cycle and from woman to woman, it is generally recognized as having a time-integrated monthly value of approximately 100 pg/ml (Riggs *et al.*, 1986; Worley, 1981; Mishell *et al.*, 1972; Fisher *et al.*, 1986). Circulating estrogen levels are positively associated with bone density (Fisher, 1986). The actions of estrogens on bone maintenance are not well understood but appear to involve several mechanisms. One action is to increase vitamin D levels, which in turn increase absorption of calcium from the gut, which would enhance new bone formation. After the menopause, estrogen levels fall to below 40 pg/ml. Estrogen deficiency appears to be associated with an increase in bone turnover and in more bone being reabsorbed than new bone being formed. The net result is bone loss. The well-recognized postmenopausal loss of bone mass can be arrested by estrogen replacement therapy (ERT) (Nachtigall *et al.*, 1979; Christiansen *et al.*, 1980; Hammond and Maxon, 1982; Judd *et al.*, 1983; Richelson *et al.*, 1984; Ettinger *et al.*, 1985). It is not yet known whether the fall in bone density in premenopausal women with low estrogen levels such as amenor-rheic athletes, anoretic women, and those with other hypoestrogenic conditions such as Turner's syndrome can be fully arrested or reversed with estrogen replacement therapy (Aitken et al., 1973; Hall, 1987).

1.2. The Effect of Dietary Fat and Dietary Fiber Intake on Circulating Estrogen Status

Vegetarians are known to be at decreased risk for hormonally dependent breast cancer (Carroll, 1986; Reed et al., 1987). A large number of epidemiologic and dietary-practice studies have been conducted to investigate this observation. At the present time, it is generally believed that the modest lowering of circulating estrogen status among vegetarians plays an important role in their lowered breast cancer frequency (Goldin et al., 1981; Armstrong et al., 1981). This reduction in circulating estrogen status among vegetarians appears to result from both decreased total dietary fat consumption and increased dietary fiber consumption. Vegetarians are known to excrete several times as much fecal estrogens as nonvegetarians (Phillips, 1985; Ingram et al., 1987; Longcope et al., 1987). Recent controlled experiments with premenopausal women have shown that estrogen status is positively associated with total dietary fat and negatively associated with dietary fiber (Goldin, 1986).

1.3. Bone Density: Measurement and Premenopausal Loss

Osteoporosis and its attendant fractures are a direct function of trabecular bone density of the most susceptible sites, namely, the vertebrae, head of the femur, and wrist. The techniques that have been used for measuring bone density include single-photon absorptiometry (SPA), dual-photon absorptiometry (DPA), quantitative computed tomography (QCT), and recently, x-ray absorptiometry (XRA) (Genant and Cann, 1984; Cann et al., 1985; Reinbold et al., 1986; Goodwin, 1987; Burgess et al., 1987). Single-photon absorptiometry has been useful for measuring changes in cortical bone, especially of the midradius. However, SPA cannot discriminate between trabecular and cortical bone. In contrast, dual-photon absorptiometry may be used to estimate both trabecular and cortical bone. However, DPA does not measure trabecular bone directly, and trabecular bone estimates from DPA are obtained after processing of integrated cortical and trabecular bone density measurements. In contrast, QCT is able to provide precise and specific measurement of trabecular bone: QCT and DPA have been the recognized "gold standards" for obtaining measurements of bone density. Until recently, trabecular bone mass was thought to reach a peak near the age of 30 (Cann et al., 1985). However, studies by our group and by other research teams have shown a linear pre-menopausal bone loss from the late teens (Fisher et al., 1986; Pacifici et al., 1987; Riggs et al., 1986).

Using the following studies we critically evaluated the hypothesis that pre-menopausal changes in bone density (and therefore changes in osteoporosis risk) are largely caused by changes in circulating estrogen levels. Changes in estrogen status, in turn, may be influenced by exercise programs and dietary practices.

2. MATERIALS AND METHODS

All participants for these studies provided informed consent to the protocols, which had been reviewed and approved by the Clinical Investigation Committee of the Pennsylvania State University College of Medicine. All participants were Caucasian. Data were collected from Pennsylvania State collegiate women athletes and from questionnaires distributed to all women registrants for a Diet Pepsi 10-km National Championship Footrace. Of the approximately 700 women entrants, we received completed questionnaires from 260 women (37% response rate). Completion of the questionnaires was entirely voluntary and anonymous. All of the University athletes studied were on one or more of the following Pennsylvania State University Collegiate Sport Teams: basketball, cross-country, field hockey, lacrosse, track, or volleyball. The data presented in Table II were obtained by review of medical records of Penn State collegiate female athletes who had sustained a bone fracture.

We recruited the following three groups of women for the cross-sectional study: eumenorrheic collegiate women athletes, oligomenorrheic collegiate women athletes, and eumenorrheic sedentary collegiate controls. All subjects for the cross-sectional studies were matched with respect to age, height, and previous nonmenstrual medical histories. Exclusion criteria include any history of smoking, recreational drug use, previous pregnancies, eating disorders, and use of oral contraceptives within the last 6 months or cumulative oral contraceptive use of greater than 6 months. All subjects enrolled in this portion of the study were between 90% and 115% of their ideal body weights. Training regimens of the women athletes were similar and consisted of 0.5 hr/day individual conditioning, 2 hr/weekday in practice, and generally one collegiate game per week. The collegiate sedentary control subjects stated that they participated in less than 1 hr/week of any exercise activity.

Each participant in the cross-sectional study completed a 7-day dietary diary after receiving instructions on correct recording of food and liquid intake. No subject consumed dietary supplements during the course of the study. A NUTRITIONIST II computer program was employed to determine daily nutrient intake (Adelman et al., 1984). Percentage body fat and lean body mass were quantitated by immersion density techniques and skinfold measurements (Allen et al., 1956). The two techniques yielded similar results ($r = 0.91$). Trabecular bone density was determined by quantitative single-energy computerized tomography of L1–L3 lumbar vertebral bodies (Reinhold et al., 1986; Elasser and Reeve, 1980; Buchanan et al., 1987). Data analyses were performed with SAS software (SAS Institute, Inc., Cary, North Carolina) on an IBM 4381 computer (IBM, Camp Hill, Pennsylvania). Data are presented as means and S.E.M.; means were compared with Student–Newman–Keuls and the Bonferroni multiple-range tests.

3. RESULTS

Our retrospective study was composed of three sections to evaluate the effect of differences in menstrual status on bone maintenance and remodeling among different groups of women athletes. From an initial study of recreational runners, we observed that the most common reason for significant interruption of a running program was a bone-related injury. After separating the respondents into two groups, those who had been injured and those who had not, we observed that those who had been injured were much more likely to have had irregular menses and much less likely to be using oral contraceptives than those who had not been injured (Table I). The physical, marital, menstrual, and running histories of the two groups of respondents were very similar (Lloyd et al., 1986).

In the second phase of this study, we performed a review of the medical histories of Pennsylvania State University women athletes to evaluate the relationship between bone fractures and menstrual status. Of the 32 collegiate athletes with x-ray-documented fractures, 22 had regular menses, and ten had irregular menses. The two groups of women athletes with fractures did not differ from one another or from the population of uninjured Pennsylvania State University women athletes with respect to age, height, weight, number of years in competitive sports, or menarche. However, the frequency of stress fractures in women with irregular menses was nearly four times that of the group with the regular menses; and when all fractures were compared, the fracture rate for the irregular group was nearly three times the rate for the regular group (Table II). The distribution of injuries by sport was similar within the two groups of subjects. The frequency of soft tissue injuries was the same for both groups.

In the third phase of this study, we surveyed a large group of women runners participating in a national championship 10-km race. Of our 260 respondents, 80 reported a bone-related injury during their running program, and 180 reported having never been injured. The two groups did not differ with respect to age, height, weight, menarche, pariety, age started running, number of days per week, and speed of running (Lloyd et al., 1986). The two groups had significant differences with respect to duration of running, miles per run, oral contraceptive use, and menstrual history. The injured group had been running longer, ran more miles per run, and was more likely to have irregular or absent menses and not to have been using oral contraceptives. We performed logistic regression analysis of our data using injury as the dependent variable to determine whether risk factors acted in concert or independently. The risk of bone injury was found to be dependent on oral contraceptive use, menstrual history, and years of running. That is, runners who did not use oral contraceptives or had irregular menses or had been running for longer periods of time had a higher risk of bone-related injury (Table III). Each of these associations was found to be independent of the others.

Although we used three diverse populations in these studies—participants in a local

TABLE I. Patterns of Menses and Oral Contraceptive (OC) Use[a]

Runner category	Menses[b]		Uses OCs[c]	
	Regular	Irregular	Yes	No
Interrupted by injury	2	7	1	9
Other runners	69	11	30	59

[a]Chi-square P values: menses, $P < 0.001$; oral contraceptives, $P < 0.24$.
[b]Eleven respondents did not answer the question regarding menses.
[c]One respondent did not answer the question regarding oral contraceptives.

TABLE II. A Retrospective Study of Medical Records of All Women Athletes on
PSU Teams to Evaluate the Relationship between Bone Fractures and Menstrual
Status: Characteristics of All 32 Athletes with Documented Fractures

	Regular ($n = 22$)	Irregular ($n = 10$)	P
Age (yr)	20.1	19.4	NS
Height (inches)	66.2	65.9	NS
Weight (lb)	132	128	NS
Menarche	13.0	13.7	NS
All fractures	9% (15/165)	24% (10/42)	<0.025
Stress fractures	4% (7/165)	14% (6/42)	<0.025

race, collegiate women athletes, and contestants in a national race—we observed in
each case that the frequency of the bone-related injury positively correlated with menstrual dysfunction and negatively with oral contraceptive use. Although it was tempting
to speculate that the increased frequency of bone-related injuries among the groups with
menstrual dysfunction occurred because of their decreased circulating estrogen status,
we did not have any direct data on either bone density or estrogen status. Furthermore, it
was possible that variables other than menstrual status contributed to the differences in
reported injuries, but we did not find any other factors to account for the observed
differences.

3.1. Assessment of Fracture Risk

On the basis of our studies showing the associations between menstrual dysfunction
and increased risk of bone-related injury, it was clear that in future studies we needed to
determine bone mineral content and hormonal status in a quantitative fashion. Initially,
we evaluated a clinical population of osteoporatic patients and healthy subjects to
determine whether measurements of spinal trabecular bone could be used to establish a

TABLE III. A Study of a Large and Heterogeneous Population of Women Runners
in July 1984 Diet Pepsi National Championship 10-km Race to Confirm or
Disprove Prior Conclusions Based on Smaller Populations

	Noninjured ($n = 180$)	Injured ($n = 80$)	P
Age (yr)	31	33	NS
Height (inches)	64	64	NS
Weight (lb)	124	123	NS
Age started	27	27	NS
Days/week	4.8	5.1	NS
Minutes/mile	9.1	8.8	NS
Oral contraceptive users	35.5%	8.8%	<0.001
Regular menses	61%	44%	<0.025

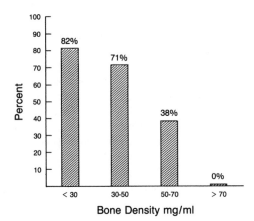

FIGURE 1. The relationship of bone density to the percentage of subjects with at least one fracture.

fracture risk threshold. We studied 96 women, of whom 65 were menopausal and 31 premenopausal. All were Caucasian, had normal levels of serum calcium, phosphate, and creatinine, and showed no evidence of an underlying endocrinopathy or neoplasm. We noted that atraumatic vertebral compression fractures were likely when bone density fell to or below 70 mg/ml (Buchanan *et al.*, 1987). We observed that the percentage of subjects with fractures escalated as bone density dropped below 70 mg/ml, as shown in Fig. 1.

3.2. Bone Density versus Age

We noted in the introduction that trabecular spinal bone density has been thought to reach a gentle peak in the mid-30s for normal women. However, our recent studies, which are shown in Fig. 2, and those by others have indicated that there is a premenopausal decrease in bone density from at least the age of 19, and the rate of bone loss accelerates after the menopause. It should be noted in Fig. 2 that the premenopausal

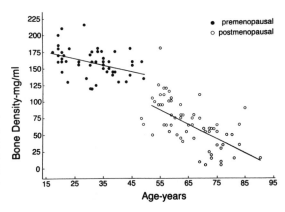

FIGURE 2. The relationship between spinal trabecular bone density and age. The premenopausal subjects were disease-free; the postmenopausal subjects had presented with osteoporosis-related complaints.

subjects were disease-free, whereas the postmenopausal subjects presented with osteoporosis-related complaints. Although longitudinal studies of trabecular bone density
of premenopausal women have not been conducted as has been done with postmenopausal women, it is probable that those premenopausal women who have low premenopausal bone density will enter their postmenopausal period with lower bone density
and be at greater risk of osteoporosis (Buchanan *et al.*, 1989).

In contrast, measurement of bone density by ^{125}I single-photon absorptiometry of
the midradius, which is composed of more than 95% cortical bone, showed no change in
bone density as a function of age in 27 normal nonosteoporotic premenopausal women,
as shown in Fig. 3. These observations suggest that changes in trabecular bone density
are dissociated from those in cortical bone and that the two bone components may
respond differentially to modulating stimuli.

3.3. Interrelationships of Diet, Athletic Activity, Menstrual Status, and Bone Density in Collegiate Women

In order to test the hypothesis that modest alteration in menstrual status, as is seen
in exercise-associated oligomenorrhea, can lead to changes in bone density and therefore
become a risk factor for the development of osteoporosis, we undertook a case-control
study to examine the impact of hormonal status, exercise practice, and dietary practices
on bone density in collegiate athletes. A summary of the physical, menstrual, and
athletic profiles of the three groups is shown in Table IV. Because of the selection
criteria that we used, the groups did not differ with respect to age, height, or weight. The
athletes had greater lean body mass and less body fat than the sedentary controls, as
would be expected. The mean menarche of the oligomenorrheic athletic group was
delayed, and the number of menses since menarche varied among the three groups, as
would be expected. Although the two athletic groups differed with respect to hours per
day and days per week of their exercise regimens, their total exercise time per week was
not different. The mean lumbar trabecular bone densities of the three groups are shown
in Fig. 4. The eumenorrheic athletes had an increase in bone density over the sedentary

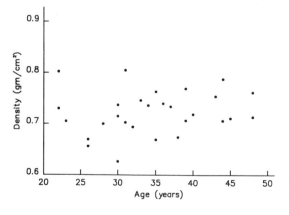

FIGURE 3. The relationship between cortical bone density and
age.

TABLE IV. Physical, Menstrual, and Athletic Profiles of the Three Study Groups

	Control (n = 12)	Eumenorrheic (n = 10)	Oligomenorrheic (n = 6)
Age (years)	19.6 ± 0.2	18.9 ± 0.3	18.8 ± 0.5
Height (cm)	163.7 ± 0.5	168.9 ± 2.0	167.4 ± 2.1
Weight (kg)	59.9 ± 1.5	61.1 ± 2.0	58.4 ± 2.7
Lean body mass (kg)	45.7 ± 1.0	49.9 ± 1.4	49.4 ± 2.2*
Body fat (%)	23.6 ± 1.2	18.2 ± 0.8*	15.6 ± 1.2*
Menarche (years)	12.2 ± 0.3	13.1 ± 0.2	15.2 ± 0.7*
Total menses	88.8 ± 5.1	68.9 ± 4.9†	18.4 ± 4.7††
Exer. (h/d)	0	2.9 ± 0.5*	2.5 ± 0.3*
Exer. (d/wk)	0	4.9 ± 0.2†	6.5 ± 0.2††

*Significantly different ($p < 0.05$) from control subjects.
† Significantly different ($p < 0.05$) from control and oligomenorrheic subjects.
†† Significantly different ($p < 0.05$) from control and eumenorrheic subjects.

controls that was statistically insignificant, and the oligomenorrheic athletes had a decrease in bone density that was different from the sedentary controls at the $P = 0.10$ level. Although the absolute change in bone density for the oligomenorrheic athletic group was modest, this decrease gains respect when the fracture threshold of about 70 mg/ml is added to the figure. The oligomenorrheic athletic group is likely to continue their premenopausal loss of bone density from a substantially lower peak bone mass than either their eumenorrheic teammates or eumenorrheic sedentary controls (Lloyd et al., 1987).

Table V is a comparison of the bone density detenninations of our study subjects to those from a similar study conducted by Marcus et al., which included sedentary eumenorrheic control subjects, eumenorrheic athletes, and, in the case of Marcus et al. (1985), amenorrheic athletes. In both studies, QCT was used to estimate trabecular spinal bone density. Although the mean ages of the subjects studied by Marcus et al. are slightly older than our study population, we note that in both studies, the eumenorrheic

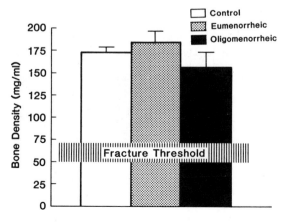

FIGURE 4. Spinal trabecular bone density values of the three study groups described in Table IV. The control group were eumenorrheic sedentary collegiate women, and the remaining two groups were collegiate athletes.

TABLE V. Comparison of Bone Density Results of Two Studies

	Lloyd et al. (1987)			Marcus et al. (1985)		
	n	Age	QCT	n	Age	QCT
Controls	12	19.6	173	20	25.0	166
Eumenorrheic athletes	10	18.9	184	6	23.8	182
Oligo- and amenorrheic athletes	6	18.8	156	11	20.0	151

athletes had an increased bone density over the eumenorrheic sedentary controls; and in both studies, the athletes with menstrual dysfunction had a substantial reduction in trabecular bone density compared either to the sedentary eumenorrheic subjects or to the eumenorrheic athletes.

3.4. Relationships among Dietary Practices, Hormonal Status, and Bone Density

The nutritional patterns of the three study groups are summarized in Fig. 5. The athletes consumed significantly more kilocalories, carbohydrate, calcium, and phosphorus than the sedentary control subjects; all groups consumed similar amounts of protein and fat. The oligomenorrheic and eumenorrheic athletes did not differ one from another in intake of any nutrients except fiber. The oligomenorrheic athletes consumed nearly twice as much dietary fiber as did the eumenorrheic athletes or the sedentary control subjects. Our finding of a difference in dietary fiber intake between the two groups of athletes was unexpected. The results of previous studies that have compared

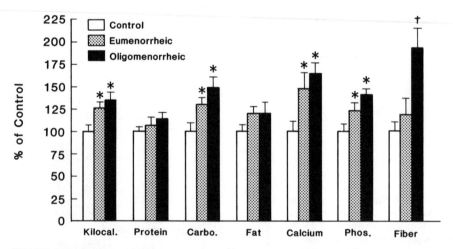

FIGURE 5. Nutritional patterns of the three study groups. The groups are described in Table IV, and the data have been normalized against the control sedentary collegiate group.

high- and low-fiber-intake populations—namely, vegetarians and nonvegetarians—have provided us with some insight on this finding. Vegetarian women who consume approximately twice as much fiber as do nonvegetarian women have markedly reduced rates of breast cancer and lowered circulating estrone and estradiol levels. In contrast, fecal excretion of estrogens by vegetarians has been found to be three to four times as great as that by nonvegetarians (Goldin *et al.*, 1982). The mechanisms responsible for depressed circulating estrogens and increased fecal excretion of estrogens among vegetarians are not known, and a variety of mechanisms including direct absorption of estrogens by fiber subspecies and alteration of gut flora, which in turn alters enterohepatic circulation of estrogens, have been proposed (Aldercreutz *et al.*, 1987). It has also been proposed that dietary fiber binds calcium and thereby affects calcium absorption (Dokkum *et al.*, 1982; Andersson *et al.*, 1983). Calcium balance studies that have been performed to examine this issue have yielded controversial results (Behall *et al.*, 1987; Kelsay, 1987; Hallbrisch *et al.*, 1986; Donangelo and Eggum, 1986).

Furthermore, unambiguous measurement of total dietary fiber is not possible at present since general agreement does not exist about the appropriate methodologies for the analysis of dietary fiber (Lanza and Butrum, 1986; Bingaham, 1987; Slavin, 1987). Therefore, a universally recognized fiber data base does not exist for computer-assisted nutrient analysis. In addition, since the term dietary fiber embraces a variety of complex chemical structures including celluloses, hemicelluloses, pectins, gums, and lignins, it is possible that different biological impacts of dietary fiber are actually mediated by specific fiber subspecies. Nonetheless, the association between increased dietary fiber intake in our oligomenorrheic athletes and their hypoestrogenic state and their decreased bone density is intriguing. The decrease in bone density in this study population could conceivably have been the result of either decreased calcium availability through binding with fiber or indirectly through the action of fiber to decrease estrogen status, which in turn has a negative effect on bone remodeling.

3.5. Impact of Dietary Practices on Menstrual Status in an Unselected Premenopausal Population

The above studies utilized recreational athletes, sedentary collegiate women, and collegiate women athletes. Although we believe that observations from these subject groups will apply to larger populations, we recognize that our very efforts to construct well-matched case-control studies may have also, unknowingly, produced biased study populations. One strategy that we developed to address this issue has been to collect menstrual history data from a population of free-living premenopausal women who are within one socioeconomic group and yet who have widely different dietary practices. Our subject population for this investigation was recruited from the premenopausal women of the Columbia Union Conference of the Seventh Day Adventist Church. The Seventh Day Adventist Church supports vegetarian practices in its dietary code, but it is well recognized that many Seventh Day Adventists are "cultural Adventists" who do not practice the dietary code. Accordingly, we were able to obtain information from both vegetarian and nonvegetarian premenopausal Seventh Day Adventist women.

Analysis of these data is shown in Tables VI and VII. The vegetarians were slightly

TABLE VI. Nutrition Profiles of the Vegetarian and Nonvegetarian Study Groups

Variable	Vegetarians ($n = 31$)	Nonvegetarians ($n = 20$)	P
Age	37.5 ± 10.2	30.3 ± 7.1	0.008
Menarche	12.4 ± 1.4	12.9 ± 1.4	0.219
Kcal	1751 ± 489	1669 ± 481	0.559
Carbohydrate (g)	250 ± 69	219 ± 72	0.130
Protein (g)	61 ± 18	69 ± 23	0.171
Total fat (g)	65 ± 30	59 ± 22	0.445
Sat. fat (g)	16.9 ± 8.7	18.3 ± 7.5	0.557
Mono. fat (g)	15.0 ± 10	13.7 ± 6.3	0.576
Polyunsat. fat (g)	12.7 ± 9.0	6.6 ± 5.3	0.009
Total fiber (g)	24.3 ± 9.9	14.6 ± 5.8	0.0002

older than our nonvegetarian respondents, but they did not differ with respect to age of menarche or intake or total calories, carbohydrates, total fat, or monosaturated fat. Not unexpectedly, the vegetarians consumed nearly twice as much dietary fiber and nearly twice as much polyunsaturated fat as the nonvegetarian respondents.

The vegetarian and nonvegetarian populations did not differ with respect to non-menstrual medical histories, but they did differ with respect to menstrual histories. Menstrual irregularity was reported by 33% of the premenopausal vegetarians and by 8% of the nonvegetarians ($P = 0.047$). To our knowledge, this is the first report of menstrual histories from a population of free-living premenopausal vegetarians. Other reports of the frequency of menstrual irregularity in nonvegetarian premenopausal populations have found that approximately 4–6% have irregular or absent menses (Pettersson et al., 1973; Bachmann and Kemman, 1982; Singh, 1981). Our findings indicate that one or more components of vegetarian diets lead to a decrease in circulating estrogen status, which can be manifest as irregular or absent menses.

4. DISCUSSION

Our results support our notion that hormonal status in premenopausal women can be modified by exercise and/or dietary practices and that such hormonal alterations may play an important role in development and maintenance of bone mass.

TABLE VII. Chi-Square Analysis of the Relationship between Dietary Practices and Menstrual History[a]

Menstrual history	Vegetarian ($n = 32$)	Nonvegetarian ($n = 39$)
Regular	24	36
Irregular	8	3
	33%	8%

[a]$P = 0.047$.

REFERENCES

Adelman, M. O., Dwyer, C. T., Woods, M., Bohn, E., and Otradovec, O. L., 1984, Computerized dietary analysis systems: A comparative view, *J. Am. Diet. Assoc.* **83**:421–427.

Adlercreutz, H., Hockerstedt, K., and Bannwart, C., 1987, Effect of dietary components, including ligans and phytoestrogens, on enterohepatic circulation and liver metabolism of estrogens and on sex hormone binding globulin (SHBG), *J. Steroid Biochem.* **27**:1135–1144.

Aitken, J. M., Hart, D. M., Anderson, J. B., Lindsay, R., Smith, D. A., and Speirs, C. F., 1973, Osteoporosis after oophorectomy for non-malignant disease in premenopausal women, *Br. Med. J.* **2**:325–328.

Allen, T. H., Peng, M. T., Chen, K. P., Huang, T. H., Chang, C., and Fang, H. S., 1956, Prediction of total adiposity from skinfolds and the curvilinear relationship between external and internal adiposity, *Metabolism* **51**:346–352.

Andersson, H., Navert, B., Bingham, S. A., Englyst, H. N., and Cummings, J. H., 1983, The effects of breads containing similar amounts of phytate but different amounts of wheat bran on calcium, zinc and iron balance in men, *Br. J. Nutr.* **50**:503–510.

Armstrong, B. K., Brown, J. B., Clarke, H. T., Crooke, D. K., Hahnel, R., Masarei, J. R., and Ratajczak, T., 1981, Diet and reproductive hormones: A study of vegetarian and nonvegetarian postmenopausal women, *J. Natl. Cancer Inst.* **67**(4):761–767.

Bachmann, G. A., and Kemmann, E., 1982, Prevalence of oligomenorrhea and amenorrhea in a college population, *Am. J. Obstet. Gynecol.* **144**:98–102.

Behall, K. M., Scholfield, D. J., Lee, K., Powell, A. S., and Moser, P. B., 1987, Mineral balance in adult men: Effect of four refined fibers, *Am. J. Clin. Nutr.* **46**:307–314.

Bingaham, S., 1987, Definitions and intakes of dietary fiber, *Am. J. Clin. Nutr.* **45**:1226–1231.

Boyce, W. J., and Vessey, M. P., 1985, Rising incidence of fracture of the proximal femur, *Lancet* **1**: 150–151.

Buchanan, J. R., Myers, C., Lloyd, T., and Greer, R. B., 1989, Early vertebral bone loss in normal premenopausal women, *J. Bone Joint Surg.* (in press).

Buchanan, J. R., Myers, C., Greer, R. B., Lloyd, T., and Varano, L., 1987, Assessment of vertebral fracture in menopausal women, *J. Bone Joint Surg.* **69A**:212–218.

Burgess, A. E., Colborne, B., and Zoffman, E., 1987, Vertebral trabecular bone: Comparison single and dual energy CT measurements with chemical analysis, *J. Comput. Assist. Tomogr.* **11**:506–515.

Cann, C. E., 1988, Quantitative CT for determination of bone mineral density: A review, *Radiology* **166**: 509–522.

Cann, C. E., Genant, H. K., Kolb, F. O., and Ettinger, B., 1985, Quantitative computed tomography for prediction of vertebral fracture risk, *Bone* **6**:1–7.

Carroll, K. K., 1986, Experimental studies on dietary fat and cancer in relation to epidemiological data, in: *Dietary Fat and Cancer,* Alan Liss, New York, pp. 231–248.

Christiansen, C., Christensen, M. S., McNair, P., Hagen, C., Stocklund, K-E., and Transbol, I., 1980, Prevention of early postmenopausal bone loss: Controlled 2-year study in 315 normal females, *Eur. J. Clin. Invest.* **10**:273–279.

Coarpron, P., 1981, Bone tissue mechanisms underlying osteoporoses, *Orthoped. Clin. North Am.* **12**(3):513–545.

Cummings, S. R., Kelsey, J. L., Nevitt, M. C., and O'Dowd, K. J., 1985, Epidemiology of osteoporosis and osteoporotic fractures, *Epidemiol. Rev.* **7**:178–208.

Dokkum, W., Wesstra, A., and Schippers, F. A., 1982, Physiological effects of fibre-rich types of bread. 1. The effect of dietary fibre from bread on the mineral balance of young men, *Br. J. Nutr.* **47**:451–460.

Donangelo, C. M., and Eggum, B. O., 1986, Comparative effects of wheat bran and barley husk on nutrient utilization in rats. 2. Zinc, calcium and phosphorus, *Br. J. Nutr.* **56**:269–280.

Elabdien, B. S. Z., Olerud, S., Karlstrom, G., and Smedby, B., 1984, Rising incidence of hip fracture in Uppsala, 1965–1980, *Acta Orthop. Scand.* **55**:284–289.

Elasser, U., and Reeve, J., 1980, Bone density measurement with computed tomography, *Br. Med. Bull.* **36:**293–296.

Ettinger, B., Genant, H. K., and Cann, C. E., 1985, Long term estrogen replacement therapy prevents bone loss and fractures, *Ann. Intern. Med.* **102:**319–324.

Fisher, E. C., Nelson, M. E., Frontera, W. R., Turksoy, R. N., and Evans, W. J., 1986, Bone mineral content and levels of gonadotropins estrogens in amenorrheic running women, *J. Clin. Endocrinol. Metab.* **62:**1232–1236.

Frandsen, P. A., and Kruse, T., 1983, Hip fractures in the County of Funen, Denmark, *Acta Orthop. Scand.* **54:**681–686.

Genant, H. K., and Cann, C. E., 1984, *Spinal Osteoporosis: Advanced Assessment Using Quantitative Computed Tomography, Spine Update,* University of California Press, San Francisco.

Goldin, B. R., Adlercreutz, H., Dwyer, J. T., Swenson, L., Warram, J. H., and Gorbach, S. L., 1981, Effect of diet on excretion of estrogens in pre- and postmenopausal women, *Cancer Res.* **41:**3771–3773.

Goldin, B. R., Adlercreutz, H., Gorbach, S. L., Warram, J. H., Dwyer, J. T., Swenson, L., and Woods, M. N., 1982, Estrogen excretion patterns and plasma levels in vegetarian and omnivorous women, *N. Engl. J. Med.* **307:**1542–1547.

Goldin, B. R., Adlercreutz, H., Gorbach, S. L., Woods, M. N., Dwyer, J. T., Conlon, T., Bohn, E., and Gershoff, S. N., 1986, The relationship between estrogen levels and diets of Caucasian American and oriental immigrant women, *Am. J. Clin. Nutr.* **44:**945–953.

Goodwin, P. N., 1987, Methodologies for the measurement of bone density and their precision and accuracy, *Semin. Nucl. Med.* **4:**293–304.

Hall, J. G., 1987, Turners syndrome: An update, *Growth, Genet. Horm.* **3:**4–8.

Hallbrisch, J., Powell, A., Carafelli, C., Reiser, S., and Prather, E. S., 1987, Mineral balances of men and women consuming high fiber diets with complex or simple carbohydrate, *J. Nutr.* **117:**48–55.

Hammond, C. B., and Maxon, W. S., 1982, Current status of estrogen therapy for the menopause, *Fertil. Steril.* **37**(1):5–25.

Johnell, O., Nilsson, B., Obrant, K., and Sernbo, I., 1984, Age and sex patterns of hip fracture changes in 30 years, *Acta Orthop. Scand.* **55:**290–292.

Judd, H. L., Meldrum, D. R., Deftos, L. J., and Henderson, B. E., 1983, Estrogen replacement therapy: Indications and complications, *Ann. Intern. Med.* **98:**195–205.

Kelsay, J. L., 1987, Effects of fiber, phytic acid and oxalic acid in the diet on mineral bioavailability, *Am. J. Gastroenterol.* **82:**983–986.

Keys, A. (ed.), 1970, Coronary heart disease in seven countries, *Circulation* **41:**I1–I198.

Klearekoper, M., Tolia, K., and Parfitt, A. M., 1981, Nutritional, endocrine, and demographic aspects of osteoporosis, *Orthop. Clin. North Am.* **12**(3):747–557.

Lanza, E., and Butrum, R. R., 1986, A critical review of food fiber analysis and data, *J. Am. Dietet. Assoc.* **86:**732–743.

Levy, R. I., and Moskowitz, J., 1972, Cardiovascular research: Decades of progress: A decade of promise, *Science* **217:**121–129.

Lipid Research Clinics Coronary Prevention Trial, 1984, Results I. Reduction in incidence of coronary artery disease, *J.A.M.A.* **251:**351–374.

Lloyd, T., Triantafyllou, S. J., Baker, E. R., Houts, P. S., Whiteside, J. A., Kalenak, A., and Stumpf, P. G., 1986, Menstrual irregularity and injury in women athletes, *Med. Sci. Sports Exercise* **18:**374–379.

Lloyd, T., Bitzer, S., Waldman, C. J., Myers, C., Ford, B. G., and Buchanan, J. R., 1987, The relationship of diet, athletic activity, menstrual status, and bone density among collegiate women, *Am. J. Clin. Nutr.* **46:**681–684.

Longcope, C., Gorbach, S., Goldin, B., Woods, M., Dwyer, J., Morril, A., and Warram, J., 1987, The effect of a low fat diet on estrogen metabolism, *J. Clin. Endocrinol. Metab.* **64:**1246–1250.

Marcus, R., Cann, C., Madvig, P., Minkoff, J., Goddard, M., Bayer, M., Martin, M., Gaudiani, L., Haskell, W., and Genant, H., 1985, Menstrual function and bone mass in elite women distance runners, *Ann. Intern. Med.* **102:**158–163.

Melton, L. J., Ilstrup, D. M., Riggs, B. L., and Beckenbaugh, R. D., 1982, Fifty-year trend in hip fracture incidence, *Clin. Orthop.* **162**:144–149.

Mishell, D. R., Thorneycroft, L. H., Nakanuta, R. M., Nagata, Y., and Stone, S., 1972, Serum estradiol in women ingesting combination oral contraceptive steroids, *Am. J. Obstet. Gynecol.* **114**:923–928.

Nachtigall, L. E., Nachtigall, R. H., Nachtigall, R. D., and Beckman, E. M., 1979, Estrogen replacement therapy I: A 10-year prospective study in the relationship to osteoporosis, *Obstet. Gynecol.* **53**(3):277–281.

Nottestad, S. Y., Baumel, J. J., Kimmel, D. B., Recker, R. R., and Heang, R. P., 1987, The proportion of trabecular bone in human vertebrae, *J. Bone Joint Surg.* **2**:221–229.

Pacifici, R., Susman, N., Carr, P. L., Birge, S. J., and Avioli, L. V., 1987, Single and dual energy tomographic analysis of spinal trabecular bone: A comparative study in normal and osteoporotic women, *J. Clin. Endocrinol. Metab.* **64**(2):209–214.

Pettersson, F., Fries, and Nillius, S. J., 1973, Epidemiology of secondary amenorrhea. I. Incidence and prevalence rates, *Am. J. Obstet. Gynecol.* **117**:80–86.

Phillips, G. B., 1985, Hyperestrogenemia, diet, and disorders of Western societies, *Am. J. Med.* **78**:363–366.

Price, T., Hesp, R., and Mitchell, R. H., 1987, Bone density in generalized osteoarthritis, *J. Rheumatol.* **13**(3):560–562.

Reed, M. J., Beranek, P. A., Cheng, R. W., McNeill, J. M., and James, V. H. T., 1987, Peripheral oestrogen metabolism in postmenopausal women with or without breast cancer: The role of dietary lipids and growth factors, *J. Steroid Biochem.* **27**:985–989.

Reinbold, W.-D., Genant, H. K., Reiser, U. J., Harris, S. T., and Ettinger, B., 1986, Bone mineral content in early postmenopausal and postmenopausal women: Comparison of measurement methods, *Radiology* **160**:469–478.

Richelson, L. S., Wahner, H. W., Melton, L. J., and Riggs, B. L., 1984, Relative contributions of aging and estrogen deficiency to postmenopausal bone loss, *N. Engl. J. Med.* **311**:1273–1275.

Riggs, B. L., and Melton, L. J., 1986, Involutional osteoporosis, *N. Engl. J. Med.* **314**:1676–1686.

Riggs, B. L., Wahner, H. W., Melton, J. L., Richelson, L. S., Judd, H. L., and Offord, K. P., 1986, Rates of bone loss in the appendicular and axial skeletons of women, *J. Clin. Invest.* **77**:1487–1491.

Singh, K. B., 1981, Menstrual disorders in college students, *Am. J. Obstet. Gynecol.* **140**:299–302.

Slavin, J. L., 1987, Dietary fiber: Classification, chemical analyses, and food sources, *J. Am. Dietet. Assoc.* **87**:1164–1171.

Worley, R. J., 1981, Age, estrogen, and bone density, *Clin. Ob. Gyn.* **24**(1):203–218.

7

Total Dietary Fiber and Mineral Absorption

DENNIS T. GORDON

1. INTRODUCTION

1.1. Dietary Fiber Hypothesis

The dietary fiber hypothesis states: "A diet that is rich in foods that contain plant cell walls is protective against a range of diseases, in particular those prevalent in affluent Western Communities." Conversely, the hypothesis implies that "in some instances a diet providing a low intake of plant cell walls is a causative factor in the etiology of the disease, and in others it provides the condition under which other etiological factors are more active" (Southgate, 1982).

This hypothesis, which represents dietary fiber as a very important component of the diet, was based on the pioneering observations of Burkitt and Trowell (1975), Burkitt et al. (1979), Trowell (1976), A. R. P. Walker, and N. S. Painter. Since these early observations, little has happened to diminish the importance of dietary fiber in health and disease. In fact, the positive image of dietary fiber continues to increase (Talbot, 1980; Committee on Diet, Nutrition, and Cancer, 1982; Life Sciences Research Office, 1987; Koop, 1988).

1.2. Total Dietary Fiber

Scientific advancements that have been accomplished to help elucidate the dietary fiber hypothesis are best found throughout the present volume and the two previous books in this series (Vahouny and Kritchevsky, 1982, 1985). One advancement has been the finding that from an analytical perspective, total dietary fiber (TDF) consists of insoluble and soluble components (Southgate, 1969, 1981; Furda, 1981; Prosky et al.,

DENNIS T. GORDON • Department of Food Science and Nutrition, University of Missouri, Columbia, Missouri 65211.

1984; Becker *et al,.* 1986). The classification of major components comprising plant cell wall TDF is shown in Fig. 1. From a physiological standpoint, the insoluble and soluble fractions appear to function differently in the gastrointestinal tract (Haber *et al.,* 1977; Eastwood *et al.,* 1983; Judd and Truswell, 1985; Leeds, 1987; Ink and Hurt, 1987). The current estimate of TDF intake is 11 to 23 g/day in the United States, and it has been recommended that the intake of TDF be increased to 20–35 g/day (Life Sciences Research Office, 1987). This same committee recommended that the TDF intake consist of 70–75% insoluble dietary fibers (IDF) and 25–30% soluble dietary fibers (SDF). Dreher (1987) has reviewed the TDF intake in different countries, which ranged from 13.9 g/day in Iceland to 45.1 g/day in Portugal.

1.3. Total Dietary Fiber and Minerals

Three factors have contributed to the scientific community's and consumer's belief and acceptance of TDF as possibly the single most healthful component in the diet. These factors are (1) real and perceived health benefits of TDF as indicated by epidemiologic, clinical, and animal studies; (2) recommendations that TDF intake can and should be increased; and (3) the lack of any strong and convincing scientific evidence that TDF causes any harmful effects. In turn, the food industry is accommodating consumer demand by incorporating more and different sources of TDF into foods. However, in the scientific community, there are concerns that high intakes of TDF (i.e., >20–35 g/day) may impair mineral absorption and nutriture (Kelsay, 1978, 1981, 1982, 1986; Harland and Morris, 1985; Toma and Curtis, 1986). In regard to the relationship between minerals and TDF, the recent Committee reporting on the physiological effects and health consequences of dietary fiber (Life Sciences Research Office, 1987) provided the following qualifying statement: "Given the possibility that there is likely to be an adaptation to any alteration in mineral availability resulting from an increased fiber intake, a moderate level of fiber intake of 20–25 g/day NDF (or insoluble fiber) does not appear to pose a problem." Long-term studies of the effect of TDF on mineral nutriture have not been done. It is interesting to note that no known pathology has ever been associated with either a deficiency or excess of TDF.

The purpose of this review is to integrate mineral nutrition with TDF consumption. More specifically, the effects of various types and amounts of TDF and its major insoluble and soluble components on mineral nutrition are presented. The apparent benefits of higher intakes of TDF appear to justify increased consumption of foods rich

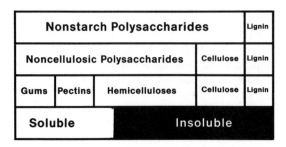

FIGURE 1. The distribution of major plant cell wall components comprising total dietary fiber. Gums and pectins are normally considered sources of soluble dietary fibers. Cellulose, lignin, and a majority of hemicellulose are insoluble. Some hemicelluloses are soluble.

in TDF. The author takes the position that although TDF does affect mineral absorption, these are not negative effects, and TDF does not impair mineral health.

2. ESSENTIAL MINERALS AND SITES OF ABSORPTION

The minerals essential for human health are indicated in Table I along with the four most common toxic elements in the human diet. From the standpoint of the interaction between TDF and mineral absorption, the minerals presented in Table I have significance for a number of reasons. Insufficient information is available as to whether TDF affects the absorption of all minerals uniformly or whether only specific minerals or groups of minerals (i.e., the transition elements, Fe, Zn, and Cu) are affected. A second consideration is the site of each mineral's absorption along the alimentary tract. As knowledge is gained that the insoluble and soluble dietary fibers act differently in the major sections of the intestine, the question can be asked: What fibers affect what minerals in what sections of the gastrointestinal tract?

Most minerals are absorbed in the small intestine (Fig. 2), with a majority of the uptake occurring in the duodenum and less along the remainder of the small intestine (Mertz, 1987). There is information to suggest that copper (Van Campen and Mitchell, 1965) and selenium are partially absorbed from the stomach. Adequate information on the role of the stomach in mineral absorption is not available. Electrolytes are exchanged between the intestinal lumen and the body along the entire intestine, with a majority of this exchange taking place in the colon. There is debate as to the magnitude and importance of the colon as a site of mineral absorption other than for the electrolytes. Absorption of calcium from the colon has received significant attention (Favurs *et al.*,

TABLE I. Essential and Toxic Minerals in the Diet along with
Their Highest Recommended Dietary Allowance (RDA) or Estimated Safe and
Adequate Daily Dietary Intake (ESI)

				Essential		
	RDA (mg/day)[a]			ESI mg/day	No exact requirement/intake established	Toxic
Ca	1200	Na		1100–3300	Ni	As[b]
P	1200	K		1875–5625	Mo	Hg
Mg	400	Cl		170–5100	As[b]	Cd
Fe	18	Cu		2.0–3.0	Co	Pb
Zn	15	Mn		2.5–5.0	Cr	
I	0.15	F		1.5–4.0	Si	
		Cr		0.05–0.2	Sn	
		Se		0.05–0.2	V	
		Mo		0.15–0.5		

[a]Maximum RDA (National Academy of Sciences, 1980).
[b]Essential in only trace amounts but toxic in higher amounts.

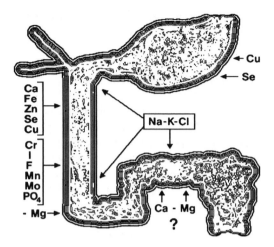

FIGURE 2. Generally accepted sites of absorption of selected essential minerals along the gastrointestinal tract. The exact magnitude and significance of copper and selenium absorption from the stomach is unknown. Calcium and magnesium absorption from the colon is only speculated. All known essential minerals are reported in Table I.

1980; Lee *et al.*, 1980; Ammann *et al.*, 1986). Because TDF has been implicated in impaired calcium utilization (McHale *et al.*, 1979; Slavin and Marlett, 1980; Godara *et al.*, 1981), possibly by decreasing absorption at sites in the small intestine, having a second area for absorption of calcium would be advantageous.

3. MINERAL BIOAVAILABILITY

The area of nutrition that has possibly received the greatest amount of attention this past decade has been nutrient bioavailability. Bioavailability is defined as that proportion of the total amount of a nutrient that is absorbed and utilized by the body (O'Dell, 1984). With minerals, the first use of this term was probably by James Fritz (Pla and Fritz, 1971), then with the Food and Drug Administration. His studies were instrumental in showing that certain forms of iron were more available to the body (e.g., ferrous sulfate) than others, which were almost totally unavailable (e.g., ferric orthophsophate).

Although the study of nutrient bioavailability remains a commendable endeavor, a major difficulty arises in obtaining a meaningful response or index that accurately reflects mineral status without conducting invasive techniques. For selected minerals, the relationship between physiological/biochemical functions and current measurements to assess status and bioavailability are reported in Table II. Because of the lack of sensitive indices, coupled with the fact that there are no overt mineral deficiencies in populations or in clinical research studies evaluating the effect of TDF on mineral nutrition, unequivocal conclusions can only be inferred as to any possible adverse affects of TDF on mineral accretion. It is interesting to note that little mention is ever made of TDF enhancing mineral utilization or bioavailability.

The bioavailability of nutrients' minerals is affected by two sets of factors, and these are listed in Table III. Those physiological changes that occur in the living system are referred to as intrinsic factors. The second set of factors, extrinsic factors, are those directly associated with the diet. Total dietary fiber is considered an extrinsic factor.

TABLE II. Selected Physiological/Biochemical Functions of Minerals and Indices to Assess Status or Bioavailability in Animals and Humans

Mineral	Physiological/biochemical function	Index of status
Ca	Bone	Bone density
Mg	Neuromuscular transmission	Serum Mg
Fe	Hemoglobin	Blood ferritin
Zn	RNA polymerase	Serum Zn
Cu	Superoxide dismutase	Serum Cu
Se	Glutathione peroxidase	Glutathione peroxidase activity

Changes in the amount or ratios of the carbohydrate, protein, and fat to TDF concentration in the diet may be important in mineral nutriture. An adequate understanding of the intrinsic and extrinsic factors that affect mineral bioavailability and long-term mineral nutriture represents a formidable challenge for the researcher.

Plant cell walls are rich in minerals (Crosby, 1978; Johnson et al., 1985). For this reason, whole-grain products and bran can contribute minerals to a food product and ultimately the human diet. This has led numerous investigators to evaluate the bioavailability of minerals from fiber sources, and these studies have been reviewed (Erdman, 1982; Frolisch, 1986). The conflicting findings suggest both that sources of TDF such as brans provide minerals with varying degrees of bioavailability and, conversely, that TDF impairs mineral bioavailability, a dilemma that should be resolved.

4. EPIDEMIOLOGIC STUDIES

Although the dietary fiber hypothesis was developed from epidemiologic studies, there was not, nor is there now, a concern about any adverse effect of TDF on mineral utilization among populations (Trowell et al., 1985). In his review on dietary fiber and mineral utilization, Walker (1987) provided an excellent discussion of the rationale that TDF should not be considered to change mineral status adversely. Walker also made the

TABLE III. Factors Affecting Nutrient–Mineral Bioavailability

Intrinsic	Extrinsic
Age	Protein
Gender	Fat
Health/disease	Carbohydrate
Pregnancy	Total dietary fiber
	Others:
	Maillard products
	Phytic acid
	Ascorbic acid

point that increased consumption of TDF and the benefits for human nutrition and health to be achieved by these changes should be viewed holistically and not in isolation.

Possibly the best long-term studies of the effects of high-TDF diets on mineral nutriture are obtained from vegetarians. These individuals have not been found to have impaired mineral status (King et al., 1981; Anderson et al., 1981; Gibson et al., 1983; Abdulla et al., 1984). Freeland-Graves et al. (1980) have reported alterations in zinc absorption among lacto-ovo-vegetarians. Vegetarians adapt to high TDF (>35 g/day) intakes, which in turn help them maintain normal mineral nutriture. The adaptation in human and animal studies caused by changes in diet is important and has significant implications in clinical studies.

5. CLINICAL STUDIES

Many clinical trials have focused on possible adverse effects of TDF on mineral absorption and nutriture with the preponderance of results suggestive of a negative influence. These studies have been reviewed (Kelsay, 1978, 1981, 1982, 1986; Harland and Morris, 1985; Toma and Curtis, 1986). Current research studies continue to support the concept of adverse affects of TDF on mineral absorption (Hallberg, 1987). However, more recent reviews on this subject (Munoz, 1986; Southgate, 1987) have begun to challenge the negative effect of TDF on mineral utilization.

McCance and Widdowson (1942) found that Ca, Mg, and P were less efficiently absorbed in subjects who consumed diets containing 40–50% of their calories from brown bread compared to subjects who consumed white bread. Since this early report, approximately one-half of all clinical studies have indicated a negative impact of TDF on minerals; the other reports indicate no change. Along with iron and zinc, the mineral that appears to be most frequently cited as being negatively affected by TDF is calcium (Walker et al., 1948; Cullumbine et al., 1950; Reinhold et al., 1976; Cummings et al., 1979), but the mechanism is not known. It would be worthwhile to clarify this possible relationship between TDF intake and calcium balance in light of the Surgeon General's Report on Nutrition and Health (Koop, 1988), which advocates increased TDF intakes and also indicates osteoporosis to be one of the major skeletal diseases in this country.

As chronicled in cited reviews, evaluations of the effect of TDF on mineral nutrition have primarily been accomplished with balance trials, a few of which have employed isotopes (Schwartz et al., 1984; Turnlund et al., 1985). The balance technique has long been used to assess nutrient requirements for humans. Most of the current recommended dietary allowances (National Academy of Sciences, 1980) for essential nutrients have been determined using balance trials. There are advantages and disadvantages of the method when used for humans or animals (Hegsted, 1976), with each species providing its own difficulties. Until better analytical techniques are developed and better methods of assessing nutrient status in animals and man become available, the balance technique will continue to be used (Beisel, 1979).

In the case of minerals, the balance method appears accurate because minerals are not degraded as are organic nutrients. Duncan (1967) made an important observation when she stated that it is wrong to use balance data to reflect absolute rather than relative change. Positive balance does not necessarily mean accretion of a mineral in the body, nor does negative balance mean mineral depletion. Inherent with the balance technique are sampling and analytical errors, which will normally overestimate positive balance results (Forbes, 1973).

To achieve good balance data may take months of steady observation to overcome changes that could be the result of adaptation alone. Isaksson and Sjogren (1967) have indicated that for calcium balance data, adaptation to a new intake level may take months to achieve in studies on calcium balance. However, reproducible balance results for calcium and phosphorus in humans have been reported (Hargreaves and Rose, 1965) in studies lasting only for months.

One of the longest and best-controlled metabolic studies evaluating the effect of TDF on mineral balance was conducted by Sandstead et al. (1978, 1979). Subjects were fed similar diets plus an additional 26 g of TDF per day for periods ranging from 4 to 8 months. The TDF provided at monthly intervals included wheat bran, corn bran, soy bean hulls, dehydrated apple powder, or dehydrated carrot powder. Among the minerals assayed in these studies were calcium, magnesium, phosphorus, zinc, copper, and iron. None were found to be adversely affected by the different sources of TDF, and the subjects were not in negative balance.

6. INSOLUBLE DIETARY FIBERS

The IDFs in the plant cell wall serve as structural components (Selvandran, 1984, 1985). Plant species, age, and growing conditions affect the amounts and ratios of the three major IDFs—cellulose, hemicellulose(s), and lignin—that are present in plant cell walls. Naturally occurring as in brans or as isolated, relatively pure polymers (i.e., cellulose), IDFs are added to foods to provide both bulk and water-holding capacity. These insoluble polymers can be expected to exhibit the same properties in the intestine.

Two theories have been suggested as to how IDF may decrease mineral utilization. First, IDF accelerates the movement of the digesta through the intestine. Although IDF has been shown to increase movement of luminal contents (Read, 1985), it has not been demonstrated that this phenomenon affects mineral absorption. The second theory suggests that IDF acts as a chelator, holding numerous metal ions and preventing their absorption (Reinhold et al., 1975; Ismail-Beigi et al., 1977; Eastwood and Kay, 1979).

Cellulose is the one source of IDF that has been frequently cited as impacting negatively on Ca nutriture (McHale et al., 1979; Slavin and Marlett, 1980; Godara et al., 1981). If cellulose affects calcium absorption, it appears to do so through a mechanism other than binding, as cellulose has no ionic charge. If decreased transit time is the explanation, then the question should be asked why other types of ID that decrease transit time do not decrease calcium absorption and why other minerals are not also affected.

7. *IN VITRO* BINDING: CATION-EXCHANGE CAPACITY

In discussing the physicochemical properties of TDF, water adsorption, water-holding capacity, and cation-exchange capacity have been considered (McConnell et al.,

1974; Eastwood and Mitchell, 1976; Rasper, 1979). The cation-exchange capacity of TDF has been defined as the number of milliequivalents of hydrogen ions held per gram TDF (mEq/g). Basically, a sample of TDF is allowed to become saturated with hydrogen ions. With sodium chloride the hydrogen ions are then stripped from the TDF and titrated. Alternatively, solutions of mono- or polycations are added to the TDF–hydrogen ion mixture, and the cations bound to the TDF are measured. This *in vitro* procedure has variation in experimental protocols among different laboratories. The cation-exchange capacity of TDF represents the foundation behind the idea that TDF adversely affects mineral nutrition via a binding or chelation mechanism. The unsubstituted uronic acid residues contained primarily in pectin and to a lesser degree in hemicellulose polymers are believed to be involved primarily in TDF cation-exchange capacity.

Eastwood and Mitchell (1976) reported that the cation-exchange capacity of dried vegetable fibers ranged from 0.6 to 2.3 mEq/g. With the highest figure measured, and assuming a TDF intake of 35 g per day, estimates by calculation are that vegetable fiber could theoretically bind 3220 mg of calcium, which is well above the maximum RDA (Table I). Calculations of this nature add support for the binding theory as a mechanism by which some kinds of TDF affect mineral absorption.

Rasper (1979) measured the cation-exchange capacity of 11 cereal and noncereal sources of TDF. Nine samples had cation-exchange values that ranged from 0.07 to 0.21 mEq/g (mean 0.10 mEq/g). Soybean hulls and peanut red skins had values of 0.68 and 0.55 mEq/g, respectively. Rasper found the following correlations between the components of the nine cereal fibers and their cation-exchange capacity: hemicelluloses 0.56; cellulose −0.49; and lignin −0.33. For the combined 11 samples, the correlation coefficients were: hemicellulose 0.46; cellulose 0.44; and lignin 0.17. The latter correlation values suggest that cellulose may bind minerals, although levels of significance were not provided. Extension and clarification of this work by Rasper would appear to be appropriate. Frolich (1986) and Dreher (1987) review numerous *in vitro* studies on the ability of various sources of TDF to bind different cations.

The number of studies in which both *in vitro* and *in vivo* binding have been investigated and compared is limited. Fernandez and Phillips (1982a,b) in comparison studies observed *in vitro* ^{59}Fe binding to various sources of TDF (in order of decreasing affinity, lignin > psyllium > cellulose > pectin) on addition to these fibers with or without ascorbate, citrate, cysteine, fructose, or ethylenediaminetetraacetic acid. Then they perfused these polymers along with ^{59}Fe into the duodenal–jejunal section of the dog small intestine. These results showed that lignin and psyllium were potent inhibitors of iron absorption, with less effect by pectin and no effect by cellulose. More *in vitro* and *in vivo* binding comparisons of this nature are needed.

8. NONTRADITIONAL SOURCES OF TOTAL DIETARY FIBER

The recommendation for increasing the amount of TDF in the human diet calls for increasing the intake of foods rich in fiber. A gray area of this recommendation is the addition of isolated fibers to the diet in processed foods. Isolated fibers include a variety of cereal and oilseed brans and purified polymers such as cellulose, guar gum, psyllium, pectin, and polydextrose, to mention only a few. Investigators have used these isolated fiber sources to help prove and understand the dietary fiber hypothesis. Specifically, these individual fiber sources are being used to identify the mechanism(s) by which TDF acts in the body.

8.1. Mineral Absorption in Rats

As mentioned, TDF has been reported to have a negative effect on calcium absorption (Marlett, 1984). How TDF specifically affects calcium nutriture is unclear. In an attempt to address the question of whether charged functional groups on TDF affect mineral absorption, three sources of conventional dietary fibers and three nonfood polymers were evaluated in the rat. Mineral balance was determined for phosphorus, calcium, magnesium, iron, zinc, and copper (Schroeder and Gordon, 1985).

Groups of rats ($n = 5$) were fed an AIN-76 diet (American Institute of Nutrition, 1972) containing cellulose, chitin, chitosan, pectin (brown N. F. pectin, CECA, Inc., St. Louis, MO), cholestyramine, or wheat bran (American Association of Cereal Chemists) at dietary concentrations of 2.5, 5, 10, 20, or 40%. These polymers were selected because they possess different charged (e.g., ionic) groups. Cellulose and chitin were neutral, and pectin had a negative charge because of its uronic acid content (28% of galacturonic units had a free carboxylic group). Chitosan is the deacylation product of chitin, and cholestyramine is a bile-acid-sequestering drug (Questran®, Mead Johnson and Co., Evansville, IN). Both of these latter two polymers contain positively charged amino groups.

Growth rates of animals with cellulose, chitin, and wheat bran were similar at dietary levels of 20% and lower. At 40% dietary fiber levels, growth was reduced because of inadequate energy intake. Pectin impaired growth at 10% concentration in the diet. Animals consuming diets with 10% or more chitosan and 20% or more cholestyramine died of septicemia during the last 10 days of the 21-day feeding period.

Mineral absorption among the groups of animals fed these different polymers at different dietary concentrations were not significantly different except for chitosan and cholestyramine. The absorption of all minerals, especially iron, was impaired in animals consuming diets with the latter two amino polymers. Percentage mineral absorption in animals consuming diets with increasing amounts of cellulose (2.5 to 40%) and 2.5% chitosan are reported in Table IV.

Three important findings resulted from this study. Calcium balance was not impaired in growing rats fed diets containing cellulose or wheat bran. In addition, the two sources of dietary fiber containing negatively changed groups, pectin and wheat bran, had no adverse effect on calcium balance or any of the other five elements. The action of the amino polymers was unique in that both chitosan and cholestyramine dramatically reduced iron utilization. We have documented this adverse effect of chitosan on iron absorption in two previous studies in which significantly lowered blood hemoglobin and liver iron concentrations were demonstrated (Gordon and Besch-Williford, 1983, 1984). Chitosan, because it is partially dissolved in the stomach acid, produces a viscous bolus in the stomach and small intestine. This physical property is believed to be responsible for reduced growth by impairment of nutrient absorption. The binding of iron by the two amino polymers is thought to take place through the formation of a coordinating complex that appears specific for iron relative to the other transition elements.

TABLE IV. Percentage Apparent Absorption of Minerals in Weanling Rats
Fed Diets Containing Cellulose at Five Different Dietary Concentrations
or 2.5% Chitosan

Dietary residue (polymer)	Dietary content (%)	Percentage absorption[a]					
		P	Ca	Mg	Fe[b]	Zn	Cu
Cellulose	2.5	73	74 (210)[c]	69	61†	43	30
	5.0	69	67 (200)	70	57†	35	22
	10.0	74	73 (255)	76	66†	44	34
	20.0	63	62 (210)	51	55†	39	25
	40.0	—	52 (220)[d]	—	—	—	—
Chitosan	2.5	56	60	48	16‡	28	16

[a] After 12 days, animals were subjected to a 60-hr balance trial. The amount of minerals ingested (dietary intake) minus minerals excreted (fecal and urine) divided by amount of minerals ingested was used to calculate percentage absorption.
[b] Values in the same column not followed by the same symbol (†, ‡) are significantly different ($P \leq 0.05$).
[c] Values in parentheses represent milligrams of calcium retained by animals during 60-hr balance period.
[d] Because growth of these animals was significantly lowered because of lower caloric density of diets and lower food intake, calcium data are only reported for comparative purposes.

An important clinical study tends to support our observations. In the Lipid Research Clinic's Coronary Primary Prevention Trial (1984) on cholesterol intervention using cholestyramine, it was observed that the participating subjects had significantly lowered circulating ferritin levels at the end of the study. The lowered serum ferritin concentrations suggested decreased body stores of iron that developed during the 7 years these individuals ingested cholestyramine. Hemoglobin concentrations did not change between the start and end of the experiment. This study has significance for two reasons. First, no other human study has been conducted over such a long period of time, 7 years. Second, the study was accomplished in men who had adequate iron stores. Even though their iron stores were reduced, it was after a 7-year period with a high polymer intake (24 g cholestyramine/day). The decrease in ferritin concentrations, although statistically significant, was not felt to be clinically significant.

Our research to date suggests that the chemical charges of conventional food sources of TDF are insufficient to cause mineral binding. Since TDF does not contain positively charged groups, the evidence further suggests that this mechanism is not responsible for decreased mineral absorption.

9. PHYTIC ACID

The classification of phytic acid as a component of TDF is an anomaly. Although phytic acid does constitute part of the plant cell wall and is not digested in the alimentary tract of animals and man (Gordon and Lee, 1982), it is not a carbohydrate. Many scientists in the area of TDF research argue that a true TDF must be a carbohydrate. Not all TDF in the human diet may be derived from plant cell walls. Along with phytic acid (or phytin), dietary polymers that may represent the sum total of TDF in the human diet are illustrated in Fig. 3.

It has long been known that phytic acid will impair zinc bioavailability (O'Dell and Savage, 1960; Davies, 1982). The discovery that zinc is essential for humans resulted from observations among Egyptian and Iranian populations found to be severely zinc deficient (Prasad, 1983). In addition to consuming diets low in zinc, individuals among these populations were found to be consuming imbalanced diets high in phytic acid. Reinhold *et al.* (1976) suggested, based on limited data, that TDF and specifically cellulose was the responsible agent for zinc deficiency observed in these populations.

The debate over whether a component of TDF in addition to phytic acid is responsible for impaired zinc nutriture will continue. However, the total action of phytic acid or TDF on other nutrients must be considered. As previously mentioned, extrinsic and intrinsic factors can affect bioavailability (Table III). In addition to the effects of these two factors, interactions can occur among minerals that can affect their bioavailability (Gordon, 1987).

Recently, it has been shown that phytic acid can enhance the bioavailability of copper (Lee *et al.*, 1988). Phytic acid has also been shown to increase the bioavailability of iron (Gordon and Chao, 1984). We chose to examine further the interactions among phytic acid, zinc, and copper.

Six groups ($n = 5$) of copper-deficient rats were fed AIN-76 diets containing 5 μg copper and 12, 30, or 270 μg Zn/g diet with or without 1% phytic acid. After 3 days, the animals were killed, and concentrations of zinc and copper in plasma were determined (Table V). With increasing amounts of dietary zinc and no dietary phytic acid, plasma copper concentrations decreased ($P \leq 0.05$). With 1% phytic acid in the diets containing

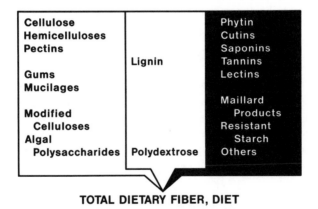

TOTAL DIETARY FIBER, DIET

FIGURE 3. Dietary components and polymers suggested to contribute to TDF. Plant cell wall carbohydrate polymers and examples of isolated cell wall polymers are indicated on the left. Lignin is a noncarbohydrate plant cell wall polymer, and polydextrose is a synthetic polymer used in foods. There is debate if these polymers should be considered as TDF. Compounds listed on the right are polymers either produced in the manufacture of foods or present as nondigestible compounds. Although these latter compounds are not considered true sources of TDF, they contribute to TDF analytical values and may affect physiological functions in the gut.

TABLE V. Effect of Increasing Dietary
Zinc Concentrations and the Presence
or Absence of Phytic Acid in the Diet
on Serum Cu and Zn Concentrations
in the Cu-Repleted Rat[a,b]

Dietary Zn (μg/g)	Dietary phytate (%)	Serum content (μg/dl)	
		Cu	Zn
12	0	43*	100*
12	1	49*	28†
30	0	32†	98*
30	1	46*	37†
270	0	22‡	102*
270	1	23‡	100*

[a] After 3 days of feeding copper-deficient rats diets that contain 5 μg Cu/g in addition to indicated Zn concentrations (Lee *et al.*, 1985).
[b] Values in the same columns not followed by the same superscript symbol are significantly different ($P \leq 0.05$).

12 or 30 μg Zn/g, plasma zinc decreased ($P \leq 0.05$). In animals consuming the diet with 270 μg Zn/g, this excess zinc overcame the inhibitory effect of the phytic acid on absorption of zinc. Information from my laboratory (Lee *et al.*, 1985) has indicated that phytic acid will impair zinc loading of the intestinal mucosal cell. Without excess Zn, the amount of zinc-binding protein, thionein, synthesized by the cell is lower. The lower concentration of this protein allows for transfer of copper into the body. The entry of zinc into the mucosal cell can be blocked with phytic acid, as illustrated in Fig. 4A. Without phytic acid, or at a high concentration of dietary zinc, the resulting protein metallothionein (thionein plus zinc and/or copper) is increased (Cousins, 1985). Both zinc and copper are bound and held by the protein, retarding the absorption of copper, as illustrated in Fig. 4B.

Two conclusions can be highlighted from these observations. When phytic acid reduces absorption of zinc, more copper can be made available to the body. The same result can occur if the diet is imbalanced with high zinc concentrations. It is agreed that a diet high in phytic acid would be detrimental to zinc nutrition. However, it is suggested that in moderate amounts, as found in a varied diet, these simultaneous series of blockages and enhancements may lead to balanced nutrient absorption.

10. SOLUBLE DIETARY FIBER

With the growing knowledge that SDF is a separate component of TDF that has its own unique physiological function(s) in the intestine, the question arises whether SDF affects mineral absorption. The properties of SDF are such that it dissolves in the stomach and small intestine. It is postulated that SDFs affect nutrient absorption or possibly in some manner regulate plasma nutrient concentrations. To illustrate this point,

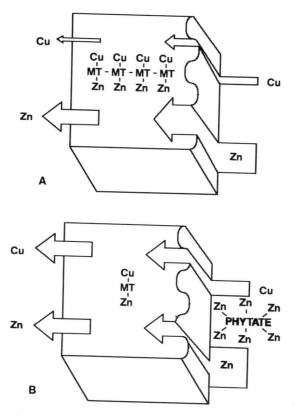

FIGURE 4. Interactions among zinc, copper, and phytic acid. Under conditions of high dietary zinc, the amount of metallothionein (MT) protein will increase, binding both zinc and copper. The net result is that less copper can be released from the mucosal cell (A). With phytic acid present to bind zinc in the intestinal lumen, less zinc enters the cell, less metallothionen is produced, and subsequently more copper can pass from the lumen into the circulatory system (B).

SDF has been repeatedly reported to be effective in lowering or maintaining blood glucose levels, especially in the diabetic (Jenkins *et al.*, 1986), and in lowering blood cholesterol levels (Chen and Anderson, 1986). Diets to lower blood glucose (Anderson, 1986) in diabetics and cholesterol in individuals with high levels (Van Horn, 1988) appear to include the incorporation of foods containing SDF.

10.1. Mechanism of Action

The mechanisms by which the actions of SDF affect nutrient absorption or regulate plasma concentrations are unknown. The actions of SDF may be physical and/or physiological within the gut (Elsenhans, 1983; Vahouny and Cassidy, 1985). Two theories are frequently mentioned for the action of SDF on nutrient absorption, and the two actions may be the same or may work synergistically. One theory suggests that the SDF is intermingling with the unstirred water layer along the luminal surface of the intestinal mucosal cells, increasing the thickness of this unstirred layer (Jenkins, 1981). The second theory proposes that the SDF in the unstirred water layer changes the composition of this diffusion barrier and restricts diffusion, thus restricting nutrient absorption (Caspary et al., 1980). With both theories, the water-holding capacity of the SDF can be related to its ability to change viscosity and possibly form gels.

Viscosity changes in the gut produced by SDF and affecting glucose absorption/tolerance have been demonstrated in humans (Gassull et al., 1976; Jenkins et al., 1978, 1980; Wolever et al., 1978; Crapo et al., 1981; Blackburn et al., 1984) animal perfusion studies (Elsenhans et al., 1980; Blackburn and Johnson, 1981), and in vitro studies (Elsenhans et al., 1980; Johnson and Gee, 1981; Edwards et al., 1987). If viscosity in the gut produced by SDF affects the absorption of certain compounds (i.e., glucose and cholesterol), the question arises whether this mechanism of action is common to any nutrient. To examine this latter possibility, the effects of viscosity produced by SDF on the luminal-to-vascular transfer of zinc were measured in the rat using a double-perfusion technique (Kim and Gordon, 1988).

10.2. Double-Perfusion Technique to Study Zinc Absorption

Details of the double-profusion technique have been described elsewhere (Smith et al., 1978; Steel and Cousins, 1985). Male weanling rats weighing 250–275 g were fed a diet containing 50 µg Zn/g for 7 days. The animals were anesthetized with phenobarbital solution, the abdominal cavity was opened, and the intestinal tract was exposed. A catheter was inserted into the superior mesenteric artery leading to the small intestine. The vascular perfusate was started through this catheter. All veins along the entire intestinal tract were ligated except those leading from the small intestine to the hepatic vein, from which the vascular perfusate was collected into a fraction collector.

At the beginning of the duodenum and at the end of the small intestine, inflow and outflow catheters were inserted, respectively. Compositions of the luminal and vascular perfusates are presented in Table VI, as are the physical conditions used during each perfusion. The total concentration of zinc in the luminal perfusate was 30 mM, consisting of zinc sulfate and 50 µCi $^{65}ZnSO_4$. On the basis of the specific activity of the vascular perfusate after it was collected, results were reported as zinc concentrations per unit time. Polymers used in this study were guar gum (Nutriloid TIC pretested guar, high viscosity) and sodium carboxymethylcellulose (Gum Ticalose, TIC pretested 5000, "R" coarse) obtained from TIC Gums, Inc., New York, NY. High methoxy pectin (HMI, Unipectin C) was supplied by Sonofi Bio-Ingredients, Germantown, WI. The B-glucan was provided by the Quaker Oats Company (Barrington, IL). The concentrations of each polymer required to produce luminal perfusates of equal viscosity are reported in Table VII.

With luminal perfusates having a viscosity of 250 cP, it was observed that the transfer of Zn into the vascular system was significantly inhibited by all polymers (Fig. 5) compared to controls with no SDF in the perfusate. Two other polymers were also investigated with similar inhibitory results at high viscosity: sodium alginate (Satialgine S1100, obtained from CECA, Sanofi, Inc., St. Louis, MO) and a second sample of B-glucan.

Two polymers, pectin and guar gum, were then evaluated at lower viscosities. With

TABLE VI. Composition and Conditions of Luminal
and Vascular Perfusates Used in Double-Perfusion
Studies in Rats

Components	Luminal perfusate	Vascular perfusate
KCl (mM)	3.9	3.9
$MgSO_4 \cdot 7H_2O$ (mM)	1.2	1.2
KH_2PO_4 (mM)	1.2	1.2
$NaHCO_3$ (mM)	2.5	2.5
$CaCl_2 \cdot 2H_2O$ (mM)	2.5	2.5
NaCl (mM)	150.0	150.0
L-Glutamine (mM)	6.0	0.6
D-Glucose (mM)	0.56	5.6
Dextran (6% soln)	—	77%
Dexamethasone (μM)	—	0.76
Zinc as $ZnSO_4$ (μM)	30.0 $(+^{65}Zn)$	15.0
Horse serum (%)	—	5.0
Flow rate (ml/min)	0.39	2.89
Norepinephrine (μg/min)	—	100
Temp (°C)	37.5	37.5
pH	7.4	7.4
Oxygen	No	Yes

a viscosity of 125 cP, zinc transfer was no different compared to a luminal viscosity of 250 cp (Fig. 6). With the pectin concentration reduced to 1% (Table VII), which gave a viscosity equal to or less than 10 cP, zinc transfer was identical to that observed in control animals with no SDF (Fig. 6). Guar gum was then tested at viscosities of 125, 40, or $\leq$10 cP (Fig. 7). Results indicate that zinc transfer was directly affected by viscosity and was unrelated to type of SDF: zinc transfer increased as viscosity decreased. When viscosity was essentially unchanged with 1% pectin or 0.1% guar gum, zinc transfer was not different from that observed in control animals with no SDF in their luminal perfusate (Figs. 6 and 7, respectively).

TABLE VII. Concentration of Soluble Dietary Fibers or Gums to Achieve
Luminal Perfusates of Equal Viscosity

Viscosity (cps)	$0 \leq 10$	40	125	250
Shear rate (sec^{-1})	23	23	46	46
Control (no SDF)	0	—	—	—
Guar gum	0.1	0.50	0.55	0.62
Sodium carboxylmethylcellulose	0.1	0.53	0.83	1.00
B-Glucan	0.1	0.46	0.69	0.80
High-methoxy pectin	1.0	1.70	2.50	3.08
Ethylenediaminetetraacetic acid[a]	0.1	—	—	—
Phytic acid[a]	0.1, 0.5	—	—	—

[a]Did not change viscosity of luminal perfusate.

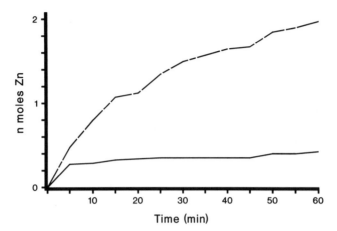

FIGURE 5. Zinc transfer from the lumen to the vascular system in rats during a double-perfusion technique over 60 min. In control animals with no SDF in their luminal perfusate (--------), zinc transfer increased with time, reaching 2 mol zinc after 60 min. This curve represents the mean data of six measurements (rats). Each of six sources of SDF, 250 cps, was tested in triplicate. The solid line (————) represents the mean data for zinc transfer from 18 animals. Deviations are not illustrated and were less than 7% of mean values.

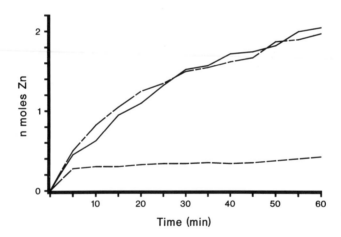

FIGURE 6. Zinc transfer from the lumen to the vascular system in rats over a double-perfusion technique for 60 min. Pectin was the source of SDF. At 125 cps (--------) zinc transfer was impaired. With 1% pectin (≤ 10 CP), zinc transfer (--————--) was equal to that observed in control animals with no SDF in their luminal perfusate (————). Each curve represents the mean value obtained from three animals. Deviations are not illustrated and were less than 7% of mean values.

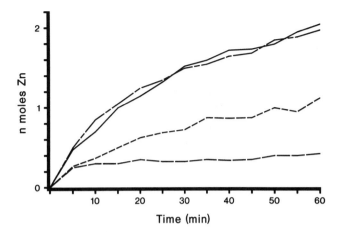

FIGURE 7. Zinc transfer from the lumen to the vascular system in rats during a double-perfusion technique over 60 min. Guar gum was the source of SDF. Zinc transfer was measured with luminal viscosities of 125 cps (———), 40 cps (-------), ≤ 10 cps (--———--), and with no added SDF (control, ———). All curves represent mean values of three animals. Deviations are not illustrated and were less than 7% of the mean values.

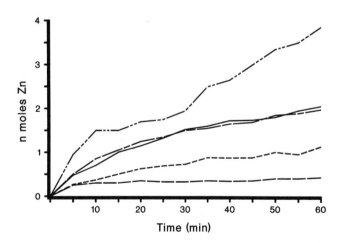

FIGURE 8. Zinc transfer from the lumen to the vascular system in rats during a double-perfusion technique over 60 min. Data represent means of analyses of four different sources of SDF evaluated at different viscosities in duplicate. Zinc transfer with luminal viscosities of 125 cps (———) 40 cps (-------), ≤10 cps (--———) and with no added SDF (———). With 0.1% EDTA in the luminal perfusate (———·--———), Zn transfer was higher than in control animals (———). No zinc transfer could be detected when phytic acid (0.5% or 1.0%) was added to the luminal perfusate (not shown). Deviations are not illustrated and were less than 7% of the mean values.

Experiments were completed in which all four of the original polymers were evaluated again at viscosities of 125, 40, and ≤10 cP. In addition, effects on zinc transfer of added phytic acid (1% or 0.5%) or 0.1% sodium ethylenediaminetetraacetic acid (EDTA) in the luminal perfusate (without SDF) were determined. Results for these experiments (Fig. 8) were identical to those previously observed and indicated in Fig. 5, 6, and 7. Phytic acid at both concentrations completely inhibited the transfer of zinc into the vascular system as was expected. The transfer of the zinc from the perfusate containing the zinc–EDTA was greater than observed in the control animals (Fig. 8). Apparently this complex (zinc–EDTA) is more easily absorbed through the cellular absorption and transport process than is free zinc (Oestreicher and Cousins, 1982).

11. SUMMARY

The consumption of foods rich in TDF should not be associated with impaired mineral absorption and long-term mineral status. In surveys of populations consuming high amounts of TDF, e.g., Third World populations and vegetarians, gross deficiencies in mineral nutrition have not been noted. If mineral status is low among these groups, it is most likely caused by the inadequacy or imbalance of the diet and not by the TDF. The key word is interaction, which should be interpreted in dietary imbalances that produce nutrient deficiencies.

There are no strong data to support the concept that TDF inhibits mineral absorption through a binding chelation mechanism. Limited data suggest that positively charged groups on polymers such as chitosan and cholestyramine will decrease iron absorption in humans and animals. Because TDF does not contain positively charged groups, future research should be directed at the possible role of protein consumed along with TDF and the combination of effects on mineral nutrition.

Phytic acid is acknowledged as a potent chelator of zinc. However, its association with zinc and its propensity to lower Zn bioavailability may enhance the absorption of other elements, notably copper and iron. The importance of interactions among nutrients, including TDF, will gain additional attention in the scientific community.

Soluble and insoluble dietary fiber function differently in the intestine. Insoluble fibers accelerate movement through the intestine. Soluble dietary fibers appear to regulate blood concentrations of glucose and cholesterol, albeit by some unknown mechanism. Increased viscosity produced by the SDF in the intestine may provide an explanation of how this class of polymers affects plasma glucose, cholesterol, and other nutrients.

Employing a double-perfusion technique in the rat, we demonstrated that viscosity produced by SDF will delay transfer of zinc into the circulatory system. This delayed absorption should not be interpreted as decreased utilization. A great deal of additional research is required to prove the importance of luminal viscosity produced by SDF on slowing nutrient absorption or regulating blood nutrient homeostasis.

Increased intake of TDF in the total human diet appears desirable. A dietary intake of 35 g/day should not be considered to have a negative effect on mineral absorption. It is important to educate people that an intake of more than 35 g TDF/day may cause an imbalance in the diet that can adversely affect mineral utilization.

ACKNOWLEDGMENTS. Appreciation is given to Dr. George V. Vahouny (deceased), who was intense, a great competitor in and out of science, and who gave the author inspiration. Portions of this work were supported by the University of Missouri Agricultural Station and by a grant from the University of Missouri Institutional Biomedical Research Support Grant RR 07053 from the National Institutes of Health. Contribution of the Missouri Agricultural Experiment Station, Journal Series No. 10747.

REFERENCES

Abdulla, M., Aly, K. O., Anderson, I., Asp, N. G., Birkhed, D., Denker, I., Johonsson, C. G., Jagestad, M., Kolar, K., Nair, B. M., Nilsson-Ekle, P., Norden, A., Rossner, S., Svensson, S., Akersson, B., and Ockerman, P. A., 1984, Nutrient intake and health status of lacto vegetarians: Chemical analysis of diets using the duplicate portion sampling technique, *Am. J. Clin. Nutr.* **40:** 325–338.

American Institute of Nutrition, 1977, Report of the AIN *Ad Hoc* Committee on Standards for Nutritional Studies, *J. Nutr.* **107:**1340–1348.

Ammann, P., Rizzoli, R., and Fleisch, H., 1986, Calcium absorption in rat large intestine *in vivo:* Availability of dietary calcium, *Am. J. Physiol.* **251**(14):G14–G18.

Anderson, B. M., Gibson, R. S., and Sabry, J. H., 1981, The iron and zinc status of long-term vegetarian women, *Am. J. Clin. Nutr.* **34:**1042–1048.

Anderson, J. W., 1986, Treatment of diabetes with high fiber diets, in: *CRC Handbook of Dietary Fiber in Human Nutrition* (G. A. Spiller, ed.), CRC Press, Boca Raton, FL, pp. 349–359.

Becker, H. G., Steller, W., Feldheim, W., Wisker, E., Kulikowski, W., Suckow, P., Meuser, F., and Seibel, W., 1986, Dietary fiber and bread: Intake, enrichment, determination and influence on colonic function, *Cereal Foods World* **31:**306–310.

Beisel, W. R., 1979, Metabolic balance studies—their continuing usefulness in nutritional research, *Am. J. Clin. Nutr.* **32:**271–274.

Blackburn, N. A., and Johnson, I. T., 1981, The effect of guar gum on the viscosity of the gastrointestinal contents and on glucose uptake from the perfused jejunum in the rat, *Br. J. Nutr.* **46:**239–246.

Blackburn, N. A., Redfern, J. S., Jarjis, H., Holgate, A. M., Hanning, I., Scorpello, J. H. B., Johnson, I. T., and Read, N.W., 1984, The mechanism of action of guar gum in improving glucose tolerance in man, *Clin. Sci.* **66:**329–336.

Burkitt, D. P., and Trowell, H. C., 1975, Refined carbohydrate foods and disease, in: *Some Implications of Dietary Fiber,* Academic Press, London.

Burkitt, D. P., Walker, A. R. P., and Painter, N. S., 1979, Dietary fiber and disease, *J.A.M.A.* **229:** 1068–1074.

Caspary, W. F., Elsenhans, B., Sufke, U., Plok, M., Blume, R., Lembcke, B., and Creutzfeldt, W., 1980, Effect of dietary fiber on absorption and motility, in: *Frontiers of Hormone Research,* Vol. 7 (T. J. B. van Wimersma Greidanus, ed.), S. Karger, Basel, pp. 202–217.

Chen, W. J. L., and Anderson, J. W., 1986, Hypochlolesterolemic effects of soluble fibers, in: *Dietary Fiber Basic and Clinical Aspects* (G. V. Vahouny and D. Kritchensky, eds.), Plenum Press, New York, pp. 275–286.

Committee on Diet, Nutrition and Cancer, 1982, *Diet, Nutrition, and Cancer,* Assembly of Life Sciences, National Research Council, National Academy Press, Washington.

Cousins, R. J., 1985, Absorption, transport and hepatic metabolism of copper and zinc: Special reference to metallothionein and ceruloplasmin, *Physiol. Res.* **65:**238–309.

Crapo, P. A., Insel, J., Sperling, M., and Kolterman, O. G., 1981, Comparison of serum glucose, insulin and glucagon responses to different types of complex carbohydrate in noninsulin-dependent diabetic patients, *Am. J. Clin. Nutr.* **34:**184–190.

Crosby, L., 1978, Fiber-standardized sources, *Nutr. Cancer* **1**:15–26.

Cullumbine, H., Basnayake, V., Lemottee, J., and Wickramanayake, T. W., 1950, Mineral absorption on rice diets, *Br. J. Nutr.* **4**:101–111.

Cummings, J. H., Southgate, D. A. T., Branch, W. J., Wiggins, H. S., Houston, H., Jenkins, D. J. A., Jivraj, T., and Hill, M. J., 1979, The digestion of pectin in the human gut and its effect on calcium absorption and large bowel function, *Br. J. Nutr.* **41**:477–485.

Davies, N. T., 1982, Effect of phytic acid on mineral availability, in: *Dietary Fiber in Health and Disease* (G. V. Vahouny and D. Kritchevsky, eds.), Plenum Press, New York, pp. 105–116.

Dreher, M. L., 1987, *Handbook of Dietary Fiber. An Applied Approach*, Marcel Dekker, New York, pp. 124–134, 323–355.

Duncan, D. L., 1967, Some aspects of the interpretation of mineral balances, *Proc. Nutr. Soc.* **26**:102–106.

Eastwood, M. A., and Kay, R. M., 1979, An hypothesis for the action of fiber along the gastrointestinal tract, *Am. J. Clin. Nutr.* **32**:364–367.

Eastwood, M. A., and Mitchell, W. D., 1976, Physical properties of fiber: A biological evaluation, in: *Fiber in Human Nutrition* (G. A. Spiller and R. J. Amen, eds.), Plenum Press, New York, pp. 109–129.

Eastwood, M. A., Robertson, J. A., Brydon, W.G., and MacDonald, D., 1983, Measurement of water-holding properties of fibre and their faecal bulking ability in man, *Br. J. Nutr.* **50**:539–547.

Edwards, C. A., Blackburn, N. A., Craigen, L., Davison, P., Tomlin, J., Sugden, K., Johnson, I. T., and Read, N. W., 1987, Viscosity of food gums determined *in vitro* related to their glycemic actions, *Am. J. Clin. Nutr.* **46**:72–77.

Elsenhans, B., 1983, Pharmacology of carbohydrate gelling agents, in: *Delaying Absorption as a Therapeutic Principle in Metabolic Diseases* (W. Creutzfeldt and U. R. Folsch, eds.), Thieme-Stratton, New York, pp. 29–43.

Elsenhans, B., Sufke, U., Blume, R., and Caspary, W. F., 1980, The influence of carbohydrate gelling agents on rat intestinal transport of monosacchardies and neutral amino acids *in vitro, Clin. Sci.* **59**:373–380.

Elsenhans, B., Zenker, D., and Caspary, W. F., 1984, Guaran effect on rat intestinal absorption. A perfusion study, *Gastrointerology* **86**:645–653.

Erdman, J. W., 1982, Bioavailability of trace minerals from cereals and legumes, *Cereal Chem.* **58**:21–26.

Favurs, M. J., Kathpalia, S. C., Fredric, L. C., and Mond, A. E., 1980, Effects of dietary calcium and 1,25 dihydroxyvitamin D_3 on colon calcium active transport, *Am. J. Physiol.* **238**(1):G75–G78.

Fernandez, R., and Phillips, S.F., 1982a, Components of fiber bind iron *in vitro, Am. J. Clin. Nutr.* **35**:100–106.

Fernandez, R., and Phillips, S. F., 1982b, Components of fiber impair iron absorption in the dog, *Am. J. Clin. Nutr.* **35**:107–112.

Forbes, G. B., 1973, Another source of error in the metabolic balance method, *Nutr. Rev.* **31**:297–300.

Freeland-Graves, J. H., Ebarget, M. L., and Hendrickson, P. J., 1980, Alterations in zinc absorption and salivary sediment zinc after a lacto-ovo-vegetarian diet, *Am. J. Clin. Nutr.* **33**:1757–1766.

Frolich, W., 1986, Bioavailability of minerals from cereals, in: *CRC Handbook of Dietary Fiber in Human Nutrition* (G. A. Spiller, ed.), CRC Press, Boca Raton, FL, pp. 173–191.

Furda, I., 1981, Simultaneous analysis of soluble and insoluble dietary fiber, in: *The Analyses of Dietary Fiber in Food* (W. P. T. James and O. Theander, eds.), Marcel Dekker, New York, pp. 163–172.

Gassull, M. A., Goff, D. V., Haisman, P., Hackaday, T. D. R., Jenkins, D. J. A., Jones, K., and Leeds, A. R., 1976, The effect of unavailable carbohydrate gelling agents in reducing the post-prandial glycemia in normal volunteers and diabetics, *J. Physiol. (Lond.)* **259**:52P–53P.

Gibson, R. S., Anderson, B. M., and Sabry, J. H., 1983, The trace mineral status of a group of postmenopausal vegetarians, *J. Am. Diet. Assoc.* **82**:246–249.

Godara, R., Kaur, A. P., and Bhat, C. M., 1981, Effect of cellulose incorporation in a low fiber diet on fecal excretion and serum levels of calcium, phosphorus and iron in adolescent girls, *Am. J. Clin. Nutr.* **34**:1083–1086.

Gordon, D. T., 1987, Interactions among iron, zinc, and copper, in: *AIN Symposium Proceedings, Nutrition '87* (O. A. Levander, ed.), The American Institute of Nutrition, Bethesda, MD, pp. 27–31.

Gordon, D. T., and Besch-Williford, C., 1983, Chitin and chitosan: Influence on element absorption in rats, in: *Unconventional Sources of Dietary Fiber* (I. Furda, ed.), American Chemical Society, Washington, DC, pp. 155–184.

Gordon, D. T., and Besch-Williford, C., 1984, Action of amino polymers on iron status, gut morphology, and cholesterol levels in the rat, *Chitin, Chitosan and Related Enzymes* (J. P. Zikakis, ed.), Academic Press, Orlando, FL, pp. 97–117.

Gordon, D. T., and Chao, L. S., 1984, Relationship of components in wheat bran and spinach to iron bioavailability in the anemic rat, *J. Nutr.* **114:**526–535.

Gordon, D. T., and Lee, D., 1982, The digestion of phytate in man, *Inst. Food Tech. Prog. Abstr.* 117.

Haber, G. B., Heaton, K. W., Murphy, D., and Burroughs, L. F., 1977, Depletion and distribution of dietary fibre, *Lancet* **2:**679–682.

Hallberg, L., 1987, Dietary fibre and mineral absorption. Effects on satiety, plasma glucose and serum-insulin, *Scand. J. Gastroenterol.* **22**(Suppl. 129):66–67.

Hargreaves, T., and Rose, G. A., 1965, The reproducibility of the balance method in man as applied to calcium and phosphorus, *Clin. Sci.* **28:**537–542.

Harland, B. F., and Morris, E. R., 1985, Fibre and mineral absorption, in: *Dietary Fibre Perspectives Reviews and Bibliography* (A. B. Leeds, ed.), John Libbey, London, pp. 72–82.

Hegsted, D. M., 1976, Balance studies, *J. Nutr.* **106:**307–311.

Ink, S. L., and Hurt, H. D., 1987, Nutritional implications of gums, *Food Technol.* **41:**77–82.

Isaksson, B., and Sjogren, B., 1967, A critical evaluation of the mineral and nitrogen balances in man, *Proc. Nutr. Soc.* **26:**106–116.

Ismail-Beigi, F., Faraji, B., and Reinhold, J. G., 1977, Binding of zinc and iron to wheat bread, wheat bran and their components, *Am. J. Clin. Nutr.* **30:**1721–1725.

Jenkins, D. J. A., 1981, Slow release carbohydrate and the treatment of diabetes, *Proc. Nutr. Soc.* **40:** 227–235.

Jenkins, D. J. A., Wolever, T. M. S., Leeds, A. R., Gassull, M. A., Haisman, P., Dilawori, J., Goff, D. V., Metz, G. L., and Alberti, K. G. M. M., 1978, Dietary fibres, fibre analogues, and glucose tolerance: Importance of viscosity, *Br. Med. J.* **1:**1392–1394.

Jenkins, D. J. A., Wolever, T. M. S., Taylor, R. H., Barker, H. M., and Fielden, H., 1980, Exceptionally low blood glucose response to dried beans: Comparison with other carbohydrate foods, *Br. Med. J.* **281:**578–580.

Jenkins, D. J. A., Wolever, T. M. S., Jenkins, A. L., and Taylor, R. H., 1986, Dietary fiber, gastrointestinal, endocrine and metabolic effects: Lente carbohydrates, *Dietary Fiber Basic and Clinical Aspects* (G. V. Vahouny and D. Kritchensky, eds.), Plenum Press, New York, pp. 69–80.

Johnson, C. D., Berry, M. F., and Weaver, C. M., 1985, Soybean hulls as an iron source for bread enrichment, *J. Food Sci.* **50:**1275–1277.

Johnson, I. T., and Gee, J. M., 1981, Effect of gel forming gums on the intestinal unstirred layer and sugar transport *in vitro, Gut,* **22:**398–403.

Judd, P. A., and Truswell, A. S., 1985, Dietary fibre and blood lipids in man, in: *Dietary Fibre Perspectives Reviews and Bibliography* (A. R. Leeds, ed.), John Libby, London, pp. 23–39.

Kelsay, J. L., 1978, A review of research on effects of fiber intake on man, *Am. J. Clin. Nutr.* **31:**142–159.

Kelsay, J. L., 1981, Effect of diet fiber level on bowel function and trace mineral balances of human subjects, *Cereal Chem.* **58:**2–5.

Kelsay, J. L., 1982, Effects of fiber in mineral and vitamin bioavailability. *Dietary Fiber In Health and Disease* (G. V. Vahouny and D. Kretchevsky, eds.), Plenum Press, New York, pp. 92–103.

Kelsay, J. L., 1986, Update on fiber and mineral aavailability, in: *Dietary Fiber Basic and Clinical Aspects* (G.V. Vohauny and D. Kritchevsky, eds.), Plenum Press, New York, pp. 361–372.

Kim, S., and Gordon, D. T., 1988, The effect of viscosity on ^{65}Zn transfer from the small intestine of the rat using a double perfusion technique, *Fed. Proc.* **2:**A1418.

King, J. C., Stein, T., and Coyle, M., 1981, Effect of vegetarianism on the zinc status of pregnant women, *Am. J. Clin. Nutr.* **34**:1049–1055.

Koop, C. E., 1988, *The Surgeon General's Report on Nutrition and Health,* Public Health Service Publication No. 88-50210, U.S. Department of Health and Human Services, Washington.

Lee, D. B. N., Walling, M. W., Gafter, U., Silis, V., and Coburn, J. W., 1980, Calcium and inorganic phosphate transport in rat colon, *J. Clin. Invest.* **65**:1326–1331.

Lee, D., Schroeder, J. J., Hess, R. L., and Gordon, D. T., 1985, Interaction of phytic acid and zinc on intestinal metallothionein levels and Cu bioavailability, *Fed. Proc.* **44**:1674.

Lee, D., Schroeder, J. III, and Gordon, D. T., 1988, Enhancement of Cu bioavailability in the rat by phytic acid, *J. Nutr.* **118**:712–717.

Leeds, A. R., 1987, Dietary fibre: Mechanisms of action, *Int. J. Obesity* **11**(Suppl. 1):3–7.

Life Sciences Research Office, Federation of American Societies for Experimental Biology, 1987, *Physiological Effects and Health Consequences of Dietary Fiber,* Prepared for Center for Food Safety and Applied Nutrition, Food and Drug Administration, Department of Health and Human Services, Washington.

Lipid Research Clinics Primary Prevention Trial, 1984, Results. I. Reduction in incidence of coronary heart disease, *J.A.M.A.* **251**:351–374.

Marlett, J. A., 1984, Dietary fiber and mineral bioavailability, *Intern. Med. Special.* **5**:99–114.

McCance, R. A., and Widdowson, E. M., 1942, Mineral metabolism of healthy adults on white and brown bread diataries, *J. Physiol. (Lond.)* **101**:44–85.

McConnell, A. A., Eastwood, M. A., and Mitchell, W. D., 1974, Physical characteristics of vegetable foodstuffs that could influence bowel function, *J. Sci. Food Agr.* **25**:1457–1464.

McHale, M., Kies, C., and Fox, H. M., 1979, Calcium and magnesium nutritional status of adolescent humans fed cellulose or hemicellulose supplements, *Food Sci.* **44**:1412–1417.

Mertz, W., 1987, *Trace Elements in Human and Animal Nutrition,* 5th ed., Vol. 1 and 2, Academic Press, New York.

Munoz, J. M., 1986, Overview of the effects of dietary fiber on the utilization of minerals and trace elements, in: *CRC Handbook of Dietary Fiber in Human Nutrition* (G. A. Spiller, ed.), CRC Press, Boca Raton, FL, pp. 193–200.

National Academy of Sciences, 1980, *Recommended Dietary Allowances,* 9th rev. ed., National Academy Press, Washington.

O'Dell, B. L., 1984, Bioavailability of trace elements, *Nutr. Rev.* **42**:301–308.

O'Dell, B. L., and Savage, J. E., 1960, Effect of phytic acid on zinc availability, *Proc. Soc. Exp. Biol. Med.* **103**:304–306.

Oestreicher, P., and Cousins, R. J., 1982, Influence of intraluminal constituents on zinc absorption by isolated, vascularly perfused rat intestine, *J. Nutr.* **112**:1978–1982.

Pla, G. W., and Fritz, J. C., 1971, Collaborative study of the hemoglobin repletion test in chicks and rats for measuring availability of iron, *J. Assoc. Off. Anal. Chem.* **54**:13–17.

Prasad, A. S., 1983, Clinical, biochemical and nutritional spectrum of zinc deficiency in human subjects: An update, *Nutr. Rev.* **14**:197–208.

Prosky, L., Asp, Y. G., Furda, I., DeVries, J. W., Schweizer, T. F., and Harland, B. F., 1984, Determination of total dietary fiber in foods, food products and total diets: Interlaboratory study, *J. Assoc. Off. Anal. Chem.* **67**:1044–1052.

Rasper, V. F., 1979, Chemical and physical characteristics of dietary cereal fiber, in: *Dietary, Fibers, Chemistry, and Nutrition* (G. E. Inglett and S. I. Falkehug, eds.), Academic Press, New York, pp. 93–115.

Read, N. W., 1985, Dietary fiber and bowel transit, in: *Dietary Fiber Basic and Clinical Aspects* (G. V. Vahouny and D. Kritchevsky, eds.), Plenum Press, New York, pp. 81–100.

Reinhold, J. G., Ismail-Beigi, F., and Faraji, B., 1975, Fiber vs. phytate as determinant of the availability of calcium, zinc and iron in breadstuffs, *Nutr. Rep. Int.* **12**:75–85.

Reinhold, J. G., Faradhi, B., Abadi, P., and Ismail-Beigi, F., 1976, Decreased absorption of calcium, magnesium, zinc and phosphorus by humans due to increased fiber and phosphorus consumption as wheat bread, *J. Nutr.* **106**:493–503.

Sandstead, H. H., Munoz, J. M., Jacob, R. A., Klevay, L. M., Rech, S. J., Logan, G. M., Dintzis, F. R., Inglett, G. I., and Shuey, W. C., 1978, Influence of dietary fiber on trace element balance, *Am. J. Clin. Nutr.* **31**:S180–S184.

Sandstead, H. H., Klevay, L. M., Jacob, R. A., Munoz, J. M., Logan, G. M., Jr., Reck, S. J., Dintzis, F. R., Inglett, G. E., and Shusey, W. C., 1979, Effects of dietary fiber and protein level on mineral element metabolism, in: *Dietary Fibers Chemistry and Nutrition* (G. E. Inglett and S. I. Falkehag, eds.), Academic Press, New York, pp. 147–156.

Schroeder, J. J., and Gordon, D. T., 1985, Mineral absorption in rats fed diets containing dietary fibers (residues) ranging in levels from 2.5 to 40%, *Fed. Proc.* **44**:1500.

Schwartz, R., Spencer, H., and Welsh, J. J., 1984, Magnesium absorption in human subjects from leafy vegetables, intrinsically labelled with stable ^{26}Mg, *Am. J. Clin. Nutr.* **39**:571–576.

Selvendran, R. R., 1984, The plant cell wall as a source of dietary fiber: Chemistry and structure, *Am. J. Clin. Nutr.* **39**:320–337.

Selvendran, R. R., 1985, Developments in the chemistry and biochemistry of pectic and hemicellulosic polymers, *J. Cell. Sci. [Suppl.]* **2**:51–88.

Slavin, J. L., and Marlett, J. A., 1980, Influence of refinished cellulose on human bowel function and calcium and magnesium balance, *Am. J. Clin. Nutr.* **33**:1932–1939.

Smith, K. T., Cousins, R. J., Silbon, B. L., and Failla, M. L., 1978, Zinc absorption and metabolism by isolated, vascularly perfused rat intestine, *J. Nutr.* **108**:1849–1857.

Southgate, D. A. T., 1969, Determination of carbohydrates in foods. II. Unavailable carbohydrates, *J. Sci. Food Agr.* **20**:331–335.

Southgate, D. A. T., 1981, One of the southgate methods for unavailable carbohydrate in the measurement of dietary fiber in the analysis of dietary fiber in food (W. P. T. James and O. Theander, eds.), Marcel Dekker, New York, pp. 1–20.

Southgate, D. A. T., 1982, Definitions and terminology of dietary fiber. in: *Dietary Fiber in Health and Disease* (G. V. Vahouny and D. Kritchevsky, eds.), Plenum Press, New York, pp. 1–7.

Southgate, D. A. T., 1987, Minerals, trace elements and potential hazards, *Am. J. Clin. Nutr.* **45**:1256–1266.

Steel, L., and Cousins, R. J., 1985, Kinetics of zinc absorption by luminally and vascularly perfused rat intestine, *Am. J. Physiol.* **248**:G46–G53.

Talbot, J. M., 1980, *The Role of Dietary in Divesticular Disease and Colon Cancer,* United States Food and Drug Administration, Washington.

Toma, R. B., and Curtis, D. J., 1986, Dietary fiber: Effect on mineral bioavailability, *Food Technol.* **46**:111–116.

Trowell, H., 1976, Definition of dietary fiber and hypotheses that it is a protective factor in certain diseases, *Am. J. Clin. Nutr.* **29**:417–427.

Trowell, H., Burkitt, D., and Heaton, K. (eds.), 1985, *Dietary Fibre, Fibre-Depleted Foods and Disease,* Academic Press, London.

Turnland, J. R., King, J. C., Gong, B., Keyes, W. R., and Michel, M. C., 1985, a Stable isotope study of copper absorption in young men: Effect of phytate and α-cellulose, *Am. J. Clin. Nutr.* **42**:18–23.

Vahouny, G. V., and Cassidy, M. M., 1985, Dietary fibers and absorption of nutrients, *Proc. Soc. Exp. Biol. Med.* **180**:432–446.

Vahouny, G. V., and Kritchevsky, D. (eds.), 1982, *Dietary Fiber in Health and Disease,* Plenum Press, New York.

Vahouny, G. V., and Kritchevsky, D. (eds.), 1985, *Dietary Fiber—Basic and Clinical Aspects,* Plenum Press, New York.

Van Campen, D., and Mitchell, E. A., 1965, Absorption of Cu^{64}, Zn^{65}, Mo^{19}, and Fe^{59} from ligated segments of the rat gastrointestinal tract, *J. Nutr.* **86**:120–124.

Van Horn, L. V., Lui, K., Parker, D., Emidy, L., Liao, U. Y. L., Pan, W. H., Giumetti, D., Hewitt, J., and Stampler, J., 1988, Serum lipid response to oat product intake with a fat-modified diet, *J. Am. Diet. Assoc.* **86:**759–764.

Walker, A. R. P., 1987, Dietary fibre and mineral metabolism, *Mol. Aspects Med.* **9:**69–87.

Walker, A. R. P., Fox, F. W., and Irving, J. T., 1948, Studies in human mineral metabolism I. The effect of bread rich in phytate phosphorus on the metabolism of certain mineral salts with special references to calcium, *Biochem. J.* **42:**452–462.

Wolever, T. M. S., Taylor, R., and Goff, D. V., 1978, Guar: Viscosity and efficacy, *Lancet* **23:**1381.

8

Effects of Fiber on Vitamin Bioavailability

JUNE L. KELSAY

1. INTRODUCTION

The effects of fiber on vitamin bioavailability were previously reviewed (Kelsay, 1982). At that time it was concluded that available data indicated that fiber would not likely interfere with vitamin absorption and bioavailability except possibly in situations where the naturally occurring vitamin might be bound to fiber in plant cell walls.

In this chapter the role of fiber in vitamin bioavailability is reconsidered in view of more recent experimental studies.

2. VITAMIN A AND CAROTENE

Results of human studies conducted in the first half of the century indicated that carotene was less available from vegetables than as pure carotene in oil, and this was later attributed to the presence of fiber (Kelsay, 1982). Barnard and Heaton (1973) found that in human subjects lignin given with a test meal containing 5000 I.U. of vitamin A per kilogram body weight had no effect on serum rise in vitamin A after the meal. However, Kasper *et al.* (1979) reported that when wheat bran, cellulose, pectin, guar flour, carob bean flour, or carrageenan was given with a test meal containing 300,000 I.U. of vitamin A palmitate, the area under the curve for serum concentration of vitamin A for the 9 hr afterward was significantly increased by all fiber sources. These two studies were conducted because there had been reports of binding of bile acids by fiber, and it was postulated that fiber might also affect bioavailability of fat-soluble vitamins through changes in intestinal absorption. Both studies involved unphysiological amounts of vitamin A.

JUNE L. KELSAY • Carbohydrate Nutrition Laboratory, Beltsville Human Nutrition Research Center, Agricultural Research Service, U.S. Department of Agriculture, Beltsville, Maryland 20705.

Results of earlier studies showed no effect of methylcellulose plus minimal doses of vitamin A given by stomach tube on growth of vitamin-A-depleted rats (Ellingson and Massengale, 1952) or of pectin on vitamin A liver stores (Phillips and Brien, 1970); however, a mixture of methylcellulose, agar, pectin, and wheat grain as 10% or 20% of the diet resulted in a modest decrease in liver stores of vitamin A (Gronowska-Senger *et al.*, 1979). Keck and Monte (1983) found that 3% lignin added to a synthetic purified diet had no effect on rat growth or vitamin A liver stores. Omaye and Chow (1983) fed three different breads and two cereals as 20% of the diet; there was no effect on rat growth, food consumption, internal organ weights, or general health, but one of the cereals resulted in lower plasma vitamin A values. Omaye and Chow (1984) also reported that there was no effect of 5% or 20% wheat bran fed for 56 days on weight gain, food intake, and food efficiency of rats; plasma vitamin A values declined at 6 weeks, but the effect was reversed at 8 weeks. These studies with animals indicate that if there is a decrease in vitamin A bioavailability because of fiber, the effect is small, and the animals likely become adjusted with time.

3. VITAMIN E

There have been several fiber and vitamin E studies conducted on rats. A vitamin-E-supplemented diet containing 10% pectin that was fed for 56 days resulted in lower body weight as well as decreased plasma and red blood cell vitamin E and increased hemolysis (deLumen *et al.*, 1982). Feeding 20% of the diet as bread or cereal for 30 days (three different breads and two cereals) did not affect vitamin E plasma levels (Omaye and Chow, 1983). In rats fed 5% or 20% wheat bran for 56 days, plasma vitamin E levels declined at 5 weeks but reverted to prestudy levels at 8 weeks (Omaye and Chow, 1984). Rats fed 6% or 8% pectin for 8 weeks had lower body weights, lower liver vitamin E levels, and higher hemolysis rates than those fed 0 or 3% pectin (Schaus *et al.*, 1985). From these studies on rats it appears that intakes of pectin at a level of 6% or more of the diet decrease vitamin E bioavailability, but fiber-containing breads or cereals as 20% of the diet have only transient or no effects.

4. ASCORBIC ACID

Keltz *et al.* (1978) fed humans a control diet supplemented with 14 g hemicellulose, cellulose, or pectin. Pectin significantly decreased and hemicellulose appeared to increase ascorbic acid absorption as measured by urinary excretion of the vitamin. This study indicates that different forms of fiber may have different effects of ascorbic acid bioavailability.

5. THIAMINE

In an early study on humans by Daum *et al.*(1951), urinary thiamine after a breakfast containing 100 g cereal was no different from that after a breakfast with bacon and eggs, indicating that the cereal had no effect on thiamine absorption. However, Saito and Yoshida (1987) found that in humans the addition of pectin or guar gum in a test meal resulted in increased urinary thiamine; glucomannan and rice decreased urinary thiamine; and cellulose or fiber from wood pulp had no effect. Ellingson and Massengale (1952) reported that weight gain was not affected in thiamine-depleted rats given methyl-cellulose along with low doses of thiamine hydrochloride by stomach tube. Ranhotra *et al.* (1985) found that thiamine in whole wheat bread or enriched white bread was not as available to thiamine-deficient rats as was thiamine mononitrate not added to bread, as determined by erythrocyte transketolase activity and thiamine liver response. These studies suggest that carbohydrates other than fiber in food products may affect thiamine bioavailability.

6. RIBOFLAVIN

Earlier work (Holman, 1954) indicated that in undernourished children riboflavin was not as well absorbed from whole wheat bread as from enriched white bread, as determined by urinary riboflavin. However, it was later reported that bran, cellulose, or cabbage fiber in the diet enhanced the absorption of riboflavin, as indicated by urinary content following a load dose of riboflavin with breakfast (Roe *et al.*, 1978). Omaye and Chow (1983) found that rats fed 20% of the diet from three breads or two cereals had significantly lower red blood cell riboflavin with two of the cereals and one of the breads than when a fiber-free diet was fed. The effect of fiber on riboflavin absorption remains questionable because of differences in protocol and paucity of data. It appears that it is of prime importance whether the vitamin is associated with the fiber in the food source. Bioavailability is more likely to be decreased when the vitamin is bound to the fiber in the food source than when the fiber and the riboflavin are given as separate supplements.

7. NIACIN

When a breakfast containing cereal was compared with a breakfast containing bacon and eggs, urinary niacin was not significantly different (Daum *et al.*, 1951). However, in undernourished children niacin was not as well absorbed from whole wheat bread as from enriched white bread as determined by urinary niacin (Holman, 1954). Carter and Carpenter (1982a) administered to human subjects test doses of pure niacin, wheat bran extracts (in which niacin was known to be bound), and alkali-treated bound niacin (which would free the niacin). As determined by urinary metabolites of niacin, the bound niacin not treated with alkali was least available. Carter and Carpenter (1982b) also studied the availability to rats of niacin in foods by adding test foods at two levels. Eight samples of mature cooked cereals gave assay values of 35% of their total niacin content as determined by chemical assay. Alkali-cooked cereals, steamed sweet corn, beans (red panameno), and liver gave values close to their total niacin content.

It appears conclusive that the niacin in some foods, such as cereals, is bound and therefore not available unless freed by the proper alkali treatment. The binding agent is likely cellulose or hemicellulose, but niacin was also bound to a peptide or glycoprotein (Mason *et al.*, 1973).

8. PANTOTHENIC ACID

One human study was conducted in which 20 g of wheat bran or corn bran, 10 or 20 g of cellulose, or 10 or 20 g hemicellulose was added to the diet. After 6 days there was no effect of any of the fibers on blood pantothenic acid levels (Trimbo *et al.*, 1979). Therefore, there is no evidence that fiber interferes with pantothenic acid availability.

9. FOLIC ACID

Four studies have been conducted to determine the bioavailability of folic acid as affected by fiber content of test meals given to human subjects. Rise in serum folate was lower when maize meal, polished rice, or whole wheat bread was fed than when pteroylglutamic acid alone was fed and was lowest with whole wheat bread (Colman *et al.*, 1975). Recovery of pteroyl monoglutamic acid in urine was not affected by whole wheat bread or cellulose (Russell *et al.*, 1976). As determined by urinary excretion, bioavailability of folic acid from seven common Indian foods appeared not to be related to fiber content of these foods (Babu and Srikantia, 1976). Consuming wheat bran or small white beans had no effect on serum and urinary levels of heptaglutamyl folic acid, the form representative of polyglutamate in foods (Keagy *et al.*, 1988); however, wheat bran accelerated absorption of monoglutamyl folic acid.

In chicks the addition of 3% cellulose, pectin, lignin, sodium alginate, or wheat bran did not affect plasma and liver response curves to monoglutamate (Ristow *et al.*, 1982). In rats, the addition of 9% or 17% cellulose, 5% or 9% xylan, 5% or 9% pectin, 11% or 20% white wheat bran, 9% or 17% red wheat bran, 29% cooked dried beans, or 25% extracted cooked dried beans had no effect on the slope of the liver folic acid response to pteroylglutamic acid (Keagy and Oace, 1984).

The results of studies of effect of fiber on folic acid availability indicate that fiber itself may not be important. The study by Colman *et al.* (1975), however, indicated that folic acid in foods is not as available as when the free vitamin is fed. Other carbohydrates in the diet may affect folic acid bioavailability.

10. VITAMIN B-12

In rats the inclusion of 40–50% cellulose or 5–15% pectin during vitamin B-12 depletion resulted in increased excretion of methylmalonic acid in urine and vitamin B-12 in feces, indicating decreased vitamin B-12 utilization (Cullen and Oace, 1978). Feeding 5% pectin, cellulose, lignin, alginate, wheat bran, xylan, or guar gum to vitamin-B-12-deprived rats resulted in decreased vitamin B-12 utilization (as determined by increased methylmalonic acid excretion) only by xylan, guar gum, and pectin. The greatest effect in these two studies was caused by the ingestion of the soluble fibers, which are digestible by intestinal bacteria (Cullen and Oace, 1980).

11. VITAMIN B-6

When human subjects were fed soy beans or beef as a source of 60% of the vitamin B-6 in the diet, there was greater fecal vitamin B-6 and less urinary vitamin B-6 as 4-pyridoxic acid (4PA) with soy beans (Leklem et al., 1980a). Bioavailability of vitamin B-6 was 5–10% less from whole wheat bread than from vitamin-B-6-enriched white bread as determined by fecal vitamin B-6 and urinary 4PA (Leklem et al., 1980b). The addition of 15 g wheat bran to the diet resulted in a modest decrease in vitamin B-6 bioavailability as measured by fecal vitamin B-6, urinary 4PA, and plasma vitamin B-6 and pyridoxal phosphate (Lindberg et al., 1983). Vitamin B-6 bioavailability was less from whole wheat bread and peanut butter than from tuna, as determined by urinary vitamin B-6 and 4PA (Kabir et al., 1983). These studies were done by the same group of investigators, who reported that the availability of vitamin B-6 to humans was inversely related to the amount of glycosylated vitamin B-6 (Kabir et al., 1983). Vitamin B-6 in wheat, corn, and rice bran were judged to be unavailable to humans as determined by total urinary vitamin B-6 (Kies et al., 1984)

In rats, bioavailability of vitamin B-6 was lower when a fortified rice-base cereal was fed than when nonfat dry milk was fed as determined by bioassay after a vitamin B-6 depletion period (Gregory, 1980). In another rat study there was no effect of 10% cellulose, pectin, or a mixed fiber source (containing cellulose, pectin, and lignin) on liver vitamin B-6 (Nguyen et al., 1981), but when 5% cellulose, pectin, or bran was fed to chicks, bran resulted in decreased vitamin B-6 availability as indicated by growth and feed consumption. There was no effect of cellulose, pectin, or lignin, but the absorption rate of B-6 vitamers perfused in rat jejunal sections was decreased by homogenized carrot (Nguyen et al., 1983). Also in rats 20% wheat bran or 7% cellulose did not affect urinary vitamin B-6 or vitamin B-6 turnover (Hudson et al., 1988). Pyridoxine-β-glucoside was about 40% as available to rats as was free pyridoxine as measured by growth rate, plasma PLP concentration, erythrocyte aspartate aminotransferase activity, and urinary 4PA (Trumbo et al., 1988). These studies indicate that the decreased vitamin B-6 availability from some foods may be related more to the occurrence of vitamin B-6 in the glycosylated form than to the presence of fiber as such. However, Shultz and Slattery (1988) reported that lignin bound 29% of vitamin B-6 (as pyridoxine) in vitro.

12. SUMMARY AND CONCLUSIONS

Studies dealing with the effects of fiber on vitamin bioavailability were reviewed. Results of some studies using responses to test meals showed that fiber resulted in increased vitamin availability. This was found in studies on serum vitamin A, urinary thiamin, and urinary riboflavin responses to fiber. However, the protocol in each of these studies was so different that results cannot really be compared.

There is some evidence that soluble fibers such as pectin and guar gum decrease the availability of vitamin E and vitamin B-12 to rats. This effect may be related to the degradability of the soluble fibers by intestinal bacteria, but the mechanism of the fiber effect is not known.

In some studies on rats the effect of fiber depended on the level of fiber intake. Examples of this were seen in the effect of pectin on vitamins E and B-12, the effect of mixed fiber on vitamin A, and the effect of cellulose on vitamin B-12.

It appears that vitamins bound to fiber or to some other carbohydrate or glycoprotein in food are not as available as the vitamin given in pure form. Examples of this are carotene in carrots, niacin in cereals, thiamin in bread, and vitamin B-6 in the glycosylated form. Fiber may not always be the agent that reduces availability of certain vitamins from foods.

Reported decreases of vitamin bioavailability because of the presence of fiber or related components in foods have been small in most cases and likely would not affect vitamin nutritional status when vitamin intakes are adequate.

REFERENCES

Babu, S., and Srikantia, S. G., 1976, Availability of folates from some foods, *Am. J. Clin. Nutr.* **29:** 376–379.

Barnard, D. L., and Heaton, K. W., 1973, Bile acids and vitamin A absorption in man: The effects of two bile acid-binding agents, cholestyramine and lignin, *Gut* **14:**316–318.

Carter, E. G. A., and Carpenter, K. J., 1982a, The bioavailability for humans of bound niacin from wheat bran, *Am. J. Clin. Nutr.* **36:**855–861.

Carter, E. G. A., and Carpenter, K. J., 1982b, The available niacin value of foods for rats and their relation to analytical values, *J. Nutr.* **112:**2091–2103.

Colman, N., Green, R., and Metz, J., 1975, Prevention of folate deficiency by food fortification. II. Absorption of folic acid from fortified staple foods, *Am. J. Clin. Nutr.* **28:**459–464.

Cullen, R. W., and Oace, S. M., 1978, Methylmalonic acid and vitamin B-12 excretion of rats consuming diets varying in cellulose and pectin, *J. Nutr.* **108:**640–647.

Cullen, R. W., and Oace, S. M., 1980, Impact on B-12 status of pectin and six dietary fibers in rats, *Fed. Proc.* **39:**785.

Daum, K., Tuttle, W. W., Kisgen, R., Imig, C. J., Bongey, R., and Martin, C., 1951, Effect of different types of breakfasts on thiamin and niacin excreted by men, *J. Am. Diet. Assoc.* **27:**298–301.

deLumen, B. O., Lubin, B., Chiu, D., Reyes, P., and Omaye, S. T., 1982, Bioavailability of vitamin E in rats fed diets containing pectin, *Nutr. Res.* **2:**73–83.

Ellingson, R. C., and Massengale, O. N., 1952, Effect of methylcellulose on growth response of rats to low vitamin intakes, *Proc. Soc. Exp. Biol. Med.* **79:**92–94.

Gregory, J. F. III, 1980, Bioavailability of vitamin B-6 in nonfat dry milk and a fortified rice breakfast cereal product, *J. Food Sci.* **45:**84–86.

Gronowska-Senger, A., Chudy, D., and Smaczny, E., 1979, The content of fibre in the diet and the utilization of vitamin A by the organism, *Roczniki Panstowowego Zakladu Higieny* **30:**553–558 [Polish]; cited 1980, *Nutr. Abstr. Rev.* **50:**805.

Holman, W. I. M., 1954, Studies on the nutritive value of bread and on the effect of variations in the extraction rate of flour on the growth of undernourished children, in: *Medical Research Council Special Report Series No. 287,* HMSO, London, pp. 92–118.

Hudson, C. A., Betschart, A. A., and Oace, S. M., 1988, Bioavailability of vitamin B-6 from rat diets containing wheat bran or cellulose, *J. Nutr.* **118:**65–71.

Kabir, H., Leklem, J. E., and Miller, L. T., 1983, Comparative vitamin B-6 bioavailability from tuna, whole wheat bread and peanut butter in humans, *J. Nutr.* **113:**2412–2420.

Kasper, H., Rabast, U., Fassl, H., and Fehle, F., 1979, The effect of dietary fiber on the postprandial serum vitamin A concentration in man, *Am. J. Clin. Nutr.* **32:**1847–1849.

Keagy, P. M., and Oace, S. M., 1984, Folic acid utilization from high fiber diets in rats, *J. Nutr.* **114:** 1252–1259.

Keagy, P. M., Shane, B., and Oace, S. M., 1988, Folate bioavailability in humans: Effects of wheat bran and beans, *Am. J. Clin. Nutr.* **47**:80–88.

Keck, B. E., and Monte, W. C., 1983, Effects of lignin on vitamin A absorption in rats, *J. Appl. Nutr.* **35**:73–79.

Kelsay, J. L., 1982, Effects of fiber on mineral and vitamin bioavailability, in: *Dietary Fiber in Health and Disease* (G. V. Vahouny and D. Kritchevsky, eds.), Plenum Press, New York, pp. 91–103.

Keltz, F. R., Kies, C., and Fox, H. M., 1978, Urinary ascorbic acid excretion in the human as affected by dietary fiber and zinc, *Am. J. Clin. Nutr.* **31**:1167–1171.

Kies, C., Kan, S., and Fox, H. M., 1984, Vitamin B-6 availability from wheat, rice, and corn brans for humans, *Nutr. Rep. Int.* **30**:483–491.

Leklem, J. E., Shultz, T. D., and Miller, L. T., 1980a, Comparative bioavailability of vitamin B-6 from soybeans and beef, *Fed. Proc.* **39**:558.

Leklem, J. E., Miller, L. T., Perera, A. D., and Peffers, D. E., 1980b, Bioavailability of vitamin B-6 from wheat bread in humans, *J. Nutr.* **110**:1819–1828.

Lindberg, A. S., Leklem, J. E., and Miller, L. T., 1983, The effect of wheat bran on the bioavailability of vitamin B-6 in young men, *J. Nutr.* **113**:2578–2586.

Mason, J. B., Gibson, N., and Kodicek, E., 1973, The chemical nature of the bound nicotinic acid of wheat bran: Studies of nicotinic acid-containing macromolecules, *Br. J. Nutr.* **30**:297–311.

Nguyen, L. B., Gregory, J. F. III, and Damron, B. L., 1981, Effects of selected polysaccharides on the bioavailability of pyridoxine in rats and chicks, *J. Nutr.* **111**:1403–1410.

Nguyen, L. B., Gregory, J. F. III, and Cerda, J. J., 1983, Effect of dietary fiber on absorption of B-6 vitamers in a rat jejunal perfusion study, *Proc. Soc. Exp. Biol. Med.* **173**:568–573.

Omaye, S. T., and Chow, F. I., 1983, High fiber breads and breakfast cereals: Effect on rat growth and selected vitamin bioavailability, *Nutr. Rep. Int.* **28**:295–304.

Omaye, S. T., and Chow, F. I., 1984, Effect of hard red spring wheat bran on the bioavailability of lipid-soluble vitamins and growth of rats fed for 56 days, *J. Food Sci.* **49**:504–506.

Phillips, W. E. J., and Brien, R. L., 1970, Effect of pectin, a hypocholesterolemic polysaccharide, on vitamin A utilization in the rat, *J. Nutr.* **100**:289–292.

Ranhotra, G., Gelroth, J., Novak, F., and Bohannon, F., 1985, Bioavailability for rats of thiamin in whole wheat and thiamin-restored white bread, *J. Nutr.* **115**:601–606.

Ristow, K. A., Gregory, J. F. III, and Damron, B. L., 1982, Effects of dietary fiber on the bioavailability of folic acid monoglutamate, *J. Nutr.* **112**:750–758.

Roe, D. A., Wrick, K., McLain, D., and Van Soest, P., 1978, Effects of dietary fiber sources on riboflavin absorption, *Fed. Proc.* **37**:756.

Russell, R. M., Ismail-Beigi, F., and Reinhold, J. G., 1976, Folate content of Iranian breads and the effect of their fiber content on the intestinal absorption of folic acid, *Am. J. Clin. Nutr.* **29**:799–802.

Saito, Y., and Yoshida, K., 1987, Effect of dietary fibre on urinary thiamin excretion in humans, *Hum. Nutr., Food Sci. Nutr.* **41F**:63–70.

Schaus, E. E., deLumen, B. O., Chow, F. I., Reyes, P., and Omaye, S. T., 1985, Bioavailability of vitamin E in rats fed graded levels of pectin, *J. Nutr.* **115**:263–270.

Shultz, T. D., and Slattery, C. W., 1988, Binding of pyridoxine *in vitro* by natural and purified fibers, *FASEB J.* **2**:A441.

Trimbo, S., Kathman, J., Kies, C., and Fox, H. M., 1979, Pantothenic acid nutritional status of adolescent humans as affected by dietary fiber and bran, *Fed. Proc.* **38**:556.

Trumbo, P. R., Gregory, J. F. III, and Sartain, D. B., 1988, Incomplete utilization of pyridoxine-β-glucoside as vitamin B-6 in the rat, *J. Nutr.* **118**:170–175.

9

Dietary Fiber and Lipid Absorption

MICHIHIRO SUGANO, IKUO IKEDA, KATSUMI IMAIZUMI, and Y.-F. LU

1. INTRODUCTION

The regulation of lipid absorption by dietary means is a reasonable approach in reducing plasma lipids, particularly cholesterol. Among various dietary factors that interfere with lipid absorption, certain dietary fibers have a substantial hypolipidemic effect through interaction with lipids in the intestine (Kay, 1982; Kritchevsky, 1985; Miettinen, 1987; Vahouny, 1982, 1985a; Vahouny and Cassidy, 1985, 1986). Although the lipid-lowering effect of dietary fibers is rather moderate compared to synthetic compounds such as cholestyramine or colestipol, they usually exert no or otherwise negligible side effects (Cassidy *et al.*, 1986, Furda, 1983). In addition, dietary fiber is relatively easily accepted as a constituent of regular meals. It has been shown that in low-fat diets recommended for the reduction of serum cholesterol, the effect of nonfat components such as protein, carbohydrate, and fiber of the diet on cholesterol lowering may become more important than the type of fat consumed (American Heart Association, 1986; Grundy, 1986; Report of the National Cholesterol Education Program, 1988).

In spite of numerous studies on the effect of dietary fiber on plasma lipid profiles (Kay, 1982; Kritchevsky, 1985; Miettinen, 1987; Vahouny, 1982, 1985a; Vahouny and Cassidy, 1985, 1986), only limited information is available regarding the interaction of the fibers with different types of dietary fats. We report here that the effect of dietary fiber on cholesterol absorption is modified by the the type of dietary fat. The results of studies searching new fiber sources with hypocholesterolemic activity and those related to the metabolic memory in respect to dietary fiber and cholesterol metabolism are also described.

MICHIHIRO SUGANO, IKUO IKEDA, KATSUMI IMAIZUMI, and Y.-F. LU • Laboratory of Nutrition Chemistry, Kyushu University School of Agriculture 46-09, Fukuoka 812, Japan.

2. MODE OF LIPID ABSORPTION: AN OVERVIEW

2.1. Absorption of Triglyceride

Figure 1 illustrates the proposed pathway by which triglyceride is digested and absorbed in the intestine (Carey *et al.*, 1983; Friedman and Nylund, 1980; Tso, 1985). Dietary fats are first emulsified in the stomach by the peristalsis. It has been believed that a portion of roughly emulsified triglycerides, in particular those containing short-chain fatty acids occurring in milk, is hydrolyzed by the lipase present in the stomach. However, in our regular meals, the magnitude of stomach hydrolysis of dietary fats seems to be negligible, if any occurs. Thus, the interference of dietary fiber with fat absorption may not be significant in the stomach except for the delaying effect on stomach emptying.

Roughly emulsified lipids are then transferred to the duodenum, where they finely emulsified by the aid of bile salts and are partially hydrolyzed by the pancreatic lipase, yielding free fatty acids and 2-monoglyceride. The hydrolysis products soon leave the emulsion and dissolve in bile juice. The mixed micelles thus formed then cross the unstirred water layer and reach the surface of the absorptive cells, where the micelles are broken and the hydrolysis products are absorbed by diffusion, mainly at the upper jejunum. Once free fatty acids and monoglyceride are absorbed, they are resynthesized to triglyceride and excreted as chylomicrons into the intestinal lymph. Chylomicron

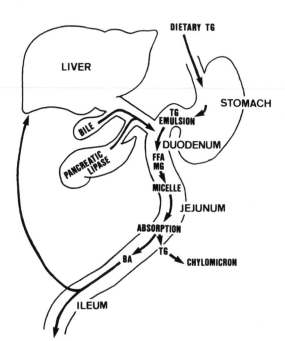

FIGURE 1. Digestion and absorption of dietary triglyceride. TG, triglyceride; MG, 2-monoglyceride; FFA, free fatty acid; BA, bile acid.

triglyceride is distributed as free fatty acids first to the extrahepatic tissues through the action of lipoprotein lipase in the capillary endothelia.

2.2. Absorption of Cholesterol

Figure 2 shows how cholesterol is absorbed from the intestine (Gibson, 1984; Grundy, 1983; Stange and Dietschy, 1985). Cholesterol is absorbed almost similarly to triglyceride. Most of dietary cholesterol is free, and a small portion exists in an esterified form. Esterified cholesterol is hydrolyzed by the cholesterol esterase in the pancreatic juice before it is absorbed. Nonesterified cholesterol (free cholesterol) is then dissolved in the mixed micelles and absorbed mainly in the upper jejunum as in the case of triglyceride, but animal studies show that a detectable amount of cholesterol is also absorbed in the ileum (Ikeda and Sugano, 1983; Oku and Sugano, 1986).

There is a controversy in respect to the mode of cholesterol uptake by the intestinal cells. Although it seems likely that the bulk of cholesterol is absorbed by the diffusion, data supporting the possibility of active transport are also available (Watanabe *et al.*, 1981).

Most of the free cholesterol absorbed is esterified in the cells. Cellular acyl-CoA:cholesterol acyltransferase (ACAT) appears to be responsible for the esterification reaction (Suckling and Stange, 1985), but pancreatic cholesterol esterase also appears to participate in the formation of cholesterol esters (Gallo *et al.*, 1984). Esterified choles-terol is then incorporated into chylomicrons and transported via the lymph. Most of

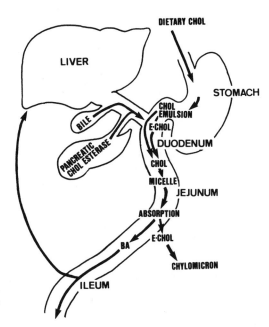

FIGURE 2. Absorption of cholesterol. CHOL, free cholesterol; E·CHOL, es-terified cholesterol; BA, bile acid.

dietary cholesterol is taken up by the specific receptor in the liver as chylomicron remnants (Brown and Goldstein, 1986).

2.3. Interference of Dietary Fiber with Lipid Absorption

Since numbers of dietary fibers are available, and since the physicochemical properties of individual dietary fibers differ considerably from each other, the mode by which the dietary fiber interacts with lipid absorption may differ depending on the type of fiber.

Most dietary fibers have no functional groups in their molecule and appear to interfere nonspecifically with bulk-phase diffusion of the lipids mainly through their water-holding capacity (Vahouny, 1982, 1985a). In addition, water-soluble viscous fibers such as guar gum may have a residual effect in limiting transmural transport of lipids (Tasman-Jones et al., 1982). On the other hand, certain dietary fibers have functional groups, and they bind more directly with bile acids and restrict the availability of bile acids for lipid absorption, like bile acid sequestrants (Story, 1986; Vahouny, 1985b; Vahouny et al., 1987). Since the bile acid is an essential component of lipid absorption (Borgstrom et al., 1985), dietary fibers that reduce the availability of bile acids in the intestine are most effective in reducing lipid absorption. Pectin and possibly guar gum are examples of these fibers. Chitosan may also belong to this group (Furda, 1983; Sugano et al., 1980). The modification of gastric emptying or a morphofunctional change in the gastrointestinal tract by the dietary fiber may also influence the lipid absorption (Cassidy et al.,1986; Tasman-Jones et al., 1982).

3. EFFECTS OF DIETARY FIBERS ON LIPID ABSORPTION

3.1. How Dietary Fibers Influence Absorption of Cholesterol

Vahouny and his co-workers reported a series of experiments in which the effects of various dietary fibers on cholesterol and fat absorption were studied in thoracic lymph-cannulated rats. After feeding diets containing various dietary fibers for several weeks ("chronic" study), they measured lymphatic absorption of cholesterol and fatty acid. An example of the results is shown in Fig. 3. Thus, dietary fibers commonly reduced absorption not only of cholesterol but also of triglyceride (Vahouny et al., 1980). It is worth mentioning that cellulose is more effective in reducing cholesterol absorption than cholestyramine or pectin, although this was not necessarily duplicated in the latter study (Vahouny et al., 1988). Cellulose usually does not reduce serum cholesterol, but its water-holding capacity and the related effect on intestinal transit time may explain in part the diminished cholesterol and triglyceride absorption (Vahouny et al., 1980). The closely associated drop in the uptake of both lipids may imply an interference in intestinal bulk-phase diffusion in general. In another study, they observed that chitosan, a glucosamine polymer, is as effective as cholestyramine in reducing cholesterol and lipid absorption in acute (an emulsion containing fat, cholesterol, and fiber was administered before measuring the absorption) and chronic studies (Fig. 4) (Vahouny et al., 1983).

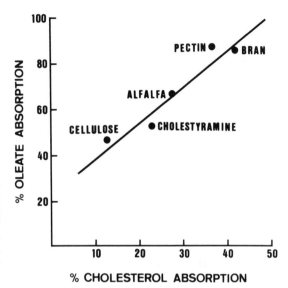

FIGURE 3. Correlation of the effects of chronic (6-week) feeding of semipurified diets containing cholestyramine or various dietary fiber sources on lymphatic absorption of administered cholesterol and triolein. (After Vahouny et al., 1980.)

The type of dietary fats also influences cholesterol absorption. As shown in Table I, the degree of cholesterol absorption tended to be lower with fish oils containing eicosapentaenoic acid than with corn oil rich in linoleic acid (Chen et al., 1987). In addition, the apparent absorption of fish oil fatty acids was markedly low compared to that of corn oil. The absorption rate of cholesterol appeared to be low when it was fed with cocoa butter as opposed to corn oil or palm oil (Table II) (I. Ikeda and M. Sugano, unpublished data). In these studies, however, the absorption rate of both lipids did not necessarily parallel each other, indicating a participation of the non-bulk-phase absorption process.

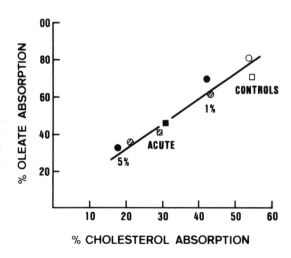

FIGURE 4. Correlation of the effects of cholestyramine and chitosan on the simultaneous lymphatic absorption of oleic acid and cholesterol. Acute studies: □, control; ■, cholestyramine; ▨, chitosan. Chronic studies: ○, control; ●, cholestyramine; ◯, chitosan. (After Vahouny et al., 1983.)

TABLE I. Absorption of Cholesterol and Fatty Acids
in Thoracic Lymph Cannulated Rats[a]

Groups	Cholesterol recovery (%)	Fatty acid recovery relative to corn oil
Oleic acid	51.0 ± 14.1^{bc}	—
Corn oil	55.4 ± 5.8^{b}	100^{b}
Menhaden oil	42.6 ± 9.3^{cd}	56.6 ± 15.9^{c}
FOC[e]	39.0 ± 9.4^{d}	47.1 ± 16.4^{c}

[a] After Chen et al. (1987). Emulsions containing sodium taurocholate, labeled cholesterol, fat, and albumin were administered in the duodenum, and lymph was collected for 24 hr.
[b-d] Values not sharing a same letter are significantly different at $P < 0.05$.
[e] FOC, fish oil concentrate.

These data strongly suggest that the effect of dietary fibers on lipid absorption is predictably modified by the type of dietary fats. Thus, it seems of interest to study the interaction of dietary fibers and fats on lipid absorption in order to know the most effective and favorable combinations for lowering plasma lipids, in particular cholesterol.

3.2. Interaction of Dietary Fibers and Fats on Cholesterol Absorption

We have studied the interaction of dietary fibers and fats on cholesterol and triglyceride absorption in the rat model using various combinations of typical types of fibers and fats under an "acute" condition. The fibers and fats chosen were all representative sources in respect to structure and function. Thus, as dietary fibers, cellulose (a water-insoluble fiber), guar gum (a water-soluble fiber), and chitosan (as a fiber with a functional group) were chosen. The dietary fats examined were safflower oil (a polyunsaturated fat), high-oleic safflower oil (a monoene fat), and palm oil (a saturated fat). Fatty acid compositions of these fats are shown in Table III.

Male Sprague–Dawley rats weighing 290 to 340 g were subjected to cannulation of

TABLE II. Absorption of Cholesterol and Fatty Acids
in Thoracic Lymph Cannulated Rats[a]

Groups	Cholesterol recovery (%)	Apparent fatty acid (%)	
Oleic acid	59.4 ± 2.7^{b}	99.6 ± 15.4^{b}	
Corn oil	55.4 ± 2.4^{b}	79.7 ± 5.5^{bc}	(100)
Palm oil	55.4 ± 3.6^{b}	65.6 ± 4.5^{c}	(83.0)
Cocoa butter	39.1 ± 5.6^{c}	50.3 ± 3.0^{c}	(63.3)

[a] After I. Ikeda and M. Sugano (unpublished data). See Table I for experimental details. Values in parentheses denote recovery relative to corn oil.
[bc] Values not sharing a same letter are significantly different at $P < 0.05$.

TABLE III. Fatty Acid Compositions of Dietary Fats

Fats	Fatty acids (weight %)			
	16:0	18:0	18:1	18:2
Palm oil	46.4	6.2	36.8	10.6
High-oleic safflower oil	11.5	4.8	67.1	16.6
Safflower oil	11.3	2.7	11.7	74.2

the left thoracic lymphatic channel, and an indwelling catheter was placed in the stomach for later administration of an emulsified meal (Chen et al., 1987). After surgery, animals were placed in restraining cages in a warm recovery room and allowed free access to drinking water containing 5% glucose and 0.9% NaCl. On the next morning, after collection of the blank lymph, each animal was administered 3 ml of a test emulsion, and the lymph was collected for 24 hr. The test emulsions contained 300 mg sodium taurocholate, 52 mg fatty-acid-free bovine serum albumin, 1 μCi [4-14C]cholesterol, 25 mg cholesterol, 200 mg fat, and 50 mg dietary fiber. The emulsion containing guar gum was prepared by homogenization in a Teflon homogenizer immediately before administration. The other emulsions were prepared by sonication. The difference in the methods for preparing emulsions depended on the nature of fibers, and essentially no difference was observed in the rate of cholesterol absorption related to the difference in the methods for preparation. The lymph was collected periodically and analyzed for radioactivities and fatty acid compositions.

The lymph flow rate was generally comparable among different fats, but it tended to be slightly higher in the safflower oil and high-oleic safflower oil groups (mean values 6.93 to 8.62 and 6.21 to 8.62 ml/hr, respectively) than in the palm oil group (mean values 5.62 to 7.33 ml/hr). There was no effect of dietary fibers on the lymph flow rate.

Figure 5 shows the time course of cholesterol absorption, and Table IV summarizes the radioactive cholesterol recovered for 24 hr. There was an interaction of dietary fibers and fats with cholesterol absorption. Chitosan was more effective in lowering the cholesterol absorption than guar gum or cellulose, and the diminishing effect was considerably greater in the combination with monounsaturated or polyunsaturated fats than with saturated fat. The effect of guar gum was not influenced by the fat type, and the cholesterol absorption rate was comparable to that with cellulose except for the combination with palm oil; in this combination, guar gum was considerably more effective in lowering cholesterol absorption than was cellulose, although the difference was not significant. Cholesterol absorption was in general highest with cellulose among the fibers examined, but the type of dietary fats appeared to influence cholesterol absorption.

In respect to the influence of dietary fats, the lymphatic recovery of cholesterol tended in general to be higher with increasing saturation of dietary fats. Thus, more cholesterol was absorbed when fibers were mixed with palm oil, and less with safflower oil, the difference between these two oils being marked when cellulose or chitosan, but not guar gum, was the fiber source.

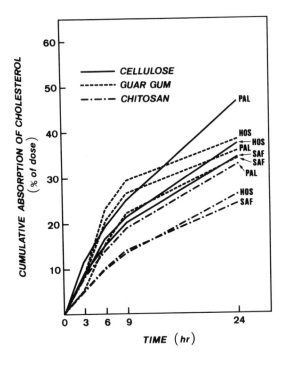

FIGURE 5. Cumulative absorption of [4-^{14}C]cholesterol into the thoracic duct lymph of rats intragastrically administered fat emulsions containing sodium taurocholate, labeled cholesterol, albumin, and various fibers. Each point represents the mean of five to six rats. PAL, palm oil; HOC, high-oleic safflower oil; SAF, safflower oil.

TABLE IV. Absorption of Cholesterol and Triglyceride in Thoracic
Lymph-Cannulated Rats[a]

Fibers—Fats	Cholesterol recovery (%)	Apparent triglyceride absorption (%)
Cellulose—PAL	44.9 ± 1.8^b	$85.5 \pm 2.8^{b-d}$
—HOS	39.5 ± 2.1^b	102 ± 7.2^b
—SAF	34.5 ± 4.1^{bc}	111 ± 7.3^b
Guar gum—PAL	35.1 ± 1.9^b	67.0 ± 4.0^{cd}
—HOS	38.3 ± 3.6^b	$75.9 \pm 8.0^{b-d}$
—SAF	34.3 ± 1.6^{bc}	77.0 ± 11.6^{bd}
Chitosan—PAL	33.9 ± 1.4^{bc}	67.8 ± 5.8^{cd}
—HOS	26.3 ± 3.7^{bc}	65.7 ± 8.3^{cd}
—SAF	24.4 ± 1.8^c	60.7 ± 6.6^c

[a]Values are mean $\pm$ S.E. of five or six rats. Emulsions containing sodium taurocholate, fat, labeled cholesterol, albumin, and fiber were administered intragastrically, and lymph was collected for 24 hr. PAL, palm oil, HOS, high-oleic safflower oil; SAF, safflower oil.
[b-d]Values not sharing a same letter are significantly different at $P < 0.05$.

In addition, as Fig. 5 shows, there was a considerable difference in the time course of cholesterol absorption depending on the type of dietary fibers. Thus, the effect of guar gum on the rate of cholesterol absorption appeared to be somewhat specific; the initial rate of cholesterol absorption was considerably faster in guar gum than in the other two fibers, whereas the rate at the later stage was rather slow. Figure 6 summarizes the periodic cholesterol absorption patterns in order to show the fiber-dependent difference. In rats given guar gum, more than 30% of cholesterol recovered for 24 hr was absorbed from 3 to 6 hr compared to less than 20% in rats fed other dietary fibers, whereas the absorption rate between 9 and 24 hr was 25–30% compared to more than 40% in rats fed cellulose or chitosan. This was not caused by a change in the lymph flow rate because it was almost linear during the 24 hr of lymph collection. This observation may be related to a specific property of guar gum that the maximum output rate of triglyceride in the distal but not proximal intestine was five times higher in rats fed guar gum than in those fed cellulose (Imaizumi and Sugano, 1986; Imaizumi *et al.*, 1982).

3.3. Interaction of Dietary Fibers and Fats on Triglyceride Absorption

The rate of triglyceride absorption is shown in Fig. 7. In general, chitosan again reduced the fat absorption most effectively, guar gum somewhat less, and cellulose the least. The diminishing effect of chitosan was not influenced by the fat source. The absorption was, however, characteristic of the type of fats. Safflower oil and high-oleic safflower oil were absorbed completely when combined with cellulose (see Table IV), whereas lymphatic absorption of these unsaturated fats was less than 80% with guar gum and 70% with chitosan.

Again, in rats given guar gum, the relative rate of fatty acid absorption was high

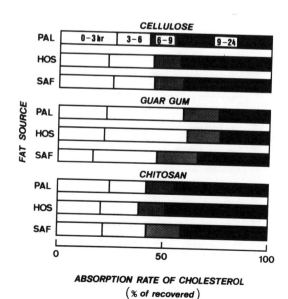

FIGURE 6. Relative rate of absorption of [4-14C]cholesterol into the thoracic duct lymph of rats. Means of five to six rats. See Fig. 5.

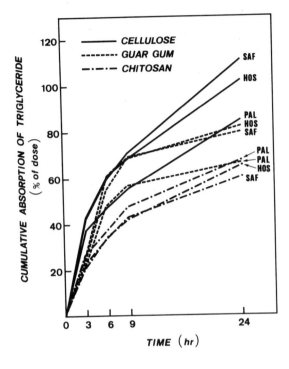

FIGURE 7. Cumulative absorption of triglyceride into the thoracic duct lymph of rats. See Fig. 5.

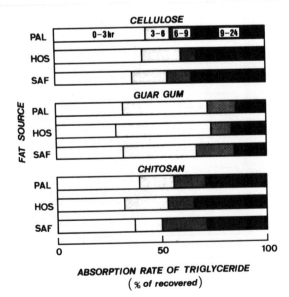

FIGURE 8. Relative rate of absorption of triglyceride into the thoracic duct lymph. See Fig. 5.

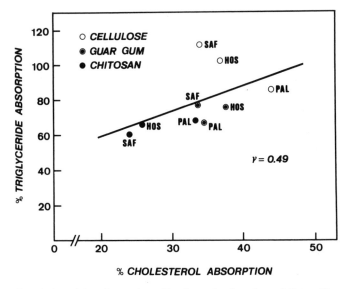

FIGURE 9. Correlation of the effects of combinations of various fat and dietary fiber sources on lymphatic absorption of administered cholesterol and triglyceride. Each point represents the mean of five to six rats.

during the initial phase and low during the later phase as in the case of cholesterol, but the bias was more evident with fatty acid than with cholesterol (Fig. 8).

As shown in Fig. 9, the degree of triglyceride absorption was roughly parallel to that of cholesterol, consistent with the observations on fat emulsions containing various dietary fibers in a common fat source (Vahouny *et al.*, 1980, 1983). However, relatively large deviations suggest possible participation of non-bulk-phase absorption of these lipids, in particular cholesterol (Vahouny *et al.*, 1988).

Table V shows fatty acid compositions of lymph lipids recovered for 24 hr. Al-

TABLE V. Fatty Acid Compositions of Lymph Lipids Recovered for 24 hr[a]

Fats—fibers	Fatty acids (weight %)						
	16:0	16:1	18:0	18:1	18:2	18:3	20:4
PAL—Cellulose	31.0	2.3	6.9	25.7	20.6	2.2	5.4
—Guar gum	32.4	1.3	6.9	28.5	20.4	1.4	4.9
—Chitosan	31.5	1.0	6.9	26.9	21.5	6.6	5.5
HOS—Cellulose	17.2	2.7	5.9	41.6	22.2	1.0	4.4
—Guar gum	16.3	0.9	6.2	42.1	22.9	1.3	6.0
—Chitosan	17.4	1.2	7.0	35.6	25.1	2.0	6.3
SAF—Cellulose	16.3	2.1	5.9	13.5	49.8	2.3	5.0
—Guar gum	15.2	1.5	6.0	13.0	54.5	1.3	4.4
—Chitosan	18.0	2.6	7.0	13.4	46.8	1.1	6.0

[a]Mean of five to six rats. PAL, palm oil; HOS, high-oleic safflower oil; SAF, safflower oil.

though the composition generally reflected the type of fats administered, there was a slight but detectable difference in the percentage of individual major fatty acids related to the type of dietary fibers. Thus, when high-oleic safflower oil was the fat source, the percentage of oleate tended to be lower in rats given chitosan than in rats given other fibers. Also, there appeared to be a dietary fiber-dependent difference in the percentage of linoleic acid when safflower oil was administered. These results may indicate that the absorption of individual fatty acids in dietary fats is specifically modified by the type of dietary fiber.

The lymphatic recovery of representative fatty acids in each dietary fat was calculated, and the results are shown in Fig. 10. The absorption of oleic and linoleic acid was

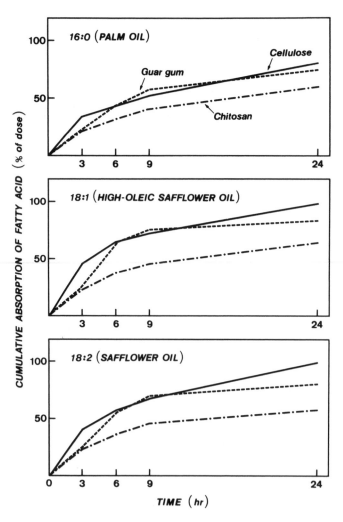

FIGURE 10. Cumulative absorption of representative fatty acids in dietary fats into the thoracic duct lymph of rats. Each point represents the mean of five to six rats.

almost quantitative when cellulose was used as a dietary fiber source. Guar gum and especially chitosan decreased absorption of these unsaturated fatty acids. Thus, absorption of oleate and linoleate was only 60% of the dose in rats administered chitosan and approximately 80% for guar gum. The absorption of palmitic acid was approximately 80% for cellulose and guar gum and 60% for chitosan. These data indicate that the magnitude of the interference of dietary fibers with the absorption of individual fatty acids is not necessarily constant.

4. NEW FIBER SOURCES WITH A HYPOCHOLESTEROLEMIC ACTIVITY

4.1. Chitosans with Different Viscosity

Although chitosan is not a novel hypocholesterolemic fiber (Sugano *et al.*, 1980), we have examined the relationship between the hypocholesterolemic effect and structure using chitosans with different viscosities (Sugano *et al.*, 1988b). When chitosans with widely different viscosity (17 to 1620 cP) were added at the 5% level to a cholesterol-enriched diet, the serum cholesterol levels of rats did not substantially elevate and stayed at the levels normally encountered in rats fed a cholesterol-free diet (Fig. 11). This observation is useful because the physicochemical properties, particularly molecular weight and the degree of the deacetylation, differ variably depending on the source of chitosan. Thus, it appears that commercially available chitosans are all effective as a

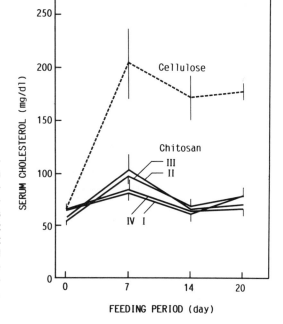

FIGURE 11. Time course of the effects of chitosan (5% level) with various viscosities on serum cholesterol levels of rats fed cholesterol-enriched diets. I to IV: chitosans with different viscosity. Each point represents mean ± S.E. of eight rats. Cholesterol levels in rats fed chitosan are all significantly different ($P < 0.05$) from the corresponding values for rats fed cellulose. (After Sugano *et al.*, 1988b.)

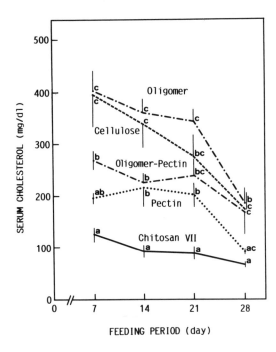

FIGURE 12. Time course of the effects of chitosan, glucosamine oligomer, and pectin (4% level) on serum cholesterol levels of rats fed cholesterol-enriched diets. Each point represents the mean ± S.E. of six rats. Values in the same time point not sharing a same letter are significantly different at $P < 0.05$. (After Sugano *et al.*, 1988b.)

soruce of hypocholesterolemic fiber. However, as shown in Fig. 12, the water-soluble glucosamine oligomer (viscosity 1.1 cP) did not have serum-cholesterol-lowering activity when it was added at the 4% level. Chitosan was still effective in lowering serum cholesterol at a lower (2%) dietary level.

4.2. Fruit Pomaces

When rats were fed cholesterol-enriched diets containing apple or grape pomaces at the 10% level, these pomaces, in particular the latter, exerted a potent hypocholesterolemic effect (Fig. 13) (Sugano *et al.*, 1988a). Although the effect was somewhat lower than that of pectin, they did not have an untoward effect on growth of rats, in contrast to pectin. The neutral detergent fiber content was 33.0% and 49.5% for apple and grape pomaces, respectively, most of which was insoluble fiber according to the method of Asp *et al.* (1983). Thus, byproducts of the fruit and vegetable juice industries may be useful as a source of hypocholesterolemic dietary fiber (Block and Lanza, 1987; Takebe *et al.*, 1982; Porzio and Blake, 1983; Ting and Rouseff, 1983).

4.3. Dietary Fiberlike Materials from Soybean Protein Isolate

Regarding new sources of hypocholesterolemic fibers, we recently found that the protease-resistant fraction of soybean protein isolate exerts an amazing hypocholesterolemic effect (Table VI) (Sugano *et al.*, 1988c). Soybean protein isolate was exhaustively digested with microbial proteases, and the digestible fraction was removed.

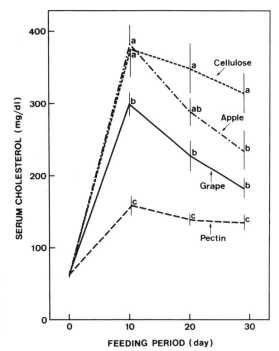

FIGURE 13. Time course of the effect of apple and grape pomaces (10% level) on serum cholesterol levels of rats fed cholesterol-enriched diets. Each point represents the mean ± S.E. of eight rats. Values in the same time point not sharing a common letter are significantly different at $P < 0.05$. (After Sugano et al., 1988a.)

The undigested fractions contained relatively large proportions of nonproteinous materials such as carbohydrates (7.5%) and saponins (4.6%). Although saponin may in part be responsible for the cholesterol-lowering effect, the undigested peptides themselves behaved much like dietary fiber and significantly reduced serum cholesterol by increasing both fecal neutral and acidic steroids. Soybean protein has been shown to reduce cholesterol absorption and increase fecal steroid excretion in relation to casein (Nagata et al., 1982; Vahouny et al., 1984).

TABLE VI. Effects of the Undigested Fraction of Soybean Protein Digests on Serum Cholesterol of Rats[a]

	Groups		
Cholesterol	Soybean protein	Digested fraction	Undigested fraction
Serum (mg/dl)	340 ± 22[b]	523 ± 50[c]	99.4 ± 6.6[d]
Liver (mg/g)	69.5 ± 2.7[b]	77.1 ± 4.1[b]	7.70 ± 0.97[c]

[a] After Sugano et al. (1988c). Means ± S.E. of eight rats. Rats were fed cholesterol-enriched diet containing the microbial protease hydrolysis products at the nitrogen level equivalent to that of the 20% soybean protein diet. Feeding period, 28 days.
[b-d] Values not sharing a same letter are significantly different at $P < 0.05$.

5. CONSEQUENCE OF FIBER FEEDING: DOES EARLY FIBER FEEDING INFLUENCE CHOLESTEROL HOMEOSTASIS IN LATER LIFE?

Several studies involved in the consequences of early dietary manipulations to the serum cholesterol level in adult life indicate the possible existence of an imprinting effect such as metabolic memory, although this is still controversial (Beynen *et al.*, 1985; Haln and Kirby, 1973; Reiser and Siedelman, 1972; Subbiah, 1983). We have studied whether dietary fibers given during the weaning period could influence the response of plasma cholesterol to later cholesterol challenge.

Male Wistar newborn rats were preweaned at age of 16 days and, under the experimental protocol shown in Table VII, were fed purified diets containing 7% cellulose, guar gum, pectin, or polydextrose or 2% cholestyramine for 2 weeks. The concentration of serum cholesterol was lower in rats fed water-soluble fibers at this stage (Table VIII). Rats were then fed commercial rat chow for 4 weeks and received a cholesterol-enriched purified diet for 1 week. The serum cholesterol level tended to be higher in rats previously fed pectin, polydextrose, or cholestyramine. Thus, rats that previously experienced hypocholesterolemic fiber diets were apt to have a high serum cholesterol level when they were later challenged with a cholesterol-enriched diet. It is likely that the suppression of steroid absorption during the weaning period may contribute adversely to the high concentration of serum cholesterol on feeding cholesterol in later life.

Cassidy *et al.* (1986) observed that dietary fibers such as cellulose and guar gum caused varying degrees of damage to the mucosal surface of the intestinal tract when they were fed to the postweaned rats, although rats fed the zero-fiber or -cellulose diets did not develop the normal pattern of intestinal villi, whereas those fed pectin were comparable to the control animals (Tasman-Jones, 1981). Thus, more studies are needed before recommending application of dietary fiber as a component of the weaning diet.

6. CONCLUSION

Studies described here showed that the effect of dietary fibers on cholesterol and fatty acid absorption differs predictably depending on the type of dietary fat simultaneously given. In most cases, saturated fat gave a high level of cholesterol absorption, and for effective reduction of cholesterol absorption by dietary fibers, choice of unsaturated fats may be recommended. However, in order to reduce the absorption of triglyceride, choice of the effective fiber rather than the type of fats was mandatory; when

TABLE VII. Experimental Protocol

Phases	I	II	III
Days of age	16–30	30–58	58–65
Diets	Fiber-supplemented purified diets	Commercial rat chow	Cholesterol-enriched chow

TABLE VIII. Serum Cholesterol Levels
of Rats Fed Dietary Fibers[a]

Groups	Cholesterol (mg/dl)	
	Days 30	Days 65
Basal	104 ± 5[b]	227 ± 20[b]
Cellulose	103 ± 3[b]	241 ± 22[bd]
Guar gum	95 ± 3[bc]	238 ± 13[bd]
Pectin	88 ± 2[cd]	288 ± 20[cd]
Polydextrose	82 ± 4[d]	298 ± 16[cd]
Cholestyramine	103 ± 3[b]	326 ± 31[c]

[a] After Lu et al. (1989). Mean ± S.E. of six rats. For experimental conditions, see Table VII.
[b–d] Values not sharing a same letter are significantly different at $P < 0.05$.

chitosan was fed, the absorption of triglyceride was apparently independent on the type of fat source. Alternatively, when cellulose or guar gum was fed, there was a considerable fat-type-dependent difference in the absorption. The difference in the effect of individual dietary fibers on the periodic rate of lipid absorption should also taken into consideration for the effective reduction of the absorption.

Several lines of studies searching for new sources of dietary fibers confirmed the effectiveness of chitosan. Pomaces from the fruit juice industries may be useful as a new source of dietary fiber. The undigested fractions of soybean protein behaved like dietary fiber. There is a possibility that similar types of fractions may be obtained from other vegetable proteins or even from animal proteins.

Use of dietary fiber as a constituent of the weaning diet may be restricted because the early experience of reduction of serum cholesterol by dietary fibers may sensitize the response to dietary cholesterol when cholesterol is fed in later life.

Finally, an appropriate application of dietary fibers seems valuable for effective reduction of cholesterol absorption and hence serum cholesterol level.

ACKNOWLEDGMENTS. The authors wish to acknowledge the valuable assistance of Y. Tomari, Y. Yamada, and K. Yoshida. Portions of these studies were supported by a Grant-in-Aid from the Ministry of Education, Science and Culture of Japan and Research Committee of Soy Protein

REFERENCES

American Heart Association, 1986, Dietary guidelines for healthy American adults: A statement for physicians and health professionals by the Nutrition Committee, American Heart Association, Circulation 74:1465A–1468A.
Asp, N.-G., Johansson, C.-G., Hallmer, H., and Siljestrom, M., 1983, Rapid enzymatic assay of insoluble and soluble dietary fiber, J. Agr. Food Chem. 31:467–482.
Beynen, A. C., De Bruijne, J. J., and Katan, M. B., 1985, Treatment of young rats with cholestyramine or a hypercholesterolemic diet does not influence the response of serum cholesterol to dietary cholesterol in later life, Atherosclerosis 58:149–157.

Block, G., and Lanza, E., 1987, Dietary fiber sources in the United States by demographic group, *J. Natl. Cancer Inst.* **79**:83–91.

Borgstrom, D., Barrowman, J. A., and Lindstrom, M., 1985, Role of bile acids in intestinal lipid digestion and absorption, in: *Sterols and Bile Acids* (H. Danielsson and J. Sjovall, eds.), Elsevier Science, Amsterdam, pp. 405–425.

Brown, M. S., and Goldstein, J. L., 1986, A receptor-mediated pathway for cholesterol homeostasis, *Science* **232**:34–47.

Carey, M. L., Small, D. M., and Bliss, C. H., 1983, Lipid digestion and absorption, *Annu. Rev. Physiol.* **45**:651–677.

Cassidy, M. M., Fitzpatrick, L. R., and Vahouny, G. V., 1986, The effect of fiber in the postweaning diet on nutritional and intestinal morphological indices in the rat, in: *Dietary Fiber, Basic and Clinical Aspects* (G. V. Vahouny and D. Kritchevsky, eds.), Plenum Press, New York, pp. 229–251.

Chen, I. S., Hotta, S. S., Ikeda, I., Cassidy, M. M., Sheppard, A. J., and Vahouny, G. V., 1987, Digestion, absorption and effects on cholesterol absorption of menhaden oil, fish oil concentrate and corn oil by rats, *J. Nutr.* **117**:1676–1680.

Friedman, H. I., and Nylund, B., 1980, Intestinal fat digestion, absorption, and transport. A review, *Am. J. Clin. Nutr.* **33**:1108–1139.

Furda, I., 1983, Aminopolysaccharides—their potential as dietary fiber, in: *Unconventional Sources of Dietary Fiber* (I. Furda, ed.), American Chemical Society, Washington, DC, pp. 105–122.

Gallo, L. L., Clark, S. B., Myers, S., and Vahouny, G. V., 1984, Cholesterol absorption in rat intestine: Role of cholesterol esterase and acyl coenzyme A : cholesterol acyltransferase, *J. Lipid Res.* **25**:604–612.

Gibson, J. C., 1984, Clinical and experimental method for the determination of cholesterol absorption, in: *Laboratory and Research Methods in Biology and Medicine*, Vol. 10 (J. A. Story, ed.), Alan R. Liss, New York, pp. 157–190.

Grundy, S. M., 1983, Absorption and metabolism of dietary cholesterol, *Annu. Rev. Nutr.* **3**:71–96.

Grundy, S. M., 1986, Cholesterol and coronary heart disease: A new era, *J.A.M.A.* **256**:2849–2858.

Haln, P., and Kirby, L., 1973, Immediate and later effects of premature weaning and of feeding a high fat or high carbohydrate diet to weaning rats, *J. Nutr.* **103**:690–696.

Ikeda, I., and Sugano, M., 1983, Some aspects of mechanism of inhibition of cholesterol absorption by β-sitosterol, *Biochim. Biophys. Acta* **732**:651–658.

Imaizumi, K., and Sugano, M., 1986, Dietary fiber and intestinal lipoprotein secretion, in: *Dietary Fiber, Basic and Clinical Aspects* (G. V. Vahouny and D. Kritchevsky, eds.), Plenum Press, New York, pp. 287–308.

Imaizumi, K., Tominaga, A., Mawatari, K., and Sugano, M., 1982, Effect of cellulose and guar gum on the secretion of mesenteric lymph chylomicrons in meal-fed rats, *Nutr. Rep. Int.* **26**:263–269.

Kay, R. M., 1982, Dietary fiber, *J. Lipid Res.* **23**:221–242.

Kritchevsky, D., 1985, Lipid metabolism and coronary heart disease, in: *Dietary Fibre, Fibre-Depleted Foods and Disease* (H. Trowell, D. Burkitt, and K. Heaton, eds.), Academic Press, London, pp. 305–315.

Lu, Y.-F., Imaizumi, K., Sakono, M., and Sugano, M., 1989, Effects of dietary fibers in early weaning on later response of serum and fecal steroid levels to high-cholesterol diet in rats, *Nutr. Res.* **9**:345–352.

Miettinen, T. A., 1987, Dietary fiber and lipids, *Am. J. Clin. Nutr.* **45**:1237–1242.

Nagata, Y., Ishiwaki, N., and Sugano, M., 1982, Studies on the mechanism of antihypercholesterolemic action of soy protein and soy protein-type amino acid mixture in relation to casein counterparts in rats, *J. Nutr.* **112**:1614–1625.

Oku, H., and Sugano, M., 1986, Site of fat absorption and localization of 3-hydroxy-3-methylglutaryl coenzyme A (HMG-CoA) reductase activity along the small intestine of rats, *Agr. Biol. Chem.* **50**:699–705.

Porzio, M. A., and Blake, J. R., 1983, Washed orange pulp: Charcterization and properties, in:

Unconventional Sources of Dietary Fiber (I. Furda, ed.), American Chemical Society, Washington, DC, pp. 191–204.

Reiser, R., and Siedelman, Z., 1972, Control of serum cholesterol homeostasis by the cholesterol in the milk of the suckling rat, *J. Nutr.* **102**:1009–1016.

Report of the National Cholesterol Education Program Expert Panel on Detection, 1988, Evaluation and treatment of high blood cholesterol in adults, *Arch. Intern. Med.* **148**:36–69.

Stange, E. F., and Dietschy, J. M., 1985, Cholesterol absorption and metabolism by the intestinal epithelium, in: *Sterols and Bile Acids* (H. Danielsson and J. Sjovall, eds.), Elsevier Science, Amsterdam, pp. 405–425.

Story, J. A., 1986, Modification of steroid excretion in response to dietary fiber, in: *Dietary Fiber, Basic and Clinical Aspects* (G. V. Vahouny and D. Kritchevsky, eds.), Plenum Press, New York, pp. 253–264.

Subbiah, M. T. R., 1983, Decreased atherogenic response to dietary cholesterol in pigeons after stimulation of cholesterol catabolism in early life, *J. Clin. Invest.* **71**:1509–1513.

Suckling, K. E., and Stange, E. F., 1985, Role of acyl-CoA:cholesterol acyltransferase in cellular cholesterol metabolism, *J. Lipid Res.* **26**:647–671.

Sugano, M., Fujikawa, T., Hiratsuji, Y., Nakashima, K., Fukuda, N., and Hasegawa, Y., 1980, A novel use of chitosan as a hypocholesterolemic agent in rats, *Am. J. Clin. Nutr.* **33**:787–793.

Sugano, M., Nakano-Morita, M., Yoshida, K., and Imai, J., 1988a, Influence of dietary apple and grape pomace on serum cholesterol level of the rat, *J. Jpn. Soc. Food Sci. Technol.* **35**:242–245.

Sugano, M., Watanabe, S., Kishi, A., Izume, M., and Ohtakara, A., 1988b, Hypocholesterolemic action of chitosans with different viscosity in rats, *Lipids* **23**:187–191.

Sugano, M., Yamada, Y., Yoshida, K., Hashimoto, Y., Matsuo, T., and Kimoto, M., 1988c, The hypocholesterolemic action of the undigested fraction of soybean protein in rats, *Atherosclerosis* **72**:115–122.

Takebe, K., Makino, J., Konuma, T., Nakamura, T., Tsutsui, M., Osoino, K., Kudoh, M., Masuda, M., Tamasawa, N., and Ochiai, S., 1982, [Basic and clinical studies on apple fibre—Influence of the fibre on bile acid, blood glucose, plasma insulin, and plasma lipoprotein lipid,] *Newest Med.* **37**:2268–2274.

Tasman-Jones, C. T., 1981, Effects of dietary fiber on the structure and function of the small intestine, *Top. Gastroenterol.* **1981**:67–74.

Tasman-Jones, C. T., Owen, R. L., and Jones, A. L., 1982, Semipurified dietary fiber and small bowel morphology in rats, *Dig. Dis. Sci.* **27**:519–524.

Ting, S. V., and Rouseff, R. L., 1983, Dietary fiber from citrus waste: Characterization, in: *Unconventional Sources of Dietary Fiber* (I. Furda, ed.), American Chemical Society, Washington, DC, pp. 205–219.

Tso, P., 1985, Gastrointestinal digestion and absorption of lipid, *Adv. Lipid Res.* **3**:71–96.

Vahouny, G. V., 1982, Dietary fiber, lipid metabolism and atherosclerosis, *Fed. Proc.* **41**:2801–2896.

Vahouny, G. V., 1985a, Dietary fibers. Aspects of nutrition, pharmacology and pathology, in: *Nutritional Pathology, Pathobiochemistry of Dietary Imbalance* (H. Sidransky, ed.), Marcel Dekker, New York, pp. 207–320.

Vahouny, G. V., 1985b, inhibition of cholesterol absorption by natural products, in: *Drugs Affecting Lipid Metabolism, VIII* (D. Kritchevsky, W. L. Holmes, and R. Paoletti, eds.), Plenum Press, New York, pp. 265–279.

Vahouny, G. V., and Cassidy, M. M., 1985, Dietary fibers and absorption of nutrients, *Proc. Soc. Exp. Biol. Med.* **180**:432–446.

Vahouny, G. V., and Cassidy, M. M., 1986, Dietary fiber and intestinal adaptation, in: *Dietary Fiber, Basic and Clinical Aspects* (G. V. Vahouny and D. Kritchevsky, eds.), Plenum Press, New York, pp. 181–209.

Vahouny, G. V., Roy, T., Gallo, L. L., Story, J. A., Kritchevsky, D., and Cassidy, M. M., 1980, Dietary fibers III. Effects of chronic intake on cholesterol absorption and metabolism in the rat, *Am. J. Clin. Nutr.* **33**:2182–2191.

Vahouny, G. V., Satchithanandam, S., Cassidy, M. M., Lightfoot, F. B., and Furda, I., 1983, Comparative effects of chitosan and cholestyramine on lymphatic absorption of lipids in the rat, *Am. J. Clin. Nutr.* **38:**699–705.

Vahouny, G. V., Chalcarz, W., Satchithanandam, S., Adamson, I., Klurfeld, D. M., and Kritchevsky, D., 1984, effect of soy protein and casein intake on intestinal absorption and lymphatic transport of cholesterol and oleic acid, *Am. J. Clin. Nutr.* **40:**1156–1164.

Vahouny, G. V., Khalafi, R., Satchithanandam, S., Watkins, D. W., Story J. A., Cassidy, M. M., and Kritchevsky, D., 1987, Dietary fiber supplementation and fecal bile acids, neutral steroids and divalent cations in rats, *J. Nutr.* **117:**2009–2015.

Vahouny, G. V., Satchithanandam, S., Chen, I., Tepper, S. A., Kritchevsky, D., Lightfoot, F. G., and Cassidy, M. M., 1988, Dietary fiber and intestinal adaptation: Effects on lipid absorption and lymphatic transport in the rat, *Am. J. Clin. Nutr.* **47:**201–206.

Watanabe, M., Oku, T., Shidoji, Y., and Hosoya, N., 1981, A new aspect on the mechanism of intestinal cholesterol absorption in rat, *J. Nutr. Sci. Vitaminol.* **27:**209–218.

10

Macronutrient Absorption

BARBARA OLDS SCHNEEMAN

1. INTRODUCTION

Several reports suggest that various sources of dietary fiber can slow the process of digestion and absorption of macronutrients. It is likely that the physical properties of fiber sources such as particle size, viscosity, water-holding capacity, gel formation, and bile acid binding capacity are important in determining the effect of a fiber source on nutrient absorption. The effects of fiber on gastrointestinal function have recently been reviewed (Schneeman, 1987; Vahouny and Cassidy, 1985), and several mechanisms have been suggested by which the physical properties of fiber slow digestion and absorption. The gastrointestinal tract is illustrated in Fig. 1; experimental evidence suggests that sources of fiber are likely to affect the function of each organ involved with the digestion and absorption of nutrients.

2. MECHANISMS TO SLOW DIGESTION AND ABSORPTION

The food consumed can be held in the stomach and metered into the small intestine. The rate at which the stomach empties will regulate the rate of nutrient absorption from the small intestine. Viscous polysaccharides but not insoluble fiber sources such as cellulose have been reported to delay gastric emptying (Schwartz *et al.*, 1982; Leeds *et al.*, 1979). Because viscous polysaccharides can form a gel matrix, they are likely to trap nutrients in the matrix and delay their emptying from the stomach. Within the small intestine the digestible components of the diet are broken down by hydrolysis, and nutrients are absorbed through the mucosal cells. *In vitro* data indicate that various fiber sources can inhibit the activity of pancreatic enzymes that digest carbohydrates, lipids, and proteins (Schneeman and Gallaher, 1986a). The mechanisms for inhibiting digestive enzyme activity are not clearly established, but in some nonpurified fiber sources,

---

BARBARA OLDS SCHNEEMAN • Department of Nutrition, University of California, Davis, California 95616.

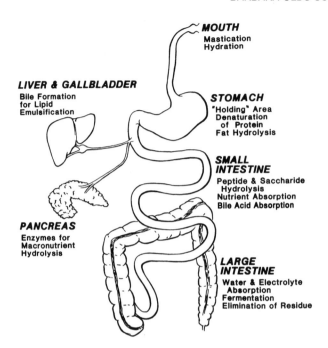

FIGURE 1. The gastrointestinal tract.

specific enzyme inhibitors may exist (Gallaher and Schneeman, 1986b). It is difficult to assess the physiological importance of this inhibition because an excess of digestive enzyme activity is secreted in response to a meal. In patients with pancreatic insufficiency the amylase, trypsin, chymotrypsin, and lipase activity available from pancreatic replacement treatment was significantly reduced when the patients were given a meal containing pectin or wheat bran, which suggests that it is possible for these fiber sources to reduce digestive enzyme capacity significantly in the small intestine (Isaksson *et al.*, 1984). In addition to inhibition of digestive enzyme activity, the presence of plant cell wall matrix in a food can provide a physical barrier to digestion (Collier and O'Dea, 1982; Wong and O'Dea, 1983; Snow and O'Dea, 1981). Grinding of the fiber source to a very fine particle size may disrupt the cell wall structure sufficiently to make digestible nutrients more available for hydrolysis.

The physical characteristics of the intestinal contents will be changed by the physical properties of the fiber sources in the diet. The bulk or amount of material in the small intestine will increase because the fiber is not digestible and hence remains during the transit of digesta through the small intestine (Schneeman, 1982). The volume of the intestinal contents can increase because of the water-holding capacity of the fiber source. Sandberg *et al.* (1981, 1983) reported that addition of wheat bran or pectin to a low-fiber meal increased the volume of ileostomy fluid by about 20–30%. In addition, animal data indicate that a greater dry and wet weight of intestinal contents is associated with the addition of a fiber supplement to experimental diets (Schneeman, 1982; Poksay and Schneeman, 1983). The presence of viscous polysaccharides in the fiber source will increase the viscosity of the contents and in particular of the aqueous phase of the intestinal contents from which nutrients are absorbed (Blackburn and Johnson, 1981; Elsenhans *et al.*, 1980; Johnson and Gee, 1981). An increase in the bulk, volume, or viscosity of the intestinal contents is likely to slow diffusion of enzymes, substrates, and nutrients to the absorptive surface, all of which can lead to a slower appearance of nutrients in the plasma following a meal.

The experimental evidence suggests that through a variety of mechanisms, certain fiber sources, especially those containing viscous polysaccharides, can slow the process of digestion and absorption, although total nutrient absorption is not necessarily reduced. Because of its effect on the rate of absorption, a greater proportion of nutrients from a diet high in fiber will undoubtedly be absorbed from the lower half of the small intestine. This pattern of nutrient absorption is likely to contribute to the physiological responses to various fiber sources. For example, the rate of nutrient absorption will affect the pattern of hormone release in response to diet (Jenkins, 1978) and the rate of nutrient delivery to the tissues. Evidence also exists that the presence of nutrients in the ileum can influence food intake, gastric emptying, and the composition and size of chylomicrons (Wu *et al.*, 1980; see Chapter 16 by Read *et al.*). Sigleo *et al.* (1984) have demonstrated that chronic feeding of fiber supplements will alter the morphology of the small intestine. Both the distribution of nutrient absorption in the small intestine as well as the influence of bulk in the small intestine on intestinal cell renewal are likely to contribute to this response (Johnson, 1988). The implications of a change in morphology on nutrient absorption from the small intestine are discussed in more detail by Cassidy *et al.* (Chapter 5).

3. SPECIFIC MACRONUTRIENTS

3.1. Lipids

Several of the effects of dietary fiber on digestion and absorption, as outlined above, are likely to interfere with the emulsification, lipolysis, and absorption of fat digestion products. Within the intestinal contents lipids in the aqueous phase that are available for digestion and absorption are present in micelles. *In vitro* studies have demonstrated that certain dietary fibers can bind the components of micelles. Vahouny and Cassidy (1985) pointed out that guar gum bound 36% of sodium taurocholate, 22% of lecithin, 23% of cholesterol, 23% of monolein, and 33% of fatty acids from a micellar solution. In contrast, wheat bran and cellulose absorbed much lower percentages of micellar components. Gallaher and Schneeman (1986a) measured the *in vivo* bile acid and phospholipid binding ability of certain dietary fibers and cholestyramine by determining bile acid and phospholipid distribution within the small intestinal contents following a test meal. Cholestyramine, which is a bile-acid-binding resin, bound bile acids and phospholipids in the small intestinal contents and reduced the concentration of bile acids and phospholipids in the aqueous phase of the intestinal contents by about 50% compared to a fiber-free control. Among the dietary fibers tested, guar gum and lignin but not cellulose, wheat bran, or oat bran bound or sequestered bile acids in the small intestinal contents. Oat bran but not cellulose, wheat bran, guar gum, or lignin bound phospholipids in the intestinal contents; however, guar gum lowered the concentration of phospholipids in the aqueous phase of the intestinal contents because it increased the volume of the intestinal contents by about 30–50%. The results of this study demonstrated that the total amount of phospholipid and bile acid in the aqueous phase are significant predictors of the amount of lipid solubilized in the aqueous phase and available for absorption from the small intestine. Consequently, the interaction of dietary components with bile acids and phospholipids will contribute significantly to their effect on the rate and site of lipid absorption. In a recent study (Ebihara and Schneeman, 1989) we have shown that konjac mannan, guar gum, and chitosan can bind or sequester bile acids in the small intestinal contents. Chitosan can also bind phospholipids, whereas konjac mannan and guar gum dilute the concentration of phospholipids because they increase the total volume of intestinal contents.

Among these studies it is interesting to note the effect of these various dietary treatments on estimators of lipid absorption. In the first study only cholestyramine significantly lowered the amount of lipid that was solubilized in the intestinal contents. In the second study, [^{3}H]cholesterol and [^{14}C]triglyceride were included in the test diet to estimate lipid disappearance from the small intestine, and both konjac mannan and guar gum, but not chitosan, delayed the isotope disappearance at 2 hr after the test meal. The results suggest that in the case of a drug, cholestyramine, the binding of micellar components is sufficiently strong to have a significant impact on lipid solubilization. However, in the case of dietary fibers the interactions are weaker, and the absolute amount of lipid solubilized during digestion and absorption does not necessarily change. Under these conditions the effect of the fiber on the physical properties of the intestinal contents becomes a significant factor in affecting the rate of lipid absorption. Specifically, konjac mannan and guar gum both increase the viscosity of the intestinal contents, whereas a material such as chitosan is insoluble and unlikely to be viscous at the pH of the intestinal contents. Although all three have bile acid binding capacity, only the viscous polysaccharides were able to delay the disappearance of lipid from the small intestine. It has been demonstrated *in vitro* that viscosity caused by guar gum will slow the diffusion of micelles (Phillips, 1986), and this factor can have an important effect on the rate of lipid absorption. Studies in patients with ileostomies also support the observation that viscous fiber sources are more likely to reduce lipid absorption. Sandberg *et al.* (1983) reported that addition of pectin to a low-fiber meal given to ileostomy patients increased the amount of fat in the ileostomy fluid by 96%. In contrast, the amount of fat in the ileostomy fluid was not changed when wheat bran was the fiber added to the meal (Sandberg *et al.*, 1981). In patients with pancreatic insufficiency who are receiving pancreatic enzyme replacements, pectin appeared to cause a malabsorption of fat (Isaksson *et al.*, 1984).

The current experimental evidence supports the view that dietary fiber will slow the digestion and absorption of fats by binding components of the micelle and by increasing viscosity and volume of the intestinal contents, which can slow diffusion. Additional experimental evidence suggests that certain dietary fibers may inhibit the activity of lipase, which hydrolyzes triacylglycerides to free fatty acids, monoglycerides, and diglycerides. *In vitro* data indicated that lipase activity in human pancreatic juice or human duodenal samples can be inhibited by xylan, cellulose, pectin, and guar gum (Schneeman and Gallaher, 1986a). Lipolytic activity measured in intestinal samples is not lowered by feeding rats test meals containing fiber supplements (Schneeman, 1982); however, this enzyme has been assayed under conditions that favor maximal enzyme activity, which does not necessarily reflect physiological availability of digestive activity. Lipase activity was reduced in duodenal samples after meals containing pectin and wheat bran in patients who were receiving pancreatic enzyme replacement therapy (Isaksson *et al.*, 1984). It is likely that specific fibers may reduce the availability of the enzyme for hydrolyzing triacylglyceride within the intestinal contents (Schneeman and Gallaher, 1986a). Gallaher and Schneeman (1985) reported that a diet high in cellulose (20% by weight) delayed the disappearance of labeled triolein but not labeled cholesterol from the small intestine. The results indicated that the high cellulose content of the diet interfered with triolein breakdown but not with overall lipid absorption. Recently, Lairon *et al.* (1985a,b) reported that an inhibitor of pancreatic lipase is present in wheat bran and wheat germ. The characteristics of this inhibitor suggest that it may be active in the small intestine and capable of slowing the digestion of triacylglycerides.

Vahouny *et al.* (1988; Vahouny, 1982) have demonstrated that addition of fiber supplements to rat diets can slow the appearance of lipids in the lymph. In a recent study they demonstrated that during the initial 4-hr period following the infusion of lipid, the appearance of oleic acid was reduced in rats that had been adapted to diets containing cholestyramine, psyllium, guar gum, pectin, and a mixed fiber preparation, whereas cellulose and alfalfa did not lower the percentage of oleic acid that appeared in the lymph at 4 hr (Vahouny *et al.*, 1988). However, when recovery of oleic acid in lymph was measured in 24-hr samples, there were no differences among animal groups except for the psyllium-treated group. The results indicated that, except in the psyllium group, fatty acid absorption was delayed but not impaired by the fiber treatment. Cholesterol recovery in the lymph was also reduced by the fiber treatments. However, an actual impairment of cholesterol absorption appeared to occur from the fiber treatment because lymphatic recovery of cholesterol in the mixed-fiber, pectin, guar gum, and psyllium groups at 24 hr was still lower. The ability of the soluble forms of fiber to slow fatty acid absorption and to interfere with cholesterol absorption undoubtedly contributes to the effect of these fiber sources on plasma lipid levels.

Although considerable evidence exists to suggest that certain dietary fibers can slow digestion and absorption of fats by their effects on the physical characteristics of the intestinal contents, by binding micellar components, or by interfering with digestive enzyme activity, overall fat digestibility does not appear to be severely restricted by fiber supplementation. In reviewing the literature, Vahouny and Cassidy (1986) reported that apparent fat digestibility was changed less than 4% because of a high-fiber diet. What may be of greater interest than fecal fat excretion is the fact that dietary fibers that increase viscosity and inhibit lipolytic activity are likely to slow the disappearance of lipid and result in more lipid absorption from the lower half of the small intestine. We are currently investigating the effect of lipid absorption from the lower half of the small intestine on lipid metabolism.

A limited number of studies have been conducted in human subjects to determine whether the effect of soluble fiber sources on lipid digestion and absorption will attenuate the appearance of triacylglycerides in the plasma following consumption of a meal. In general, the use of plasma triacylglyceride appearance has not resulted in a clear-cut picture of the effect of fiber on lipid absorption. Anderson *et al.* (1980) reported that plasma triacylglycerides did not increase postprandially in subjects given a high-carbo-hydrate, high-fiber meal. However, the low fat content of this meal undoubtedly contrib-uted to the lack of postprandial increase. In contrast, Jenkins (1978) reported that both pectin and guar gum raise the serum triacylglycerides above control during the first 3 hr after consuming a Lundh test meal. We have also reported a larger increase in postpran-dial triacylglycerides in female subjects in response to a test meal when the meal was supplemented with guar gum and oat bran (Redard *et al.*, 1988). It is possible that, even if the absorption of lipid were delayed by viscous fiber sources, the clearance of triacylglycerides may be slowed because of a modification in insulin release caused by the fiber source. Jenkins (1978) also suggested that the interaction of pectin and guar gum might enhance absorption of fatty acids via the portal vein, which could also contribute to the postprandial response. It is interesting to note that treatment with cholestyramine, which binds micellar components, has recently been reported to en-hance postprandial triglyceridemia in type IIa patients (Weintraub *et al.*, 1987).

3.2. Carbohydrate

Some similarity in the mechanisms that are likely to slow digestion and absorption exists between lipids and carbohydrates. For example, viscosity caused by a fiber supplement has been shown to be an important factor in blunting the plasma glucose response in tolerance tests (Jenkins, 1978; Jenkins *et al.*, 1978; Leeds *et al.*, 1979). With respect to carbohydrate absorption, viscosity can delay glucose absorption by slowing gastric emptying (Leeds *et al.*, 1979) and through interference with glucose diffusion in the intestinal contents (Ebihara and Kiriyama, 1983). The glucose-flattening ability of various fiber supplements is correlated with their viscosity (Ebihara *et al.*, 1981; Jenkins, 1978). In a recent study we reported that the disappearance of ^{14}C-labeled starch from the small intestine can be delayed by the addition of guar gum but not by wheat bran to a fiber-free diet (Tinker and Schneeman, 1989). An earlier report demonstrated that addition of pectin to a diet also delayed the disappearance of starch from the small intestine, suggesting that viscosity may be important in slowing absorp-tion of complex carbohydrates from the small intestine (Forman and Schneeman, 1982).

It is feasible that the digestibility of complex carbohydrates can be reduced by the presence of fiber. Legumes have been reported to contain amylase inhibitors, which could slow the hydrolysis of starch in the small intestine (Gallaher and Schneeman, 1986b). Inhibition of amylase in human pancreatic or duodenal fluid by wheat bran, xylan, cellulose, guar gum, and psyllium has been reported (Schneeman and Gallaher, 1986a). This inhibition of amylase may slow the release of glucose or maltose units from starch but does not appear to impair starch availability. In whole foods such as rice or wheat, the bran layers that contain fiber can serve as a barrier to digestive enzymes, reducing the digestibility of the starch in the kernel (Snow and O'Dea, 1981; Wong and O'Dea, 1983).

3.3. Protein

As with carbohydrate and fat, it is likely that the absorption of the digestion products of proteins is slowed by the increased bulk, volume, and viscosity caused by certain fiber sources in the gastrointestinal tract. In ileostomy patients addition of pectin to a low-fiber diet increased nitrogen output by 30%, whereas addition of wheat bran did not change nitrogen output from the ileostomy (Sandberg *et al.*, 1981, 1983). This observation indicates that pectin, a viscous polysaccharide, can interfere with the digestion and absorption of either dietary or endogenous protein in the intestine. In certain nonpurified sources of fiber such as wheat bran, oat bran, or other cereals, which are a source of dietary protein, the cell wall layers, which contain fiber, can prevent penetration by digestive enzymes, and the protein associated with the fiber source will have a lower digestibility. In addition, trypsin and/or chymotrypsin, two pancreatic proteases, have been reported to be inhibited *in vitro* by alfalfa, safflower meal, soybean meal, xylan, cellulose, pectin, and carrageenan (Schneeman and Gallaher, 1986a). Many cereals and legumes contain pancreatic protease inhibitors that can decrease protein digestibility (Gallaher and Schneeman, 1986b; Schneeman and Gallaher, 1986b). These inhibitors are often inactivated by heat treatment; however, some inhibitor activity can potentially survive normal processing conditions and remain active in the gut. If the diet provides sufficient protein of high nutritional quality, animal studies have indicated that there is an adaptation to the presence of inhibitors in the diet, typically reflected in a greater synthesis of pancreatic enzymes (Schneeman and Gallaher, 1986b). When protein quality or quantity is poor, the animal is not capable of adapting, which is often reflected in poor growth. The pancreatic response of humans to the dietary presence of pancreatic protease inhibitors is not fully understood; however, it is likely that the human pancreas is capable of adapting by increasing the synthesis or secretion of pancreatic enzymes.

The effect of various fiber sources on protein digestibility and utilization has been comprehensively reviewed by Gallaher and Schneeman (1986c). Their review indicates that protein digestibility as estimated from fecal nitrogen excretion can be decreased by addition of fiber to experimental animal diets; however, this observation is confounded by the method of adding fiber to the diet and by the age of the animal being studied. In human subjects, although certain fibers increase fecal nitrogen excretion, overall nitrogen balance typically remains positive. All of the observations taken together suggest that with consumption of a high-fiber diet, protein digestibility will be reduced slightly because of lower digestibility of the protein that is associated with the fiber source, some inhibition of digestive enzymes, and/or less efficient absorption of amino acids. If protein intake is adequate and high-quality protein sources are available, nitrogen balance will not be compromised by the presence of fiber in the diet. In contrast, the high bulk and low protein quality of certain diets could make it difficult to meet protein and amino acid requirements, especially in growing children.

4. CONCLUSIONS

The human diet is typically omnivorous, and it is reasonable to assume that the physiology of the gastrointestinal tract reflects adaptations to utilize efficiently the nutrients from this dietary pattern. Studies on gastrointestinal function in which the absorption of a single nutrient or purified compounds has been measured usually indicate that the gastrointestinal tract has a very high capacity to digest and absorb carbohydrates, lipids, and proteins. Some of the excess capacity available in the gut may exist to compensate for the lower efficiency of nutrient absorption that is likely to occur when a mixture of digestible and nondigestible compounds, such as dietary fiber, is present. It seems reasonable to suggest that digestion and absorption of macronutrients in the presence of dietary fiber reflect normal physiological function. The presence of fiber in the gut probably has an important function in maintaining the gastrointestinal system by regulating the rate and site of nutrient absorption. Because the physical characteristics of fiber sources within the gut will determine the rate and site of nutrient absorption, it is important to examine the relationship between the rate and site of nutrient absorption and the subsequent metabolism of the absorbed compounds.

ACKNOWLEDGMENTS. This work has been supported in part by NIH grant DK 20446. The author is grateful to B. Diane Richter, Daniel Gallaher, Paul Davis, Kiyoshi Ebihara, Carol Redard, and Lesley Fels Tinker for their excellent research contributions.

REFERENCES

Anderson, J. W., Chen, W.-J. L., and Sieling, B., 1980, Hypolipidemic effects of high-carbohydrate, high-fiber diets, *Metabolism* **29**:551–558.

Blackburn, N. A., and Johnson, I. T., 1981, The effect of guar gum on the viscosity of the gastrointestinal contents and on glucose uptake from the perfused jejunum in the rat, *Br. J. Nutr.* **46**:239–246.

Collier, G., and O'Dea, K., 1982, Effect of physical form of carbohydrate on the postprandial glucose, insulin, and gastric inhibitory polypeptide responses in type 2 diabetes, *Am. J. Clin. Nutr.* **36**:10–14.

Ebihara, K., and Kiriyama, S., 1983, Plasma glucose-flattening activity of dietary fiber, in: *Dietary Fiber: Approach in Food Science and Nutrition* (S. Innami and S. Kiriyama, eds.), Shinohara Shoten, Tokyo, pp. 119–131.

Ebihara, K., and Schneeman, B. O., 1989, Interaction of bile acids, phospholipids, cholesterol, and triglyceride with dietary fibers in the small intestine of rats, *J. Nutr.* **119**:1100–1106.

Ebihara, K., Masuhara, R., Kiriyama, S., and Manabe, M., 1981, Correlation between viscosity and plasma glucose- and insulin-flattening activities of pectins from vegetables and fruits in rats, *Nutr. Rep. Int.* **23**:985–992.

Elsenhans, B., Sufke, U., Blume, R., and Caspary, W. F., 1980, The influence of carbohydrate gelling agents on rat intestinal transport of monosaccharides and neutral amino acids *in vitro, Clin. Sci.* **59**: 373–380.

Forman, L. P., and Schneeman, B. O., 1982, Dietary pectin's effect on starch utilization in rats, *J. Nutr.* **112**:528–533.

Gallaher, D., and Schneeman, B. O., 1985, Effect of dietary cellulose on site of lipid absorption, *Am. J. Physiol.* **249**:G184–G191.

Gallaher, D., and Schneeman, B. O., 1986a, Intestinal interaction of bile acids, phospholipids, dietary fibers, and cholestyramine, *Am. J. Physiol.* **250**:G420–G426.

Gallaher, D., and Schneeman, B. O., 1986b, Nutritional and metabolic response to plant inhibitors of digestive enzymes, in: *Advances in Experimental Medicine and Biology, Vol. 199: Nutritional and Toxicological Significance of Enzyme Inhibitors in Foods* (M. Friedman, ed.), Plenum Press, New York, pp. 167–184.

Gallaher, D., and Schneeman, B., 1986c, Effect of dietary fiber on protein digestibility and utilization, in: *Handbook of Dietary Fiber in Human Nutrition* (G. A. Spiller, ed.), CRC Press, Boca Raton, FL, pp. 143–164.

Isaksson, G., Lundquist, I., Akesson, B., and Ihse, I., 1984, Effects of pectin and wheat bran on intraluminal pancreatic enzyme activities and on fat absorption as examined with the triolein breath test in patients with pancreatic insufficiency, *Scand. J. Gastroenterol.* **19**:467–472.

Jenkins, D. J. A., 1978, Action of dietary fiber in lowering fasting serum cholesterol and reducing postprandial glycemia: Gastrointestinal mechanisms in: *International Conference on Atherosclerosis* (L. A. Carlson, ed.), Raven Press, New York, pp. 173–182.

Jenkins, D. J. A., Wolever, T. M. S., Leeds, A. R., Gassull, M. A., Haisman, P., Dilawari, J., Goff, D. V., Metz, G. L., and Alberti, K. G. M. M., 1978, Dietary fibres, fibre analogues, and glucose tolerance: Importance of viscosity, *Br. Med. J.* **1**:1392–1394.

Johnson, I. T., and Gee, J. M., 1981, Effect of gel-forming gums on the intestinal unstirred layer and sugar transport *in vitro, Gut* **22**:398–403.

Johnson, L. R., 1988, Regulation of gastrointestinal mucosal growth, *Physiol. Rev.* **68**:456–502.

Lairon, D., Lafont, H., Vigne, J. L., Nalbone, G., Leonardi, J., and Hauton, J. C., 1985a, Effects of dietary fibers and cholestyramine on the activity of pancreatic lipase *in vitro, Am. J. Clin. Nutr.* **42**: 629–638.

Lairon, D., Borel, P., Termine, E., Grataroli, R., Chabert, C., and Hauton, J. C., 1985b, Evidence for a proteinic inhibitor of pancreatic lipase in cereals, wheat bran and wheat germ, *Nutr. Rep. Int.* **32**: 1107–1113.

Leeds, A. R., Bolster, N. R., Andrews, R., and Truswell, A. S., 1979, Meal viscosity, gastric emptying, and glucose absorption in the rat, *Proc. Nutr. Soc.* **38**:44A.

Phillips, D. R., 1986, The effect of guar gum in solution on diffusion of cholesterol mixed micelles, *J. Sci. Food Agr.* **37**:548–552.

Poksay, K. S., and Schneeman, B. O., 1983, Pancreatic and intestinal response to dietary guar gum in rats, *J. Nutr.* **113**:1544–1549.

Redard, C. L., Schneeman, B. O., and Davis, P. A., 1988, Differences in postprandial lipemia due to gender and dietary fiber, *FASEB J.* **2**:A1418.

Sandberg, A. S., Andersson, H., Hallgren, B., Hasselblad, K., Isaksson, B., and Hulten, L., 1981, Experimental model for *in vivo* determination of dietary fibre and its effect on the absorption of nutrients in the small intestine, *Br. J. Nutr.* **45**:283–294.

Sandberg, A. S., Ahderinne, R., Andersson, H., Hallgren, B., and Hulten, L., 1983, The effect of citrus pectin on the absorption of nutrients in the small intestine, *Hum. Nutr. Clin.* **37C**:171–183.

Schneeman, B. O., 1982, Pancreatic and digestive function, in: *Dietary Fiber in Health and Disease* (G. V. Vahouny and D. Kritchevsky, eds.), Plenum Press, New York, pp. 73–83.

Schneeman, B. O., 1987, Dietary fiber and gastrointestinal function, *Nutr. Rev.* **45**:129–132.

Schneeman, B. O., and Ebihara, K., 1988, Bile acid binding by dietary fibers, *FASEB J.* **2**:A1418.

Schneeman, B. O., and Gallaher, D., 1986a, Effects of dietary fiber on digestive enzymes, in: *Handbook of Dietary Fiber in Human Nutrition* (G. A. Spiller, ed.), CRC Press, Boca Raton, FL, pp. 305–312.

Schneeman, B. O., and Gallaher, D., 1986b, Pancreatic response to dietary trypsin inhibitor: Variations among species, in: *Advances in Experimental Medicine and Biology*, Vol. 199: *Nutritional and Toxicological Significance of Enzyme Inhibitors in Foods* (M. Friedman, ed.), Plenum Press, New York, pp. 185–187.

Schwartz, S. E., Levine, R. A., Singh, A., Scheidecker, J. R., and Track, N. S., 1982, Sustained pectin ingestion delays gastric emptying, *Gastroenterology* **83**:812–817.

Sigleo, S., Jackson, M. J., and Vahouny, G. V., 1984, Effects of dietary fiber constituents on intestinal morphology and nutrient transport, *Am. J. Physiol.* **246**:G34–G39.

Snow, P., and O'Dea, K., 1981, Factors affecting the rate of hydrolysis of starch in food, *Am. J. Clin. Nutr.* **34**:2721–2727.

Tinker, L., and Schneeman, B. O., 1989, The effects of guar gum and wheat bran on the disappearance of ^{14}C-labeled starch from the rat gastrointestinal tract, *J. Nutr.* **119**:403–408.

Vahouny, G. V., 1982, Dietary fibers and intestinal absorption of lipids, in: *Dietary Fiber in Health and Disease* (G. V. Vahouny, and D. Kritchevsky, eds.), Plenum Press, New York, pp. 203–227

Vahouny, G. V., and Cassidy, M. M., 1985, Dietary fibers and absorption of nutrients, 1985, *Proc. Soc. Exp. Biol. Med.* **180**:432–446.

Vahouny, G. V., and Cassidy, M. M., 1986, Effect of dietary fiber on intestinal absorption of lipids, in: *Handbook of Dietary Fiber in Human Nutrition* (G. A. Spiller, ed.), CRC Press, Boca Raton, FL, pp. 121–128.

Vahouny, G. V., Satchithanandam, S., Chen, I., Tepper, S. A., Kritchevsky, D., Lightfoot, F. G., and Cassidy, M. M., 1988, Dietary fiber and intestinal adaptation: Effect on lipid absorption and lymphatic transport in the rat, *Am. J. Clin. Nutr.* **47**:201–206.

Weintraub, M. S., Eisenberg, S., and Breslow, J. L., 1987, Different patterns of postprandial lipoprotein metabolism in normal, type IIa, type III, and type IV hyperlipoproteinemic individuals, *J. Clin. Invest.* **79**:1110–1119.

Wong, S., and O'Dea, K., 1983, Importance of physical form rather than viscosity in determining the rate of starch hydrolysis in legumes, *Am. J. Clin. Nutr.* **37**:66–70.

Wu, A. L., Clark, S. B., and Holt, P. R., 1980, Composition of lymph chylomicrons from proximal or distal rat small intestine, *Am. J. Clin. Nutr.* **33**:582–589.

11

Physiological Effects of Fiber

C. A. EDWARDS

1. INTRODUCTION

Ingestion of a diet rich in dietary fiber may result in slower gastric emptying, delayed absorption in the small intestine, faster colonic transit, and an increased stool output. These effects, however, are dependent on the type of dietary fiber ingested. The plant polysaccharides that make up dietary fiber have a variety of chemical structures and physical properties. The soluble polysaccharides form viscous solutions in the intestinal lumen and delay absorption of nutrients in the small intestine. Many of these soluble polysaccharides are extensively fermented in the proximal colon and have minimal effects on stool output. In contrast, the insoluble polysaccharides act as indigestible solids and have little effect on small bowel physiology. However, they are more resistant to bacterial degradation and have a greater influence on stool output and colonic transit. This chapter discusses the mechanisms by which dietary fiber influences the functions of the gastrointestinal tract.

2. REDUCTION IN ABSORPTION BY VISCOUS POLYSACCHARIDES

The addition of viscous polysaccharides to carbohydrate meals reduces postprandial hyperglycemia (Jenkins *et al.*, 1978), indicating an impairment of carbohydrate absorption. However, although ingestion of viscous polysaccharides with xylose reduced the blood levels of xylose, the amount of xylose excreted by the kidneys was unchanged, suggesting that the rate of absorption was slowed but that the total amount of xylose absorbed was not reduced (Jenkins *et al.*, 1978). Studies of ileostomy effluent in rats and in man have shown that more fat is egested from the stoma after pectin ingestion (Isaksson *et al.*, 1983; Sandberg *et al*, 1983), indicating a malabsorption of lipids.

C. A. EDWARDS • Sub-Department of Human Gastrointestinal Physiology and Nutrition, Royal Hallamshire Hospital, Sheffield S10 2JF, England; *present address:* Wolfson Gastrointestinal Unit, Western General Hospital, Edinburgh EH4 2XU, Scotland.

2.1. Gastric Emptying

In general, viscous polysaccharides slow the emptying of liquids from the stomach, and this effect was thought to be responsible for the reduced rate of absorption of rapidly absorbed substances (Holt *et al.*, 1979; Schwartz and Levine, 1980). However, there is no correlation between the rate of gastric emptying and postprandial plasma glucose levels after administration of guar gum (Blackburn *et al.*, 1984a), which suggests that the action of viscous polysaccharides on gastric emptying may not be the dominant factor in reducing postprandial glycemia. Indeed, at low concentration some viscous polysaccharides, such as locust bean gum (1%), accelerate gastric emptying but still depress postprandial glycemia (Edwards *et al.*, 1987).

2.2. Small Intestinal Absorption

Viscous polysaccharides reduce the absorption of nutrients in the small intestine independent of their effects on gastric emptying. They reduce the absorption of glucose and amino acids from jejunal everted sacs *in vitro* (Johnson and Gee, 1981; Elsehans *et al.*, 1984), from perfused (Blackburn and Johnson, 1981; Elsehans *et al.*, 1984) or tied intestinal loops (L. Craigen, N. A. Blackburn, and N. W. Read, unpublished data) in the anethetised rat, and from the small intestine of the conscious pig (Rainbird *et al.*, 1984) and man (Blackburn *et al.*, 1984a). There is little evidence for a direct action of the viscous polysaccharides on the intestinal epithelium. The reduced rate of absorption is probably the result of a reduction in the mixing and movement of intraluminal contents caused by their increased viscosity.

2.2.1. Effect of Viscous Polysaccharides on Intraluminal Mixing

Two mechanisms bring nutrients into contact with the epithelium where they are absorbed. First, intestinal contractions create turbulence and convective currents, which mix the luminal contents and bring material from the center of the lumen close to the epithelium (Macagno *et al.*, 1982), and second, nutrients have to diffuse across the thin relatively unstirred layer of fluid lying adjacent to the epithelium (Blackburn *et al.*, 1984a). Viscous polysaccharides could reduce intestinal absorption by reducing mixing and/or diffusion. Viscous polysaccharides have been shown to increase the apparent thickness of the unstirred layer (Johnson and Gee, 1981; Flourie *et al.*, 1984). The unstirred layer, however, is not an anatomic reality but a functional concept that has been invented to explain the changes in absorption that occur under stirred or less stirred conditions (Dietschy *et al.*, 1971). An increase in unstirred layer thickness could imply a decrease in the diffusion coefficient for the test solute across an unstirred layer of unchanged thickness (Holzheimer and Winne, 1986), or it could imply a reduction in luminal convection causing an actual increase in the unstirred zone.

Studies that claim to show an effect of viscous polysaccharides on diffusion have usually measured the movement of glucose out of dialysis bags (Taylor *et al.*, 1980; Jenkins, 1981) or across diffusion chambers (Ebihara *et al.*, 1981) under stirred conditions and have not eliminated the effect of convection (Lucas, 1984). We have shown that the conductivity (and hence ion mobility) of unstirred electrolyte solutions is unchanged by adding guar to increase the viscosity, whereas the time taken for two electrolyte solutions to mix under the influence of a constant rotary force is proportional to their viscosity (Edwards *et al.*, 1988). We also designed a model to simulate the effect of intestinal contractions on absorption. This consisted of a dialysis tube filled with 10% glucose and 0.9% saline. The tube was anchored at each end in a perspex trough containing distilled water. Intestinal contractions were simulated using two pairs of cylindrical paddles positioned at each end of the dialysis tube, driven by a motor so that each pair alternatively moved together until they occluded 1 cm of each end of the tube at a rate of either 36 or 72 contractions per minute. The appearance of glucose in the external solution was monitored over 90 min. Our results showed that increasing the contraction rate with nonviscous solutions in the dialysis tubing caused a significant increase in the rate of movement of glucose into the external solution (Fig. 1). This effect was inhibited by incorporation of 1% guar gum into the dialysis tubes (Fig. 1).

These observations suggest that the reduction of absorption of small molecules by viscous polysaccharides is probably caused by the viscous luminal contents resisting the convection effects of intestinal contractions rather than by a decrease in diffusion coefficients. This may not be the case with large molecules or complexes such as micelles, however, where reductions in diffusion may play a larger role (Phillips, 1986).

In addition to limiting the access of digested nutrients to the epithelial surface, this reduction in intraluminal mixing would also reduce the interaction between food and enzymes and impair disruption of solid foods. These effects would tend to delay absorption, reducing nutrient levels in the blood and resulting in a greater delivery of nutrients to more distal sites in the small intestine. The reduced access of nutrients to the intestinal epithelium and the change in site of absorption could change the pattern of hormonal release and modulate neurohumoral reflexes responsible for gastrointestinal function.

2.2.2. Effect of Luminal Secretions on Viscosity

Because the reduction in intestinal absorption by soluble fibers is related to their viscous properties, it is important to consider the effects of intestinal secretions on the viscosity of ingested plant polysaccharides. A mixture of xanthan and locust bean gum

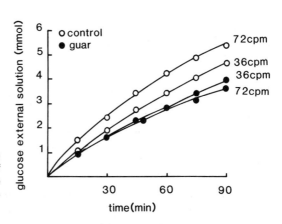

FIGURE 1. Effect of guar gum (1%) on the movement of glucose out of a dialysis bag under different paddle contraction rates (36 and 72 contractions per minute).

(X/LBG; 1% w/v) is three times as viscous as the same concentration of guar gum. However, when fed to human volunteers in a glucose drink, these polysaccharides were equally potent in reducing postprandial hyperglycemia. This apparent discrepancy between the higher viscosity of X/LBG and its effects on blood glucose was explained when we simulated the effects of intestinal secretions on the viscous properties of the gums. The viscosity of X/LBG was reduced by dilution with saline or by acidification to pH 1 and reneutralization to a much greater extent than the viscosity of guar (Table I), and the final viscosities of the two gums were very similar after these treatments (Table I; Edwards *et al.*, 1987). Thus, preingestion viscosity measurements may be misleading if used to predict the action of viscous polysaccharides on postprandial glycemia.

2.2.3. Effect of Viscous Polysaccharides on Intestinal Propulsion

The degree of absorption from the small intestine depends on both the efficiency of digestion and epithelial transport and the length of time that food remains in the small intestine—the small intestinal contact time or residence.

Until recently, most investigators working in this area considered absorption to be so rapid in relation to contact time that the latter was not an important factor in determining the degree of absorption. This belief probably derived from the pioneering studies of Borgstrom and his colleagues (1957), who used small intestinal aspiration in human volunteers to demonstrate that over 90% of the components of a liquid meal were absorbed by a point 100 cm from the pylorus. The hypothesis is challenged by observations in people with ileostomies fashioned from the terminal ileum for diseases involving the colon; these studies show that up to 10% of ingested fat, protein, and absorbable carbohydrate is voided from the ileal stomas, and these amounts can be increased by agents that accelerate small bowel transit and reduced by loperamide, which slows small bowel transit (King and Hill, 1981; Holgate and Read, 1983). Moreover, data obtained by aspiration of the contents of a carbohydrate meal from the ileum of intact human volunteers showed that up to 10% of ingested carbohydrate escaped absorption (Stephen *et al.*, 1983) and that carbohydrate absorption was markedly reduced when transit was accelerated by metoclopramide and increased when transit was delayed by infusion of lipid into the ileum (Holgate and Read, 1983).

Viscous polysaccharides tend to delay mouth-to-cecum transit (Jenkins *et al.*, 1978; Blackburn *et al.*, 1984a; Brown *et al.*, 1988). This effect is partly related to the delay in gastric emptying, but there is also a direct effect on small intestinal transit (Blackburn *et al.*, 1984a; Bueno *et al.*, 1981; Brown *et al.*, 1988). Experiments in rats have shown that the greatest effects of guar on transit were seen in the stomach and ileum (Brown *et al.*, 1988). Transit through the upper portion of the small intestine remains unchanged (Leeds, 1982; Brown *et al.*, 1988), probably because the viscosity in the upper small intestine is reduced by digestive secretions, but by the time the remnants of the meal reach the ileum, most of this fluid has been reabsorbed, and intestinal contents regain their viscosity. The rate of cecal filling (or ileal emptying) is also reduced by ingestion of viscous polysaccharides (Spiller *et al.*, 1987; Brown *et al.*, 1988).

In theory, the delay in the transit of a meal down the small intestine caused by viscous polysaccharides could reduce the rate of nutrient absorption by limiting the spread of luminal contents along the intestine and hence the area of mucosa in contact

TABLE I. Effect of Acidification with a Solution of 0.1 mM HCl, 54 mM NaCl and Reneutralization with a Solution of 120 mM NaHCO$_3$, 5 mM KCl, and 30 mM NaCl on the Viscosity of Guar and a Xanthan/Locust Bean Gum Mixture

Concentration	Initial, 1%	Acidified, 0.89%	Saline-diluted control, 0.89%	Reneutralized, 0.68%	Saline-diluted control, 0.68%	Acid-saline-diluted control, 0.68%
			Viscosity of gum solutions[a]			
G	2396 ± 51	1939 ± 78*	1714 ± 160	1209 ± 51**	1227 ± 38	1305 ± 53
X/LBG	9922 ± 524	1294 ± 142+	4474 ± 595	2138 ± 68+***	1391 ± 78	1009 ± 53

[a]Mean ± S.E.M.; $n = 5$; G, guar; X, xanthan; LBG, locust bean gum. *$P < 0.05$, +$P < 0.001$, significantly different from saline control at same concentration. **$P < 0.05$, ***$P < 0.001$, significantly different from acid-saline control.

with the nutrients. Although we reduced peak plasma glucose levels to a small extent in normal volunteers by limiting the intestinal spread of a glucose drink with occluding balloons, the reduction in blood glucose caused by adding guar gum to a drink of glucose was not associated with any change in the distribution of a radiolabeled drink in the upper small intestine (Blackburn *et al.*, 1984b). If viscous polysaccharides do not reduce the contact area of nutrients with the small intestine, a slower overall small bowel transit would be expected to increase absorption of nutrients. Presumably the decrease in intraluminal mixing by viscous polysaccharides is sufficient to reduce absorption despite the longer transit time.

Although viscous polysaccharides slow small bowel transit, nonnutrient meals containing viscous polysaccharides produce a highly active contractile motor pattern in the small intestine, consisting of clusters of contractions propagated over long distances (Schemann and Erhlein, 1986; Welch and Worlding, 1986) (Fig. 2). Studies of isolated rat small intestine mounted in a Trendelenburg apparatus (Weems and Seygal, 1981;

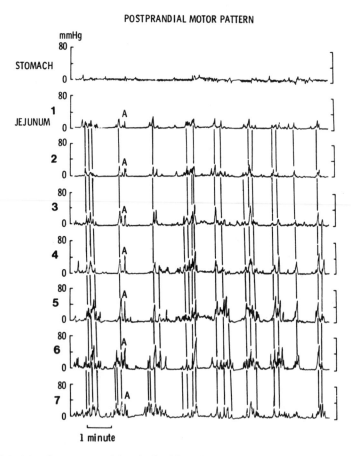

FIGURE 2. Jejunal pressure activity of a healthy volunteer after ingestion of a guar gum meal. Vertical lines joining traces indicate propagated contractions. A is an artifact. (I. McL. Welch and J. Worlding, unpublished data.)

Fisher *et al.*, 1986) in which waves of propulsive contractions are stimulated by a hydrostatic load showed that when the viscosity of the luminal contents was increased with guar gum, the frequency of peristalsis was unchanged, but the volume of contents moved by each wave was decreased and the duration of each peristaltic wave was prolonged (B. Murray, R. D. Rumsey, C. A. Edwards, and N. W. Read, unpublished data) (Fig. 3). These data suggest that although intestinal propulsion is stimulated by viscous meals, the resistive properties of the luminal contents predominate, and the movement of contents along the intestine is slowed.

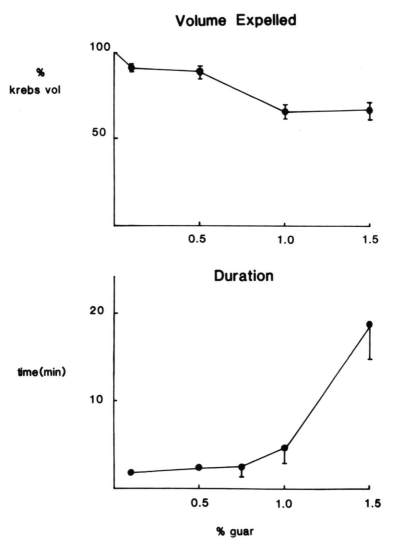

FIGURE 3. Effect of intraluminal guar gum on the propagative motility of isolated segments of rat ileum *in vitro*. Top indicates the reduction in volume of fluid propelled compared with intra-luminal Krebs solutions. Bottom shows the increase in the duration of each peristaltic wave.

by viscous meals, the resistive properties of the luminal contents predominate, and the movement of contents along the intestine is slowed.

Viscous polysaccharides thus appear to reduce absorption in the small intestine, acting as antimotility agents reducing both intraluminal mixing and propulsion. Absorption is reduced probably because the effects on intraluminal mixing overcome the effects of delayed small intestinal transit.

3. NONVISCOUS POLYSACCHARIDES

Although the ingestion of purified nonviscous fibers has very little effect on postprandial glycemia, the ingestion of unpurified plant material results in less pronounced hyperglycemia than would be expected (Jenkins *et al.*, 1980). This may result in part from the presence of starches that are resistant to degradation by pancreatic enzymes (Englyst and Cummings, 1985) or of enzyme inhibitors (Leiner and Kakade, 1980; Mistunaga, 1974). The major mechanism, however, is probably the sequestration of the absorbable carbohydrate within the cellular matrix of the plant, which restricts the access of pancreatic enzymes (Wong *et al.*, 1985). The glycemic indices of plant foods and their rates of hydrolysis by pancreatic enzymes are markedly increased if they are ground before ingestion (O'Dea *et al.*, 1981). A similar effect can be observed if a variety of different plant foods are chewed thoroughly before they are swallowed (Read *et al.*, 1986).

4. COLONIC FUNCTION

The action of fiber on fecal output depends on events occurring in the colon. The rich colonic bacterial flora is capable of fermenting many of the plant polysaccharides, but there is considerable interindividual variation in the range of polysaccharides that can be fermented, and some polysaccharides such as cellulose are fermented much more slowly than others such as pectin. Lignification of the plant material increases its resistance to bacterial degradation (Cummings, 1982). In general, those fibers that are poorly digested are more efficient stool bulkers than the fermentable fibers.

Fiber increases stool bulk by its physical presence but perhaps more importantly by the extent to which it absorbs and retains water. However, the action of fiber in stool bulking cannot be predicted on the basis of its water-holding capacity (WHC) *in vitro*. In fact, there tends to be an inverse relationship between WHC and stool output (Cummings, 1984). This is because most of the fibers with a large WHC are soluble polysaccharides that are extensively fermented by the colonic bacteria, losing their WHC in the colon (McBurney *et al.*, 1985). There is, however, a good correlation between the WHC of the ethanolic extracts of fibers exposed to bacterial degradation and their effects on stool output (McBurney *et al.*, 1985).

The influence of colonic fermentation on the action of fiber in the colon was recently demonstrated in our laboratory when the degradation of three plant polysaccharides (ispaghula, guar and xanthan) by fecal bacterial from normal volunteers was compared with their effects on stool output in the same volunteers (Tomlin and Read,

1988a). There was considerable variation in the ability of the subjects to ferment the fibers, but the comparison of fermentability with bowel function supported McBurney's observations in that those fibers that retained their structure were associated with an increase in stool bulk. However, there was no significant change in either transit time or frequency after ingestion of these polysaccharides. In contrast, complete degradation of a polysaccharide was associated with a decrease in transit time, an increase in frequency, but no change in stool bulk. This suggests that the relationship between stool bulk and transit time is indirect, and each may be influenced independently by different components of fiber.

The mechanism by which fermentation accelerates transit is unclear. Fermentation of fiber results in the production of gas and short-chain fatty acids. Gas may distend the bowel and cause propulsion (Chauve *et al.*, 1976). The effects of short-chain fatty acids on transit have not been fully studied, but propionic and *n*-butyric acids have been shown to stimulate contractions in muscle strips taken from the middle and distal colon of the rat (Yajima, 1985). Another possible explanation is the release of bile acids or fatty acids previously adsorbed onto the fiber in the small intestine when the fiber is fermented (Vahouney *et al.*, 1981). These are transformed by bacterial action into substances with laxative properties such as deoxycholic acid and hydroxy fatty acids. We have shown that deoxycholic acid infused into the rectum of normal volunteers at low concentration (3 mM) causes large rectal contractions that are associated with an intolerable urge to defecate (Edwards *et al.*, 1989) (Fig. 4).

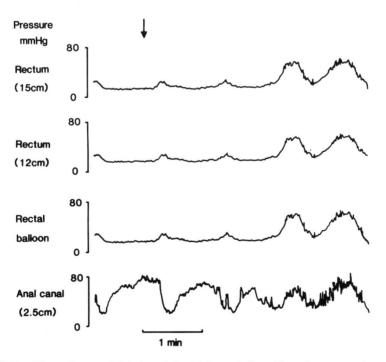

FIGURE 4. Effect of a rectal infusion of 3 mM deoxycholic acid on the anorectal pressure activity of a healthy volunteer. Arrow indicates the onset of an intolerable urge to defecate.

An additional mechanism by which fiber (especially particulate fiber) may increase stool output and decrease transit is the interaction of the edges of the fiber particles with multimodal mucosal receptors that cause reflex secretion (Beubler and Juan, 1978) and propagated motor patterns. This may explain why plastic particles (2 mm × 2 mm; 15 g/day) fed to normal volunteers were as effective as the same amount wheat fiber in increasing stool output and decreasing transit time (Tomlin and Read, 1988b). Similarly, polystyrene particles (2 mm) fed to dogs induced propagated colonic motor patterns similar to those seen with wheat bran (Cherbut and Ruckebusch, 1984).

5. CONCLUSION

The action of fiber on intestinal function is dependent on the type of polysaccharide ingested. Viscous polysaccharides are the most effective in reducing absorption and delaying transit in the small intestine, and this is probably because of their antimotility effects, the increased viscosity of the luminal contents resisting the convective and propulsive movements caused by intestinal contractions.

In the colon, fibers that resist bacterial degradation and retain most of their physical structure are the best stool bulkers, probably because of their water-holding capacity. However, colonic fermentation of plant polysaccharides appears to accelerate colonic transit. The mechanism of this action is unclear and needs further study. The most effective bulk laxatives appear to be those that are partially fermented but retain sufficient of their structure to hold water and increase stool bulk.

REFERENCES

Beubler, E., and Juan, H., 1978, PGE-released blood flow and transmucosal water movement after mechanical stimulation of the rat jejunal mucosa, *Naunyn Schmiedebergs Arch. Pharmacol.* **305:** 91–95.

Blackburn, N. A., and Johnson, I. T., 1981, The effect of guar gum on the viscosity of the gastrointestinal contents and on glucose uptake from the perfused jejunum in the rat, *Br. J. Nutr.* **46:**239– 246.

Blackburn, N. A., Redfern, J. S., Jarjis, M., Holgate, A. M., Hanning, I., Scarpello, J. H. B., Johnson, I. T., and Read, N. W., 1984a, The mechanism of action of guar gum in improving glucose tolerance in man, *Clin. Sci.* **66:**329–336.

Blackburn, N. A., Holgate, A. M., and Read, N. W., 1984b, Does guar gum improve postprandial hyperglycaemia in humans by reducing small intestinal contact area? *Br. J. Nutr.* **52:**197–204.

Borgstrom, B., Dahlquist, A., and Lundh, G., 1957, Studies of intestinal digestion and absorption in humans, *J. Clin. Invest.* **36:**1521–1536.

Brown, N. J., Worlding, J., Rumsey, R. D. E., and Read, N. W., 1988, The effect of guar gum on the distribution of a radiolabelled meal in the gastrointestinal tract of the rat, *Br. J. Nutr.* **59:**223–231.

Bueno, L., Praddaude, F., Fioramonti, J., and Ruckebusch, Y., 1981, Effect of dietary fibre on gastrointestinal motility and jejunal transit time, *Gastroenterology* **80:**701–707.

Chauve, A., Devroede, G., and Bastin, E., 1976, Intraluminal pressures during perfusion of the human colon *in situ*, *Gastroenterology* **70:**336–344.

Cherbut, C., and Ruckebusch, Y., 1984, Modifications de l'electromyogramme du colon chez le chien, *Gastroenterol. Clin. Biol.* **8:**995–959.

Cummings, J. H., 1982, Polysaccharide fermentation in the human colon, in: *Colon and Nutrition* (H. Kasper and H. Goebell, eds.) MTP Press, Lancaster, pp. 91–103.

Cummings, J. H., 1984, Constipation, dietary fibre and the control of large bowel function, *Postgrad. Med. J.* **60:**811–819.

Dietschey, J. M., Sallee, V. L., and Wilson, F. A., 1971, Unstirred water layers and absorption across the intestinal mucosa, *Gastroenterology* **61:**932–943.

Ebihara, K., Masuhara, R., and Kiriyamam, S., 1981, Major determinants of plamsa glucose. Flattening activity of a water soluble dietary fibre. Effect of konjac mannan on gastric emptying and intra-luminal glucose diffusion, *Nutr. Rep. Int.* **23:**1145–1156.

Edwards, C. A., Blackburn, N. A., Craigen, L., Davison, P., Tomlin, J., Sugden, K., Johnson, I. T., and Read, N. W., 1987, Viscosity of food gums determined *in vitro* related to their hypoglycaemic actions, *Am. J. Clin. Nutr.* **46:**72–77.

Edwards, C. A., Johnson, I. T., and Read, N. W., 1988, Do viscous polysaccharides reduce absorption by inhibiting diffusion or convection? *Eur. J. Clin. Nutr.* **42:**307–312.

Edwards, C. A., Brown, S., Baxter, A. J., Bannister, J. J., and Read, N. W., 1989, Effect of bile acid on anorectal function in man, *Gut* **30:**383–386.

Elsenhans, B., Zenker, D., Caspary, W. F., and Blume, R., 1984, Guaran effect on rat intestinal absorption. A perfusion study, *Gastroenterology* **86:**645–653.

Englyst, H., and Cummings, J. H., 1985, Digestion of the polysaccharides of some cereal foods in the human small intestine, *Am. J. Clin. Nutr.* **42:**778–787.

Fisher, S., Murray, B. E., and Rumsey, R. D. E., 1986, Factors affecting the volume expelled from isolated segment of rat small intestine during the peristaltic reflex, *J. Physiol. (Lond.)* **378:**15P.

Flourie, B., Vidon, N., Florent, C. H., and Bernier, J. J., 1984, Effect of pectin on jejunal glucose absorption and unstirred layer thickness in normal man, *Gut* **25:**936–941.

Holgate, A. M., and Read, N. W., 1983, The relationship between small bowel transit time and absorption of a solid meal. Influence of metaclopramide, magnesium sulphate, and lactulose, *Dig. Dis. Sci.* **28:**812–819.

Holt, S., Heading, R. C., Carter, D. C., Prescott, L. F., and Tothill, P., 1979, Effect of gel-forming fibre on gastric emptying and absorption of glucose and paracetomol, *Lancet* **1:**636–639.

Holzheimer, G., and Winne, D., 1986, Influence of dietary fibre and intraluminal pressure on absorption and pre-epithelial diffusion resistance (unstirred layer) in rat jejunum *in situ*, *Naunyn Schmiedebergs Arch. Pharmacol.* **334:**514–524.

Jenkins, D. J. A., Wolever, T. M. S., Taylor, R. H., Ghafari, H., Jenkins, A. L., Barker, H., and Jenkins, M. J. A., 1980, Rate of digestion of foods and postprandial glycaemia in normal and diabetic subjects, *Br. Med.J.* **2:**14–17.

Isaksson, G., Asp, N.-G., and Ihse, I., 1983, Effect of dietary fibre on pancreatic enzyme activities of ileostomy evacuates and on excretion of fat and nitrogen in the rat, *Scand. J. Gastroenterol.* **18:** 417–423.

Jenkins, D. J. A., 1981, Slow release carbohydrate and the treatment of diabetes, *Proc. Nutr. Soc.* **40:** 227–235.

Jenkins, D. J. A., Wolever, T. M. S., Leeds, A. R., Gassull, M. A., Haisman, P., Dilawari, J., Goff, D. V., Metz, G. L., and Albert, K. G. M. M., 1978, Dietary fibres, fibre analogues and glucose tolerance: Importance of viscosity, *Br. Med .J.* **1:**1392–1394.

Johnson, I. T., and Gee, J. M., 1981, Effect of gel-forming food gums on the intestinal unstirred layer and sugar transport *in vitro*, *Gut* **22:**398–403.

King, R. F. G. J., and Hill, G. H., 1981, Effect of loperamide and codeine phosphate on ileostomy output, in: *Diarrhoea—New Insights* (N. W. Read, ed.), Janssen Research Foundation, Beerse, Belgium, pp. 207–213.

Leeds, A. R., 1982, Modification of intestinal absorption by dietary fibre and fibre components, in: *Dietary Fibre in Health and Disease* (G. V. Vahouney and D. Kritchevsky, eds.), Plenum Press, New York, pp. 53–71.

Leiner, I. E., and Kakade, M. L., 1980, Protease inhibitors, in: *Toxic Constituents of Food Plantstuffs*, 2nd ed. (I. E. Leiner, ed.), Academic Press, New York, pp. 7–71

Lucas, M. L., 1984, Estimation of sodium chloride diffusion coefficient in gastric mucins, *Dig. Dis. Sci.* **29:**336–345.

Macagno, E. C., Christensen, J., and Lee, C. I., 1982, Modelling the effect of wall movements on absorption in the intestine, *Am. J. Physiol.* **243:**G541–550.

McBurney, M. I., Horvath, P. J., Jeraci, J. L., and Van Soest, P. J., 1985, Effect of *in vitro* fermentation using faecal inoculum on the water holding capacity of dietary fibre, *Br. J. Nutr.* **53:**17–24.

Mistunaga T., 1974, Some properties of protease inhibitors in wheatgrain, *J. Nutr. Sci. Vitaminol.* **20:** 153–159.

O'Dea, K., Snow, P., and Nestel, P., 1981, Rate of starch hydrolysis in vitro as a predictor of metabolic responses to complex carbohydrate *in vivo, Am. J. Clin. Nutr.* **34:**1991–1993.

Phillips, D. R., 1986, The effect of guar gum in solution on diffusion of cholesterol mixed micelles, *J. Sci. Food. Agr.* **37:**548–552.

Rainbird, A. L., Low, A. G., and Zebrowsky, T., 1984, Effect of guar gum on glucose and water absorption from isolated jejunal loops in conscious growing pigs, *Br. J. Nutr.* **52:**489–498.

Read, N. W., Welch, I. McL., Austen, C. J., Barnish, C., Bartlett, C. E., Baxter, A. J., Brown, G., Comptom, M. E., Hume, K. E., Storie, I., and Worlding, J., 1986, Swallowing food without chewing: A simple way to reduce postprandial glycaemia, *Br. J. Nutr.* **55:**43–7.

Sandberg, A. S., Ahderinne, R., Andersson, H., Hallgren, B., and Hulten, L., 1983, The effect of citrus pectin on the absorption of nutrients in the small intestine, *Hum. Nutr. Clin. Nutr.* **37:**171–183.

Schemann, M., and Erhlein, H. J., 1986, Postprandial patterns of canine jejunal motility and transit of luminal contents, *Gastroenterology* **90:**991–1000.

Schwartz, S. E., and Levine, G. D., 1980, Effect of dietary fibre on intestinal glucose absorption and glucose tolerance in rats, *Gastroenterology* **79:**833–836.

Spiller, R. C., Brown, M. L., and Phillips, S. F., 1987, Emptying of the terminal ileum in intact humans. Influence of meal residue and ileal motility, *Gastroenterology* **92:**724–729.

Stephen, A. M., Haddad, A. C., and Phillips, S. F., 1983, Passage of carbohydrate into the colon. Direct measurements in humans, *Gastroenterology* **85:**589–596.

Taylor, R. H., Wolever, T. M. S., Jenkins, D. J. A., Ghafari, H., and Jenkins, M. J. A., 1980, Viscosity and glucose diffusion: Potential for modification of absorption in the small intestine, *Gut* **21:**A452.

Tomlin, J., and Read, N. W., 1988a, The relationship between bacterial degradation of viscous polysaccharides and stool output in human beings, *Brit. J. Nutr.* **60:**467–475.

Tomlin, J., and Read, N. W., 1988b, The effect of inert plastic particles on colonic function in human volunteers, *Brit. Med. J.* **297:**1175–1176.

Vahouny, G. V., Tombes, R., Cassidy, M. M., Kritchevsky, D., and Gallo, L. L., 1981, Dietary fibers. VI. Binding of fatty acids and monolein from mixed micelles containing bile salts and lecithin, *Proc. Soc. Exp. Biol. Med.* **166:**12–16.

Weems, W. A., and Seygal, G. E., 1981, Fluid propulsion by cat intestinal segments under conditions requiring hydrostatic work, *Am. J. Physiol.* **240:**G147–G156.

Welch, I. McL., and Worlding, J., 1986, The effect of ileal infusion of lipids on the motility pattern in humans after ingestion of a viscous, non-nutrient meal, *J. Physiol. (Lond.)* **378:**12P.

Yajima, T., 1985, Contractile effect of short chain fatty acids on the isolated colon of the rat, *J. Physiol. (Lond.)* **368:**667–678.

Wong, S., Traianedes, K., and O'Dea, K., 1985, Factors affecting the rate of hydrolysis of starch in legumes, *Am. J. Clin. Nutr.* **42:**38–43.

12

Fiber Metabolism and Colonic Water

ELIZABETH F. ARMSTRONG, W. GORDON BRYDON, and MARTIN EASTWOOD

1. INTRODUCTION

The physiology of the colon is difficult to investigate, partly because structure and function vary between species and findings in experimental animals are not necessarily applicable to man (Phillips, 1988). In addition, differences in ileal effluent that are dependent on diet and biliary secretions and bacterial metabolism in the colon result in feces that can vary both qualitatively and quantitatively (Eastwood *et al.*, 1984).

The colon has an important role in water, mineral, and nutrient salvage. The vagus nerve runs along the esophagus to the midtransverse colon. The parts that are innervated by this nerve are involved in nutrient absorption to the cecum. In contrast, defecation is controlled by the lumbar–sacral outflow. In normal man the colon absorbs 1–2 liters of water and up to 200 mmole of sodium and chloride ions. The maximum capacity of absorption of the human colon when exposed to infusions of isotonic saline is 5 to 6 liters per day. Mucosal bicarbonate exchanges for luminal chloride. Fiber (complex carbohydrates) is degraded by bacteria to produce carbon dioxide, hydrogen, methane, and short-chain fatty acids, the major anionic components of fecal water (Eastwood and Brydon, 1985). In man urea is metabolized within the colon to ammonia to provide a nitrogen source.

In man there is a slow progression of ileal chyme throughout the colon, wherein there is a conversion of a viscous liquid to a plasticinelike stool, which can be expressed after voluntary decision.

In the rat the cecal contents are liquid, but by the transverse colon, well-formed, pellety stools have developed.

ELIZABETH F. ARMSTRONG, W. GORDON BRYDON, and MARTIN EASTWOOD • Wolfson Gastrointestinal Laboratories, Department of Medicine, University of Edinburgh, Western General Hospital, Edinburgh EH4 2XU, Scotland.

The colonic absorption of intestinal contents results in the ileal water content being reduced from 1 to 2 liters to between 40 and 250 g per day (Eastwood et al., 1984). The output of feces varies from person to person and over time in the same person (Wyman et al., 1976), yet the percentage of water in the stool remains constant between 70% and 80%.

2. STOOL BULKING

The single most important factor that dictates stool weight in the Western world is dietary fiber (Eastwood and Brydon, 1985), but the mechanism whereby this is moderated has not been fully clarified. It has been suggested that the laxative effect of dietary fiber is the result of the products of bacterial fermentation. Hellendoorn (1978) suggested that the fermentation of unabsorbed food carbohydrate gives rise to a range of end products with different physiological effects. He suggested that gas formation produced softer and more voluminous stools and hence accelerated intestinal transit time, and that short-chain fatty acid production increased stool weight by accelerating intestinal transit. Individual responses to fiber were not to be regarded as constant, as fermentation leads to fluctuation in the concentration and type of end products. This hypothesis became less tenable when it was shown that short-chain fatty acids (SCFA) are rapidly absorbed from the human rectum. It has been calculated that 20 g of fiber a day yields approximately 100 mmole of short-chain fatty acids in the colon, of which 80 to 95 mmole is absorbed or further metabolized. The amount of short-chain fatty acids (SCFAs) that cross the colonic mucosa is important in the overall control of electrolyte absorption from the colonic lumen. The movement of large amounts of ions will materially affect several epithelial transport processes (Cummings and Branch, 1986).

The water-holding capacity (WHC) of dietary fiber is important in explaining the fecal bulking effect. This fecal bulking action, however, only applies to the residue of fiber left after bacterial fermentation. Finely milled bran not only has a lower WHC for ingested fiber than coarse bran but also has less bulking activity (Broadribb and Groves, 1978; Smith et al., 1981). However, for nonlignified fruit and vegetable fibers, WHC has an inverse relationship to the eventual fecal bulking effect (Stephen and Cummings, 1979). Since there is extensive fermentation of this type of fiber by bacteria, such fermentation leads to substantial generation of short-chain fatty acids, which would indicate that these acids are not an important dictate of stool weight.

It has been shown that the dry weight of stool has two major components, residual fiber and bacteria (Stephen and Cummings, 1980a). When stool weight is increased by a fermented fiber, e.g., cabbage or carrot fiber, the increase in stool weight corresponds to an increase in bacterial mass. However, this does not invariably happen. For example, when extensively fermented fibers such as pectin pass through the colon it would be predicted that there would be an increase in both bacteria in the stool and in stool weight. However, pectin does not increase stool weight (Stephen and Cummings, 1980b). When stool weight is increased by coarse wheat bran, it is the water-holding capacity of the residual wheat bran that affects the increase in weight (Smith et al., 1981).

3. CECAL, BACTERIAL METABOLISM, AND STOOL WEIGHT

Further information about the relationship between diet, bacteria, stool weight, and colonic water can be obtained by studies of the cecum. There are substantial differences in colonic anatomy between mammals (Van Soest, 1982). In some herbivores the cecal fermentation organ products are expressed as feces and the animals indulge in coprophagy. This is an important soure of nutrition in the rat, hare, and lemming. In these animals there is a stratification or streaming of intestinal contents such that two distinct kinds of feces are expressed. One form of stool is liquid and is enriched in microbial protein, voided at night, and is selected for coprophagy. The other voided stool is pellety and low in energy. This enables these small animals to consume low-energy-containing fiber sources, e.g., grass, and readily to convert the fiber to available energy sources. In these cases the colon acts as a form of rumen (Van Soest, 1982). The rat, which is an omnivore, has a capacious cecum. In contrast, man, who is an omnivore of very wide dietary preferences, has a small cecum relative to the size of the colon. This implies that the colon is not an important source of nutrition, certainly to Western man. A common consequence of adding fiber to the diet of rats is enlargement of the cecum; this is believed to be part of the normal physiological adjustments to increasing fermentable fiber in the diet (Walters *et al.*, 1986; Rowland *et al.*, 1986; Tulung *et al.*, 1987).

There is no relationship between cecal size and stool weight. Those fibers that are extensively fermented appear to be associated with an increased cecal but not fecal weight. Those fibers that result in fecal bulking do not appear to be associated with an increase in cecal weight (Walters *et al.*, 1986; Sakata, 1987).

4. COLONIC WATER AND FIBER

The WHC of dietary fiber can be measured by a variety of methods, including centrifugation (McConnell *et al.*, 1974), filtration (Robertson and Eastwood, 1981a), column chromatography, and the ability of fiber to hold on to water when subjected to an osmotic or suction pressure (Robertson and Eastwood, 1981a). These methods give different values for WHC because each method measures different phases of water. Water can occupy three phases in relation to fiber (Robertson and Eastwood, 1981b; Eastwood *et al.*, 1983):

1. Tightly bound water, which can be removed only with difficulty, e.g., by freeze drying, which results in irreversible changes in the fiber polymeric structure.
2. Water held in the interstices of the fiber, which can only removed by pressure but not by centrifugation.
3. Loosely associated or free water, which is readily removed by filtration.

The amount of water held by a fiber will depend on the fiber source, the mode of preparation of the fiber, and the method of measurement. Wheat bran holds 2 to 6 g of water per gram of fiber, whereas fruits and vegetables vary from 8 to 30 g water in each gram (McConnell *et al.*, 1974). These estimates come from the centrifugation method, which tends to give a greater measurement for WHC than the suction method (Robertson and Eastwood, 1981b).

Human feces usually consist of approximately 75% water, of which probably 80% to 90% is either bound by bacteria or by the undigested fiber (Stephen and Cummings, 1980a). The percentage of water in rat fecal pellets is less than that of human feces. This may result from differences in sodium absorption or be because smaller fecal pellets dehydrate more easily than larger stools.

5. FIBER METABOLISM AND COLONIC AND CECAL WATER

Water is passively absorbed from the colon along osmotic gradients created mainly by the movement of sodium (Phillips and Giller, 1973). It has been estimated that about 2 atm (89 mosm kg^{-1}) osmotic difference exists across the colonic mucosa (McBurney et al., 1985). This present study has attempted to simulate the removal of water from colonic contents using dialysis membranes and polyethylene glycol to create an osmotic gradient of 2 atm over 72 hr. This gives a measure of the water that was strongly bound to the colonic contents. By this technique Canadian red spring wheat bran has a WHC of 4.3 g/g water, and high-methoxylated pectin has a WHC of 16.4 g/g dry weight.

The influence of wheat bran (Canadian red spring wheat) and pectin of a high-methoxylated type has recently been studied in this laboratory in relation to the distribution of cecal and fecal water. These fiber sources, one of which is nonfermentable and the other totally fermentable, have been studied when added to a fiber-free diet.

Ten percent of wheat bran supplement resulted in a 90% increase in fecal weight. This would be water held by undigested wheat bran fiber (Nyman and Asp, 1982). Pectin (10%) increased the dry stool weight by about 20% but doubled the output of bacteria as judged by the excretion of diamino pimelic acid (DAPA).

Although the total water content of the feces from rats fed either wheat bran or pectin was similar, this water was distributed in different ways (Fig. 1). The bound water per gram dry feces from rats fed pectin was greater than in those fed wheat bran. This may be in part because of an increased percentage of bacteria in the feces, since bacteria are about 80% water, which is mainly intracellular and therefore unavailable for absorption from the colon. Dietary pectin had a higher WHC prior to fermentation and held water more strongly than wheat bran fiber. Therefore, if about 25% of pectin had survived digestion (Hove and King, 1979; Nyman and Asp, 1982) in the present study, it would also hold water in the feces. McBurney et al. (1985) have shown that pectin has a high WHC in vitro but is completely fermented and that the resultant stool bulk is primarily from microbial organic matter.

Wheat bran did not increase the water held per gram dry feces in both the in vivo and in vitro measurements. There are reports that demonstrate that bran-supplemented diets do not increase fecal water content, and therefore the WHC of bran is no greater than that of the other fecal components (Slavin et al., 1985). Spiller et al. (1986) found that fecal water only increased with high levels (66 g per day or about 10%) of hard red wheat bran compared to the basal diet and that this may have been related to faster mean transit times. In the present study, wheat bran had a higher WHC before it was consumed by the rat compared to the residual WHC in the feces. Since wheat bran contained 38%

dietary fiber and up to 40% of this may have been fermented (Stephen and Cummings, 1980a; Nyman and Asp, 1982), this might lead to a reduction in the WHC of the residual material in the feces.

The feces from rats fed bulk laxatives from gum karaya and ispaghula sources have been shown (using centrifugation) to have four to six times higher WHC than feces from rats fed enriched wheat bran or barley products (Nyman and Asp, 1987). These fibers are at the most 30% fermented, and therefore the residual WHC reflects their original WHC. In contrast, increasing the amount of cellulose in the diet can cause a reduction in fecal moisture (Kies et al., 1984). Therefore, the ideal fecal bulking fiber should not only be largely undigested but also have resistant hydrophilic properties. Using a mixed-fiber diet of vegetables and cereals, Kurpad and Shetty (1986) have shown that the WHC of unfermented dietary fiber is of greater importance to fecal bulking in man than bacterial proliferation.

In general, cecal contents had a higher water content than fresh feces but a similar bound water (Fig. 1). The bound water of cecal contents might have been expected to be higher than that of the corresponding feces if fiber was incompletely digested. Although pectin caused cecal enlargement and produced feces with a higher water content than the feces from rats fed the fiber-free diet, the animals excreted formed pelleted stools. It has been reported that diets that cause cecal distention can lead to diarrhea (Elsenhans et al., 1981). An increase in osmotically active material following fermentation may have led

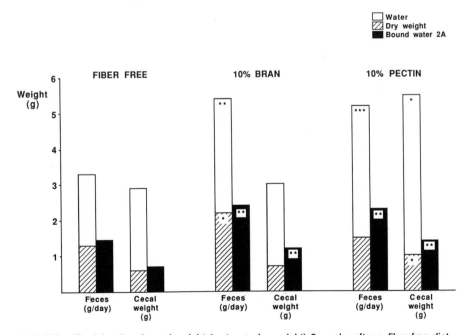

FIGURE 1. Total fecal and cecal weight (water + dry weight) 3 weeks after a fiber-free diet, alone or supplemented with 10% wheat bran or 10% pectin. The bound water is bound to the dry matter after dialysis at 2 atm.

to a greater retention of water, thereby increasing the cecal contents (Leegwater *et al.*, 1974). The cecum is the most important site for the production and absorption of SCFA. Demigne and Remesy (1985) have suggested that the concentration difference across the cecal wall is not favorable for absorption of sodium and that the cecum is a minor site for the absorption compared to the jejunum and the colon. So it may be that SCFA are readily absorbed from the cecum without a net movement of water. Sakata (1987) has also suggested that the cecum is not a major site for water absorption in the rat.

The water bound to colonic contents, measured by the suction method at 2 atm, can be considered to be unavailable for absorption by the colon, whereas the relatively free water removed by 2-atm pressure is considered available for absorption. The ability of the colonic contents to bind water against the dehydrating forces of the colon is important. However, there are a number of factors that may have contributed to the less strongly bound, or relatively free, water content in feces:

1. Fecal pellets from rats fed wheat bran were bulkier than those from pectin-fed rats. Increased fecal volume is believed to decrease transit time (Van Soest *et al.*, 1983). The longer the residence time of digesta in the colon, the longer there is for the absorption of water.
2. The absorption of 100 mmole SCFA is associated with the absorption of 360 ml water (Caspary *et al.*, 1981); therefore, increased production of SCFA results in a greater absorption of water from the colon.
3. The colon is a secretory organ, and different fibers may influence secretion into the gut in different ways. Hydroxy fatty acids produced by bacteria from undigested host fatty acids (Ammon and Phillips, 1973; Spiller *et al.*, 1985) and bile acids (Mekhjian *et al.*, 1971) have been shown to increase colonic secretion.

Wheat bran has a moderate capacity to bind water, but since it can be tolerated in large quantities in the diet and is only partially fermented, it contributes considerably to the total fecal water output (Fig. 1). However, other factors apart from the direct WHC of fiber, such as transit time, colonic secretion (including mucus), and SCFA, may contribute to fecal moisture and hence fecal bulking. Pectin increased the bacterial fraction of feces, which is an important water-holding component, but was unable to increase fecal bulk to the same extent as wheat bran, demonstrating that the bacterial fraction of feces may have made an important qualitative but not quantitative contribution to fecal bulking.

REFERENCES

Ammon, H. V., and Phillips, S. F., 1973, Inhibition of colonic water and electrolyte absorption by fatty acids in man, *Gastroenterology* **65**:744–749.

Broadribb, A. G. M., and Groves, C., 1978, Effect of bran particle size on stool weight, *Gut* **19**:60–63.

Caspary, W. F., Lembcke, B., and Elsehans, B., 1981, Bacterial fermentation of carbohydrates within the gastrointestinal tract, *Clin. Res. Rev.* **1**:107–117.

Cummings, J. H., and Branch, W. J., 1986, Fermentation and the production of short chain fatty acids in the human large intestine, in: *Dietary Fiber* (G. V. Vahouny and D. Kritchevsky, eds.), Plenum Press, New York, pp. 131–150.

Demigne, C., and Remesy, C., 1985, Stimulation of absorption of volatile fatty acids and minerals in the cecum of rats adapted to a very high fiber diet, *J. Nutr.* **115**:53–60.

Eastwood, M. A., and Brydon, W. G., 1985, Physiological effects of dietary fiber on the alimentary tract, in: *Dietary Fiber, Fiber-Depleted Foods and Disease* (H. Trowell, D. Burkitt, and K. Heaton, eds.), Academic Press, London, pp. 105–132.

Eastwood, M. A., Robertson, J. A., Brydon, W. G., and MacDonald, D., 1983, Measurement of water-holding properties of fibre and their faecal bulking ability in man, *Br. J. Nutr.* **50**:539–547.

Eastwood, M. A., Brydon, W. G., Baird, J. D., Elton, R. A., Smith, J. H., and Pritchard, J. L., 1984, Fecal weight and composition, serum lipids, and diet among subjects aged 18–80 years not seeking health care, *Am. J. Clin. Nutr.* **40**:628–634.

Elsenhans, B., Blume, R., and Caspary, W. F., 1981, Long-term feeding of unavailable carbohydrate gelling agents. Influence of dietary concentration and microbiological degradation on adaptive responses in the rat, *Am. J. Clin. Nutr.* **34**:1837–1848.

Hellendoorn, E. W., 1978, Fermentation as the principle cause of the physiological activity of indigestible food residue, in: *Topics in Dietary Fiber Research* (G. A. Spiller, ed.), Plenum Press, New York, pp. 127–168.

Hove, E. L., and King, S., 1979, Effects of pectin and cellulose on growth, feed efficiency, and protein utilisation, and their contribution to energy requirement and cecal VFA in rats, *J. Nutr.* **109**:1274–1278.

Kies, C., Sanchez, V. E., and Fox, H. M., 1984, Cellulose supplementation of a nutritionally complete, liquid formula diet: Effect on gastrointestinal tract function of humans and faecal fibre recovery, *J. Food Sci.* **49**:815–837.

Kurpad, A. V., and Shetty, P. S., 1986, Effects of antimicrobial therapy on faecal bulking, *Gut* **27**:55–58.

Leegwater, D. C., De Groot, A. P., and Van Kalmthout-Kuyper, M., 1974, The aetiology of caecal enlargement in the rat, *Food Cosmet. Toxicol.* **12**:687–697.

McBurney, M. I., Horvath, P. J., Jeraci, J. L., and Van Soest, P. J., 1985, Effect of *in vitro* fermentation using human faecal inoculum on the water-holding capacity of dietary fibre, *Br. J. Nutr.* **53**:17–24.

McConnell, A. A., Eastwood, M. A., and Mitchell, W. D., 1974, Physical characteristics of vegetable foodstuffs that could influence bowel function, *J. Sci. Food Agr.* **25**:1457–1464.

Mekhjian, K. S., Phillips, S. F., and Hofmann, A. F., 1971, Colonic secretion of water and electrolytes induced by bile acids: Perfusion studies in man, *J. Clin. Invest.* **50**:1569–1577.

Nyman, M., and Asp, N.-G., 1982, Fermentation of dietary fibre components in the rat intestinal tract, *Br. J. Nutr.* **47**:357–366.

Nyman, M., and Asp, N.-G., 1987, Chemistry and fermentative breakdown of dietary fibre in bulk-laxatives compared to some fibre containing foods, *Scand. J. Gastroenterol.* **22**(22):52–54.

Phillips, S. F., 1988, Physiology and pathophysiology of the large intestine and anal canal, in: *Diseases of the Colon, Rectum and Anal Canal* (J. B. Kirnser and R. G. Shorter, eds.), Williams & Wilkins, Baltimore, pp. 23–46.

Phillilps, S. F., and Giller, J. G., 1973, The contribution of the colon to electrolyte and water conservation in man, *J. Lab. Clin. Med.*, **81**:733–746.

Robertson, J. A., and Eastwood, M. A., 1981a, A method to measure the water-holding properties of dietary fibre using suction pressure, *Br. J. Nutr.* **46**:247–255.

Robertson, J. A., and Eastwood, M. A., 1981b, An investigation of the experimental conditions which could affect water-holding capacity of dietary fibre, *J. Sci. Food. Agr.*, **32**:819–825.

Sakata, T., 1987, Short-chain fatty acids and water in the hindgut contents and faeces of rats after hindgut bypass surgery, *Scand. J. Gastroenterol.* **22**:639–648.

Slavin, J. L., Nelson, N. L., McNamara, E. A., and Cashmere, K., 1985, Bowel function of healthy men consuming liquid diets with and without dietary fibre, *J. Parent. Ent. Nutr.* **9**:317–321.

Smith, A. N., Drummond, E., and Eastwood, M. A., 1981, The effect of coarse and soft Canadian red spring and French soft wheat bran on colonic motility in patients with diverticular disease, *Am. J. Clin. Nutr.* **34**:2460–2463.

Spiller, R. C., Brown, M. L., and Phillips, S. F., 1985, Segmental colonic function in experimental steatorrhea-decreased capitance of the proximal colon, *Gut* **26:**A1136–1137.

Spiller, G. A., Story, J. A., Wong, L. G., Nunes, J. D., Alton, M., Petro, M. S., Furumoto, E. J., Whittam, J. H., and Scala, J., 1986, Effect of increasing levels of hard wheat fiber on fecal weight, minerals and steroids and gastrointestinal transit time in healthy young women, *J. Nutr.* **116:**778–785.

Stephen, A. M., and Cummings, J. H., 1979, Water holding by dietary fibre *in vitro* and its relationship to faecal output in man, *Gut* **20:**722–729.

Stephen, A. M., and Cummings, J. H., 1980a, The microbial contribution to human faecal mass, *J. Med. Microbiol.* **3:**45–56.

Stephen, A. M., and Cummings, J. H., 1980b, Mechanism of action of dietary fibre in the human colon, *Nature* **284:**283–284.

Van Soest, P. J., 1982, in: *Nutrition Ecology of the Ruminant,* O. & B. Books, Oregon, pp. 199.

Walters, D. J., Eastwood, M. A., Brydon, W. G., and Elton, R. A., 1986, An experimental design to study colonic fibre fermentation in the rat: The duration of feeding, *Br. J. Nutr.* **55:**465–479.

Wyman, J. B., Heaton, K. W., Manning, A. P., and Wicks, A. C. B., 1976, The effect on intestinal transit and the faeces of raw and cooked bran in different doses, *Am. J. Clin. Nutr.* **29:**1476–1479.

13

Activities of Polysaccharide-Degrading Bacteria in the Human Colon

ABIGAIL A. SALYERS

1. INTRODUCTION

Polysaccharides in the human diet that are not digested and absorbed in the stomach and small intestine enter the colon, where they are available for fermentation by the complex and numerous population of bacteria that resides there. Most of the polysaccharides that reach the colon are plant cell wall polysaccharides, but some starch may also reach the colon (Englyst and Cummings, 1985). Plant polysaccharides are extensively degraded during passage through the colon and probably provide a major part of the carbon and energy required by colonic bacteria (Van Soest, 1978). Dietary polysaccharides are not the only type of carbohydrate that enters the colon. Goblet cells in the mucosa produce copious amounts of glycoprotein mucin, and the constant sloughing of cells from the mucosa of the small and large intestine releases tissue mucopolysaccharides into the lumen (Lennarz, 1980). Mucopolysaccharides are also found in meat.

Since at least one of the numerically predominant genera of colonic bacteria, *Bacteroides,* can ferment both host and plant polysaccharides (Salyers *et al.,* 1977a,b), it is necessary to take host products into account when considering polysaccharide breakdown by colon bacteria. One consequence of bacterial fermentation of host products is that fermentation of amino sugars produces ammonia as well as short-chain fatty acids and gases. It is thus possible that ammonia in the colon comes not only from proteolysis but also from fermentation of host mucins and mucopolysaccharides. In fact, one effect of dietary fiber could be to lessen the reliance of colon bacteria on host products and thus reduce ammonia levels in the colon. Another possible consequence of bacterial digestion of glycoprotein mucins is that it might reduce the protective capacity

ABIGAIL A. SALYERS • Department of Microbiology, University of Illinois, Urbana, Illinois 61801.

of the mucin lining of the colon and increase the exposure of the mucosa to bacteria and toxic substances.

2. ASSESSING THE EXTENT TO WHICH HOST POLYSACCHARIDES ARE DIGESTED BY COLONIC BACTERIA

To assess the importance of host product fermentation, it is necessary to determine to what extent host polysaccharides are actually utilized in the colon. In the case of plant polysaccharides, an estimate of the extent of bacterial digestion could be made by comparing the amount of polysaccharide ingested with the amount excreted in feces (Van Soest, 1978). Results of this analysis showed that most plant polysaccharides were extensively digested in the colon. Such a comparison is technically much more difficult in the case of mucopolysaccharides and glycoprotein mucins, and no detailed comparison of production versus excretion of host polysaccharides has ever been attempted.

The closest approximation to such a study was made by Vercellotti et al. (1977). They measured measured mucin sugars (glucosamine, galactosamine, and fucose) in a high-molecular-weight fraction from intestinal contents from accident victims. Levels of these mucin sugars were high in ileal contents but low in contents taken from all parts of the colon. Only soluble polysaccharides were considered in this study, and the percentage of total carbohydrate accounted for in the compositional analysis was low. Nonetheless, these results indicate that at least some host polysaccharides are digested in the colon.

Assessment of the extent to which host polysaccharides are fermented by colon bacteria is complicated not only by methodological difficulties but also by the fact that the amount of host polysaccharide produced is probably influenced by the composition of the host's diet. For example, fiber in the diet affects the rate of turnover of mucosal cells (Vahouny and Cassidy, 1986). Fiber may also affect the amount of glycoprotein mucin secreted by the goblet cells.

Still another factor that needs to be kept in mind is that the amounts of various host polysaccharides in the colon may be less important than their solubility and accessibility. Solubility is known to be an important factor in fermentation of dietary polysaccharides (Ehle et al., 1982). Soluble plant polysaccharides such as pectin and some hemicelluloses are more readily fermented by colon bacteria than insoluble hemicelluloses and cellulose. Mucopolysaccharides such as chondroitin sulfate are quite soluble and are rapidly fermented by bacteria. Low concentrations of these substances might be fermented preferentially, even if higher levels of other less easily fermented compounds were available. Bacterial preference for the amino-sugar-containing mucopolysaccharides could also be enhanced because these polysaccharides can serve as a source of nitrogen as well as carbon (A. Salyers, unpublished data). The highly branched glycoprotein mucins are relatively soluble but may be less easily degraded because of their high viscosity and because of the number of different sugars and carbohydrate linkages they contain.

A more direct way of assessing bacterial utilization of polysaccharides is to incubate fecal suspensions with different polysaccharide substrates and measure the rate and extent of digestion. Hoskins and Boulding (1981) have used this approach to show that there are bacteria in the colon that can degrade glycoprotein mucins. Although studies of this sort are very useful for gaining an overview of the metabolic capabilities of the microflora as a whole, they do not necessarily prove that the organisms in the fecal suspension were actually degrading the polysaccharide being tested unless the bacteria responsible for the digestion can only utilize that one type of polysaccharide. Bacteria such as the *Bacteroides,* which can use a variety of dietary and host polysaccharides and can adapt quickly to new substrates, will simply switch from whatever they were utilizing in the colon to whatever is provided in the assay mixture. To assess the *in vivo* activities of such bacteria, it will be necessary to develop methods for making direct measurements on colon contents that will not allow bacteria to change substrates. The sections below summarize some of our recent work on developing methods for determining what polysaccharides are being used by *Bacteroides* in the colon, with a particular emphasis on the question of whether host products are significant sources of carbon and energy for these bacteria.

3. ABILITY OF COLONIC *BACTEROIDES* TO USE HOST POLYSACCHARIDES AS A SOURCE OF CARBOHYDRATE

It is easy to show that a number of *Bacteroides* strains could utilize tissue mucopolysaccharides such as chondroitin sulfate and hyaluronic acid (Salyers *et al.,* 1977a), because these polysaccharides are commercially available in quantities sufficient for bacteriological testing. However, it is not possible at present to obtain adequate quantities of undegraded human or animal goblet cell mucin. Thus, when we surveyed *Bacteroides* strains for the ability to utilize mucin, we had to use substrates with similar structures, such as porcine gastric mucin and bovine submaxillary mucin, as substitutes. Similarly, Hoskins and Boulding (1981) used commercially available mucins rather than human goblet cell mucin to look for mucin-degrading activity in human feces. Only a few strains of *Bacteroides* could utilize porcine gastric mucin or bovine submaxillary mucin as a sole source of carbohydrate (Salyers *et al.,* 1977a). Nor did any of the other genera of colon anaerobes we surveyed exhibit the ability to degrade mucin (Salyers *et al.,* 1977b).

This does not necessarily rule out goblet cell mucin as a carbohydrate source for *Bacteroides* in the colon. First, human goblet cell mucin may be more digestible than porcine gastric mucin or hog submaxillary mucin. Second, many of the strains that could not grow to high numbers when mucin was the only substrate could utilize individual mucin sugars such as L-fucose and N-acetylhexosamines (Salyers *et al.,* 1977a,b), and they produce glycosidases that could cleave some linkages in mucin. It is possible that these bacteria carry out only a partial digestion of the carbohydrate portion of goblet cell mucin. Partial digestion would have been scored as a negative in our survey but would have been registered in the approach used by Hoskins and Boulding (1981). More recently, Roberton and Stanley (1982) used a more sensitive and quantitative approach to show that *Bacteroides fragilis* can partially degrade glycoprotein mucins.

Demonstrating that a certain species of bacteria isolated from the human colon can ferment a particular polysaccharide when grown in laboratory medium only indicates the metabolic capabilities of that species. It does not prove that this species of bacteria actually utilizes the polysaccharide in the colon because in the colon there may be other, preferred substrates available. Similarly, the fact that an organism grows faster in laboratory medium on one polysaccharide than another does not predict which polysaccharide will be used in the colon, where differences in concentrations of the polysaccharides or association with other compounds (e.g., in the plant cell wall matrix) may affect the organism's choice.

4. ASSESSING POLYSACCHARIDE UTILIZATION IN THE GASTROINTESTINAL TRACTS OF GERM-FREE MICE

Our first approach to determining what polysaccharides are most important as carbohydrate sources for *Bacteroides in vivo* has been to generate mutants that are deficient in the ability to utilize particular polysaccharides or groups of polysaccharides and to test these mutants for the ability to compete with the isogenic wild-type parent strain in the gastrointestinal tracts of germ-free mice. Germ-free mice are mice that have been delivered by caesarean section and reared in a completely sterile environment. These animals have no resident microflora and can be colonized with a defined mixture of bacterial strains. We colonized the mice with a mixture of wild type and mutant and then took fecal samples at intervals and determined the ratio of wild type to mutant in these samples. If the mutation is deleterious *in vivo*, the wild type will outcompete the mutant and will soon become the predominant strain in the feces. If the mutation is not deleterious, the wild type and the mutant will maintain their original balance.

A summary of the results of these experiments is given in Table I. Surprisingly, very few mutations that affected polysaccharide utilization had any effect *in vivo;* i.e., the mutant was not outcompeted by the wild type. Two of the mutations that were deleterious *in vivo* were mutations that interfered with the organism's ability to utilize the mucopolysaccharide chondroitin sulfate. One of these mutants, which was unable to utilize chondroitin sulfate because it could not utilize hexosamines, would also have been deficient in the ability to utilize glycoprotein mucins. However, mutants that were unable to utilize L-fucose (a sugar that is found exclusively in glycoprotein mucins) were not outcompeted by wild type. Thus, the deleterious effect of the hexosamine-minus mutation *in vivo* seems to be associated with inability to use chondroitin sulfate rather than mucins.

These results support the hypothesis that host mucopolysaccharides are a source of carbohydrate for *Bacteroides* in the intestinal tract. However, it should be noted that there was one chondroitin-sulfate-minus mutant that was not outcompeted by the wild type *in vivo*. We now know that this mutant differs from the mutant that was outcompeted in that it is unable to induce any of the enzymes needed for chondroitin sulfate utilization. The mutant that was outcompeted *in vivo* responds to chondroitin sulfate by producing all the enzymes but cannot gain carbon and energy from chondroitin sulfate. The inability to ignore chondroitin sulfate *in vivo*, which causes it to expend energy on proteins that do not confer any benefit, may be the source of its inability to compete with wild type *in vivo*. However, there may be some as yet undiscovered consequence of this mutation that is the real cause of its effect *in vivo*.

Two other mutations affected survival *in vivo*. One of these mutations interfered with utilization of uronic acids, and the other interfered with the ability to utilize galactose. Interestingly, both of these mutations caused the organisms to be deficient in utilization of two or more types of polysaccharide, whereas the other mutations interfered with utilization of only one type of polysaccharide. With the exception of one of the chondroitin-sulfate-minus mutants, all of the mutants that were deficient in utilization of only one type of polysaccharide were able to compete with the wild type. This finding supports the hypothesis that *Bacteroides* are scavengers that use small amounts of a variety of polysaccharides. For such organisms, loss of the ability to use a single polysaccharide would not be particularly deleterious because they could fall back on other carbohydrate sources.

Experiments such as those described above are useful for gaining a general idea of what polysaccharides might be important *in vivo*, but results must be interpreted with great caution. First, the murine gastrointestinal tract may differ in subtle ways from the human gastrointestinal tract. Second, the absence of the rest of the microflora in germ-free mice colonized with one or two strains may make it easier for mutants to survive. Ideally, we want to be able to test conclusions drawn from such studies by measuring activities of bacteria directly in human fecal specimens. Such a procedure should not

TABLE I. Importance of Different Polysaccharide Substrates in a Simple *in Vivo* Model System[a]

Mutant phenotype[b]	Polysaccharides affected[c]	Outcome of competition[d]
Amylose⁻	Amylose Amylopectin Pullulan Maltodextrins	Not outcompeted
Galactose⁻	Arabinogalactan Raffinose, stachyose Goblet cell mucin	Outcompeted
PGA⁻, galacturonate⁺	Pectin	Not outcompeted
PGA⁻, galacturonate⁻	Pectin Mucopolysaccharides	Outcompeted
CS⁻, hexosamine⁻	Chondroitin sulfate Goblet cell mucin	Outcompeted
CS⁺/⁻, hexosamine⁺	Chondroitin sulfate	Outcompeted
CS⁺/⁻, hexosamine⁺[e]	Chondroitin sulfate	Not outcompeted
L-Fucose⁻	Goblet cell mucin	Not outcompeted

[a]Mutants of *B. thetaiotaomicron* that are defective in the ability to utilize one or more polysaccharides were competed with wild type *B. thetaiotaomicron* in the intestinal tracts of germ-free mice (Salyers et al., 1988; Salyers and Pajeau, 1989).
[b]PGA, polygalacturonic acid; CS, chondroitin sulfate.
[c]Polysaccharides the mutant is unable to use or uses less efficiently.
[d]A mutant that is not outcompeted by wild type is equally able to grow *in vivo*. A mutant that is outcompeted by wild type has a diminished ability to grow *in vivo*.
[e]Regulatory mutant.

involve growth of the fecal bacteria because *Bacteroides* polysaccharide-degrading enzymes are regulated (Salyers, 1984). Thus, growth on laboratory medium allows the bacteria to synthesize new enzymes that are appropriate for the growth medium and to lose enzymes that were synthesized during growth in the colon.

We decided to take advantage of the fact that the polysaccharide-degrading enzymes and membrane proteins involved in polysaccharide uptake are produced at high levels only when *Bacteroides* are grown on the inducing polysaccharide and to use regulated proteins as indicators of what the bacteria were doing in the colon. Since we had antiserum that detected a chondroitin-sulfate-induced outer membrane protein, we decided to use this antiserum to probe a Western blot containing bacteria taken directly from the ceca of germ-free mice colonized with *Bacteroides*. The results indicated that although there was evidence for a low level of induction, the bacteria growing in the animal were not producing this chondroitin-sulfate-induced protein at fully induced levels (Salyers *et al.*, 1988). This finding indicates that chondroitin sulfate is not the major source of carbohydrate for *Bacteroides* growing in the animal's gastrointestinal tract but does not rule out the possibility that chondroitin sulfate is being utilized at low levels. Taken together with the results of the competition experiments, this finding supports the hypothesis that *Bacteroides* are utilizing low levels of a variety of polysaccharides *in vivo* rather than relying mainly on a single source of carbon and energy.

5. ASSESSING POLYSACCHARIDE UTILIZATION IN HUMAN COLON CONTENTS

We have used an antibody probe to determine if an α-glucosidase activity that was detected in a bacterial fraction from human feces was being produced by *Bacteroides vulgatus* (McCarthy *et al.*, 1988). We thought that starch might be a substrate for this species because, unlike the other major *Bacteroides* species, *B. vulgatus* has a relatively narrow substrate spectrum. Starch is the polysaccharide that supports the best growth in laboratory medium. The *B. vulgatus* α-glucosidase is induced at least 20-fold when bacteria are growing on starch as their sole source of carbohydrate. Thus, if the enzyme was being produced at induced levels in the colon, this would indicate that *B. vulgatus* was utilizing starch. Our results indicate that the α-glucosidase detected in feces was not being produced by *B. vulgatus*. Moreover, since we should have been able to detect the *B. vulgatus* α-glucosidase if it was fully induced, our results indicate that *B. vulgatus* was not using high enough levels of starch in the colon to fully induce this enzyme.

Previously, we had used properties of *Bacteroides* polygalacturonases to show that polygalacturonic acid is probably not a major source of carbon and energy for *Bacteroides* in the colon (McCarthy and Salyers, 1986). Polygalacturonic acid, like starch, is very rapidly fermented by *Bacteroides* and supports good growth of the organisms in laboratory medium. Our results show why it is risky to extrapolate from the results of experiments involving growth of bacteria in laboratory medium to what the organisms do *in vivo*. So far the results of our studies of enzymes in human feces do not contradict the hypothesis formulated on the basis of the competition studies in germ-free mice, i.e., that *Bacteroides* utilize a variety of polysaccharides in the colon rather than relying on one or two as a major source of carbon and energy. However, to obtain conclusive proof of this hypothesis, we will have to test for utilization of other polysaccharides, especially those such as arabinogalactan that may be more abundant in the colon than starch and polygalacturonic acid. Also, we will have to rule out the possibility that for some reason induction of enzymes never reaches *in vivo* the high levels that are attainable in laboratory medium.

Although our results to date do not provide conclusive proof of *Bacteroides* polysaccharide utilization patterns *in vivo*, we think the use of antibody probes to detect expression of polysaccharide-regulated proteins in colon contents is a promising approach to directly monitoring bacterial activities *in vivo*. If successful, this approach could be applied to determining whether *Bacteroides* use host products in the human colon and how polysaccharide utilization patterns are affected by changes in the host's diet.

ACKNOWLEDGMENTS. This work was supported by grant AI 17876 from the National Institutes of Health.

REFERENCES

Ehle, F. R., Robertson, J. B., and Van Soest, P. J., 1982, Influence of dietary fiber on fermentation in the human large intestine, *J. Nutr.* **112:**158–166.

Englyst, H. N., and Cummings, J. S., 1985, Digestion of the polysaccharides in some cereal foods in the human small intestine, *Am. J. Clin. Nutr.* **42:**778–787.

Hoskins, L. C., and Boulding, E. T., 1981, Mucin degradation in the human colonic ecosystems, *J. Clin. Invest.* **67:**163–172.

Lennarz, W. J. (ed.), 1980, *The Biochemistry of Glycoproteins and Proteoglycans,* Plenum Press, New York.

McCarthy, R. E., and Salyers, A. A., 1986, Evidence that polygalacturonic acid is not an important substrate for *Bacteroides* species growing in the human colon, *Appl. Environ. Microbiol.* **52:**9–11.

McCarthy, R. E., Pajeau, M. P., and Salyers, A. A., 1988, Importance of starch as a carbohydrate source for *Bacteroides vulgatus* growing in the human colon, *Appl. Environ. Microbiol.* **54:**1911–1916.

Roberton, A. M., and Stanley, R. A., 1982, *In vitro* utilization of mucin by *Bacteroides fragilis, Appl. Environ. Microbiol.* **43:**325–330.

Salyers, A. A., 1984, *Bacteroides* of the human lower intestinal tract, *Annu. Rev. Microbiol.* **38:**293–313.

Salyers, A. A., and Pajeau, M. P., 1989, Importance of dietary and host polysaccharides as substrates for *Bacteroides thetaiotaomicron* growing in the intestinal tracts of exgermfree mice, *Appl. Environ. Microbiol.* (in press).

Salyers, A. A., Vercellotti, J. R., West, S. E. H., and Wilkins, T. D., 1977a, Fermentation of mucin and plant polysaccharides by strains of *Bacteroides* from the human colon, *Appl. Environ. Microbiol.* **33:**319–322.

Salyers, A. A., West, S. E. H., Vercellotti, J. R., and Wilkins, T. D., 1977b, Fermentation of mucin and plant polysaccharides by anaerobic bacteria from the human colon, *Appl. Environ. Microbiol.* **34:**529–533.

Salyers, A. A., Pajeau, M. P., and McCarthy, R. E., 1988, Importance of mucopolysaccharides as substrates for *Bacteroides thetaiotaomicron* growing in the intestinal tracts of exgermfree mice, *Appl. Environ. Microbiol.* **54:**1970–1976.

Vahouny, G. V., and Cassidy, M. M., 1986, Dietary fiber and intestinal adaptation, in: *Dietary Fiber:*

Basic and Clinical Aspects (G. V. Vahouny and D. Kritchevsky, eds.), Plenum Press, New York, pp. 181–210.

Van Soest, P. J., 1978, Dietary fibers: Their definition and nutritional aspects, *Am. J. Clin. Nutr.* **31:** S12–S20.

Vercellotti, J. R., Salyers, A. A., Bullard, W. S., and Wilkins, T. D., 1977, Breakdown of mucin and plant polysaccharides in the human colon, *Can. J. Biochem.* **55:**1190–1196.

14

The Influence of Dietary Fiber on Microbial Enzyme Activity in the Gut

IAN R. ROWLAND and ANTHONY K. MALLETT

1. CHARACTERIZATION OF THE GUT MICROFLORA

The mammalian gut microflora is a highly complex community of microorganisms that exhibits a large degree of species diversity. This diversity, coupled with the cumbersome procedures necessary to identify the predominant (anaerobic) components of the flora (Holdeman and Moore, 1975), makes the analysis and characterization of the flora by conventional microbiological methods extremely difficult and time consuming. Thus, studies of dietary modification of the flora need to be based on simpler, quicker methods. Measurement of microbial enzyme activities in suspensions of gut luminal contents or feces provides such a method and has been used to assess the influence of diet (Goldin and Gorbach, 1976; Rowland et al., 1985), exogenous microorganisms (Goldin et al., 1980), and disease (Mastromarino et al., 1976) on the gut microflora. By choosing enzymes that are known to participate in the detoxification of foreign compounds or in their activation to toxic, mutagenic, or carcinogenic metabolites, it is possible to gain insight into the likely consequences for the health of the host animal of the changes induced.

Some of the enzymes that can be used in this way are summarized in Table I together with some examples of the toxicological consequences of their actions. Many of the enzymes listed have a broad specificity, metabolizing a wide range of compounds of similar basic structure. For example, β-glycosidase hydrolyzes a variety of plant glycosides to release aglycones, which are toxic (e.g., cyanide from amygdalin; Hill et al., 1980), mutagenic (e.g., quercetin from rutin; Brown and Dietrich, 1979), or carcinogenic (e.g., methylazoxymethanol from cycasin; Laqueur and Spatz, 1968). Simi-

IAN R. ROWLAND and ANTHONY K. MALLETT • The British Industrial Biological Research Association, Carshalton, Surrey SM5 4DS, United Kingdom.

TABLE I. Bacterial Enzymes of Toxicological Importance

Enzyme	Substrate (example)	Consequence of reaction for host
β-Glycosidase	Amygdalin	Cyanide toxicity
β-Glucuronidase	Diethylstilbestrol glucuronide	Increased retention of steroids in body
Nitroreductase	Dinitrotoluene	Tumor induction
Azoreductase	Benzidine dyes	Induction of bladder tumors
Nitrate reductase	Nitrate	Methemoglobinemia
Urease	Urea	Ammonia toxicity

larly, β-glucuronidase hydrolyzes hepatic glucuronic acid conjugates of many toxic xenobiotics, e.g., benzo(a)pyrene (Renwick and Drasar, 1976), that are secreted into the gut via the bile. The enzyme is known to be important in the induction of colon tumors in rats by dimethylhydrazine. The bacterial hydrolysis of these polar conjugates results in the regeneration of the (usually) less polar parent molecule, which can be reabsorbed, returning to the liver. This enterohepatic recirculation leads to increased retention of foreign compounds in the body, thus potentiating their pharmacological or toxicological activities (Williams *et al.*, 1965). The role of these and other bacterial enzymes in toxicity of xenobiotics has been reviewed in detail elsewhere (Williams, 1972; Goldman, 1982; Rowland and Walker, 1983; Simon and Gorbach, 1987).

In our laboratory, these bacterial enzymes (Table I) are usually assayed by incubating buffered suspensions of freshly collected cecal contents or feces under anaerobic conditions with model chromogenic substances (Wise *et al.*, 1982; Coates *et al.*, 1988).

2. EFFECTS OF DIETARY FIBER ON GUT MICROFLORA

The term dietary fiber covers a wide range of indigestible plant material of complex chemical constitution (Selvendran, Chapter 1, this volume). The source of the fiber and any processing that it has undergone markedly influence its physicochemical characteristics and its physiological effects on the intestinal tract. Even purified forms of fiber can vary in chemical composition with concomitant changes in physiological effects on the gut. For example, pectins derived from different sources vary in their degree of methoxylation, which influences their gelling capacity and their degree of fermentability by intestinal bacteria. It is not surprising, therefore, that the mammalian gut microflora

TABLE II. Possible Effects of Dietary Fiber on Gut Microflora

1. Bulking effect of undergraded fiber may dilute gut contents and decrease enzyme activities.
2. Fiber may inhibit, directly or indirectly, bacterial metabolism.
3. Fiber may increase level of nutrients in colon by altering digestion or by adsorbing nutrients in duodenum or ileum.
4. Fermentation of fiber components may provide energy for bacterial growth and/or metabolism.
5. Products from fermented fiber may alter physicochemical conditions in gut, e.g., pH.

is influenced in different ways and to different extents by the various forms of dietary fiber. These are summarized in Table II and discussed below.

2.1. Bulking Effect of Fiber on Gut Contents

Many types of dietary fiber have the capacity to cause bulking of the gut contents (apparent as increased fecal mass in humans) by the physical presence of the fiber or as a consequence of the ability of the fiber or its breakdown products to absorb water (Leegwater *et al.*, 1974; Cummings, 1982). In other cases, fecal bulking results from increased microbial mass following the utilization of fiber as a substrate for bacterial growth in the large intestine (Cummings *et al.*, 1981). Finally, there are nonfermentable fibers, such as crystalline cellulose, that have only minimal fecal bulking properties. They provide little or no energy for bacterial metabolism or growth, and their presence in the gut usually results in decreases in the concentration of bacteria. For example, when added to a purified diet at levels of up to 40% by weight, cellulose causes a dose-related decrease in bacterial numbers and cecal enzyme activities concomitant with the increase in weight of cecal contents (Mallett *et al.*, 1983; Shiau and Chang, 1983).

Although not all types of dietary fiber cause bulking of gut contents (in humans, pectin and guar gum are notable exceptions), it is a common feature with many fibers and underlies many of the other effects of fiber on the flora outlined in Table II.

2.2. Inhibition of Bacterial Metabolism by Fiber

With some types of dietary fiber the decrease in enzyme activities is markedly in excess of that expected simply from dilution of the bacteria by bulking of the gut contents. Such is the case with the algal polysaccharide carrageenan. Carrageenan is a polymer of sulfated galactose and anhydrogalactose and is not degraded by the gut microflora (Hawkins and Yaphe, 1965). When fed to rats in a purified rodent diet at the level of 5% by weight, the polysaccharide caused a striking reduction in activity of all enzymes studied, particularly nitroreductase activity, which was decreased by over 95% (Rowland *et al.*, 1983b; Mallett *et al.*, 1984, 1985). By comparison, cellulose at the same dietary level had no effect on enzyme activities. The decreases in the activities, which were common to all three major forms of the algal polysaccharide (iota, lambda, and kappa), were associated with reductions in bacterial numbers, with all the main types of bacteria being affected (Mallett *et al.*, 1985). *In vitro* studies suggest that carrageenans have no direct effects on viability of bacterial cells, and it is possible that the mechanism of inhibition is via an influence on immunologic status of the host animal. In rats fed 5% dietary *i* or *k* carrageenan, an increased titer of biliary IgA was seen (Mallett *et al.*, 1985; Table III). The IgA is the predominant immunoglobulin secreted into the gut, where it may function as a barrier against bacterial invasion (Beinenstock, 1979).

2.3. Increase in Nutrient Levels in Large Bowel

Carboxymethylcellulose is a modified cellulose used as a thickening agent in many processed foods (Glicksman, 1969). It has considerable water-holding capacity and,

TABLE III. Influence of Carrageenans on Rat Cecal Microflora
and Biliary Antibodies

	Diet[a]		
	Fiber-free	i-Carrageenan (5%)	k-Carrageenan (5%)
Cecal contents	1.0 ± 0.1	2.1 ± 0.5***	2.7 ± 0.4***
Bacterial/g cecal contents ($\log_{10}$)	11.18 ± 0.12	9.88 ± 0.21***	9.72 ± 0.20***
Biliary IgA/ml bile	360 ± 81	300 ± 79	269 ± 128
Biliary agglutination response ($\log_2$ dilution)	2.6 ± 0.8	3.9 ± 0.5*	4.3 ± 0.9**

[a]Values shown are means ± S.D. ($n = 8$). Asterisks indicate that result is significantly different from control (fiber-free): *$P < 0.05$; **$P < 0.01$; ***$P < 0.001$. Rats were fed purified diets shown for 4 weeks.

when fed (5% w/w) in a purified diet to rats, induces large increases in weight of cecal contents (Mallett *et al.*, 1984; Table IV). Unlike cellulose or carrageenan, however, activities of rat cecal bacterial enzymes are generally increased, in the case of β-glucuronidase by eightfold over the control, fiber-free diet group. Carboxy-methylcellulose would not appear to be a substrate for bacterial fermentation (Frawley *et al.*, 1964) and so cannot be providing energy for increased bacterial metabolism in this way. A more likely explanation is that the ion-exchange properties of the polysaccharide allow it to adsorb nutrients in the small intestine, preventing their absorption from the

TABLE IV. Influence of Carboxymethylcellulose (CMC)
on Rat Cecal Enzyme Activities

	Enzyme activity (μmole/hr per cecum)[a]	
Enzyme	Fiber-free diet	5% (w/w) CMC diet
β-Glucosidase	15 (9–29)	123*** (43–270)
β-Glucuronidase	14 (10–33)	41** (33–144)
Urease	79 (35–466)	422*** (146–591)
Azoreductase	4.7 (3–8)	11** (7–19)
Nitroreductase	1.1 (0.6–1.7)	1.8** (1.5–2.4)
Nitrate reductase	11 (5–17)	51*** (38–128)
Cecal contents (g)	1.7 (1.4–2.1)	8.1*** (5.4–11.0)

[a]Results shown are medians for six rats (range in parentheses) fed diets for 4 weeks. Those marked with asterisks differ significantly from control (fiber-free): **$P < 0.01$; ***$P < 0.001$ (Mann–Whitney U test).

gut and transporting them to the cecum where they can be utilized by bacteria as a source of energy for increased growth and metabolism.

It is feasible that other plant polysaccharides may act, at least in part, in a similar fashion to carboxymethylcellulose by providing nutrients indirectly to the cecal microflora. By decreasing transit time, for example, some fibers may reduce the time available for digestive enzymes to act, thus allowing partially digested material to reach the lower bowel. Some fibers may stimulate mucin secretion in the gut. Mucin is known to be an important nutrient source for many intestinal organisms (McCarthy and Salyers, 1988), and increased levels in the large intestine might be expected to increase biomass and metabolism.

2.4. Fermentation of Dietary Fiber

Energy derived from the microbial degradation of highly fermentable fibers such as pectins and hemicelluloses can be channeled into bacterial growth and metabolism. In contrast to carboxymethylcellulose, which increased the activity of all the enzymes studied, fermentable fibers like pectin tend to have more specific effects.

Pectins consist predominantly of polymerized galacturonic acid with galactose, arabinose, and rhamnose subunits. The degree of methylation of the carboxyl groups of the galacturonic acid backbone differs considerably in different types of plant material yielding polysaccharides with different gelling properties and susceptibility to fermentation. We have used a pectin derived from citrus fruits, which has a high methoxyl content with approximately 70% of the carboxyl groups methylated.

When fed to rats at a level of 5% in a purified diet, such pectin caused a modest increase in cecal size and had little or no effect on bacterial numbers (Table V) or on the proportions of the major types of organisms in the cecum (Mallett, 1982). Of the bacterial enzymes studied, most were not greatly affected by pectin feeding, although β-glucuronidase was increased, particularly when higher dietary pectin levels (10%) were fed (DeBethizy et al., 1983; Rowland et al., 1983a). The most marked effect associated with dietary pectin (5% w/w) was a six- to ten-fold increase in the rate of reduction of

TABLE V. Effects of Pectin on Cecal Bacteria and Enzyme Activities in Rats[a]

	Control (fiber-free diet)	Pectin (50g/kg diet)
Cecal contents (g)	1.01 ± 0.04	1.73 ± 0.11***
Bacteria ($\times 10^{10}$/g cecal contents)	3.72 ± 0.25	3.69 ± 0.36
β-Glucuronidase	14.5 ± 1.5	27.1 ± 1.9***
β-Glucosidase	8.0 ± 1.0	4.9 ± 0.6*
Azoreductase	1.9 ± 0.1	1.5 ± 0.1*
Nitroreductase	1.4 ± 0.1	1.3 ± 0.2
Nitrate reductase	8.0 ± 1.1	47.4 ± 4.7***

[a]Enzyme activities (μmole/hr per gram cecal contents) are shown as means ± S.E. for at least six rats fed the diet for 4 weeks. Asterisks indicate that the result was significantly different from the control (*$P < 0.05$; ***$P < 0.001$).

nitrate to nitrite by cecal contents (Table V; Wise *et al.*, 1982). This increase in nitrate reductase activity was dependent on the degree of methoxylation of the polysaccharide, since the effect was diminished when low (30%) methoxyl pectin was fed and was absent in rats given the nonmethoxylated pectic acid (Conning *et al.*, 1984).

Presumably, the changes induced by pectin are associated with the microorganisms that are proficient at fermenting the fiber, although no numerical differences in bacterial species composition were seen in these studies (Wise *et al.*, 1982). An alternative

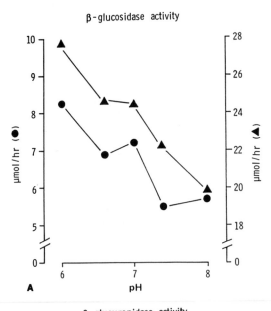

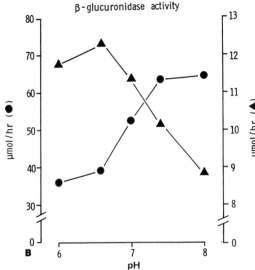

FIGURE 1. Activity of bacterial enzymes in samples of human feces (▲) or rat cecal contents (●) incubated with buffers of different pH. Enzyme activity is expressed as μmole product formed/hr per 10^{11} bacteria, and results shown are means for three humans and three rats. A, β-glucosidase; B, β-glucuronidase; C, nitrate reductase.

explanation may be the specific induction of enzymes, such as those involved in nitrate reduction, responsible for the efficient anaerobic degradation and assimilation of carbohydrate in the gut ecosystem (Hassan and Hall, 1975; Ota, 1982).

As might be expected, fibers such as pectin that appear to depend on fermentation for their effects on bacterial enzyme activities tend to exhibit species-specific effects, presumably because of the differences in gut flora composition between animals (see Section 3.1).

2.5. Changes in Physicochemical Conditions in the Gut

Fermentation of polysaccharides by bacteria in the large intestine generates short-chain fatty acids, which can lead to a lowering of luminal pH (Cummings *et al.*, 1986). Thus, individuals consuming a vegan diet and therefore having a high intake of dietary fiber have a lower fecal pH (6.6) than when eating a mixed "Western" diet (pH 7.2; van Dokkum *et al.*, 1983). Similarly, in rats, dietary fiber has been shown to decrease colonic pH (Jacobs and Lupton, 1986). In addition, it has been suggested that colonic pH is an important factor in cancer of the large bowel, since colon cancer patients have a higher fecal pH than healthy individuals (Thornton, 1981), although whether this is a predisposing factor or one that develops after the onset of the disease has not been established (Samelson *et al.*, 1985). A change in hydrogen ion concentration in the gut milieu could have important sequelae for bacterial metabolism in the hind gut, either by altering the expression of bacterial enzyme activities or, over a longer period of time, modifying the composition of the flora itself. We have recently investigated the first of

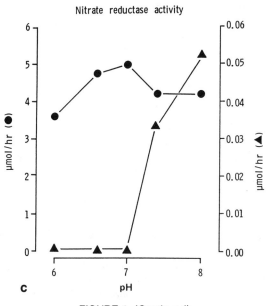

FIGURE 1. (*Continued*)

these alternatives by studying the effect of altering the pH (between 6.0 and 8.0) during the assay of enzyme activities in cecal suspensions from the rat (A. K. Mallett and I. R. Rowland, unpublished observations, 1988). The three enzymes studied exhibited distinct pH profiles with β-glucuronidase showing an increase and β-glycosidase an overall decrease in activity as pH of incubation increased. In contrast, nitrate reduction was optimal at pH 7.0 (Fig. 1). It is therefore difficult to generalize about the influence of luminal pH on microbial metabolism in the large intestine. However, the low activity of cecal β-glucuronidase at acidic pH does at least correlate with the observed lower incidence of dimethylhydrazine-induced colon tumors in rats with low fecal pH (Samelson *et al.*, 1985). As mentioned above (Section 1), dimethylhydrazine is a colon carcinogen that is dependent on β-glucuronidase for its action.

3. SPECIES DIFFERENCES IN RESPONSE TO FIBER

3.1. Interspecies Differences in Microflora

There are marked differences between animal species in the microbial composition of the cecal or fecal flora, presumably related to differences in the biotic and abiotic conditions (e.g., pH, supply of endogenous nutrients, peristalsis) in the intestinal tract (reviewed by Drasar, 1988). This microbial variability is reflected in interspecies differences in microbial enzyme activity in the gut. We have studied reductive and hydrolytic enzyme activities associated with the gut flora of six animal species: rat, mouse, guinea pig, hamster, marmoset, and man (Rowland *et al.*, 1986). Significant differences in microbial metabolic activity were apparent, although none of the species had consistently higher enzyme activities than any other. Furthermore, none of the laboratory animals, including the marmoset, possessed a profile of gut flora enzyme activities similar to that of man, indicating that extreme care should be taken when extrapolating data on microbial xenobiotic metabolism from animals to man.

3.2. Effect of Fiber in Different Animal Species

In view of the differences between the gut microfloras of animal species discussed above, it seems likely that the influence of dietary fiber on microbial activity would also vary with species, particularly for those fibers that exert their effects after fermentation. The response of the rat gut flora to dietary pectin has already been detailed (Section 2.4). When fed in a purified diet to mice, pectin elicited a similar response, having little or no effect on most enzyme activities apart from β-glucuronidase, which was increased slightly, and nitrate reductase, which increased by about sevenfold, by comparison to mice fed the fiber-free control diet (Rowland *et al.*, 1983a). In contrast, when fed to hamsters, pectin had no effect on reduction of nitrate or on hydrolysis of glucuronides (Rowland *et al.*, 1983a).

Recently, we have extended our studies of the influence of dietary pectin on the gut microflora to man (Mallett *et al.*, 1988). Six volunteers consumed a mixed, free-choice diet (control) and then the same diet with additional high-methoxyl pectin (18 g/day) for a period of 3 weeks before returning to the control diet for a further 3 weeks. Microbial enzyme activities were measured in fecal samples collected during the final week of each dietary period. The results, summarized in Table VI, show that dietary pectin decreased the activities of β-glucosidase and β-glucuronidase and had no significant effect on nitrate reductase. Thus, the human gut flora responds differently to dietary pectin than does the flora of the rat and mouse.

It is possible that the differences in response to pectin of the rat and human may be a consequence of the basal diet fed. In the human study the volunteers ate a mixed diet containing 16–21 g fiber, which was then supplemented with pectin. The basal diet used in the rat studies was a much simpler purified diet containing no fiber. When incorporated into a stock laboratory diet (which contained approximately 16% fiber), pectin did not significantly increase nitrate reductase activity in the rat (Wise *et al.*, 1986). However, there is also good evidence that the human flora is intrinsically unable to respond to dietary pectin. It is possible to maintain the human gut microflora in association with the rat by administering a suspension of human feces to germ-free animals under isolator conditions (Mallett *et al.*, 1987). In these circumstances, pectin fed at 5% (w/w) in a purified diet did not elicit a significant increase in nitrate reductase activity.

3.3. Species Differences in Response to pH Changes in the Gut

Changes in bacterial enzyme activity in rat cecal contents in response to alterations in the pH of the incubation medium have been discussed in Section 2.5. When these studies were extended to human fecal samples, differences in pH profile of some enzymes were seen (A. K. Mallett and I. R. Rowland, unpublished observations, 1988). For example, whereas the rat cecal flora showed optimal reduction of nitrate at pH 7, nitrate reductase activity in human feces was undetectable below pH 6.6 and then increased with increasing pH. The most pronounced difference between the species was seen with β-glucuronidase, the activity of which increased with pH in rat cecal contents but was inversely related to pH in human fecal suspensions (Fig. 1B).

It is clear, therefore, that although fermentable dietary fiber may have similar

TABLE VI. Effect of Dietary Pectin on Bacterial Numbers
and Enzyme Profile in Human Feces

	Activity (μmole/hr per g feces)[a]	
	Control (n = 24)	+ Pectin[b] (n = 12)
β-Glucosidase	51.6 ± 19.2	33.9* ± 18.8
β-Glucuronidase	34.7 ± 20.4	17.4* ± 8.3
Nitrate reductase	5.8 ± 4.2	4.3 ± 2.2
Nitroreductase	0.8 ± 0.3	0.8 ± 0.3
Bacterial concentration ($\log_{10}/g$)	11.2 ± 0.3	11.0 ± 0.2

[a]Results shown are means ± S.E.; asterisks indicate that a result is significantly different from control (*$P < 0.01$).
[b]During pectin period (3 weeks), volunteers supplemented their normal (control) diet with 18 g pectin/day.

effects on pH of the gut contents in rats and man, the consequences of such changes on foreign compound metabolism in the gut may be very different in the two species.

4. CONCLUSIONS

The studies described above demonstrate that dietary fiber can modify, in a variety of ways, enzyme activities associated with the gut microflora. Since the enzymes studied are chosen on the basis of their known involvement in the metabolism of potentially toxic compounds ingested by man, it can be concluded that fiber may alter the response of an individual to the toxic or carcinogenic effects of foreign compounds subject to microbial biotransformation. That such an effect can occur has been demonstrated for pectin and nitrate toxicity, since rats fed dietary pectin exhibit a greater susceptibility to methemoglobinemia after nitrate exposure as a consequence, presumably, of the increased capacity of the animals to reduce nitrate to nitrite when compared to animals fed a fiber-free diet (Wise *et al.*, 1982).

The effects of fiber on the gut flora are very diverse, even when purified sources of plant material are used. Naturally occurring fiber in the human diet comprises mixtures of fermentable and nonfermentable material from various sources and consequently would be expected to have even more complex effects on bacteria in the large bowel.

Finally, it must not be forgotten that there are pronounced differences between animal species in the response of their gut microfloras to dietary fiber, which makes it difficult to extrapolate results from laboratory animals to man. In this regard, particular care should be exercised when drawing conclusions from colon carcinogenesis studies in rats. Not only does the rat flora respond to some fibers differently from man, but pH changes in the gut induced by fiber may engender very different effects on bacterial enzyme activities in the two species. Nevertheless, the degree of nutritional control that can be exercised in laboratory animals makes them valuable tools in the study of diet–toxicity interactions. It may be necessary, therefore, to put more effort into devising animal models, such as the human-flora-associated rat (Mallett *et al.*, 1987), that combine the ease of dietary manipulation of laboratory animals with greater relevance to man.

ACKNOWLEDGMENT. We are grateful to the Ministry of Agriculture Fisheries and Food for financial support.

REFERENCES

Beinenstock, J., 1979, The physiology of the local immune response, in: *Immunology of the Gastrointestinal Tract* (P. Asquith, ed.), Churchill Livingstone, London, pp. 1–13.

Brown, J. P., and Dietrich, P. S., 1979, Mutagenicity of plant flavonols in the *Salmonella*/microsome test: Activation of flavonol glycosides by mixed glycosidases from rat cecal bacteria and other sources, *Mutat. Res.* **66**:223–240.

Coates, M. E., Drasar, B. S., Mallett, A. K., and Rowland, I. R., 1988, Methodological considerations for the study of bacterial metabolism, in: *Role of the Gut Flora in Toxicity and Cancer* (I. R. Rowland, ed.), Academic Press, London, pp. 1–21.

Conning, D. M., Mallett, A. K., and Nicklin, S., 1984, Novel toxicological aspects of gums and stabilisers, in: *Gums and Stabilisers for the Food Industry,* Vol. 2 (G. O. Phillips, D. J. Wedlock, and P. A. Williams, eds.), Pergamon Press, Oxford, pp. 389–404.

Cummings, J. H., 1982, Consequences of the metabolism of fiber in the human large intestine, in: *Dietary Fiber in Health and Disease* (G. V. Vahouny and D. Kritchevsky, eds.), Plenum Press, New York, pp. 9–22.

Cummings, J. H., Stephen, A. M., and Branch, W. J., 1981, Implications of dietary fiber breakdown in the human colon, in: *Banbury Report 7—Gastrointestinal Cancer: Endogenous Factors* (W. R. Bruce, P. Correa, M. Lipkin, S. R. Tannenbaum, and T. D. Wilkins, eds.), Cold Spring Harbor Laboratory, New York, pp. 71–81.

Cummings, J. H., Englyst, H. N., and Wiggins, H. S., 1986, The role of carbohydrates in lower gut function, *Nutr. Rev.* **44:**50–54.

DeBethizy, J. D., Sherril, J. M., Rickert, D. E., and Hamm, T. E., 1983, Effects of pectin-containing diets on hepatic macromolecular covalent bonding of 2,6-dinitro-(H^3)toluene in Fischer 344 rats, *Toxicol. Appl. Pharmacol.* **69:**369–376.

Drasar, B. S., 1988, The bacterial flora of the intestine, in: *Role of the Gut Flora in Toxicity and Cancer* (I. R. Rowland, ed.), Academic Press, London, pp. 23–38.

Frawley, J. P., Wiebe, A. K., and Klug, E. D., 1964, Studies on the gastro-intestinal absorption of purified sodium carboxymethylcellulose, *Food Cosmet. Toxicol.* **2:**539–543.

Glicksman, M., 1969, *Gum Technology in the Food Industry,* Academic Press, New York.

Goldin, B. R., and Gorbach, S. L., 1976, The relationship between diet and rat fecal bacterial enzymes implicated in colon cancer, *J. Natl. Cancer Inst.* **57:**371–375.

Goldin, B. R., Swenson, L., Dwyer, J., Sexton, M., and Gorbach, S. L., 1980, Effect of diet and *Lactobacillus* supplements on human fecal bacterial enzymes, *J. Natl. Cancer Inst.* **64:**255–262.

Goldman, P., 1982, Role of the intestinal microflora, in: *Metabolic Basis of Detoxication* (W. B. Jakoby, J. R. Bend, and J. Caldwell, eds.), Academic Press, New York, pp. 323–337.

Hassan, M., and Hall, J. B., 1975, The physiological function of nitrate reduction in *Clostridium perfringens, J. Gen. Microbiol.* **87:**120–128.

Hawkins, W. W., and Yaphe, W., 1965, Carrageenan as a dietary constituent for the rat: Faecal excretion, nitrogen absorption, and growth, *Can. J. Biochem.* **43:**479–485.

Hill, H. Z., Backer, R., and Hill, G. J., 1980, Blood cyanide levels in mice after administration of amygdalin, *Biopharmaceut. Drug Dispos.* **1:**211–220.

Holdeman, L. V., and Moore, W. E. C., 1975, *Anaerobe Laboratory Manual,* 3rd ed., VPI Laboratory, Blacksburg, VA.

Jacobs, L. R., and Lupton, J. R., 1986, Relationship between colonic luminal pH, cell proliferation, and colon carcinogenesis in 1,2-dimethylhydrazine treated rats fed high fiber diets, *Cancer Res.* **46:**1727–1734.

Laqueur, G. L., and Spatz, M., 1968, Toxicology of cycasin, *Cancer Res.* **28:**2262–2267.

Leegwater, D. E., De Groot, A. P., and Van Kalmthout-Kuyper, M., 1974, The aetiology of caecal enlargement in the rat, *Food Cosmet. Toxicol.* **12:**687–697.

Mallett, A. K., 1982, Effect of dietary pectin on the metabolic activity of the rat hind gut microflora, *Chem. Ind.* 984–987.

Mallett, A. K., Wise, A., and Rowland, I. R., 1983, Effect of dietary cellulose on the metabolic activity of the rat caecal microflora, *Arch. Toxicol.* **52:**311–317.

Mallett, A. K., Wise, A., and Rowland, I. R., 1984, Hydrocolloid food additives and rat caecal microbial enzyme activities, *Food Chem. Toxicol.* **22:**415–418.

Mallett, A. K., Rowland, I. R., Bearne, C. A., and Nicklin, S., 1985, Influence of dietary carrageenans on microbial biotransformation activities in the cecum of rodents and on gastrointestinal immune status in the rat, *Toxicol. Appl. Pharmacol.* **78:**377–385.

Mallett, A. K., Bearne, C. A., Rowland, I. R., Farthing, M. J. G., Cole, C. B., and Fuller, R., 1987, The use of rats associated with human faecal flora as a model for studying the effects of diet on the human gut microflora, *J. Appl. Bacteriol.* **62:**39–46.

Mallett, A. K., Rowland, I. R., Bearne, C. A., Flynn, J. C., Fehilly, B. J., Udeen, S., and Farthing, M.

J. G., 1988, Effects of dietary supplements of apple pectin, wheat bran or fat on the enzyme activity of the human faecal flora, *Microbiol. Ecol. Health Dis.* **1**:23–29.

Mastromarino, A. J., Reddy, B. S., and Wynder, E. L., 1976, Metabolic epidemiology of colon cancer: Enzymic activity of fecal microflora, *Am. J. Clin. Nutr.* **29**:1455–1460.

McCarthy, R. E., and Salyers, A. A., 1988, The effects of dietary fibre utilization on the colonic microflora, in: *Role of the Gut Microflora in Toxicity and Cancer* (I. R. Rowland, ed.), Academic Press, London, pp. 295–313.

Ota, A., 1982, Phosphorylation coupled to nitrate reduction, *Int. J. Biochem.* **14**:341–346.

Renwick, A. G., and Drasar, B. S., 1976, Environmental carcinogens and large bowel cancer, *Nature* **263**:654–655.

Rowland, I. R., and Walker, R., 1983, The gastro-intestinal tract in food toxicology, in: *Toxic Hazards in Food* (D. M. Conning and A. Lansdown, eds.), Croom-Helm, London, pp. 183–274.

Rowland, I. R., Mallett, A. K., and Wise, A., 1983a, A comparison of the activity of five microbial enzymes from rats, mice, and hamsters, and response to dietary pectin, *Toxicol. Appl. Pharmacol.* **69**:143–148.

Rowland, I. R., Mallett, A. K., Wise, A., and Bailey, E., 1983b, Effect of dietary carrageenan and pectin on the reduction of nitro-compounds by rat caecal microflora, *Xenobiotica* **13**:251–256.

Rowland, I. R., Mallett, A. K., and Wise, A., 1985, The effect of diet on the mammalian gut flora and its metabolic activities, *CRC Crit. Rev. Toxicol.* **16**:31–103.

Rowland, I. R., Mallett, A. K., Bearne, C. A., and Farthing, M. J. G., 1986, Enzyme activities of the hindgut microflora of laboratory animals and man, *Xenobiotica* **16**:519–523.

Samelson, S. L., Nelson, R. L., and Nyhus, L. M., 1985, Protective role of faecal pH in experimental colon carcinogenesis, *J. R. Soc. Med.* **78**:230–233.

Shiau, S.-Y., and Chang, G. W., 1983, Effects of dietary fiber on fecal mucinase and β-flucuronidase in rats, *J. Nutr.* **113**:138–144.

Simon, G. L., and Gorbach, S. L., 1987, Intestinal flora and gastrointestinal function, in: *Physiology of the Gastrointestinal Tract*, 2nd ed. (L. R. Johnson, ed.), Raven Press, New York, pp. 1729–1747.

Thornton, J. R., 1981, High colonic pH promotes colorectal cancer, *Lancet* **1**:1081–1083.

van Dokkum, W., deBoer, B. C. J., van Fassen, A., Pikaar, N. A., and Hermus, R. J. J., 1983, Diet, faecal pH and colorectal cancer, *Br. J. Cancer* **48**:109–110.

Williams, R. T., 1972, Toxicologic implications of biotransformation by intestinal microflora, *Toxicol. Appl. Pharmacol.* **23**:769–781.

Williams, R. T., Millburn, P., and Smith, R. L., 1965, The influence of enterohepatic circulation on toxicity of drugs, *Ann. N.Y. Acad. Sci. (U.S.A.)* **123**:110–124.

Wise, A., Mallett, A. K., and Rowland, I. R., 1982, Dietary fibre, bacterial metabolism and toxicity of nitrate in the rat, *Xenobiotica* **12**:111–118.

Wise, A., Mallett, A. K., and Rowland, I. R., 1986, Effect of mixtures of dietary fibers on the enzyme activity of the rat caecal microflora, *Toxicology* **38**:241–248.

15

The Effects of α-Amylase-Resistant Carbohydrates on Energy Utilization and Deposition in Man and Rat

GEOFFREY LIVESEY

1. INTRODUCTION

Current nutritional guidance in virtually all developed countries (Truswell, 1987) includes the recommendation to reduce fat intake and to replace the food energy from fat with complex carbohydrates. Among the reasons for this guidance is evidence that complex carbohydrates, especially those derived from the plant cell wall, have a lower energy value and also reduce the metabolizable energy from protein and fat. The complex carbohydrates, principally dietary fiber and resistant starches, have also attracted interest from food producers in view of their potential use as constituents in reduced-energy products. The putative efficacy of complex carbohydrates as an aid to slimming and as a possible adjunct to the clinical management of obesity, whether effective or not (FASEB, 1987), has focused attention on their energy value. Studies on the energy value of dietary fiber, however, are few in number. Nevertheless, it is clear that some types of fiber make a considerable contribution to the energy value of the diet (Harley et al., 1989; Davies et al., 1987; Livesey and Davies, 1988). Some, by contrast, add little energy to the diet since they cause substantial loses to feces of protein and fat, whereas others contribute appreciable amounts of energy as a result of their fermentation with production of short-chain fatty acids for absorption and oxidation in the host, this simultaneously with little effect on the losses of protein and fat. It is generally accepted at present, however, to assume a low (i.e., zero) energy contribution from dietary fiber or unavailable carbohydrates from natural sources and even with manufactured products

GEOFFREY LIVESEY • AFRC Institute of Food Research, Norwich Laboratory, Norwich NR4 7UA, United Kingdom.

enriched in dietary fiber. This position is increasingly less tenable and coming under challenge.

It is traditional to assess the energy value of food components as apparent metabolizable energy values (Atwater, 1910; Merrill and Watt, 1973; Paul and Southgate, 1978), the general caloric conversion factors for fat and protein being 9 and 4 kcal/g, respectively. It is by virtue of the method used for their determination that they are apparent metabolizable energy values (see below). The value for dietary fiber (commonly taken to be zero as implied by the British system of food energy assessment and increasingly elsewhere in the world, e.g., the United States) is, however, a partial digestible energy value, which arises from the work of Southgate and Durnin (1970). Before we progress further, these different energy terms (apparent and partial) need further description. Moreover, use of a common calculation method for the estimation of the partial energy values is desirable; a suggested method will be described. Furthermore, it was unavailable carbohydrate rather than dietary fiber that was inferred to have a low energy contribution to the diet. There is some need, therefore, to consider these terms also.

2. DIETARY FIBER OR UNAVAILABLE CARBOHYDRATE

Dietary fibers were once collectively termed unavailable carbohydrate both to distinguish this dietary component from available carbohydrate by chemical analysis (Southgate, 1969) and for the purpose of assessing their contribution to the energy value of the diet (Southgate and Durnin, 1970). It is known that analytical values for dietary fiber may sometimes include resistant starch (Englyst et al., 1987). This starch complicates the issue of definitions. It is resistant to digestion in the small intestine and can therefore also be considered as unavailable carbohydrate. According to the physiological definition of dietary fiber as given by Trowell (1976)—the sum of lignin and polysaccharides that are not digested by endogenous secretions of the human gastrointestinal tract—it would also be appropriate to include resistant starch within the term dietary fiber.

Whether one does or does not requires some consideration of the objective for advising more widespread consumption of the type of diet with more dietary fiber. The recommendation derives from the fiber hypothesis, which postulates a protective effect of a particular type of diet, one rich in foods containing plant cell wall materials. The protective component of this diet was suggested to be to the indigestible plant cell wall material and was termed dietary fiber (Trowell, 1976). On this basis, therefore, resistant starch, which does not originate from the plant cell wall, would be excluded from a definition of dietary fiber. A single principal or proximate component of the diet with a protective effect has, however, never been firmly elucidated, probably because several components are involved, some perhaps being causative (e.g., fat) and some protective (e.g., plant cell wall polysaccharide).

The (re)discovery of a dietary component, resistant starch, that is utilized in a manner similar to plant cell walls by itself provides no new insight into the dietary fiber hypothesis other than to add to the list of candidates that could influence the development of disease. There has not yet been sufficient investigation of resistant starch to know whether it has protective effects. Indeed, this is true also of nonstarch polysaccharides. There is at present, therefore, neither ground to eliminate resistant starch from the dietary component to be called dietary fiber nor ground to include it. This leaves the term dietary fiber undefined and possibly redundant, whereas a precise definition of unavailable carbohydrate is possible and is useful for the purpose of assessing food energy values; it is dietary carbohydrate that resists digestion in the small intestine. To summarize, unavailable carbohydrate certainly includes resistant starch, whereas dietary fiber, although being predominantly composed of unavailable carbohydrate, may be considered not to include resistant starch.

3. APPARENT OR PARTIAL ENERGY VALUES FOR UNAVAILABLE CARBOHYDRATE

When devising a method for estimating the energy value of the whole diet, Atwater (1910) found it possible to sum the caloric values he attributed to the individual proximate component of the diet. In doing so he investigated several diets. Each composition represented an individual experiment from which he determined nutrient balances for fat, nitrogen (as an index of protein), and carbohydrate (by difference). These balances provided estimates of apparent digestibility that, when multiplied by the corresponding heat of combustion value for these proximate constituents, gave *apparent digestible energy values*. Since no loss of energy from carbohydrate or fat to urine was suspected, each apparent digestible energy value was also considered to be the apparent metabolizible energy value (using current terminology, or "physiological fuel value" to use past terminology). For proteins, a correction was made for the loss to urine of energy that was associated with the loss of nitrogen. All food energy assessment systems in current usage with food tables apply *apparent metabolizable energy values* for protein, fat, and carbohydrate (Périsse, 1983). As discussed below, the situation with unavailable carbohydrate is more complex.

It is well known that unavailable carbohydrate undergoes fermentation in the large bowel of man to give rise to volatile fatty acids and that these acids are absorbed and oxidized for energy. Also well known is that unavailable carbohydrate may have effects on the digestibility of protein and fat. When Southgate and Durnin (1970) carried out energy balance investigations with more than one dietary composition, a question that concerned them was the effect unavailable carbohydrate had on the energy value of the whole diet. It was logical for them to obtain a regression analysis of the energy value of the diets plotted against the intake of unavailable carbohydrate. Both of these quantities had been normalized for gross energy intake. The slope of such a plot was such that unavailable carbohydrate appeared to contribute no energy to the diet. Subsequently, this has been taken to indicate that the energy value of unavailable carbohydrate is zero. A further development has been the notion that any carbohydrate that is unavailable as carbohydrate, i.e., escapes digestion in the small intestine, is likely to have no energy value.

For sugar replacements this has been shown to be wrong in several instances (Cooley and Livesey, 1987; van Es *et al.*, 1986; Grupp and Siebert, 1978), and it is becoming apparent that this is also true for fiber isolates (see below). The important point here is that the regression analysis of Southgate and Durnin (1970), which pointed towards a zero energy contribution to the diet, is, in fact, a *partial digestible energy value*, i.e., the change in energy value of the diet per gram increase in unavailable carbohydrate intake, whereas Atwater (1910) derived apparent energy values from single diets using nutrient balances. Though energy balances were performed, this was used in the final analysis to validate his method of calculating the energy value of the whole diet. This difference in approach does not seem to be generally considered, but it is important. For example, using the data of Southgate and Durnin (1970) to obtain an apparent digestible energy value for unavailable carbohydrate, one finds it to be 3.0 (S.D. 0.6, $n = 9$ dietary groups) kcal/g, a value considerably greater than zero, which is the correspondingly partial digestible energy value derived from the same study.

The question arises as to which, apparent or partial, energy value should be used in the assessment of the energy values of foods. This question did not arise before because available carbohydrate, protein, and fat all have partial digestible energy values close or equal to their apparent digestible energy values. This is because these materials, being predominantly digested and absorbed in the small intestine, have relatively little effect on colonic metabolism and appear to interact with each other to a relatively minor extent. A question of whether it would be inconsistent to use apparent energy values for protein, fat, and available carbohydrate together with partial energy values for unavailable carbohydrate does not, therefore, arise.

The reason the partial digestible energy value for unavailable carbohydrate is less than the apparent value is that the former also accounts for interactions between unavailable carbohydrate intake and the losses of protein and fat to feces. In the terms of Kleiber (1975), it should be stated that the "partial digestibility of unavailable carbohydrate for protein" is negative when unavailable carbohydrate lowers the apparent digestibility of protein. Similarly, "the partial digestibility of unavailable carbohydrate for fat" is negative when unavailable carbohydrate lowers the apparent digestibility of fat. When no additional losses of protein and fat to feces occur, the apparent and partial digestible energy values will be identical within experimental error. However, when the partial digestibilities of unavailable carbohydrate for protein and fat are large and negative, for example, when the additional losses of energy to feces exceed that contained in the unavailable carbohydrate that is eaten, the partial digestible energy value of unavailable carbohydrate will also be negative. With partial digestible energy values but not with apparent digestible energy values, therefore, a value is possible that lies outside the range between 0 and its heat of combustion (about 4 kcal/g). The important point here is that although unavailable carbohydrate may contribute a small amount to the overall gross energy intake, it may have a relatively large effect on the digestible energy value of the whole diet.

4. WHICH FORMULA TO USE FOR THE CALCULATION OF PARTIAL DIGESTIBLE ENERGY VALUES?

Several different formulas can be used to calculate partial digestible energy values from raw energy-balance data. Among those that are correct in principle, it is important to choose the one that derives partial digestible energy values without introduction of error. A procedure can be said to differ from another if it makes different assumptions. One that is correct in principle will derive the correct partial digestible energy value when accurate (and cogent) energy-balance data are used. In practice some error exists in all experimental data, and this error is magnified with every procedure for calculating the partial digestible energy value. The extent of this magnification depends on the procedure and the assumption it makes (Livesey, 1989). A useful procedure that makes only small magnification of measurement errors is to plot digestibility (or meta-bolizability) of dietary gross energy against the proportion of gross energy intake attributed to the substance of interest. The absolute value of the slope (which is usually negative for unavailable carbohydrates) gives the partial indigestibility of the substances for energy. The difference between 1 and the absolute value of the slope gives, therefore, partial digestibility for energy. Finally, the product of the heat of combustion of the substance and its partial digestibility for energy gives the partial digestible energy value of the substance. The procedure is simple, and its use should eliminate the use of those procedures that are incorrect in principle or those that can give inaccurate energy values because of large magnification of measurement errors. The procedure should enable, more easily than previously possible, the comparison of energy values between different research groups who have, until now, adopted a wide range of different procedures to assess their experimental data.

5. PREDICTION OF FOOD ENERGY VALUES

Because the work of Southgate and Durnin (1970) provided data consistent with a zero partial digestible energy value for unavailable carbohydrate, the British system of food energy assessment (that is 4, 9, 3.75 kcal/g for protein, fat, and available carbohydrate as monosaccharide) seemed to be justified. More information is available now on which to test this system. The energy values of 17 diets have been predicted using the British system and compared with experimentally determined values (Livesey, 1988). The Atwater general system of food energy assessment (i.e., 4, 9, 4 kcal/g protein, fat, and carbohydrate by difference) was also investigated on these diets. Figure 1 shows that the Atwater system progressively overestimates, whereas the British system progressively underestimates, the energy value of whole diets with increasing unavailable carbohydrate content. Interestingly, both systems are adequate when only small amounts of unavailable carbohydrate are present in the diet, an observation that is consistent with this component being the major modifier of the digestibility of dietary gross energy. By contrast with these general factor systems, the Atwater specific system of food energy assessment, which was adopted and developed by Merrill and Watt (1973), performs better than either of the general systems at high intakes of unavailable carbohydrate (Göranzon et al., 1983; Göranzon and Forsum, 1987).

The data in Fig. 1 are consistent with an energy value for unavailable carbohydrate being between 0 and 4 kcal/g, at about 2 kcal/g. I deliberately omitted to state that this (2 kcal/g) is a partial or apparent energy value. In fact it is neither. It is merely a combination of the apparent energy value of the unavailable carbohydrate and the error or bias in the Atwater and British caloric conversion factors. This leads me to a problem that is a point in principle rather than one of great quantitative significance. The use of

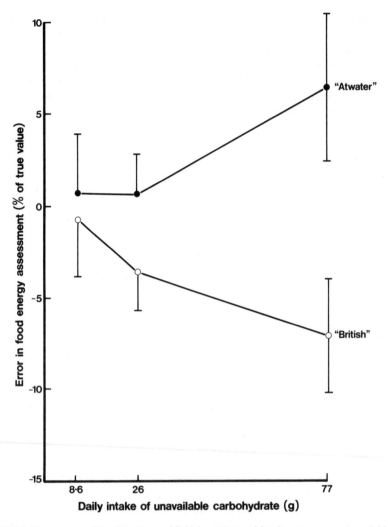

FIGURE 1. Mean error of the Atwater and British systems of food energy assessment. Values are mean ±S.D. for seven diets with a mean intake of 8.6 g, seven diets with a mean intake of 26 g, and three diets with a mean intake of 77 g unavailable carbohydrate per day.

apparent energy values for unavailable carbohydrate when calculating the energy value of the whole diet would be appropriate only if modification to the utilization of dietary protein and fat did not occur as a consequence of increased consumption of unavailable carbohydrate. Since this modification often occurs, apparent energy values would be inappropriate. The use of partial energy values for unavailable carbohydrate would be appropriate if the caloric conversion factor for protein and fat (and carbohydrate by difference) applied in principle to diets of no unavailable carbohydrate content. However, this is not strictly the case: they apply only to diets with an unavailable carbohydrate content similar to those in the experimental diets of Atwater (1910) and of South-gate and Durnin (1970). Marginally higher energy values for protein and fat should be adopted if both exact reasoning is used and partial energy values for unavailable carbo-hydrate are applied. This is a digression, however. What presently gives us concern are the energy values of unavailable carbohydrate and effects on body fat.

6. THE ENERGY VALUES OF UNAVAILABLE CARBOHYDRATE

Although the partial energy value suggested by the data of Southgate and Durnin (1970) was 0 kcal/g, values suggested by Göranzon et al. (1983) and Göranzon and Forsum (1987) are higher, >2 kcal/g. By contrast, Miller and Judd (1984) considered that the additional energy lost to feces as fat and protein may be more than is obtained from the unavailable carbohydrate ingested, which implies negative energy values. Attempts have been made to estimate the partial digestible energy value of unavailable carbohydrate using measurements published in a number of studies (Table I). Values range between >2 kcal/g and <−4kcal/g, with the highest and lowest values being significantly different from zero. It seems, therefore, that the partial digestible energy value of unavailable carbohydrate cannot be considered as constant independent of its source, at zero or at any other value.

Further analysis of the published energy-balance data from studies based mostly on

TABLE I. Partial Digestible Energy Values
for Unavailable Carbohydrate

Source	kcal/g[a]	Species
Mixed diets	From −3.0 to +2.4	Man
Cereal diet, mainly barley	−4.8	Man
Fruit and vegetable diet	−4.7	Man
Oat bran	−3.8	Man
Cereal diet, mainly wheat	−3.1	Man
Psyllium	1.4	
Wheat bran	1.4	
Guar gum	2.4	Rat
Gum arabic	3.3	Rat
Cellulose (Solka-floc)	0	Rat
Cellulose (Solka-floc)	0	Man
Ispagula husk	approx. +3.1	Man
Retrograde resistant pea starch	2.9	Rat
Retrograde resistant corn starch	3.6	Rat

[a]Data based either on experiments in the author's laboratory or on calculations using information in the literature (Southgate and Durnin, 1970; Göranzon et al., 1983; Judd, 1982; Kelsay et al., 1978; Stevens et al., 1987; Davies et al., 1987; Harley et al., 1989; Slavin and Marlett, 1980; Livesey et al., 1990).

whole foods as a source of fiber suggested that variability in the energy value is statistically related to two things (G. Livesey, unpublished data): (1) the apparent digestibility of unavailable carbohydrate such that at low values for this, the partial digestible energy value is very low and negative and at high apparent digestibilities the partial digestible energy value is higher than zero and approaching the apparent digestible energy value; (2) the level of intake of unavailable carbohydrate, so that generally higher values are associated with higher intakes; this relationship may be circumstantial, however, since experiments with sources of unavailable carbohydrate that cause substantial loss of energy to feces are likely to be the less readily tolerated ones, whereas those that cause relatively little energy loss may be readily tolerated and, therefore, can be ingested at higher levels.

These two generalizations apply to diets containing unavailable carbohydrate mostly from food sources that are not artificially enriched. Supplements of unavailable carbohydrates tend to give a different pattern of results, primarily because of a relative lack of effect on the losses of fat to feces and a diminished effect on the losses of protein. For example, wheat bran, cellulose (Solkafloc), and the unavailable carbohydrate of isphagula husk are calculated to have partial digestible energy values of 1.4, -0.2, and 3.1 kcal/g, values that compare well with apparent energy values of about 1.6 (34% apparent digestibility), -0.2 (-7% apparent digestibility), and >3.2 kcal/g (>80% apparent digestibility), respectively. [These values, estimated by the author, are based on data published by Stevens et al. (1987), Slavin and Marlett (1980), Prynne and Southgate (1979), and Nyman et al. (1986).] Viscous gums, nevertheless, may enhance losses of protein and fat to feces when mixed into a dry diet, which explains the difference between the apparent and partial digestible energy value for guar gum in the rat (Livesey and Davies, 1988).

The distinction between isolated fiber sources and whole-food diets may have some relevance for sugar replacement with alternative carbohydrate. It might be expected, by analogy with the isolates of unavailable carbohydrates, that the sugar replacers have partial digestible energy values not much below their apparent digestible energy values. This would be true of lactitol (van Es et al., 1986).

Finally, a word of caution may be made here: dietary energy values of gums have sometimes been estimated using animal growth assays. Such values can be described as neither partial nor apparent energy values and in any event may be unreliable, as shown with experiments with gum arabic (Harley et al., 1989).

7. THE INFLUENCE OF UNAVAILABLE CARBOHYDRATE ON FAT DEPOSITION

Unavailable carbohydrate may potentially influence energy deposition in one of three ways: impairment of food intake, causation of additional losses of energy to feces, and elevation of the rate of energy expenditure. There is a literature on the effects of unavailable carbohydrates on appetite and food intake, a subject that is still in its infancy, to which the reader can be referred (FASEB, 1987; Stevens et al., 1987; Blundell and Burley, 1987). The effect of unavailable carbohydrates on obesity has been addressed on several occasions, but mostly with inconclusive findings (FASEB, 1987).

The experimental problems of reliably assessing the effects of unavailable carbohydrate on the development and regression from obesity may not be well appreciated. It is hardly surprising that there are many apparently conflicting findings, some suggesting small beneficial effects and others no beneficial effect of fiber. Such observations, as made until now, are not really conflicting (or contradictory as suggested by FASEB, 1987). Mostly, too little unavailable carbohydrate is ingested in these studies to have a measurable effect. Moreover, control over food intake is difficult but crucial to obtain. One difficulty may be that subjects aim for an ideal rate of weight loss irrespective of the composition of dietary intake, so that the rate of weight loss controls a subject's food intake. A major factor may be that isolates of unavailable carbohydrates are often used to elevate the intake of this dietary fraction, whereas larger effects might be expected with the unavailable carbohydrate presented with whole foods (compare, for example, the difference in effects on digestibility of dietary energy in Table I). A further difficulty with studies on weight loss is that the losses of body fat may be balanced, in part, by a gain of body water, particularly that associated with enlarged intestinal masses and larger weight of digesta in the large bowel.

In the absence of sufficiently reliable data from experiments with human subjects, observations in laboratory animals take on greater than usual importance. Interpretation of such data in terms of possible implications for humans relies on the basic principles of energetics being similar across relevant species. Although this is essentially so, some caution may be warranted because there is still little experience worldwide of animal and human energetics related to the effects of unavailable carbohydrates. We have undertaken a number of energy-balance studies in the laboratory rat to investigate their effects on food intake, their partial and apparent digestible energy values when fed as supplements to a basal ration, and their effect on the gain of lean tissue and the deposition of fat. For this we use 28-day energy-balance studies.

Consistent with observations in humans (Krotkiewski, 1984), we find guar gum to depress food intake markedly when animals are fed a single meal with access to food limited to only 1.5 hr each day or when food intake is *ad lib*. In both instances the decrease is acute; on the first day of supplementation food intake may be halved when 9% of the weight of the food provided is the gum. Further decreases are not seen with time; rather, food intake recovers towards more normal values, suggesting that the animals either learn to eat more of the supplemented ration or adapt to tolerate a larger dose of the gum. The observation raises the question of whether human subjects who respond acutely to a higher-fiber diet with a decreased intake of food would in the longer term also show a diminished response and return toward more "usual" food intakes.

A notable observation in our studies with rats is that the partial digestible energy values for unavailable carbohydrate give little indication of what to expect of fat deposition. Normally, as one increases energy intake, one expects more body fat to be deposited. We observed this effect with two different but fully available carbohydrates, sucrose and corn starch. The unavailable carbohydrate guar gum, by contrast, supplies substantial amounts of partial digestible energy (2.5 kcal/g) and apparent digestible energy (4.1 kcal/g) when fed as 9% w/w supplements to a basal ration, yet the effect on fat deposition is to decrease it! The decrease can be expressed as net fat deposition (in excess of that for animals fed only the basal ration) per gram of gum ingested. This expression, net energy for fat deposition, of course is negative because less fat is deposited when guar is fed. With juvenile animals fed single meals per day, a value of -2 kcal fat/g guar gum has been observed. With similar animals permitted access to food all day (self-restricted by the effects of guar gum on food intake), the value is -3 kcal fat/g guar gum. In some experimental conditions, no effect has been observed on total fat deposition, and this is associated with low rates of fat deposition relative to the gains of lean tissue in control groups of animals not receiving the gum supplement. Our data indicated that when fat deposition is decreased by guar gum, it is not because energy is being used to enhance the gain of lean body mass. Also, it needs to be noted that this lack of effect in the gain of lean body mass occurs even though guar gum enhances the losses of nitrogen to feces. A possible implication of our observations with guar gum is that unavailable carbohydrate of this type may not aid loss of body fat (cf. slimming) because this is not a condition of fat gain, it may help decrease the development or regain of body fat deposits. Presently, we have data that point toward possible decreased fat deposition with other gums also, but guar gum has so far shown the largest effect. Also, cellulose (Solka-floc) appears to have a small negative effect on fat deposition in some circumstances. These effects, it needs to be stressed, are independent of food intake and losses of energy to feces. Also, it is not plausible for our observations to arise from artifacts associated with the measurement of food intake.

We recently turned our attention to two retrograde resistant starches, asking whether they resemble the available carbohydrate or the unavailable carbohydrates (Livesey *et al.*, 1990). These starches were prepared from starch isolated from pea and from corn starch. Resistance to α-amylase was induced by heat treatment, cooling, ethanol precipitation, and then drying. Both had partial digestible energy values >2.8 kcal/g and apparent digestible energy values greater than 3.5 kcal/g. Fat deposition subsequent to supplementation of a basal food ration with these isolates was enhanced to an extent equal to or greater than with sucrose. From the point of view of effects on fat deposition, it therefore seems more appropriate to classify the resistant starches apart from the nonstarch polysaccharides.

REFERENCES

Atwater, W. O. 1910, *Principles of Nutrition and Nutritive Value of Foods*, USDA Bulletin 142 (2nd rev.), U.S. Government Printing Office, Washington.
Blundell, J. E., and Burley, V. J., 1987, Satiation, satiety and the action of fibre on food intake, *Int. J. Obesity* 11(Suppl.):9–25.
Cooley, S., and Livesey, G., 1987, The metabolizable energy value of Polydextrose[R] in a mixed diet fed to rats, *Br. J. Nutr.* 57:235–243.
Davies, I. R., Johnson, I. T., and Livesey, G., 1987, Food energy values of dietary fibre components and decreased deposition of body fat, *Int. J. Obesity* 11(Suppl. 1):101–105.
Englyst, H. N., Travel, H., Southgate, D. A. T., and Cummings, J. H., 1987, Dietary fibre and resistant starch, *Am. J. Clin. Nutr.* 46:873–876.
FASEB, 1987, *Physiological Effects and Health Consequences of Dietary Fiber* (S. W. Pilch, ed.), Federation of American Societies for Experimental Biology, Bethesda.
Göranzon, H., and Forsum, E., 1987, Metabolizable energy in humans in two diets: Calculations and analysis, *J. Nutr.* 117:267–273.
Göranzon, H., Forsum, E., and Thilén, M., 1983, Calculation and determination of metabolizable energy in mixed diets to humans, *Am. J. Clin. Nutr.* 38:954–963.

Grupp, U., and Siebert, G., 1978, Metabolism of hydrogenated palatinose, an equimolar mixture of α-D-glucopyranside-1,6-sorbitol and α-D-glucopyranside-1,6-mannitol, *Res. Exp. Med.* **173**:261–278.

Harley, C. J., Davies, I. R., and Livesey, G., 1989, Digestible energy values of gums in the rat—data on gum arabic, *Food Add. Contam.* **6**:13–20.

Judd, P. A., 1982, The effects of high intakes of barley on gastrointestinal function and apparent digestibility of dry matter, nitrogen and fat in human volunteers, *J. Plant Foods* **4**:79–88.

Kelsay, J. L., Behall, K. M., and Prather, E. S., 1978, Effect of fiber from fruit and vegetables on metabolic responses of human subjects. I. Bowel transit time, number of defecations, fecal weight, urinary excretions of energy and nitrogen and apparent digestibilities of energy, nitrogen and fat, *Am. J. Clin. Nutr.* **31**:1149–1153.

Kleiber, M., 1975, *The Fire of Life,* Robert E. Krieger, New York, pp. 262–263.

Krotkiewski, M., 1984, Effect of guar on body weight, hunger rating and metabolism in obese subjects, *Br. J. Nutr.* **52**:97–105.

Livesey, G., 1988, Energy from food—old values and new perspectives, *Br. Nutr. Found. Bull.* **13**(1):9–28.

Livesey, G., 1989, Procedures for calculating the digestible and metabolizable energy values of food components making a small contribution to dietary intakes *J. Sci. Food Agric.* **48**:475–481.

Livesey, G., and Davies, I. R., 1988, Caloric value of fibre and guar gum, in: *Low-Calorie Products* (G. Birch and M. Lindley, eds.), Applied Science Publishers, London pp. 223–244.

Livesey, G., Davies, I. R., Brown, J. C., Faulkes, R. M., and Southon, S., 1990, Energy balance and energy value of α-amylase resistant corn and pea starch in the rat *Br. J. Nutr.* (in press).

Merrill, A. L., and Watt, B. K., 1973, *Energy Value of Foods, Basis and Derivation,* U.S.D.A. Handbook 74, U.S. Government Printing Office, Washington.

Miller, D. S., and Judd, P. A., 1984, The metabolizable energy values of foods, *J. Sci. Food Agr.* **35**:111–116.

Nyman, M., Asp, N. G., Cummings, J., and Wiggins, H., 1986, Fermentation of dietary fibre in the intestinal trace: Comparison between man and rat, *Br. J. Nutr.* **55**:487–496.

Paul, A. A., and Southgate, D. A. T., 1978, McCance and Widdowson's *The Composition of Foods*, 4th ed., H. M. Stationery Office, London.

Périsse, J., 1983, Heterogeneity in food composition table data, *Food Nutr. Rev.* **9**:14–17.

Prynne, C. J., and Southgate, D. A. T., 1979, The effects of a supplement of dietary fibre on faecal excretions of human subjects, *Br. J. Nutr.* **41**:494–503.

Slavin, J. L., and Marlett, J. A., 1980, Effect of refined cellulose on apparent energy, fat and nitrogen digestibilities, *J. Nutr.* **110**:2020–2026.

Southgate, D. A. T., 1969, Determination of carbohydrates in foods. II Unavailable carbohydrates, *J. Sci. Food Agr.* **20**:331–339.

Southgate, D. A. T., and Durnin, J. V. G. A., 1987, Calorie conversion factors: An experimental reassessment of the factors used in the calculation of the energy value of human diets, *Br. J. Nutr.* **24**:517–536.

Stevens, J., Levitsky, D. A., Van Soest, P. J., Roberts, J. B., Kalkwaf, H. J., and Roe, A. D., 1987, Effect of psyllium gum and wheat bran on spontaneous energy intake, *Am. J. Clin. Nutr.* **46**:812–817.

Trowell, H., 1976, Definition of dietary fibre and hypothesis that it is a protective factor in certain diseases, *Am. J. Clin. Nutr.* **29**:417–427.

Truswell, A. S., 1987, Evolution of dietary recommendations, goals and guidelines, *Am. J. Clin. Nutr.* **45**:1060–1072.

van Es, A. J. H., ReGroot, L., and Voyt, J. E., 1986, Energy balance of eight volunteers fed on diets supplemented with either laclitol or saccharose, *Br. J. Nutr.* **56**:545–554.

16

The Ileal Brake

Is It Relevant to the Action of Viscous Polysaccharides?

N. W. READ, C. P. SEPPLE, and N. J. BROWN

1. WHAT IS THE ILEAL BRAKE?

This is a term applied to the delay in gastric emptying and small intestinal transit induced by the presence of certain nutrient solutions in the ileum.

A few years ago, we demonstrated that the infusion of Intralipid® into the ileum, 205 cm from the teeth, of normal volunteers increased the time taken for a bolus of the unabsorbable disaccharide lactulose infused at the ligament of Treitz to reach the cecum and generate hydrogen gas (Read *et al.*, 1984). Infusion of a protein hydrolysate had a similar though much reduced effect, but infusion of solutions of glucose or hyperosmotic solutions failed to slow small bowel transit. This effect was confined to the ileum, since infusion of any of the nutrient solutions into the jejunum or into the colon had no effect on small bowel transit time.

At around the same time, Spiller and his colleagues, using very different methods, showed that the infusion of a partial digest of triglyceride into the distal small intestine, 170 cm from the teeth, slowed the transit of a nonabsorbable marker continuously perfused through a 30-cm segment of jejunum (Spiller *et al.*, 1984). It was this group of workers that coined the term "the ileal brake."

Further experiments carried out in human volunteers showed that infusion of lipid into the ileum reduced the frequency of jejunal contractions (Welch *et al.*, 1988; Spiller *et al.*, 1988) and also the proportion of contractions that were propagated downstream but did not apparently alter ileal contractility (Welch *et al.*, 1989). Ileal infusion of glycerol, corn starch, and lactalbumin did not influence jejunal contractility (Spiller *et al.*, 1988).

N. W. READ, C. P. SEPPLE, and N. J. BROWN • Sub-Department of Human Gastrointestinal Physiology and Nutrition, Royal Hallamshire Hospital, University of Sheffield, Sheffield S10 2JF, United Kingdom.

Ileal infusion of lipid also increased the absorption of a liquid carbohydrate meal in the whole small intestine (Holgate and Read, 1985) and increased the absorption of xylose in a perfused 30-cm segment of jejunum (Spiller *et al.*, 1984). Thus, these experiments suggested that the presence of unabsorbed fat in the distal small intestine served as a feedback mechanism to enhance absorption of nutrients by prolonging contact with the small intestinal epithelium.

Studies in animals have indicated that the lipids did not need to be hydrolyzed to interact with ileal receptors (Brown *et al.*, 1987), and the effects appeared to be directly related to the chain length of the fatty acid (Brown and Richardson, 1987): infusion of short-chain fatty acids into the ileum accelerated small bowel transit time (N. J. Brown, unpublished observations). Interestingly, the infusion of glucose solution into a jejunal loop in the rat slowed small bowel transit, although jejunal infusion of glucose did not affect transit time in humans (Brown, 1988).

2. MECHANISM OF ACTION

It is generally assumed that the ileal brake is mediated at least in part by a humoral mechanism. Spiller's work has consistently indicated that infusion of lipids (either triolein, oleic acid, medium-chain triglycerides, or a partial hydrolysate of triglyceride) into the distal small intestine was associated with an elevation of the peptides neurotensin and enteroglucagon, which are predominantly stored in the ileum (Spiller *et al.*, 1984, 1988). We could not confirm this observation; infusion of a triglyceride into the ileum did not elevate enteroglucagon and neurotensin, although infusion of triglyceride into the jejunum did (Read *et al.*, 1984). The difference in the two sets of results may depend on small differences in methodology. For example, Spiller infused their lipids 40 to 50 cm more proximal than our ileal infusion site. In any case, it seems possible to obtain a very marked delay in transit without any elevation of plasma neurotensin and enteroglucagon. In a recent paper Spiller *et al.* (1988) have also shown that blood levels of PYY are elevated during ileal infusion of lipid and that the rise in PYY, unlike the other peptides, correlated significantly with the inhibition of jejunal motor activity. PYY is found in ileal and colonic endocrine cells (Adrien *et al.*, 1985) and delays gastric emptying (Allen *et al.*, 1984) at blood levels similar to those seen after ileal lipid infusion (Spiller *et al.*, 1988) and in patients with steatorrhea (Adrien *et al.*, 1986).

Neural mechanisms are also implicated. Kinsman and his colleagues (Kinsman and Read, 1984) showed that the delay in small bowel transit time induced by ileal infusion of lipid could be blocked by an intravenous infusion of naloxone. This suggests that enkephalinergic neurons may be involved in the ileal brake.

3. ILEAL EFFECT ON GASTRIC EMPTYING

When we tried to put the ileal brake into a physiological context by studying its effect on the passage of a standard meal through the upper gastrointestinal tract, we noted that the ileal infusion of lipid not only delayed the small bowel transit of the meal but also delayed gastric emptying (Read *et al.*, 1984; Welch *et al.*, 1988a). This mechanism was distinct from the effects on small bowel transit: it could not be blocked by naloxone (Baxter *et al.*, 1987), it could not be provoked by the infusion of a protein hydrolysate into the ileum (Welch *et al.*, 1988a), and gastric emptying but not small bowel transit could be slowed by the infusion of lipid into the duodenum and jejunum (Welch *et al.*, 1988a), an observation that had been made many times before. Recent manometric studies have shown that ileal infusion of fat reduces antral contractility (Fone *et al.*, 1988) and increases isolated pyloric pressure waves (Fone *et al.*, 1988).

4. EFFECT ON SATIETY

The slowing of gastric emptying by infusion of lipid into the small intestine should, in theory, reduce the amount of food ingested from a meal. We investigated this by measuring the intake of an appetizing meal following infusion of either 50% lipid emulsion or isosmotic saline into the ileum. Infusion of lipid reduced the amount of food eaten and the total energy intake but did not affect the rate of eating (Welch *et al.*, 1985). This action of lipid was unlikely to be a postabsorptive effect because intravenous infusion of Intralipid® did not influence food intake in a similar group of volunteers (Welch *et al.*, 1985). Infusion of the same lipid solution into the upper small intestine had a slightly different effect: it not only reduced the amount of food eaten, but it also slowed the rate of eating and suppressed hunger before the meal (Welch *et al.*, 1988b). This suggests that receptors throughout the small intestine may be involved in evoking early satiety, but only those in the upper intestine have a direct effect on appetite or hunger.

5. PHYSIOLOGICAL RELEVANCE OF THE ILEAL BRAKE

5.1. Transit Time and Absorption

Several clinical observations are compatible with the ileal brake concept. Spiller and his colleagues have shown that celiac patients with steatorrhea have slow transit through the small bowel (Spiller *et al.*, 1987). Surgical resection of the ileum in rats (Kremen *et al.*, 1954; Nygaard, 1966) causes much more severe diarrhea than resection of an equivalent length of jejunum, whereas quite modest ileal resection in addition to proctocolectomy in humans leads to rapid small bowel transit and marked malabsorption of nutrients (Neal *et al.*, 1984).

Despite this, a recent study carried out in patients with terminal ileostomies showed that feeding a diet rich in fat did not cause a large increase in the amount of fat coming through the stoma, did not slow small bowel transit, and did not increase the absorption of other components of the meal (unpublished observations). It is possible that the effects of the high fat and slowing small bowel transit may be masked by the increased volume of pancreatic biliary secretions induced by the high-fat meal, since jejunal infusion of lipid markedly increases pancreaticobiliary secretion and jejunal flow (Spiller *et al.*, 1988).

5.2. Gastric Emptying and Postprandial Blood Levels

Gastric emptying can be an important factor limiting the absorption of nutrients or drugs that are rapidly transported across the intestinal epithelium. Thus, infusion of lipid into the jejunum and the ileum delays gastric emptying of a carbohydrate meal and reduces postprandial glucose and insulin levels (Welch et al., 1987). Similar effects have been observed after a drink of ethanol (McFarlane et al., 1986), providing a physiological basis for the belief that a drink of milk or a tablespoonful of olive oil before consuming alcohol can help motorists beat the breathalyzer!

Reductions in postprandial glucose and insulin can be seen when lipid is incorporated in the carbohydrate meal (Welch et al., 1987), suggesting that the judicious use of fat may help to reduce blood glucose levels in patients with diabetes mellitus. A different effect is observed when lipid is ingested as a soup 20 min before the carbohydrate meal (Cunningham and Read, 1989). Then the emptying of the carbohydrate ceases, probably until the majority of the lipid has emptied (Houghton et al., 1988), but it empties at the same rate some time later. Glucose and insulin peaks are delayed but not reduced.

5.3. Satiety and Eating Behavior

Patients who suffer from diarrhea after an ileal resection often eat an enormous amount of food. Their diarrhea can be reduced by reducing energy intake. The excessive food consumption could be explained by the lack of any brake on eating behavior from the distal small intestine.

In a recent study, we showed that feeding fatty soup slowed the gastric emptying of a carbohydrate meal fed 20 min later (Cunningham and Read, 1989). Despite this effect, we could not demonstrate any reduction in energy intake when a high-fat compared with a low-fat soup was fed to normal volunteers (Sepple and Read, 1988), questioning the role of gastric distension in human satiety. Eating a high-fat breakfast, however, reduced the intake of food at lunch 4 hr later (Sepple and Read, 1988). This result may be explained by the effect of unabsorbed fat in ileal receptors, since we would expect the residues of a meal to be in the distal small intestine by this time, and fat is more slowly absorbed than other components of the meal.

5.4. Adaptation

It seems likely that nutrient receptors in the small intestine can be up-regulated or down-regulated according to exposure. This would explain the observations that people who are habituated to a very-low-fat diet can become quite nauseated if they eat what would be otherwise considered to be a normal amount of fat, whereas some obese subjects appear to be able to consume very large amounts of food without feeling unwell. Thus, it seems likely that the acute effects of exposure of the intestine to lipid may be very much reduced by chronic exposure. In recent experiments, we have shown that putting normal volunteers on a high-glucose diet over 3 days reduces the effect of hyperosmotic glucose on gastric emptying in normal volunteers (unpublished data), whereas a high-lipid diet fed over 14 days reduces the effect of fat on gastric emptying and small bowel transit in some but not all normal volunteers (unpublished observations).

6. POSSIBLE RELEVANCE OF THE ILEAL BRAKE HYPOTHESIS TO THE ACTION OF FIBER

Incorporating viscous polysaccharides in a drink or a meal slows gastric emptying, slows small bowel transit (Blackburn *et al.*, 1984a), and reduces the rate of absorption of nutrients (Blackburn *et al.*, 1984b). It is assumed that this effect is caused by the effect. The increased luminal viscosity may impede the action of intestinal contractions and slow the movement of contents (Blackburn *et al.*, 1984b). However, this would not explain why feeding a meal containing viscous polysaccharides at breakfast time can reduce postprandial glycemia and delay gastric emptying of a meal fed at lunchtime (the second-meal effect) (Jenkins *et al.*, 1980). One explanation is that the gum in some way coats the intestinal epithelium (Blackburn and Johnson, 1981), increasing the thickness of the unstirred water layer. However, we could not demonstrate any increase in unstirred water layer after exposure of the human small intestine to guar gum (Blackburn *et al.*, 1984b).

An alternative possibility involves stimulation of the ileal receptors by unabsorbed nutrients. The reduction in movement of intestinal contents by soluble fibers must reduce the interaction between food and enzymes (Isaakson *et al.*, 1982), impair the disruption of solid foods, and limit the access of digested nutrients to the epithelial surface. These effects would tend to delay absorption, reducing nutrient levels in the blood and resulting in a greater delivery of nutrients to more distal sites in the small intestine. There is indirect evidence to support this concept: chronic ingestion of viscous polysaccharides causes the distal small intestine to adapt morphologically and functionally to become "jejunalized" (Imaizumi and Sugaro, 1986; Vahouney, 1986; Johnson and Gee, 1986; Imaizumi *et al.*, 1982), presumably in response to an increased nutrient load (Williamson, 1978). Imaizumi and colleagues (1982) found that supplementing the diets of rats with guar gum increased the secretion of radiolabeled triglyceride into the mesenteric lymph from the distal small intestine but not from the jejunum. The increased load of nutrients to the distal small intestine would stimulate the ileal brake and influence the way in which the next meal is digested. It could also explain the chronic effects of viscous polysaccharides on plasma glucose levels and the observed effects of viscous polysaccharides on energy intake and their possible use in the treatment of obesity (Heaton, 1980).

REFERENCES

Adrian, T. E., Ferri, G.-L., Bacarese-Hamilton, A. J., Fuessi, H. S., Polak, J. M., and Bloom, S. R., 1985, Human distribution and release of a putative new gut hormone, peptide YY, *Gastroenterology* **89:**1070–1077.
Adrian, T. E., Savage, A. P., Bacarese-Hamilton, A. J., Wolfe, K., Besterman, H. S., and Bloom, S. R., 1986, Peptide YY abnormalities in gastrointestinal disease, *Gastroenterology* **90:**379–384.

Allen, T. E., Fitzpatrick, M. L., Yeats, J. C., Darcy, K., Adrian, T. E., and Bloom, S. R., 1984, Effect of peptide YY and neuropeptide Y on gastric emptying in man, *Digestion* **30:**255–262.

Baxter, A. J., Edwards, C. A., Holden, S., Cunningham, K. M., Welch, I. McL., and Read, N. W., 1987, Effect of alpha₂ receptor antagonists and agonists on small bowel transit in man, *Alim. Pharmacol. Ther.* **1:**649–656.

Blackburn, N. A., and Johnson, I. T., 1981, The effect of guar gum on the viscosity of the gastrointestinal contents and on glucose uptake from the perfused jejunum in the rat, *Br. J. Nutr.* **46:**239–246.

Blackburn, N. A., Holgate, A. M., and Read, N. W., 1984a, Small intestinal contact area—another mechanism by which guar reduces postprandial hyperglycaemia in man, *Br. J. Nutr.* **52:**197–204.

Blackburn, N. A., Redfern, J. S., Jarjis, M., Holgate, A. M., Hanning, I., Scarpello, J. H. B., Johnson, I. T., and Read, N. W., 1984b, The mechanism of action of guar gum in improving glucose tolerance in man, *Clin. Sci.* **66:**329–336.

Brown, N. J., 1988, Effect of nutrient infusions into jejunal and ileal Thiry–Vella loops on stomach-to-caecum transit (SCTT) in the rat, *J. Physiol. (Lond.)* **403:**43P.

Brown, N. J., and Richardson, A., 1987, Effect of ileal infusion of fatty acids varying in chain length and saturation on stomach to caecum transit time (SCTT) in the rat, *J. Physiol. (Lond.)* **396:**20P.

Brown, N. J., Richardson, A., Rumsey, R. D. E., and Read, N. W., 1987, Effect of ileal infusion of fats, detergents, and bile acids on stomach to caecum transit time in the rat, *Gut* **28:**A1355.

Cunningham, K., and Read, N. W., 1989, Effect of incorporating lipid in different components of a standard meal on gastric emptying and postprandial plasma glucose and insulin levels, *Br. J. Nutr.* **61**(2):285–290.

Fone, D., Maddox, A., Horowitz, M., Read, N. W., Dent, J., and Collins, P., 1988, Antropyloroduodenal motor responses to terminal ileal infusion—motility correlates of the ileal brake, *Gastroenterology* **94:**A133.

Heaton, K. W., 1980, Food intake regulation and fiber, in: *Medical Aspects of Dietary Fiber* (G. A. Spiller and R. M. Kay, eds.), Plenum Press, New York, pp. 223–238.

Holgate, A. M., and Read, N. W., 1985, Effect of ileal infusion of Intralipid on gastrointestinal transit, ileal flow rate and carbohydrate absorption in humans, after a liquid meal, *Gastroenterology* **88:** 1005–1011.

Houghton, L. A., Read, N. W., Heddle, R., Horowitz, N., Collins, P. J., Chatterton, B., and Dent, J., 1988, The relationship between the motor activity of the antrum, pylorus and duodenum emptying of a solid–liquid mixed meal, *Gastroenterology* **94:**1285–1291.

Imaizumi, K., and Sugaro, M., 1986, Dietary fiber and intestinal lipoprotein secretion, in: *Dietary Fiber, Basic and Clinical Aspects* (G. V. Vahouney and D. Kritchevsky, eds.), Plenum Press, New York, pp. 287–308.

Imaizumi, K., Tominaga, A., Maivatari, K., and Sugaro, M., 1982, Effect of cellulose and guar gum on the secretion of mesenteric lymph chylomicrons in meal fed rats, *Nutr. Rep. Int.* **26:**263–269.

Isaakson, G., Lundquist, I., and Ihse, I., 1982, Effect of dietary fibre on pancreatic enzyme activity—*in vitro, Gastroenterology* **82:**918–924.

Jenkins, D. J. A., Wolever, T. M. S., Nineham, R., Sarson, D. L., Bloom, S. R., Ahern, J., Alberti, K. G. M. M., and Hockaday, T. D. R., 1980, Improved glucose tolerance four hours after guar with glucose, *Diabetologia* **19:**21–24.

Johnson, I. T., and Gee, J. M., 1986, Gastrointestinal adaptation in response to soluble non-available polysaccharides in the rat, *Br. J. Nutr.* **55:**497–505.

Kinsman, R. I., and Read, N. W., 1984, The ileal brake: A mechanism for controlling gastric emptying and small bowel transit, modulated by opiates, *Gastroenterology* **87:**335–337.

Kremen, A. J., Linner, J. H., and Nelson, C. H., 1954, An experimental evaluation of the nutritional importance of proximal and distal small intestine, *Ann. Surg.* **140:**439.

McFarlane, A., Pooley, L., Welch, I. McL., Rumsey, R. D. E., and Read, N. W., 1986, How does dietary lipid lower blood alcohol levels? *Gut* **27:**15–18.

Neal, D. E., Williams, N. F., Barker, M. C. J., and King, R. F. G. J., 1984, The effect of resection of distal ileum on gastric emptying, small bowel transit and absorption after proctocolectomy, *Br. J. Surg.* **71**:666–670.

Nygaard, K., 1966, Resection of the small intestine in rats. 1. Nutritional status and adaption of fat and protein absorption, *Acta Clin. Scand.* **132**:721.

Read, N. W., MacFarlane, A., Kinsman, R., Bates, T., Blackhall, N. W., Farrar, C. B. J., Hall, J. C., Moss, G., Morris, A. P., O'Neill, B., Welch, I., Lee, Y., and Bloom, S. R., 1984, Effect of infusion of nutrient solutions into the ileum on gastrointestinal transit and plasma levels of neurotensin and enteroglucagon in man, *Gastroenterology* **86**:274–280.

Sepple, C. P., and Read, N. W., 1988, The effect of feeding lipid on food intake and satiety in human volunteers, in: *Obesity in Europe* (P. Bjorntorp and S. Rossner, eds.), pp. 131–134.

Spiller, R. C., Trotman, I. F., Higgins, B. E., Ghodei, M. A., Grimble, G. K., Lee, Y. C., Bloom, S. R., Misiewicz, J. J., and Silk, D. B. A., 1984, The ileal brake—inhibition of jejunal motility after ileal fat perfusion in man, *Gut* **25**:365–74.

Spiller, R. C., Lee, Y. C., Edge, C., Ralphs, D. N. L., Stewart, J. S., Bloom, S. R., and Silk, D. B. A., 1987, Delayed mouth–caecum transit of a lactulose-labelled liquid test meal in patients with steatorrhoea due to partially treated coeliac disease, *Gut* **28**:1275–1282.

Spiller, R. C., Trotman, I. F., Adrian, T. E., Bloom, S. R., Misiewicz, J. J., and Silk, D. B. A., 1988, Further characterization of the 'ileal brake' reflex in man—effect of ileal infusion of partial digests of fat, protein and starch on jejunal motility and release of neurotensin, enteroglucagon, and peptide YY, *Gut* **29**:1042–1051.

Vahouney, G. V., 1986, Dietary fibre and intestinal adaptation, in: *Dietary Fiber Basic and Clinical Aspects* (G. V. Vahouney, and D. Kritchevsky, eds.), Plenum Press, New York, pp. 181–209.

Welch, I. McL., Saunders, K., and Read, N. W., 1985, The effect of ileal and intravenous infusions of fat emulsions on feeding and satiety in human volunteers, *Gastroenterology* **89**:1293–1297.

Welch, I. McL., Bruce, C., Hill, S. E., and Read, N. W., 1987, Duodenal and ileal lipid suppresses postprandial blood glucose and insulin responses, *Clin. Sci. Mol. Med.* **72**:209–216.

Welch, I. McL., Cunningham, K. M., and Read, N. W., 1988a, The ileal control of gastric emptying, *Gastroenterology* **94**:401–404.

Welch, I. McL., Sepple, C., and Read, N. W., 1988b, Comparisons of the effects on satiety and eating behaviour of infusion of lipid into the different regions of the small intestine, *Gut* **29**:306–311.

Welch, I. McL., Davison, P. A., and Read, N. W., 1988, The effect of ileal infusion of lipid emulsion on jejunal motor patterns after a nutrient and non-nutrient meal, *Am. J. Physiol.* **255**:9800–9806.

Williamson, R. C. N., 1978, Intestinal adaptation, *N. Engl. J. Med.* **298**:1393–1402.

17

Action of Dietary Fiber on the Satiety Cascade

VICTORIA J. BURLEY and JOHN E. BLUNDELL

Satiety is a complex sensation but an important one in the therapeutic use of dietary fibre. Since we aim to treat the whole patient, not just the process of absorption, a better understanding of satiety is fundamental to achieving good compliance.

Taylor (1981)

1. BACKGROUND

The present era in the late 20th century is a particularly appropriate time to consider the potential effects of dietary fiber on food consumption and body weight and on the mechanisms controlling appetite. There are good reasons for looking closely at processes of appetite control in view of certain cultural forces that currently influence people's patterns of consumption. On the one hand, in western industrial societies obesity has reached epidemic proportions (Van Itallie and Abraham, 1985; Garrow, 1982). Associated with this is a powerful cultural ideology characterized by a pursuit of slimness mainly by regulated or casual dieting. Consequently, many individuals—particularly women—perceive the need to curtail their hunger, eat selectively, and reduce energy intake. This doctrine exerts a potent effect on life styles.

On the other hand, technological advances in food manufacture have made available a wide range of products often arising from intense refining and processing. Such products are often characterized by the inclusion of components whose effect on appetite may be unknown (Blundell *et al.*, 1987). In addition, such products are typified by their

VICTORIA J. BURLEY and JOHN E. BLUNDELL • Biopsychology Group, Psychology Department, University of Leeds, Leeds LS2 9JT, United Kingdom. This chapter was prepared when *V.J.B.* was undertaking research at King's College, University of London, Kensington Campus, London, United Kingdom.

variety and high palatability. The pleasing sensory aspects of these products means that they will be well accepted and readily consumed—perhaps in excess. Accordingly, individuals may well be caught between two opposing forces—one to resist consumption and the other to promote it. People are expected to exert restraint and discretion over eating but are constantly stimulated by food signals that encourage purchase and consumption. It can be assumed that both of these forces place considerable demands on the appetite control mechanisms, and both may work against the natural biological wisdom of the body, i.e., ingesting nutrients in accordance with nutritional needs. Of course it is clearly recognized that people eat for reasons other than to achieve satisfaction of a physiological need. Given the subtle interaction between psychological and physiological factors in the control of eating (Blundell and Hill, 1986), the cultural tendencies outlined above may well lead ultimately to a fragile control over appetite.

Under these circumstances there are good reasons to examine the action of food products and nutritional components on appetite control. Since dietary fiber represents an important element of dietary intake, and one that is claimed to confer physiological benefit, there are grounds for scrutinizing closely the potential actions of fiber on the mechanisms underlying the expression of hunger and satiety. This seems especially significant in view of the fact that fiber can be either totally excluded from food products or incorporated in designated and graded amounts during manufacture. This review therefore considers the role played by fiber in the network of factors influencing the pattern of food consumption.

2. STRATEGIES FOR GETTING FIBER INTO THE DIET

A number of procedures are available for adjusting the amount of fiber consumed. The manner in which fiber is administered and its contact with food will clearly determine any observed action on appetite. In addition, it is of course understood that any effect of fiber will depend on the type of fiber, the amount, and the timing of its administration relative to meals and other activities.

2.1. Natural High-Fiber Foods

Foods containing a high proportion of fiber such as fruits, cereals, and pulses not only contain the physiologically active components—fiber itself—but may also possess certain physical characteristics and textural qualities. Consequently, fiber manipulations achieved by judicious use of natural foods may exert effects during mastication as well as influencing physiological processes in the stomach and intestine. Some researchers have proposed that these physical properties of naturally high-fiber food play a major role in their effect on satiety (Heaton, 1981).

2.2. Products Supplemented with Fiber

In certain cases it is possible for a food product such as bread to be supplemented with a high-fiber constituent. For example, a proportion of the flour in a conventional recipe could be replaced by guar gum (Ellis et al., 1988). Replacements such as this may have effects on the textural properties of the product and therefore on its sensory qualities, but the major effect on appetite mechanisms is likely to be achieved during the postingestive phase.

2.3. Fiber Isolates

In this type of manipulation fiber is administered in a capsule or tablet, and its delivery can be timed so as to be synchronous with food consumption or to be separated from periods of eating by time intervals judged to be optimal for obtaining maximal effects. Methodologically this procedure has many advantages since it is not limited, for example, by necessary procedures for incorporating fiber into food products. Theoretically any amounts of fiber, of varying types, can be delivered by this route. However, in practice social convention and physical restraint will limit the number of capsules that may be consumed. It appears that approximately 20–30 tablets per day will be needed to administer useful amounts of fiber. Clearly, since the tablets or capsules are normally swallowed with water, any effect on appetite arising from sensory impact of fiber in the mouth is minimal. The effects of fiber isolates therefore emanate exclusively from their postingestive properties. Considering the methodology of research, this route of administration is valuable for detecting effects on satiety of different types of fiber with different effects on the physiology of the gastrointestinal tract. For example, this route of delivery could be used for an evaluation of the satiety properties of soluble and insoluble types of fiber.

It is worth mentioning at this point the difference between fiber and pharmacology in the modulation of appetite. Although the delivery of fiber isolates resembles the mode of delivery of a drug, it is unlikely that a dose of fiber could ever approach the potency of chemical modulation of satiety. The intention behind the use of fiber is not to achieve a total blockade of hunger but to use fiber to augment the natural action of energy or nutrients (see below). Although occasionally fiber isolates are referred to as a drug (Pitto *et al.*, 1987) it is probably advisable, as a general principle, to separate the control of appetite by fiber and pharmaceuticals. Of course certain provisions must be included. First, fiber should be regarded not simply as a commodity that adds indigestible bulk to food but as a chemically active agent stimulating physiological processes in the gastrointestinal tract. Second, certain appetite suppressant drugs could affect food consumption by means of similar mechanisms, and there is now good evidence that serotonergic agents intensify the satiating power of food (Blundell and Hill, 1987). This appears to be an action also induced by fiber (see below).

3. MEASURING THE SATIATING POWER OF FOOD: THE SATIETY CASCADE

In the study of therapeutic control of appetite, the major voiced concern is the need to reduce food intake. This goal sometimes causes researchers to overlook the fact that food is the natural anorexic agent and it is the physicochemical properties of food that (interacting with physiological processes) confer satiating power and cause oscillations in perceived hunger. A consideration of this issue involves an examination of the states and processes that modulate the pattern of eating (profile of meals and snacks) and the motivation to eat. In previous reviews we have described and defined terms such as hunger, appetite, satiation, and satiety (Blundell, 1979; Blundell and Burley, 1987). The use of hunger and satiety can be confusing. In much of the literature on the control of eating and body weight regulation, hunger and satiety are used as if they were polar opposites. For example, the question has been posed whether obese people eat because of excessive hunger or defective satiety (Blundell, 1977). This implies that the reason must be one or the other. However, when hunger is used as a motivational construct (see Blundell and Burley, 1987), it can be seen that hunger and satiety really refer to the same process: hunger is the force impelling eating (arising from the lack of food), and satiety is resistance to eating (arising from having eaten). The strengths of hunger and satiety are therefore reciprocally related.

However, in considering hunger the conscious sensation (which is the way most people use the term), it is clear that the perception of this feeling can be assessed before eating, during eating, or after eating. Therefore, the intensity of perceived hunger can reflect the tendency to start eating (hunger construct), the willingness to stop eating (satiation), and the maintenance of inhibition over further eating (satiety). The measured sensation of hunger therefore becomes one index of the strength of satiety. (For experimental purposes hunger is normally calibrated metrically by means of visual analogue rating scales.) If hunger sensations are weak or low, then satiety will be maintained; the restoration of perceived hunger represents the dissipation or weakening of satiety. This conceptualization reflects the different status of these two terms: hunger a conscious sensation and satiety a state or process. Measuring the hunger feeling can therefore be used to evaluate the development of satiation and the durability of satiety. How do these issues help in understanding the satiating power of food?

One approach to this problem is to consider the satiety cascade. Eating food has the capacity to take away hunger, and after satiation has occurred, further eating is inhibited for a period. What mechanisms are responsible for these processes? It is likely that the mechanisms involved in terminating eating and maintaining inhibition range from those that occur when food is initially sensed to the effects of metabolites following digestion and absorption. Satiation is the process that brings eating to a halt, whereas satiety is the state that maintains inhibition over further eating. By definition, therefore, satiety is not an instantaneous event but occurs over a finite time period; it is therefore useful to distinguish different phases of satiety, which can be associated with different mechanisms. This concept is illustrated in Fig. 1. Four mediating processes are identified: sensory, cognitive, postingestive, and postabsorptive. These processes maintain inhibition over eating (and hunger) during the early and late phases of satiety.

Sensory effects are generated through the smell, taste, temperature, and texture of food, and it is likely that these factors help to bring eating to a halt (satiation) and inhibit eating (particularly of foods with similar sensory properties) in the short term (immediate postprandial phase). Such a mechanism is embodied in the idea of a sensory-specific action of food first disclosed by Le Magnen (1960) and later taken up by other researchers (see Le Magnen, 1985). Cognitive effects represent the beliefs held about the properties of foods and their presumed effect on the eater; they have been demonstrated to operate under certain circumstances (Wooley, 1972). The category identified here as postingestive processes includes a number of possible actions including gastric disten-

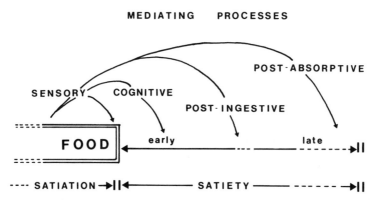

FIGURE 1. Conceptualization of the contribution of different processes to the intensity and time course of satiety.

sion and rate of gastric emptying, the release of hormones such as cholecystokinin from the duodenum, and the stimulation of physicochemically specific receptors along the gastrointestinal tract (Mei, 1985). The postabsorptive phase of satiety includes those mechanisms arising from the action of metabolites after absorption across the intestine and into the blood. This category embraces the action of chemicals such as glucose and the amino acids, which may directly act upon the brain after crossing the blood–brain barrier or which may influence the brain indirectly via neural inputs following stimulation of peripheral chemoreceptors.

The essence of the satiety cascade is the representation of the way in which the physiological processing of food after ingestion is reflected in the modulation of perceived hunger and in the inhibition over subsequent eating. This cascade has been used to evaluate the satiating effectiveness of foods in general (Blundell et al., 1987), to define the temporal bounds of satiety induced by different macronutrients (Hill and Blundell, 1988), and to understand the roles of taste and energy in the control of appetite (Blundell et al., 1988). With judicious designs of experiments and careful choice of experimental parameters, the satiety cascade can be used to determine the time periods of maximal satiety power brought about by different food components. The approximate anticipated moment of action of the mediating processes is illustrated in Fig. 1, but of course the mechanisms overlap, and their effects will be integrated to produce a conjoint action.

4. POTENTIAL ACTIONS OF FIBER

Before examining experimental evidence concerning the action of fiber, it is useful to consider how ingested fiber could influence the elements of the satiety cascade. As noted previously, the intensive chewing required for certain high-fiber foods (natural or synthesized) may deter continued consumption and prematurely terminate the eating process. Technically this would be referred to as an action on intrameal satiety (Van Itallie and Vanderveele, 1981) or satiation as it is here defined. In this way fiber could reduce food intake by creating a physical obstruction to eating (Bolton *et al.*, 1981). After eating has finished, this action of fiber in the mouth could contribute to early-phase satiety through a sensorily mediated process (Fig. 1). The legacy of the taste or "mouth feel" of fiber, in addition to the motoric action of chewing, would also be expected to influence this early stage of satiety. A further contribution to early satiety could arise from cognitively mediated processes. Attributes about fiber foods made by the eater such as the belief that "the food is filling" or "the food is nutritious and satisfying" may suppress appetite and deter any immediate resumption of eating. In this way both sensory and cognitive components of the satiety cascade could play a role in the satiety effect of fiber when it occurs as an integral part of the eaten product.

However, many researchers believe that it is in the gut that the dynamic action of fiber on physiological processes could most potently influence the expression of appetite. Fiber affects gastric distension, the rate of gastric emptying, and intestinal transit time. In addition, fiber may transport nutrients (digested or partially digested) along the gastrointestinal tract. By some of these actions the fiber content of food could activate a satiety mechanism instigated by some other component of ingested food. In addition to recruiting mechanisms involved in the passage of food along the gastrointestinal tract and the stimulation of specialized receptors in this region, there is good evidence that fiber will exert an action on satiety via postabsorptive processes. The amount of ingested fiber is known to modulate plasma glucose levels, and there are good logical reasons why the monitoring of blood glucose—by the liver or the brain—should play a role in controlling appetite. In the long-term control of body weight, fiber may also exert a role on fat, perhaps by preventing the absorption of fat or by transporting fat to sites in the intestinal tract where it may exert an inhibitory action on other physiological processes. The ileal brake phenomenon (Read *et al.*, 1984) is an example of this type of mechanism.

Consequently, it can be seen that fiber has the potential to influence all aspects of the satiety cascade discussed here (Fig. 1). Depending on the strategy used to incorporate fiber into the diet (natural, supplemented, isolates), the maximal effect may be exerted on satiation or on the early, middle, or late phase of satiety. In addition, the timing of maximal effect and the process most involved will vary with the type and amount of fiber consumed. At the present time little is known about the effect of these parameters on the processes of the satiety cascade.

5. EXPERIMENTAL STUDIES

Tables I, II, and III review briefly the available literature on dietary fiber, hunger, satiety, and food intake. Studies measuring only weight loss have not been included, as the power of fiber to aid weight reduction is reviewed elsewhere in this volume. The literature has been categorized according to the various modes of administration outlined earlier. It then becomes obvious how popular and successful are the many studies using isolated sources of dietary fiber. This mode of administration is surely the simplest and most flexible, yet still many have failed to use a placebo or control product to balance study design. It is true that a placebo product for some of the soluble fibers is extremely difficult to devise because of their unusual sensory properties, but open-ended studies do

TABLE I. Studies Using Fiber Isolates

Authors and study design	Type of fiber	Mode of administration	Hunger and food intake	Weight loss
Evans and Miller, 1975: Guar gum and methylcellulose were given for periods of 1 week each. Weighed dietary intakes were obtained over 4 consecutive weeks, and intakes during fiber weeks compared to 2 base-line weeks.	15.5 g guar gum granules or 16 g methylcellulose granules	With water, in two equal doses 30 min before meals. Total of 10 g fiber daily.	Not measured; 10% reduction in energy intake when taking fiber.	Yes
Krotkiewski, 1984: Alternating weekly treatment with guar gum or wheat bran for an average of 10 weeks. No dietary restrictions made.	Guar gum or wheat bran granulates	10 g of product twice daily with 150 ml water.	Reduced by guar gum at L and D[a]	Yes
Ryttig et al., 1985: Double-blind study of 89 slimming club members divided into fiber and placebo groups; 1200-kcal diet prescribed with placebo or fiber tablets.	Tablets of 80% cereal, 20% citrus fibers	Seven tablets four times daily 30 min before meals, with water. Total of 10 g fiber daily.	Reduced at B and L	Yes
Shearer, 1976: 1. 3-week double-blind placebo-controlled study of 73 obese dieters divided into fiber and placebo groups. 2. 4-week crossover study of 60 obese dieters on reducing diet or reducing diet plus fiber tablets.	Tablets of methylcellulose plus alginic acid. Minimum of 1.8 g fiber daily	2–4 tablets before and between meals with water	Increase in no. patients with reduced hunger and no. days patients had reduced hunger (both studies)	Not reported

(continued)

TABLE I. (*Continued*)

Authors and study design	Type of fiber	Mode of administration	Hunger and food intake	Weight loss
Durrant and Royston, 1978: 13 obese inpatients on 3-week energy-restricted diet given 0.84-MJ preloads on 2 days with the addition of methylcellulose on one day	Methylcellulose tablet	1-g tablet with 100 ml water	No effect	N/A
Hylander and Rossner, 1983: 120 obese on 3-week energy-restricted diet divided into two fiber groups and one control group.	Wheat bran or ispaghula granulate	1 (6.6-g) sachet 15 min before meals with 300 ml water	Reduced in fiber groups compared to controls at B and D	No effect of fiber
Rossner *et al.*, 1985: 60 obese women on prescribed 1400-kcal diet for up to 6 months given either fiber or placebo tablets	Mixed grain and citrus fiber	6 tablets 3 times daily 20–30 min before meals with 300 ml water (5 g fiber daily)	No effect of fiber	No effect of fiber
Gropper and Acosta, 1987: 8 obese children (6–12 years) on a weight-reducing diet were given either a high-fiber (15 g/day) supplement or a low-fiber placebo in crossover design for two consecutive 4-week periods. Energy intake was assessed by weekly 3-day diet diaries	Supplement contained corn bran, wheat flour, oat flakes, corn germ meal (180 kcal/day) Placebo content unknown (300 kcal/day)	3 suppl. daily, timing not specified; 15 g fiber daily	Hunger/satiety not measured; no effect on food intake	No effect of fiber; compliance to regimen doubtful

Study	Fiber type	Dose	Effect on hunger/satiety	Weight loss
Rossner et al., 1987: 1. 60 moderately obese women on a 1400-kcal diet were given fiber or placebo tablets in double blind design for 2 months.	Mixed grain and citrus fibers	6 tablets 3 times/day 20–30 min before meals; 5 g fiber/day	No effect of fiber	Both studies showed small improvements in weight loss with fiber
2. 45 moderately obese women on a 1600-kcal diet were given fiber or placebo tablets for 3 months. Hunger sensations (using VAS[a]) and weight change were assessed every second week of each study.	As above	As above: 7 g fiber/day (360 mg each)	No effect of fiber	
Rigaud et al., 1987: 20 healthy volunteers (10 m, 10 f) were given fiber or placebo tablets for 4 weeks in randomized, double-blind crossover study. Hunger assessments (using VAS) were made three times daily, and food intake recorded by questionnaire	As above	As above: 7.3 g fiber/day (350 mg each)	Mean weekly total hunger score signif. lower on fiber	Not reported
Pitto et al., 1987: 20 healthy volunteers were given vegetable fiber tablets for 15 days. Study was open-ended, with no placebo or control product. Food intake was not recorded. Reports of satiety were made (mode of assessment unknown)	Tablets contained cereal fiber, pea bran, pectin, and guar gum; Fibre Trim.	As above: 7.2 g fiber/day	Reports of reduced appetite and an early sensation of satiety in most subjects	N/A

[a]B, breakfast; L, lunch; D, dinner; VAS, visual analogue scales.

TABLE II. Studies Using Fiber-Supplemented Foods

Authors and study design	Type of fiber	Mode of administration	Hunger and food intake	Weight loss
Porikos and Hagamen, 1986: 50 young men (19 obese) consumed 400-kcal sandwich preloads of high or low fiber content; 30–45 min later test meals of sandwiches offered and intake recorded	Methylcellulose (inferred, not reported in article)	Sandwich preloads contained 6.6 vs. 0.4 g fiber per meal	Reduced by fiber preload. Fiber reduced size of test meal in obese only	N/A
Burley et al., 1987b: 20 nonobese females each consumed high- and low-fiber (toast and cereal) breakfasts (608 vs. 697 kcal) followed by a test meal lunch	Guar gum and wheat bran	Guar bread and bran breakfast cereal compared to white wheat bread and cornflakes. Fiber content of meals was 12 vs. 3.3 g	Similar hunger ratings after both breakfasts. Greater fullness after high-fiber breakfast. Lunchtime energy intake did not differ in spite of lower energy content of high-fiber breakfast.	N/A
Cocchi et al., 1984: 10 nonobese volunteers consumed isocaloric pasta meals of high and low fiber content. Hunger and satiety rated on a 9-point category scale	Crusca di carruba (bran)	Fiber added to pasta	Reduced hunger with high-fiber pasta meal	N/A
Ellis et al., 1985: 12 diabetic and 9 healthy volunteers consumed bread breakfasts with four different fiber	Guar gum	Guar gum replaced wheat flour by 0, 5,	Both groups found 6 and 8 g guar bread	N/A

contents and rated hunger and satiety for 2 hr after each meal		10, and 15% to make bread at four different fiber levels. Isocaloric meals containing 0, 3, 6, and 8 g guar gum given.	meals more satiating than control	
Mickelson et al., 1979: 16 overweight men on an 8-week weight loss program consumed an additional 12 slices of an energy-reduced high-fiber bread or 12 slices of an enriched low-fiber white bread	Methylcellulose	12 slices high-fiber bread adding 25.5 g crude fiber per day, compared with 1.02 g from low-fiber bread	Fewer subjects hungry at night and fewer currently hungry by end of study on high-fiber bread regimen	Yes
Quaade et al., 1988: 20 obese patients took part in a randomized single-blind crossover study of the effects of adding cellulose fiber to a VLCD formula diet (388–460 kcal/day). Duration of study was 2+2 weeks for each group. Frequent assessments of hunger and satiety were made using VAS	Cellulose	Fiber was incorporated into the formula; no control or placebo product was used; 30 g fiber/day	Hunger was significantly lower during fiber periods, but no effect of fiber on satiety	Weight loss not improved by fiber
Grimes and Gordon, 1978: 12 healthy volunteers consumed meals of white and wholemeal bread to comfortable fullness	Wheat bran	Wholemeal bread contains 8.5 g fiber/100 g vs. white bread at 2.2 g/100 g	10 out of 12 subjects consumed less wholemeal bread and therefore less energy than at the white bread meal	N/A

(continued)

TABLE II. (*Continued*)

Authors and study design	Type of fiber	Mode of administration	Hunger and food intake	Weight loss
Wilmshurst and Crawley, 1980: 7 overweight subjects consumed low-energy milky drinks with and without the addition of fiber	Guar gum	2 g of guar was added to a 200-g milky drink, increasing its viscosity	Time to maximal hunger delayed by addition of guar	N/A
Bryson *et al.*, 1980: 10 healthy volunteers consumed meals of white or wholemeal bread with butter and jam in *ad-lib* quantities. Weighed dietary intakes were obtained for 48 hr to assess the influence of the meals on subsequent intake	Wheat bran	Wholemeal bread contains 8.5 g fiber/100 g, and white bread 2.2 g per 100 g	No effect of bread type on energy or weight of meals consumed. No effect of wholemeal bread on following 24 hr	N/A
Stevens *et al.*, 1987: 12 healthy female volunteers consumed 4 types of cracker (12/day) for 2 weeks each in balanced design. Digestible energy intake was calculated for each 2-week period to assess the influence of fiber type on spontaneous food intake	Wheat bran, psyllium gum or combination giving 23 g/day compared to 4 g/day in control crackers	Fiber added to wheat crackers	No effect of fiber on appetite; digestible energy intake reduced by psyllium and bran + psyllium crackers compared to control	No
Barkeling *et al.*, 1988: 28 lean and 20 obese subjects breakfasted on yogurt meals containing fiber or a placebo product; 3 1/2 hr later an excess portion of an homogeneous dish was served, and amounts consumed assessed by a hidden scale (VICTOR). Food preference lists and hunger ratings (VAS) were completed before and after the meals	Unknown composition	Fiber or placebo stirred into yogurt; 7 g fiber	No effects of fiber on hunger or food prefs. No effects of fiber on lunchtime food intake	N/A

TABLE III. Studies Using Foods Naturally High in Fiber

Authors and study design	Type of fiber	Mode of administration	Hunger and food intake
Haber et al., 1977: 10 nonobese volunteers consumed isocaloric meals of whole apples, pureed apples, and apple juice at the same rate and rated hunger for the following 2 hr	Apple fiber	482 g apple flesh (14 g fiber), the same quantity pureed, and 469 g apple juice (0 g fiber)	Juice was less satiating than puree, and puree less satiating than fruit
Bolton et al., 1981: Healthy volunteers consumed meals of whole grapes and oranges and their isocaloric equivalents as orange or grape juice.	Fruit fiber	626 g orange flesh (15.7 g fiber) was compared to 610 ml juice (0 g fiber), and 360 g of whole grapes (4.3 g fiber) were compared to 323 ml juice (0 g fiber)	Whole-fruit meals more satiating than juice meals
Duncan et al., 1983: 10 obese and 10 lean volunteers consumed high-(HED) and low-(LED)-energy-density diets of equal palatability for 5 days. *Ad-lib* quantities of foods were offered on each diet, and volunteers asked to eat to satiety	Mixed fiber	7 g fiber per 1000 kcal from HED diet and 1 g fiber per 1000 kcal on the LED diet	Satiety was achieved on the LED diet at an energy intake approx. half that on the HED diet. Eating time was longer on LED by 33%
Burley et al., 1987a: 16 nonobese females consumed high- and low-fiber isocaloric lunches based on fiber-intact or fiber-depleted foods. Meals consisted of soup, bread, pasta dish, and dessert. Hunger and satiety ratings (VAS) were completed before and for 2 1/2 hr after lunches. All food and drink consumed until bedtime was weighed and recorded to assess the influence of the lunchtime meals on food intake	Mixed fiber	Fiber from wholemeal bread, lentil soup, beans, wholemeal pasta, and fruit: 30 g fiber vs. 3.3 g for the low-fiber lunch	Both meals had a potent satiating effect Hunger and satiety ratings similar after both meals, but 590 kcal less consumed after high-fiber lunch

little to enhance our knowledge of the potential of fiber to influence energy intake. Additionally, the precise content and composition of fiber are surely known for the commercially produced isolates but are often not clearly stated in publications.

Studies using fiber-supplemented foods (Table II) seem, on the whole, to give more consistent results. Negative or ambiguous results seem to arise here when the fiber content of the meals differs only slightly or when the energy content of the food vehicle is particularly low.

Few studies using fiber-intact or natural high-fiber foods have been conducted, possibly because of the difficulties associated with experimental design. On a weight basis fiber-intact foods are generally lower in macronutrient content than their fiber-depleted versions (e.g., whole fruit versus fruit juice). This creates problems for the experimenter who wishes to devise meals of differing fiber content but similar palatability, weight, and macronutrient content. Notwithstanding the difficulties involved in this methodology, all studies in this section reveal an effect of high-fiber foods on either satiety or energy intake. This is testimony to the robustness of these types of food to influence food intake.

6. HOW DOES FIBER ACT?

The literature surveyed in Tables I, II, and III has indicated that fiber will, if given in sufficient quantities, reduce food intake, and it seems in some cases to be sufficient to reduce body weight. How does fiber act? By influencing satiety, satiation, or both? We have conducted a number of studies specifically designed to investigate the mode of action of fiber on food intake.

In the first of these studies (Burley *et al.*, 1987b) 20 normal-weight female volunteers consumed breakfast meals of high and low fiber content (12.0 and 3.0 g fiber, respectively) on two separate occasions. Visual analogue scales were used to record hunger, fullness, desire to eat, and a measure of prospective consumption (as described by Hill *et al.*, 1984) for 2.5 hr after each meal. At this point, a tray of preweighed lunch foods was offered, and subjects were requested to eat as much or as little as they desired. The breakfast meals were both based on breakfast cereal and toast. The high-fiber breakfast meal used bread to which guar gum had been added before baking and Kellogg's Bran Flakes. The low-fiber breakfast was based on ordinary white bread and Kellogg's Corn Flakes. These meals were constructed to resemble—as far as possible—an ordinary breakfast, as commonly eaten in Britain. The fiber difference between the meals was therefore not large, and there was an energy difference between the meals of about 90 kcal.

In spite of the lower energy content of the high-fiber breakfast, there was no significant difference in visual analogue scale ratings after the two meals except for fullness ratings, which were higher after the high-fiber meal. Additionally, *ad-lib* food intake at lunchtime was similar after the two breakfasts (819 ± 307 versus 830 ± 308 kcal) despite the reduced energy content of the high-fiber breakfast. One interpretation of the results of this study is that the fiber (guar gum and wheat bran) was able to intensify the satiating power of the reduced level of available energy of the high-fiber

TABLE IV. Composition of High- and Low-Fiber Lunch Meals

High fiber		Low fiber	
200 g	Lentil soup (Heinz)	200 g	Chicken soup (Heinz)
45 g	Wholemeal bread	45 g	White bread
10 g	Butter	5 g	Butter
53 g	Wholemeal pasta (dry wt.)	60 g	White pasta (dry wt.)
165 g	Spiced red kidney beans	132 g	Bolognese Sauce
13 g	Cheddar cheese		
110 g	Tinned blackcurrants (in juice)	150 g	Fruit yogurt
40 g	Fruit yogurt		

breakfast. In other words, the fiber content of the high-fiber breakfast made up for the loss of satiating power associated with the 90-kcal deficit.

The second study was conducted with 16 female volunteers of normal weight for height. Equicaloric high- and low-fiber lunches (of equal palatability) based on fiber-intact or fiber-depleted foods were devised, with similar macronutrient contents but vastly differing amounts of fiber. See Tables IV and V for details of the meals. Subjects fasted overnight and followed the scheme of assessments as outlined in Table VI. Visual analogue scales (VAS) of hunger, desire to eat, fullness, and prospective consumption were completed by the volunteers as in the first study. Times of eating and the weights of all food and drink consumed after each test lunch and until bedtime were recorded by the subjects. These data were subsequently analyzed using a computerized version of Mc-Cance and Widdowson's food tables (Paul and Southgate, 1978), and the energy intake after each meal was calculated.

Analysis of visual analogue scale ratings indicated no significant effect of fiber on hunger, desire to eat, fullness, and prospective consumption. Both lunches had a very potent satiating effect, as they were quite substantial. However, analysis of the dietary records revealed a significant effect of fiber on subsequent energy intake. Volunteers consumed some 590 kcal less after the high-fiber lunch when compared to intake after the low-fiber meal (967 ± 532 versus 1559 ± 930, $P < 0.01$). This study, using quite large doses of fiber, has shown that fiber can intensify satiety—the maintenance of postmeal inhibition over further eating. Further work is needed to ascertain whether smaller amounts of fiber and different modes of administration would also provoke such a pronounced effect on subsequent food intake.

TABLE V. Nutrient Content of High- and Low-Fiber Lunches (by Calculation)

	Energy		Protein		Fat		Carbohydrate		Fiber
	Kcal	MJ	g	%E	g	%E	g	%E	(g)
High fiber	795	3.3	33	17	28	32	106	50	30
Low fiber	790	3.3	33	17	29	33	106	50	3.3

TABLE VI. Experimental Schedule

Time	Rating no.	Assessment[a]
11:45	1	VAS (on arrival)
Premeal		SC
	2	VAS
During lunch		
	3	VAS
	4	VAS
	5	VAS
	6	VAS
Postmeal (time from starting meal)		
1:00 60 min	7	VAS
		SC
1:30 90 min	8	VAS
		SC
2:00 120 min	9	VAS
2:30 150 min	10	VAS
		SC

[a]VAS, visual analogue scales; SC, sensations checklist.

Thus far we have demonstrated that fiber may influence satiety and also intensify the satiating power of the energy content of a meal. Will fiber also influence the process of satiation? To investigate this, open meals consisting of an excess portion of the same foods as in the previous study were presented to eight female volunteers. Subjects were requested to consume as much as they wished of all food items offered. Energy intake was calculated from weights of food selected. Visual analogue scale ratings were completed as before. Analysis of these ratings showed that the high-fiber meal reduced desire to eat and prospective consumption more than the low-fiber meal ($P < 0.03$), but significantly less energy (difference between meals 76 ± 63 kcal, $P < 0.05$,) was consumed from the high-fiber food. This small study does suggest that fiber (within fiber-intact food) can impede food intake and does indeed influence satiation, the process that brings eating to a close. Again, further work needs to be done to differentiate between the action of fiber *per se* and the qualities of fiber-intact food (lower energy content, bulkiness, etc.) on satiation.

7. FIBER AND SATIETY

The usefulness of dietary fiber for the treatment of obesity is reviewed elsewhere in this volume. However, the evidence presented here on short-term studies indicates that dietary fiber does influence energy intake. The strength of the effect varies with the amount of fiber delivered, the timing of the delivery, and the strategy used to deliver the fiber to the system. Known physiological effects of fiber suggest that it acts on various mechanisms involved in the satiety cascade, and experimental evidence indicates that fiber acts both within and between meals. What is the relevance of these effects for the potential impact of dietary fiber on obesity? The situation is depicted in Fig. 2.

The crucial step is the strength of the link between the demonstrated action of fiber on short-term energy intake and the effect on body weight achieved over months or years. The potential exists, and the key factor appears to be that of compliance. Fiber will only work, whatever the strategy, if it is ingested. Therefore, the effectiveness of dietary fiber to affect energy intake in the long term will depend on the success of methods for maintaining its intake over long periods—in other words, maintaining compliance with dietary instructions. How can good compliance be achieved? There is no magic formula, and the factors that favor compliance vary with circumstances— social and physiological. However, it is known that the palatability of food promotes consumption. Therefore, one challenge for the nutrition industry is to develop high-fiber products that are highly palatable. This may not be easy, since it will be necessary to insure that other dietary requirements are not compromised.

Fiber isolates provide one apparently convenient solution, since in theory, people could continue to eat their usual, and presumably preferred, diet while simultaneously ingesting packages of fiber. In situations where compliance can be maintained through the use of a palatable naturally high-fiber diet (Duncan et al., 1983), then effects on energy intake are encouraging. However, it is worth considering that the most effective use of fiber in the management of obesity may be not in the promotion of weight loss but as a prophylactic to prevent weight gain in the first place or to prevent weight regain after a weight loss has been achieved through some more vigorous type of therapy. By its physiological nature dietary fiber can not be expected to bring about huge energy deficits. However, fiber will act in concert with other components of food to influence the physiological mechanisms underlying satiety. At the present time surprisingly little is known about this aspect of the action of dietary fiber. Further investigations of fiber not only will reveal how it can be used more effectively but will also disclose more details about the structure and operations of satiety.

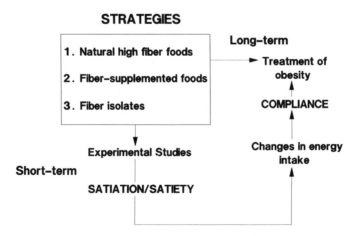

FIGURE 2. Representation of the link between the demonstrated effects of fiber on short-term energy intake and potential long-term effects on intake and body weight.

REFERENCES

Barkeling, B., Ryttig, K., and Rossner, S., 1988, VIKTOR—Objective analysis of human eating behaviour and food preferences, in: *1st European Congress on Obesity, Stockholm*, p. 136.

Blundell, J. E., 1977, Hunger and satiety in the control of food intake: Implications for the treatment of obesity, *Clin. Dietol.* **23:**257–272.

Blundell, J. E., 1979, Hunger, appetite and satiety—psychological constructs in search of identities, in: *Nutrition and Lifestyles* (M. Turner, ed.), Applied Science Publishers, London, pp. 21–42.

Blundell, J. E., and Burley, V. J., 1987, Satiation, satiety and the action of fibre on food intake, *Int. J. Obesity.* **11**(Suppl. 1)**:**9–25.

Blundell, J. E., and Hill, A. J., 1986, Biopsychological interactions underlying the study and treatment of obesity, in: *The Psychosomatic Approach: Contemporary Practise of Whole-Person Care* (M. J. Christie and P. G. Mellet, eds.), John Wiley & Sons, Chichester, pp. 115–138.

Blundell, J. E., and Hill, A. J., 1987, Serotonergic modulation of the pattern of eating and the profile of hunger–satiety in humans, *Int. J. Obesity.* **11**(Suppl. 3)**:**141–153.

Blundell, J. E., Rogers, P. J., and Hill, A. J., 1987, Evaluating the satiety power of foods: Implications for acceptance and consumption, in: *Chemical Composition and Sensory Properties of Food and Their Influence on Nutrition* (J. Solms, ed.), Academic Press, London, pp. 205–219.

Blundell, J. E., Rogers, P. J., and Hill, A. J., 1988, Uncoupling sweetness and calories: Methodological aspects of laboratory studies on appetite control, *Appetite* **11:**54–61.

Bolton, R. P., Heaton, K. W., and Burroughs, L. F., 1981, The role of dietary fibre on satiety, glucose, and insulin: Studies with fruit and fruit juice, *Am. J. Clin. Nutr.* **34:**211–217.

Bryson, E., Dore, C., and Garrow, J. S., 1980, Wholemeal bread and satiety, *J. Hum. Nutr.* **34:**113–116.

Burley, V. J., Blundell, J. E., and Leeds, A. R., 1987a, The effect of high and low fibre lunches on blood glucose, plasma insulin levels and hunger sensations, *Int. J. Obesity* **11**(Suppl. 2)**:**12.

Burley, V. J., Leeds, A. R., and Blundell, J. E., 1987b, The effect of high and low fibre breakfasts on hunger, satiety, and food intake in a subsequent meal, *Int. J. Obesity* (Suppl.) **1:**87–93.

Cocchi, M., Siniscalchi, C., Billi, G. C., Sciarretta, G., De Muti, R., and Ruffilli, E., 1984, Alimentary fibre effect on the feeling of satiety: Subjective evaluation and biochemical data modifications, *Giorn. Clin. Med.* **65:**99–108.

Duncan, K., Bacon, J. A., and Weinsier, R. L., 1983, The effect of high and low energy density diets on satiety, energy intake, and eating time of obese and non-obese subjects, *Am. J. Clin. Nutr.* **37:**763–767.

Durrant, M. L., and Royston, P., 1978, The effect of preloads of varying energy density and methyl cellulose on hunger, appetite and salivation, *Proc. Nutr. Soc.* **37:**87A.

Ellis, P. R., Apling, E. C., Leeds, A. R., Peterson, D. B., and Jepson, E. W., 1985, Guar bread and satiety: Effects of an acceptable new product in overweight diabetic patients and normal subjects, *J. Plant Foods* **6:**253–262.

Ellis, P. R., Burley, V. J., Leeds, A. R., and Peterson, D. B., 1988, A guar-enriched wholemeal bread reduces postprandial glucose and insulin responses, *J. Hum. Nutr. Diet.* **1:**77–84.

Evans, E., and Miller, D. S., 1975, Bulking agents in the treatment of obesity, *Nutr. Metab.* **18:**199–203.

Garrow, J., 1982, Does plumpness matter? *Nutr. Bull.* **7:**49–53.

Grimes, D. S., and Gordon, C., 1978, Satiety value of wholemeal and white bread, *Lancet* **2:**106.

Gropper, S. S., and Acosta, P. B., 1987, The therapeutic effect of fibre in treating obesity, *J. Am. Coll. Nutr.* **6**(6)**:**533–535.

Haber, G. B., Heaton, K. W., Murphy, B., and Burroughs, L., 1977, Depletion and disruption of dietary fibre. Effects on satiety, plasma-glucose and serum-insulin, *Lancet* **2:**679–682.

Heaton, K. W., 1981, Dietary fibre and energy intake, in: *Regulators of Intestinal Absorption in Obesity, Diabetes and Nutrition* (P. Berchtold, A. Cairella, A. Jacobelli, and V. Silano, eds.), Societa Editrice Universo, Rome, pp. 283–294.

Hill, A. J., and Blundell, J. E., 1988, Role of amino acids in appetite control in man, in: *Amino Acid Availability and Brain Function in Health and Disease* (G. Huether, ed.), Springer-Verlag, Berlin, pp. 239–248.

Hill, A. J., Magson, L. D., and Blundell, J. E., 1984, Hunger and palatability: Tracking ratings of subjective experience before, during and after the consumption of preferred and less preferred food, *Appetite* **5**:361–371.

Hylander, B., and Rossner, S., 1983, Effects of fibre intake before meals on weight loss and hunger in a weight reducing club, *Acta Med. Scand.* **213**:217–220.

Krotkiewski, M., 1984, Effect of guar gum on body-weight, hunger ratings and metabolism in obese subjects, *Br. J. Nutr.* **52**:97–105.

Le Magnen, J., 1960, Effets d'une pluralite de stimuli alimentaires sur le determinisme quantif de l'ingestion chez le rat blanc, *Arch. Sci. Physiol.* **14**:411–419.

Le Magnen, J., 1985, *Hunger*, Cambridge University Press, Cambridge.

Mei, N., 1985, Intestinal chemosensitivity, *Physiol. Rev.* **65**:211–237.

Mickelson, O., Makdani, D. D., Cotton, R. H., Titcomb, S. T., Colmey, J. C., and Gatty, R., 1979, Effects of a high fibre diet on weight loss in college-age males, *Am. J. Clin. Nutr.* **32**:1703–1709.

Paul, A. A., and Southgate, D. A. T., 1978, *McCance and Widdowson's "The Composition of Foods,"* 4th ed., MRC Special Report No. 297, HMSO, London.

Pitto, G., Sganga, F., and Mereta, F., 1987, Una nuova fibra vegetale. Studio clinico condotto su 20 soggetti voluntari sani, *Minerva Dietol. Gastroenterol.* **33**(3):235–237.

Porikos, K., and Hagamen, S., 1986, Is fibre satiating? Effects of a high fibre preload on subsequent food intake of normal-weight and obese young men, *Appetite* **7**:153–162.

Quaade, F., Vrist, E., and Astrup, A., 1988, Fibre supplementation to very low calorie diet normalises hunger and alleviates constipation, in: *1st European Congress on Obesity, Stockholm,* p. 289.

Read, N. W., McFarlane, A., Kinsman, R., Bates, T., Blackhall, N. W., Farrar, G. B. J., Hall, J. C., Moss, G., Morris, A. P., O'Neill, B., Welch, I., Lee, Y., and Bloom, S. R., 1984, Effect of infusion of nutrient solutions into the ileum on gastrointestinal transit and plasma levels of neurotensin and enteroglucagon in man, *Gastroenterology* **86**:274–280.

Rigaud, D., Ryttig, K. R., Leeds, A. R., Bard, D., and Apfelbaum, M., 1987, Effects of a moderate dietary fibre supplement on hunger rating, energy input and faecal energy output in young, healthy volunteers: A randomized, double-blind cross-over trial, *Int. J. Obesity* **11**(Suppl. 1):73–78.

Rossner, S., Von Zweigbergk, D., and Ohlin, A., 1985, Effects of dietary fibre in treatment of overweight out-patients, in: *Dietary Fiber and Obesity* (P. Bjorntorp, G. V. Vahouny, and D. Kritchevsky, eds.), Alan R. Liss, New York, pp. 69–76.

Rossner, S., Von Zweigbergk, D., Ohlin, A., and Ryttig, K. R., 1987, Weight reduction with dietary fibre supplements. Results of two double-blind studies, *Acta Med. Scand.* **222**(1):83–88.

Ryttig, K. R., Larsen, S., and Haegh, L., 1985, Treatment of slightly to moderately overweight persons. A double-blind placebo-controlled investigation with diet and fibre tablets (Dumovital), in: *Dietary Fiber and Obesity* (P. Bjorntorp, G. V. Vahouny, and D. Kritchevsky, eds.), Alan R. Liss, New York, pp. 77–84.

Shearer, R., 1976, Effects of bulk-producing tablets on hunger intensity in dieting patients, *Curr. Ther. Res.* **19**:433–441.

Stevens, J., Levitsky, D. A., VanSoest, P. J., Robertson, J. B., Kalkwarf, H. J., and Roe, D. A., 1987, Effect of psyllium gum and wheat bran on spontaneous energy intake, *Am. J. Clin. Nutr.* **46**:812–817.

Taylor, R., 1981, Report on Seminar V: Dietary fibres, in: *Regulators of Intestinal Absorption on Obesity, Diabetes and Nutrition* (P. Berchtold, M. Cairella, A. Jacobelli, and V. Silano, eds.), Sozieta Editrice Universo, Rome, p. 328.

Van Itallie, T. B., and Abraham, S., 1985, Some hazards of obesity and its treatment, in: *Recent Advances in Obesity Research* IV (J. Hirsch and T. B. Van Itallie, eds.), John Libbey, London, pp. 1–19.

Van Itallie, T. B., and Vanderweele, D. A., 1981, The phenomenon of satiety, in: *Recent Advances in*

Obesity Research III (P. Bjorntorp, M. Cairella, and A. N. Howard, eds.), John Libbey, London, pp. 278–289.

Wilmshurst, P., and Crawley, J. C. W., 1980, The measurement of gastric transit time in obese subjects using Na^{24} and the effects of energy content and guar gum on gastric emptying, *Br. J. Nutr.* **44:**1–6.

Wooley, S. C., 1972, Physiologic versus cognitive factors in short term food regulation in the obese and non-obese, *Psychosom. Med.* **34:**62–68.

18

Lente Carbohydrate or Slowly Absorbed Starch

Physiological and Therapeutic Implications

DAVID J. A. JENKINS, ALEXANDRA L. JENKINS, THOMAS M. S. WOLEVER, VLADIMIR VUKSAN, FURIO BRIGHENTI, STEPHEN CUNNANE, A. VENKET RAO, LILIAN U. THOMPSON, and ROBERT G. JOSSE

1. SUMMARY

Current dietary recommendations support the use of increased carbohydrate intakes from minimally processed or high-fiber starchy foods. A characteristic of such food is often that it is more slowly digested and absorbed than many refined and highly processed foods. The assumption that slow absorption of nutrients may have benefit is central to the fiber hypothesis and has been described as a new therapeutic principle. Other similar approaches include the use of specific digestive enzyme inhibitors and increased feeding frequency. Slow carbohydrate absorption may result in a number of sequelae, including flatter glucose and endocrine responses and lower blood lipids, reduced blood urea levels through increased NH_3 trapping in the colon, increased laxation, and enhanced colonic retrieval of short-chain fatty acids with further potential beneficial effects.

DAVID J. A. JENKINS, ALEXANDRA L. JENKINS, THOMAS M. S. WOLEVER, VLADIMIR VUKSAN, FURIO BRIGHENTI, STEPHEN CUNNANE, A. VENKET RAO, LILIAN U. THOMPSON, and ROBERT G. JOSSE • Department of Nutritional Sciences, Faculty of Medicine and Division of Endocrinology and Metabolism, St. Michael's Hospital, University of Toronto, Toronto, Ontario M5S 1A8, Canada.

2. INTRODUCTION

There is a growing consensus among agencies concerned with health to recommend consumption of diets low in saturated fat and possibly cholesterol and replacement of fat calories with starchy foods, especially those that are less processed and high in fiber (American Diabetes Association, 1987; British Diabetic Association, 1982; Canadian Diabetes Association, 1981; American Heart Association, 1982; National Academy of Sciences, 1982). A major influence in initiating this change has been the dietary fiber hypothesis of Drs. Denis Burkitt and Hugh Trowell, drawing on their experiences in Colonial Uganda (Burkitt and Trowell, 1975), and other senior scientists such as Dr. Alec Walker in South Africa. Indeed, it was work in the Colonial Service that allowed many men and women, including Arthur McHarrison in India and Cicely Williams in the Gold Coast (now Ghana), to gain valuable scientific insights, especially in the area of nutrition. Comparisons were documented between traditional and Western diets and the attendant differences in patterns of disease. It also exposed clinical scientists and nutritionists to the wisdom of ancient civilizations preserved in traditional medical practice (e.g., the ancient Indian medicine of more than 2000 years ago).

3. THE HYPOTHESIS

The overall hypothesis that has developed is that slower absorption of nutrients may confer metabolic advantages. This concept has been expanded to a proposal that slowing absorption should be regarded as a new therapeutic principle (Creutzfeldt and Folsch, 1983). Thus, reduced rates of carbohydrate absorption resulting in less potential demand for insulin may be associated with improved diabetes control (Anderson and Chen, 1979), and, in the longer term, lower blood lipids (Jenkins and Wolever, 1981). Furthermore, with slower carbohydrate absorption, increased amounts of nutrient would be delivered to the colon. Here it would serve as a substrate for bacterial fermentation generating short-chain fatty acids (SCFA), which in turn would be absorbed and may contribute to some of the benefical effects seen in carbohydrate and lipid metabolism with fiber (Cummings and Branch, 1986). The provision of increased carbohydrate to the colon, by providing an energy substrate, would also allow the entrapment of colonic NH_3, thereby possibly reducing blood urea levels from which the NH_3 is derived (Rampton et al., 1984). This may be turned to possible advantage in renal failure (Fig. 1). In addition, in liver disease, the burden of gut-derived nitrogenous products absorbed into the circulation would be reduced, with potential benefits in hepatic encephalopathy. Finally, the increased colonic bulk and nutrient supply, especially in the form of butyrate, may enhance colonic health (Cummings, 1981). Methods for achieving a reduced rate of nutrient absorption with respect to carbohydrate include (1) high dietary fiber content of foods or viscous fiber supplements; (2) enzyme inhibitors in foods or provided as pharmacological agents, e.g., *acarbose,* the glycoside hydrolase inhibitor; (3) selection of low-glycemic-index starchy foods, which by definition are foods with slower rates of digestion and absorption; and (4) increased meal frequency.

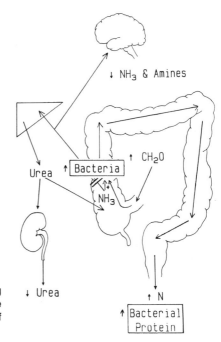

FIGURE 1. The potential effect that increasing the delivery of fermentable carbohydrate to the colon may have in reducing the filtered load of urea.

4. APPROACHES USED TO SLOW CARBOHYDRATE ABSORPTION

4.1. Dietary Fiber

Dietary fiber appears to delay absorption by a number of potential mechanisms. By contributing to food form (e.g., enveloping a grain) it may reduce the rate of digestion of the starchy interior or reduce starch granule expansion on cooking, so rendering the food less readily digestible (e.g., "parboiling") (Wolever *et al.*, 1986). Alternatively, viscous fibers (e.g., the gums—guar, tragacanth, locust bean, etc.—pectins, and mucilaginous materials) may be added to foods to reduce the rate of absorption (Jenkins *et al.*, 1978a). Under some circumstances viscous fiber may act by delaying gastric emptying (Schwartz *et al.*, 1982; Leeds *et al.*, 1981) but, more usually and predominantly, by delaying small intestinal absorption through impeding diffusion through the bulk phase and possibly increasing the thickness of the unstirred water layer (Blackburn *et al.*, 1984; Flourie *et al.*, 1984). *In vivo* dietary fiber in starchy foods has been shown in ileostomate volunteers to result in increased losses of starch and sugars in the ileostomy effluent (Jenkins *et al.*, 1987a) (Fig. 2). Although the range is relatively small and unlikely to account for the sometimes very large differences seen in glycemic effect, nevertheless, especially on high-carbohydrate diets, available carbohydrate losses to the colon are likely to constitute a very substantial proportion of the colonic load of carbohy-

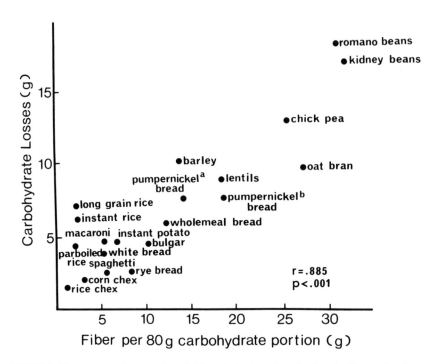

FIGURE 2. The increased losses of available carbohydrate from the terminal ileum of an ileosto-mate as foods with increasing fiber content were fed on an otherwise constant diet. (Adapted from Jenkins *et al.*, 1987a.)

drate. *In vitro* studies have demonstrated that the rate of starch digestion relates to the fiber content of foods tested (Jenkins *et al.*, 1984), and the ability of various fibers to impede diffusion has been clearly demonstrated, especially in the case of bile salts, with important implications in cholesterol metabolism (Kritchevsky and Story, 1974; Vahouny *et al.*, 1980).

The therapeutic implications of fiber have been many. Early on viscous fibres were used to reduce serum lipid levels (Miettinen and Tarpila, 1977). The reductions were seen predominantly in LDL cholesterol, with other favorable changes in lipoprotein subclasses (Jenkins *et al.*, 1979; Bosello *et al.*, 1984). Although viscous fibers increase bile acid losses to a variable extent (Kay and Truswell, 1977; Anderson *et al.*, 1984; Jenkins *et al.*, 1977), this cannot be the sole mechanism of action. Other properties related to the reduced rate of absorption, e.g., lower insulin responses with reduced stimulus to HMG-CoA reductase synthesis (Rodwell *et al.*, 1976) and possibly inhibition of cholesterol synthesis secondary to the SCFA (propionate) produced on colonic carbohydrate fermentation (Anderson *et al.*, 1984; Thacker *et al.*, 1981), are likely also to play a part.

At an early stage in the exploration of fiber, its effects on carbohydrate metabolism were exploited in the treatment of diabetes. Addition of fiber to mixed meals flattened blood glucose responses in type I and II diabetics. In metabolic studies as well as in longer-term outpatient studies, reduced urinary glucose losses and improved diabetic control were observed (Anderson and Chen, 1979; Jenkins *et al.*, 1977; Ray *et al.*, 1983; Aro *et al.*, 1981).

To a variable extent, studies with purified fiber have shown a tendency for diminution of the effect over time. The reasons for this and the degree to which this relates to noncompliance remain to be determined. In the absence of general availability of palatable formulations, noncompliance is likely to be a major problem in long-term outpatient studies.

High-fiber foods were also used to improve diabetes control and lower blood lipids (Anderson and Ward, 1979a). The effects of high-carbohydrate, high-fiber diets in the studies of Anderson and co-workers have been impressive. Insulin need was eliminated in type II diabetes controlled initially on substantial daily doses (up to 30 IU/day) (Anderson and Ward, 1979b). The diets, however, often contained several changes in addition to fiber, e.g., food form alterations, antinutrients, and higher carbohydrate, all of which, in addition to fiber, may have contributed to the final effect. Beans have proved particularly useful as high-fiber foods in improving glucose control in diabetes (Burke *et al.*, 1982) and lowering blood lipids, especially triglycerides (Albrink *et al.*, 1979).

Most recently the use of fiber sources to provide fermentable carbohydrate to the colon has been exploited. Soluble fiber supplements have been shown to result in reduced blood urea levels in patients with renal failure when given at levels of 7–15 g daily for 6–8 weeks (Rampton *et al.*, 1984). Furthermore, the presence of fiber in high-vegetable-protein diets may have been responsible for improvement of hepatic encephalopathy when patients changed to these diets from more conventional foods (DeBruin *et al.*, 1983). These phenomena presumably relate to the ability of bacteria to entrap HN_3 for their own cell protein synthesis. The NH_3 in turn is derived from blood urea that crosses the colonic mucosa to the lumen, where it is hydrolyzed by bacterial action to NH_3. It is this NH_3 that forms the substrate for nitrogen fixation and bacterial protein synthesis as long as fermentable carbohydrate is available as an energy source (Fig. 3).

The lessons learned from studies of fiber, in terms of mechanism and application, apply to many of the approaches developed for slowing carbohydrate absorption. Whenever carbohydrate absorption rate is slowed in the small intestine, a spectrum of physiological events with therapeutic potential is set in train. Likewise, whenever fermentable carbohydrate enters the colon, whether fiber, starch, or sugars, depending on the rate and degree to which it is fermented, this again will produce physiological effects with possible therapeutic implications.

4.2. Enzyme Inhibitors

Enzyme inhibitors or factors that limit the rate of nutrient digestion and absorption independent of fiber occur naturally in foods. They include antiamylases and protease inhibitors and antinutrients such as phytates, lectins, tannins, and saponins. Although cooking may destroy or reduce the levels of many of these, even residual amounts may serve to reduce the rate of carbohydrate digestion and absorption (Thompson *et al.*, 1984; Yoon *et al.*, 1983). Enzyme inhibitors may also increase the amount of available carbohydrate that escapes small intestinal digestion and enters the colon.

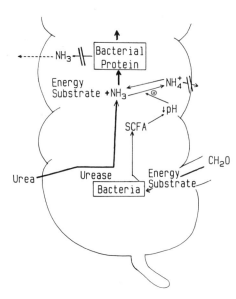

FIGURE 3. Schematic representation of nitrogen-fixing action of baceria in the human cecum. Carbohydrate provides the energy source for bacteria, which fix urea nitrogen to form their own cell proteins after first splitting the urea that crosses the bowel wall from the blood compartment. The NH_3 so formed at low pH as a result of SCFA synthesis does not pass back since in its form as NH_4^+ the bowel is much less permeable. Nitrogen is therefore trapped within the bowel and further immobilized as bacterial cell protein.

The broad application of antinutrients in the treatment or prevention of specific disorders has not yet been exploited, although their natural presence in certain high-fiber foods may be part of the reason for the beneficial effects of those foods, independent of their fiber content.

The pharmaceutical industry is, however, exploiting enzyme inhibitors of carbohydrate absorption (e.g., Acarbose, Miglitol). These have been shown to improve glycemic control in diabetes and to reduce serum triglyceride levels of those on higher-carbohydrate diets (Clissold and Edwards, 1988). Interestingly, effects have not been noted with respect to cholesterol lowering, and no reports exist on the effects of this class of drugs on nitrogen metabolism in renal failure or hepatic encephalopathy.

4.3. Lente Carbohydrate in Low-Glycemic-Index Starchy Foods

Such foods, by virtue of their food form (Haber *et al.*, 1977; Heaton *et al.*, 1989), higher amylose content (Behall *et al.*, 1989), and increased levels of fiber and antinutrients, may result in a broad range of effects. These include flatter glucose and hormone responses, lower blood lipids, increased colonic fermentation, and reduced urinary urea outputs.

Over the years, studies have noted that different starchy foods produce different glycemic responses (Crapo *et al.*, 1977; Jenkins *et al.*, 1981; Gannon and Nuttall, 1987; Arends *et al.*, 1987; Thorburn *et al.*, 1987; Brand *et al.*, 1985; Walker and Walker, 1984). It was found convenient to classify these differences in terms of a glycemic index (GI) (Jenkins *et al.*, 1981), where

$$GI = \frac{\text{Incremental blood glucose response area to a food}}{\text{Corresponding area to an equal amount of carbohydrate from a reference food}} \times 100$$

Originally glucose was the reference (Jenkins *et al.*, 1981); more recently bread has been preferred (Wolever and Jenkins, 1986). Composite lists have been devised, and in general the GI values have been found to relate well to food digestibility *in vitro* (O'Dea *et al.*, 1981; Jenkins *et al.*, 1984). In addition, a good relationship has been observed between the proportion of available carbohydrate collected from the terminal ileum of ileostomates and both the glycemic response and the *in vitro* rate of digestion of foods (Jenkins *et al.*, 1987a). Nevertheless, this colonic loss of carbohydrate is too small (at most approximately 15%) to account for the very large (100%) differences in glycemic responses between foods. The rate of small intestinal digestion of foods is therefore likely to be important in determining their glycemic effect (Torsdottir *et al.*, 1984).

Therapeutic studies with low-glycemic-index foods have demonstrated lower levels of LDL-cholesterol and triglycerides in hyperlipidemic volunteers (Jenkins *et al.*, 1987c). The lower levels of triglycerides are of interest and only appear to be seen in hypertriglyceridemic volunteers (in whom the most marked reductions in serum cholesterol are also seen, i.e., type IV and IIb phenotypes) (Fig. 4). In normal volunteers, no

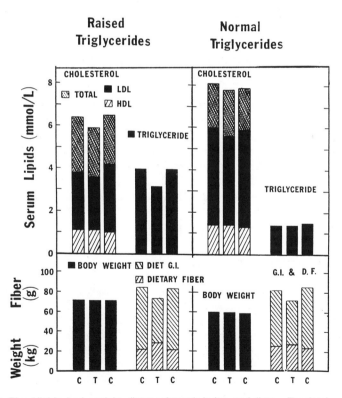

FIGURE 4. Blood lipids, body weight, dietary glycemic index, and dietary fiber intake on control, test (low glycemic index), and return to a control month in 24 type IIb or IV hyperlipidemic patients (left) and six type IIa hyperlipidemic patients (right), showing the greater lipid reduction in type IIb and IV patients on a low-GI diet. (Adapted from Jenkins *et al.*, 1978.)

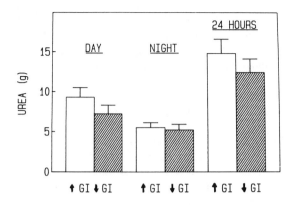

FIGURE 5. Urinary urea outputs in six healthy volunteers placed on high- and low-glycemic-index metabolic diets for 2-week periods in random order.

changes were seen in serum triglyceride levels despite significant reductions in LDL-cholesterol (Jenkins et al., 1987a,b,c).

Studies of both type I and II diabetes have demonstrated improvement in indices of glycemic control including reduced fructosamine levels and Hb_{A1C} levels (Jenkins et al., 1988; Fontvieille et al., 1988; Truswell and Brand, 1988; Collier et al., 1988). These studies achieved the reported changes without major alterations in fiber content and employed the same macronutrient profiles in both low- and high-GI limbs of the studies. During some low-GI diets the nature of the fiber (higher soluble) will have been different but unlikely to account for the magnitude of the differences seen.

Increased breath H_2 evolution during the day has provided evidence of enhanced colonic fermentation on the low-GI diet. This was associated with reduced urinary urea excretion in normal volunteers (Fig. 5) and relatively lower serum urea levels at the end of the low-GI period in type II diabetics (Jenkins et al., 1988). This further suggests that a similar spectrum of activities is shared by fermentable fiber and starch delivered to the colon.

However, more studies are required to be undertaken to confirm the effects of low-GI diets on nitrogen metabolism. Studies in patients with renal disease and raised serum urea levels may be particularly rewarding.

4.4. Meal Frequency: Nibbling versus Gorging

Finally, without altering the nature of the food or using pharmacological agents, it is possible to slow the rate of nutrient absorption by spreading the delivery of the nutrient load over a longer period of time. In effect, this is the ultimate proof that slowing small intestinal absorption per se (without alterations in fiber, food form, enzyme inhibitors, etc.) produces the observed metabolic effects.

Over 25 years ago there was considerable interest in this area. Studies of "nibbling" (six to ten meals daily) compared to "gorging" (one to three meals daily) demonstrated lower cholesterol levels, improved carbohydrate tolerance, and reduced postprandial chylomicronemia (Gwinup et al., 1963a,b; Cohn, 1964; Jagannathan et al., 1964; Fabry and Tepperman, 1970). These changes were also associated with a number

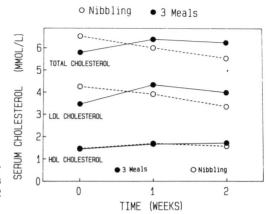

FIGURE 6. Blood lipids in one subject (DJAJ) placed on a nibbling or a three-meal feeding regimen for 2 weeks in random order.

of enzymatic changes in adipose tissue on nibbling, suggestive of depressed lipogenesis (Bray, 1972).

More recently, studies have demonstrated that sipping glucose over 3–5 hr as opposed to drinking it in a bolus resulted in a significantly reduced insulin area and a prolonged suppression of FFA with no rebound at 4 hr.

The reduced insulin area was also associated with lower postprandial serum growth hormone and glucogen levels and lower urinary total catecholamine excretion.

In a 2-week metabolic study in which 17 meals (nibbling) were compared to three meals daily, lower LDL levels were also seen after nibbling (Fig. 6). This was associated with 20% reductions in urinary C-peptide excretion and lower day-long serum insulin and C-peptide levels. When nibbling was compared to one large evening meal and two small snacks during the day, a substantial reduction in apolipoprotein B was also seen (Fig. 7). No differences were observed in glucose tolerance or fasting serum triglyceride levels in these studies of otherwise healthy volunteers.

It is suggested that reduction of insulin secretion may be an important factor in

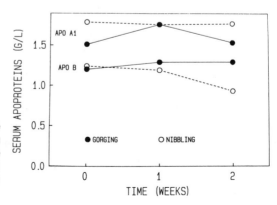

FIGURE 7. Apolipoproteins in one subject (DJAJ) placed on a nibbling (17 meals daily) or gorging (one large meal and two snacks daily) regimen for 2 weeks in random order.

reducing LDL-cholesterol levels in situations in which slowing carbohydrate absorption is associated with lower cholesterol levels. It is possible that the action of insulin in enhancing the concentration of HMG-CoA reductase in the liver is an important part of this mechanism. Possible colonic effects of lente carbohydrate would not be expected in this paradigm, although in the situation of high-fiber or low-glycemic-index diets, propionate generation may provide an additional effect (Chen *et al.*, 1984; Thacker *et al.*, 1981).

5. CONCLUSION

It is concluded that slowing starch absorption by various means has a number of possible metabolic advantages that have therapeutic implications in the treatment of diabetes and hyperlipidemia. Clinical trials and exploration of basic mechanisms are urgently required in this area, as are studies of the actions of lente carbohydrate on colonic metabolism in relation to nitrogen trapping and the local and systemic effects of SCFA generation.

ACKNOWLEDGMENTS. Studies reported by the authors were funded by NSERC, Canada.

REFERENCES

Albrink, M. J., Newman, T., and Davidson, P. C., 1979, Effect of high- and low-fiber diets on plasma lipids and insulin, *Am. J. Clin. Nutr.* **32**:1486–1491.

American Diabetes Association Policy Statement, 1987, Nutritional recommendations and principles for individuals with diabetes mellitus: 1986, *Diabetes Care* **10**:126–132.

American Heart Association Committee Report, 1982, Rationale of the diet–heart statement of the American Heart Association, *Circulation* **65**:839A–854A.

Anderson, J. W., and Chen, W. L., 1979, Plant fiber: Carbohydrate and lipid metabolism, *Am. J. Clin. Nutr.* **32**:346–363.

Anderson, J. W., and Ward, K., 1979a, Long-term effects of high carbohydrate, high fiber diets on glucose and lipid metabolism: A preliminary report on patients with diabetes, *Diabetes Care* **1**:77–82.

Anderson, J. W., and Ward, K., 1979b, High carbohydrate, high fiber diets for insulin treated men with diabetes mellitus, *Am. J. Clin. Nutr.* **32**:2312–2321.

Anderson, J. W., Story, L., Sieling, B., Chen, W. L., Petro, M. S., and Story, J., 1984, Hypocholesterolemic effects of oat-bran or bean intake for hypercholesterolemic men, *Am. J. Clin. Nutr.* **40**:1146–1155.

Arends, J., Ahrens, K., Lubke, D., and Willms, B., 1987, Physical factors influencing the blood glucose response to different breads in type II diabetic patients, *Klin. Wochenschr.* **65**:469–474.

Aro, A., Uusitupa,, M., Vontilainen, E., Hersio, K., Korhonen, T., and Siitonen, O., 1981, Improved diabetic control and hypocholesterolemic effect induced by long-term dietary supplementation with guar gum in type 2 (insulin-independent) diabetes, *Diabetologia* **21**:29–33.

Behall, K. M., Schofield, D. J., Yuheniak, I., and Canary, J., 1989, Diets containing amylose vs. amylopectin starch: Effects on metabolic variables in human subjects, *Am. J. Clin. Nutr.* **49**:337–442.

Blackburn, N. A., Redfern, J. S., Jarjis, H., Holgate, A. M., Hanning, I., Scarpello, J. H. B., Johnson,

I. T., and Read, N. W., 1984, The mechanism of action of guar gum in improving glucose tolerance in man, *Clin. Sci.* **66**:329–336.

Bosello, O., Cominacini, L., Zocca, I., Garbin, U., Ferrari, F., and Davoli, A., 1984, Effect of guar gum on plasma lipoproteins and apolipoproteins C-II and C-III in patients affected by familial combined hyperlipoproteinemia, *Am. J. Clin. Nutr.* **40**:1165–1174.

Brand, J. C., Nicholson, P. L., Thorburn, A. W., and Truswell, A. S., 1985, Food processing and the glycemic index, *Am. J. Clin. Nutr.* **42**:1192–1196.

Bray, G. A., 1972, Lipogenesis in human adipose tissue: Some effects of nibbling and gorging, *J. Clin. Invest.* **51**:537–548.

British Diabetic Association, 1982, The Nutrition Sub-committee of the British Diabetic Association's Medical Advisory Committee: Dietary recommendations for diabetics for the 1980's—a policy statement by the British Diabetic Association, *Hum. Nutr. Appl. Nutr.* **36A**:378–394.

Burke, B. J., Hartog, M., Heaton, K. W., and Hooper, S., 1982, Assessment of the metabolic effects of dietary carbohydrate and fibre by measuring urinary excretion of C-peptide, *Hum. Nutr. Clin. Nutr.* **36C**:373–380.

Burkitt, D. P., and Trowell, H. C., 1975, *Refined Carbohydrate Foods and Disease,* Academic Press, London.

Canadian Diabetes Association, 1981, Special Report Committee: Guidelines for the nutritional management of diabetes mellitus: A special report from the Canadian Diabetes Association, *J. Can. Dietet. Assoc.* **42**:110–118.

Chen, W. J. L., Anderson, J. W., and Jennings, D., 1984, Propionate may mediate the hypocholesterolemic effects of certain soluble plant fibers in cholesterol-fed rats, *Proc. Soc. Exp. Biol. Med.* **175**:215–218.

Clissold, S. P., and Edwards, C., 1988, Acarbose: A preliminary review of its pharmacodynamic and pharmacokinetic properties, and therapeutic efficacy, in: *Drugs,* ADIS Press, Auckland, New Zealand, pp. 214–243.

Cohn, C., 1964, Feeding patterns and some aspects of cholesterol metabolism, *Fed. Proc.* **23**:76–81.

Collier, G. R., Giudici, S., Kalmusky, J., Wolever, T. M. S., Helman, G., Wesson, V., Ehrlich, R. M., and Jenkins, D. J. A., 1988, Low glycaemic index starchy foods improve glucose control and lower serum cholesterol in diabetic children, *Diab. Nutr. Metab.* **1**:11–19.

Crapo, P. A., Reaven, G., and Olefsky, J., 1977, Post-prandial plasma-glucose and -insulin responses to different complex carbohydrates, *Diabetes* **26**:1178–1183.

Creutzfeldt, W., and Folsch, U. R. (eds.), 1983, *Delaying Absorption as a Therapeutic Principle in Metabolic Diseases,* Thieme-Stratton, New York.

Cummings, J. H., 1981, Short chain fatty acids in the human colon, *Gut* **22**:763–779.

Cummings, J. H., and Branch, W. J., 1986, Fermentation and the production of short-chain fatty acids in the human large intestine, in: *Dietary Fiber: Basic and Clinical Aspects* (G. V. Vahouny and D. Kritchevsky, eds.), Plenum Press, New York, pp. 131–149.

DeBrujn, K. M., Blendis, L. M., Zilm, D. H., Carlen, P. L., and Anderson, G. H., 1983, Effect of dietary protein manipulations in subclinical portal systemic encephalopathy, *Gut* **24**:53–60.

Fabry, P., and Tepperman, J., 1970, Meal frequency—a possible factor in human pathology, *Am. J. Clin. Nutr.* **23**:1059–1068.

Flourie, B., Vidon, N., Florent, C. H., and Bernier, J. J., 1984, Effect of pectin on jejunal glucose absorption and unstirred layer thickness in normal man, *Gut* **25**:936–941.

Fontvieille, A. M., Acosta, M., Rizkalla, S. W., Bornet, F., Davis, P., Letanoux, M., Tchobroutsky, G., and Slama, G., 1988, A moderate switch from high to low glycaemic-index foods for 3 weeks improves the metabolic control of type I (IDDM) diabetic subjects, *Diab. Nutr. Metab.* **1**:139–143.

Gannon, M. C., and Nuttall, F. Q., 1987, Factors affecting interpretation of postprandial glucose and insulin areas, *Diabetes Care* **10**:759–763.

Gwinup, G., Byron, R., Roush, W., Kruger, F. A., and Hamwi, G. J., 1963a, Effect of nibbling versus gorging on glucose tolerance, *Lancet* **2**:165–167.

Gwinup, G., Byron, R. C., Roush, R. W., Kruger, F. A., and Hamwi, G. J., 1963b, Effect of nibbling versus gorging on serum lipids in men, *Am. J. Clin. Nutr.* **13**:209–213.

Haber, B. G., Heaton, K. W., Murphy, D., and Burroughs, L. F., 1977, Depletion and disruption of dietary fibre: Effects on satiety, plasma-glucose, and insulin, *Lancet* **2**:679–682.

Heaton, K. M., Marcus, S. N., Emmett, P. M., and Bolton, C. H., 1989, Effect of particle size on the plasma glucose and insulin responses to test meals of wheat, maize, and oats and the rate of starch digestion *in vitro*, *Am. J. Clin. Nutr.* (in press).

Jagannathan, S. N., Connel, W. F., and Beveridge, J. M. R., 1964, Effect of gormandizing and semicontinuous eating of equicaloric amounts of formal type high fat diets on plasma cholesterol and triglyceride levels in human volunteer subjects, *Am. J. Clin. Nutr.* **15**:90–93.

Jenkins, D. J. A., and Wolever, T. M. S., 1981, Slow release carbohydrate and the treatment of diabetes, *Proc. Nutr. Soc.* **40**:227–235.

Jenkins, D. J. A., Leeds, A. R., Gassull, M. A., Houston, H., Goff, D. V., and Hill, M. J., 1976, The cholesterol lowering properties of guar and pectin, *Clin. Sci. Mol. Med.* **51**:8.

Jenkins, D. J. A., Wolever, T. M. S., Hockaday, T. D. R., Leeds, A. R., Haworth, R., Bacon, S., Apling, E. C., and Dilawari, J., 1977, Treatment of diabetes with guar gum, *Lancet* **2**:779–780.

Jenkins, D. J. A., Wolever, T. M. S., Leeds, A. R., Gassull, M. A., Dilawari, J. B., Goff, D. V., Metz, G. L., and Alberti, K. G. M. M., 1978, Dietary fibres, fibre analogues and glucose tolerance: Importance of viscosity, *Br. Med. J.* **1**:1392–1394.

Jenkins, D. J. A., Reynolds, D., Leeds, A. R., Waller, A. L., and Cummings, J. H., 1979, Hypocholesterolemic action of dietary fiber unrelated to fecal bulking effect, *Am. J. Clin. Nutr.* **32**:2430–2435.

Jenkins, D. J. A., Wolever, T. M. S., Taylor, R. H., Barker, H. M., Fielden, H., Baldwin, J. M., Bowling, A. C., Newman, H. C., Jenkins, A. L., and Goff, D. V., 1981, Glycemic index of foods: A physiological basis for carbohydrate exchange, *Am. J. Clin. Nutr.* **34**:362–366.

Jenkins, D. J. A., Wolever, T. M. S., Thorne, M. J., Jenkins, A. L., Wong, G. S., Josse, R. G., and Csima, A., 1984, The relationship between glycemic response, digestibility, and factors influencing the dietary habits of diabetics, *Am. J. Clin. Nutr.* **40**:1175–1191.

Jenkins, D. J. A., Cuff, D., Wolever, T. M. S., Knowland, D., Thompson, L. U., Cohen, Z., and Prokipchuk, E., 1987a, Digestibility of carbohydrate foods in an ileosomate: Relationship to dietary fiber, *in vitro* digestibility, and glycemic response, *Am. J. Gastroenterol.* **82**:709–717.

Jenkins, D. J. A., Wolever, T. M. S., Collier, G. R., Ocana, A., Rao, A. V., Buckley, G., Lam, Y., Mayer, A., and Thompson, L. U., 1987b, The metabolic effects of a low glycemic index diet, *Am. J. Clin. Nutr.* **46**:968–975.

Jenkins, D. J. A., Wolever, T. M. S., Kalmusky, J., Giudici, S., Giordano, C., Patten, R., Wong, G. S., Bird, J. N., Hall, M., Buckley, G., Csima, A., and Little, J. A., 1987c, Low-glycemic index diet in hyperlipidemia: Use of traditional starchy foods, *Am. J. Clin. Nutr.* **46**:66–71.

Jenkins, D. J. A., Wolever, T. M. S., Buckley, G., Lam, K. Y., Giudici, S., Kalmusky, J., Jenkins, A. L., Patten, R. L., Bird, J., Wong, G. S., and Josse, R. G., 1988, Low glycemic-index starchy foods in the diabetic diet, *Am. J. Clin. Nutr.* **48**:248–254.

Kay, R. M., and Truswell, A. S., 1977, Effect of citrus pectin on blood lipids and fecal steroid excretion in men, *Am. J. Clin. Nutr.* **30**:171–175.

Kritchevsky, D., and Story, J. A., 1974, Binding of bile salts *in vitro* by nonnutritive fiber, *J. Nutr.* **104**:458.

Leeds, A. R., Ralphs, D. N. L., Ebied, F., Metz, G., and Dilawari, J. B., 1981, Pectin in the dumping syndrome: Reduction of symptoms and plasma volume changes, *Lancet* **1**:1075–1078.

Miettinen, T. A., and Tarpila, S., 1977, Effect of pectin on serum cholesterol, fecal bile acids and biliary lipids in normolipidemic and hyperlipidemic individuals, *Clin. Chim. Acta.* **79**:471–477.

National Academy of Sciences, Committee on Diet, Nutrition and Cancer, 1982, *Diet, Nutrition and Cancer,* National Research Council, Washington.

O'Dea, K., Snow, P., and Nestel, P., 1981, Rate of starch hydrolysis *in vitro* as a predictor of metabolic responses to complex carbohydrate *in vivo*, *Am. J. Clin. Nutr.* **34**:1991–1993.

Rampton, D. S., Cohen, S. L., Crammond V. de B., Gibbons, J., Lilburn, M. F., Rabet, J. Y., Vince, A. J., Wager, J. D., and Wrong, O. M., 1984, Treatment of chronic renal failure with dietary fiber, *Clin. Nephrol.* **21**:159–163.

Ray, T. K., Mansell, K. M., Knight, L. C., Malmud, L. S., Owen, O. E., and Boden, G., 1983, Long-term effects of dietary fiber on glucose tolerance and gastric emptying in noninsulin-dependent diabetic patients, *Am. J. Clin. Nutr.* **37**:376–381.

Rodwell, V. W., Nordstrom, J. L., and Mitschelen, J. J., 1976, Regulation of HMG-CoA reductase, *Adv. Lipid Res.* **14**:1–76.

Schwartz, S. E., Levine, R. A., Singh, A., Scheidecker, J. R., and Track, N. S., 1982, Sustained pectin ingestion delays gastric emptying, *Gastroenterology* **83**:12–17.

Thacker, P. A., Salomons, M. O., Aherne, F. X., Milligan, L. P., and Bowland, J. P., 1981, Influence of propionic acid on cholesterol metabolism of pigs fed hypercholesterolemic diets, *Can. J. Anim. Sci.* **61**:969–975.

Thompson, L. U., Yoon, J. H., Jenkins, D. J. A., Wolever, T. M. S., and Jenkins, A. L., 1984, Relationship between polyphenol intake and blood glucose response of normal and diabetic individuals, *Am. J. Clin. Nutr.* **39**:745–751.

Thorburn, A. W., Brand, J. C., and Truswell, A. S., 1987, Slowly digested and absorbed carbohydrate in traditional bushfoods: A protective factor against diabetes? *Am. J. Clin. Nutr.* **45**:98–106.

Torsdottir, I., Alpsten, M., Andersson, D., Brummer, R. J. M., and Andersson, H., 1984, Effect of different starchy foods in composite meals on gastric emptying rate and glucose metabolism. I. Comparison between potatoes, rice and white beans, *Hum. Nutr. Clin. Nutr.* **38C**:329–338.

Truswell, A. S., and Brand, J. C., 1988, Low glycemic index diet in type II diabetes, *Proc. Nutr. Soc. Abstr.* **13**:150A.

Vahouny, G. V., Tombes, R., Cassidy, M. M., Kritchevsky, D., and Gallo, L. L., 1980, Dietary fibers. V. Binding of bile salts, phospholipids and cholesterol from mixed micelles by bile sequestrants and dietary fibers, *Lipids* **15**:1012.

Walker, A. R. P., and Walker, B. R., 1984, Glycaemic index of South African foods determined in rural blacks—a population at low risk to diabetes, *Hum. Nutr. Clin. Nutr.* **36C**:215–222.

Wolever, T. M. S., and Jenkins, D. J. A., 1986, The use of the glycemic index in predicting the blood glucose response to mixed meals, *Am. J. Clin. Nutr.* **43**:167–172.

Wolever, T. M. S., Jenkins, D. J. A., Kalmusky, J., Jenkins, A. L., Giordano, C., Giudici, S., Josse, R. G., and Wong, G. S., 1986, Comparison of regular and parboiled rices: Explanation of discrepancies between reported glycemic responses to rice, *Nutr. Res.* **6**:349–357.

Yoon, J. H., Thompson, L. U., and Jenkins, D. J. A., 1983, The effect of phytic acid on *in vitro* rate of starch digestibility and blood glucose response, *Am. J. Clin. Nutr.* **38**:835–842.

19

Fiber and Gastrointestinal Disease

MARTIN EASTWOOD

1. INTRODUCTION

The earlier volumes devoted to dietary fiber concentrated on the association between fiber and disease of the gastrointestinal tract. It is noteworthy that in the conference that led to the present volume, the time allocated to dietary fiber and diseases was reduced to 60% of a morning session. This suggests that no longer is there a requirement to prove that a deficiency of fiber is an important etiological factor in the genesis of a number of diseases. Yet such precise logic as Koch's hypothesis has not been applied to studies of fiber in experimental or clinical situations.

Figure 1 shows the conditions that have been suggested to be related to a deficiency of fiber.

Hiatus hernia has not been regarded as an interesting subject of study in relation to dietary fiber, largely because it is not a substantial cause of death, and most treatments are based on weight loss. Perhaps the important place fiber has here is in reducing weight and hence symptoms.

It is in the colon that the association between gastrointestinal disease and fiber has been proposed. The areas that are worth discussing are those of appendicitis, constipation, irritable bowel syndrome, diverticulosis, and carcinoma of the colon.

A distinction has to be drawn between the place of dietary fiber in the etiology of diseases and in the treatment of diseases.

2. ETIOLOGY

Bradford Hill (1977) postulated a series of "canons" analogous to Koch's postulates that had to be fulfilled in order to determine whether the relationship between an

MARTIN EASTWOOD • Wolfson Gastrointestinal Laboratories, Department of Medicine, University of Edinburgh. Western General Hospital, Edinburgh EH4 2XU, Scotland.

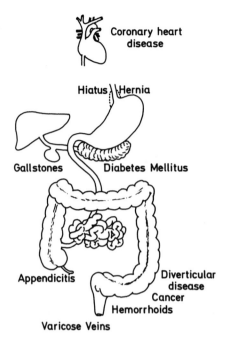

FIGURE 1. The diseases, in the original dietary fiber hypothesis by Burkitt and Trowell, thought to owe their etiology to a deficiency of dietary fiber.

environmental phenomenon and a disease was genuinely causal or merely associative. As most of the work on dietary fiber has epidemiologic overtones, it is worthwhile applying these canons.

1. The correlation is biologically plausible.
2. The correlation is strong.
3. The correlation reflects a biological gradient; i.e., the less fiber eaten the higher is the risk of disease.
4. The correlation is found consistently.
5. The correlation holds over time; i.e., the longer there is a reduced fiber content to the diet the greater is the risk of an increase in the incidence of the disease.
6. The association is confirmed by experiments.

There are few experiments that confirm a relationship between dietary fiber and intestinal disease. Most contemporary studies are devoted to the physiological effects of fiber. In the early days of the dietary fiber hypothesis there was considerable resistance to accepting the clinical relevance of dietary fiber. The work of Neil Painter (1975) sustained the validity of the hypothesis by showing that symptomatic colonic diverticulosis could respond to an enhanced fiber content in the diet. Here was a practical concept that clinicians could understand and apply.

In examining the etiology of a disease there are two aspects that have to be taken into consideration when considering why any individual succumbs to that condition. The first is the consequence of an exposure to the causal agent, if it is known, and the second is the individual's susceptibility to that causal agent. Not everyone exposed to the tuberculosis bacillus develops tuberculosis. Not everyone who smokes heavily develops lung cancer. Yet with lung cancer, the increasing exposure to cigarettes increases the chance of lung cancer. It is a frequent assumption of epidemiologic studies that exposure to the agent studied, e.g., dietary fiber, has a uniform effect on inducing that disease irrespective of the cultural and ethnic composition of the population. It is not proven that different populations have the same susceptibility to a given cause of a disease. This is palpably not the case. For example, the mortality rate from lung cancer per 100,000 population is quite different in Mexico and Belgium despite both having approximately the same cigarette consumption per adult (Le Fanu, 1987). It is possible that the tobacco smoked is different for the two populations, but there may be other factors, e.g., the blood vitamin A status and genetic susceptibilities.

The age distribution of a population also has a large influence on the development of disease. In 1911 in the United States the average citizen died 38 years before the expected life expectancy. Tuberculosis was an important cause of death (Fries, 1980). By 1950 the life expectations came within 17 years of the biological expectation, and by 1980 to 11 years. This was in part related to increased survival by the young. Cancer is rare until individuals are 50 years old. Seventy percent of all circulatory disease and cancer appears in individuals aged over 70 years. Individuals are living longer and therefore are more at risk of developing the diseases associated with being elderly. Cancer and the degenerative diseases are predominantly found in the elderly. So the older a population, the more at risk is that population.

Dietary fiber has not yet been proven to be an obligate in the diet, i.e., an essential component. It is, however, an important modulator of events in the gastrointestinal tract.

3. CONSTIPATION

If dietary intake was all important in dictating individual bowel habit, one would expect that spouses eating from the same kitchen would have similar stool weight and fecal constituents. In a survey of Edinburgh couples who had lived together between 20 and 50 years and who had eaten most of their meals together, there was concordance between the type but not amount of food they ate. There was no relationship of fecal weight between husbands and wives, nor between total fecal bile acids or individual bile acids, deoxycholic acid, lithocholic acid, total bile acid concentration, neutral sterols, and serum cholesterol. The only weak correlate was for daily fecal fat output. The only dietary constituents that were related were dietary fat and dietary fiber (Eastwood et al., 1982). Dietary fiber appears to be the only dietary constituent that affects stool weight (Eastwood et al., 1984a). It is well proven that by increasing the fiber content of the diet with wheat bran, particularly of a coarse type, the stool weight increases (Eastwood et al., 1973).

There are other important causes of constipation. Complex abnormalities in the anorectal neurological control of defecation can result in constipation or fecal incontinence. As a result of the work of Sir Alan Parks (Henry and Swash, 1985) and others (Varma and Smith, 1988), it is appreciated that abnormalities of anorectal innervation can cause refractory constipation. It has been shown that hysterectomy can result in a high prevalence of constipation (A. N. Smith, T. Taylor, and P. M. Fulton, unpublished data). The recognition of the phenomenon of anismus is also an important development (Preston and Lennard-Jones, 1985). Neither of these conditions is related to dietary fiber deficiency.

4. APPENDICITIS

Appendectomy is the most common abdominal operation carried out in the United Kingdom. The management of cases of suspected appendicitis results in the removal of some normal appendices, and this is more likely to happen in young women (Lee, 1961). It is well known that patients who have had a normal appendix removed at appendectomy and in whom no other cause is found to account for the pain have more emotional problems confirmed compared with those patients who have had an inflamed appendix removed (Ingram *et al.*, 1965). It is possible, therefore, that any increase in the incidence of appendicitis may in part result from factors unrelated to inflammation of the appendix. In a study by Vassiulas (1988), the records were examined of all 628 operations for appendectomy at the Royal Infirmary of Edinburgh in women aged between 15 and 35 years between 1975 and 1980. Of these, 368 of the appendices removed showed evidence of inflammation. In 184 cases normal histology was found. In 75 cases a diagnosis other than appendicitis was found, e.g., ovarian cyst. The case notes of those individuals were compared with admissions for suicide who were also admitted to the same hospital in that enclosed geographical area. Among those in whom a normal appendix had been removed, the number of subjects admitted for suicide either in the 6 months before or after the operation was three times greater than in patients from whom an inflamed appendix had been removed (14.7% compared with 4.6%). Rates of other psychiatric contact were also significantly higher.

When comparing the incidence of problems that possibly have a functional element to them, e.g., appendicitis, irritable bowel syndrome, and symptomatic diverticulosis, factors other than diet are important. It is possible that in rural communities, which are more common in Africa than in Britain, community support is substantially greater in a village community than in an urban spread.

5. IRRITABLE BOWEL SYNDROME

The irritable bowel syndrome presents in two forms: (1) diarrhea without abdominal pain and (2) alternating diarrhea and constipation accompanied by abdominal pain.

The constipated form of the irritable bowel syndrome is characterized by abdominal pain, usually, though not always, in the left iliac fossa or the right iliac fossa. The passage of fragmented stool accompanied by mucus but not blood relieves this pain.

These symptoms also occur in many other conditions, and therefore the diagnosis of the irritable bowel syndrome must, by necessity, be by exclusion of other important causes of these particular symptoms. The differential diagnosis for diarrhea of colonic origin includes diverticular disease, cancer of the rectum and colon, bacterial infection, viral infection, and Crohn's disease. This differential diagnosis can be particularly demanding in the diarrheal form of the irritable bowel.

There are many views on the etiology of the irritable bowel syndrome, including psychiatric, environmental, and purely organic causes.

A further problem is that of finding a logical description of irritable bowel syndrome. At the moment there is no method of checking by some yardstick whether a doctor is right or wrong in his notion that a given patient has the irritable bowel syndrome (Card et al., 1984). The problem is also made more difficult by the fact that the condition seems to be self-limiting (Harvey et al., 1987).

Since the characterization of the irritable bowel syndrome as an entity, consistent emphasis has been placed on the personality that is found with many patients suffering from this condition. Tension, anxiety, guilt, and resentment have been suggested to be significant factors in the evolution of the irritable bowel syndrome. Chaudhary and Truelove (1962) suggested that stress was a significant feature in 80% of the patients they studied.

There is a body of opinion that believes that specific food intolerance contributes to the symptoms of the irritable bowel syndrome. In one study significant increases in rectal prostaglandin E_2 production were found after food challenge (Jones et al., 1982). Other workers have suggested that gastrointestinal infection may be a long-term contributor to the symptoms of the irritable bowel, attributable to changes in bowel flora. Chronic alcohol abuse has long been recognized as the cause of nonulcer dyspepsia and diarrhea. Alcohol abuse has been recorded in 1% of the general population, 10% of hospital admissions, and 1% of medical referrals to a gastrointestinal unit (Harvey et al., 1983; Garman and Kellett, 1979).

A holistic view of the irritable bowel syndrome sees the problem being derived from three sources (Eastwood et al., 1987):

1. Physical causes such as deficiency of dietary fiber, lactase deficiency, or undiagnosed or early organic disease such as Crohn's disease.
2. Adverse life events that lead to feelings of hopelessness and hence anxiety.
3. The patient's personality.

These factors form an etiologic triangle; the shape of the triangle is dependent on the relative strengths of the factors and make-up of each particular patient. Indeed, the shape of the triangle may vary in any one patient along with an alterations in the relative contributions over a period of time.

The irritable bowel syndrome is a common condition and yet poorly classified. In a number of studies, patients have been found to have very similar fecal weights and fecal constituents to a control group. In a study of patients in Edinburgh, all of the group of irritable bowel syndrome patients had normal breath hydrogen tests in response to ingested lactose, which is in accordance with the suggestion that hypolactasia plays only a minor role in the etiology of irritable bowel syndrome in a Scottish population (Eastwood et al., 1984b).

The irritable bowel syndrome has been said to be a chronic relapsing condition. Harvey and his colleagues (1987) showed that the management and outlook of patients with irritable bowel is better than is generally believed. Treatment consists of a simple explanation of the condition and adequate treatment with high-fiber bulking agents and antispasmodics. Approximately 85% of patients become virtually symptom-free in the short term, and approximately 67% remain virtually symptom-free in the long term. The response to treatment is best in men, particularly those with predominant constipation and with a short history.

6. DIVERTICULAR DISEASE OF THE LARGE INTESTINE

Diverticulosis is an acquired condition rarely found before the age of 40 (Parks, 1969). The frequency of colonic diverticula increases with advancing age, reaching 40% by the age of 65 (Parks, 1968). The colonic diverticula are most commonly found in the sigmoid colon as two parallel rows running longitudinally along the mesenteric and antimesenteric taeniae. Because the diverticula are usually hidden within the fat of the appendices epiploica, it is difficult to see the diverticula at operation. The usual manner of diagnosis is by barium enema. It is almost certain that in parallel with any changes in diet during this century there has been an increased exposure of the population to diagnostic procedures using the barium enema. Therefore, the percentage of colons found with diverticula at the present time must be seen against the lack of colons exposed to barium enemas in the distant past. The disease tends to localize in the sigmoid colon.

The mechanical properties, burst strength, width at burst, percentage elongation, and stress elongation of the colon have been compared in colons from Africans and Europeans, and differences were found (Watters et al., 1985) in that African patients had significantly greater stretch capacity in the distal colon than Europeans, but there were no differences between Europeans with and without diverticula.

The circular muscle layer in the colonic taeniae coli is greatly thickened in diverticulosis. A deficient fiber content to the diet leads to a constipated scibelous stool that requires effort to extrude in those individuals who are constipated. Therefore, the association with the amount of fiber in the diet is attractive. However, the muscle cells in diverticular disease are normal; neither hypertrophy nor hyperplasia is present.

The elastin content of the taeniae coli is doubled compared to normals (Whiteway and Morson, 1985). Because of the elastin laid down between the muscle cells, the normal fascicular pattern of the taeniae coli is distorted. Contractions in the elastin may result in shortening of the taeniae, leading to the concertinalike corrugation of the circular muscle. This permanent structural change would appear to be resistant to change by therapy.

The submucosa of the colon consists of a series of layers of collagen fibers, each 0.5 to 2.0 μm thick. The fibers between each layer are codirectionally oriented. The mean diameter of the collagen fibers is 60 nm, and the mean fibril count per unit area is 159 ± 58 μm^2 (Thomson et al., 1986). If the ultrastructure of the constituent collagen fibers in the isolated submucosa of colons is examined, it is shown that collagen fibrils in the left colon become smaller and more tightly packed than those in the right colon with increasing age. This difference is accentuated in diverticular disease. It is possible that these differences are developmental, the right colon being of midgut and left colon of hindgut origin. Studies relating the ultrastructure of the various animal and human tissues to their function have shown that tensile strength and elasticity are related to the diameter of the fibrils that constitute the collagen fibers (Parry et al., 1978). This suggests that the structure changes result from the different mechanical stresses to which the connective tissue of each region is subjected. It is possible that the accentuation of the differences in collagen structure in the right and left colon in patients with diverticular disease may reflect the abnormally high sigmoid colonic intraluminal pressures in such cases. These changes, which are normally associated with aging, are more pronounced in diverticular disease. This may represent premature aging for whatever cause (Thomson *et al.*, 1987).

Colorectal cancer and diverticular disease of the colon are common diseases in clinical practice. Both of them, it is suggested, result from a deficiency of dietary fiber (Painter, 1975). If this was to be the case, one would expect overlap in the prevalence of the two conditions. In a study of the barium enemas of patients with colorectal cancer and age-matched controls, it was found that among men 39.6% of patients with colorectal cancer have diverticulosis compared with 50% of the control group. In the women 39.3% of patients with colorectal cancer had diverticular disease compared with 36% of the controls. There appeared to be no significant etiologic association between the two conditions (McCallum *et al.*, 1988).

In general diverticulosis of the colon is a benign condition. The most common complication of diverticulosis is symptomatic diverticulosis. The symptoms of this are indistinguishable from those of the irritable bowel syndrome. It is this condition that responds most readily to the giving of coarse wheat bran. The origin of the wheat bran is less important than its physical state. A comparison of bran from a Canadian red spring wheat and that of a French soft wheat had exactly the same effect provided the bran was coarse. The increase in stool weight and the reduction in colonic pressure and symptoms were the same with the two treatments (Smith *et al.*, 1981). If the bran is highly processed, then the fiber is rendered impotent, and there is no value of this preparation in the treatment of diverticulosis (Ornstein *et al.*, 1981).

On the other hand it is noteworthy that ispaghula, while increasing stool weight, increases colonic pressure (Eastwood *et al.*, 1978). Because of this there has been a widespread belief that an increased amount of dietary wheat bran fiber should be accepted as the standard management of symptomatic, uncomplicated diverticular disease (Royal College of Physicians, 1980). The Scottish Hospital inpatient statistics for diverticular disease have been analyzed for a 15-year period (Au *et al.*, 1988). The last 10 years fell within the era when wheat bran was widely prescribed for symptomatic diverticular disease. The object of this study was to determine whether the popularity of fiber treatment had influenced the course of diverticular disease in Scotland. The rate of hospital admissions for diverticulosis had not decreased in the period following the introduction of a high-fiber diet in therapy. The rate of admission was highest in the elderly over 75 years, particularly in females. Females under 45 years are the only group so far to register a fall in admission rates in the last 5 years of the survey period. The colectomy rate is unchanged throughout and is higher for younger male patients. Therefore, there is no absolute evidence that the popularity of the high-fiber diets has influenced the problem of diverticular disease as it occurs in Scotland. It is possible that a longer time interval than 10 years is necessary to exhibit a change. It is also possible that diverticulosis is essentially a problem of the very old and that a high-fiber diet will not alter the anatomic changes associated with frailty and advanced years.

7. CANCER OF THE LARGE BOWEL

Cancer of the large bowel is common in economically developed countries and is rarely found in rural Africa (Burkitt, 1971). The incidence of colonic cancer even correlates with the gross national product per capita. This may indicate in a general way the importance of environmental factors, but no specific carcinogen for the colon has been identified. The African high-fiber diet is based on unprocessed plant foods and is usually low in fat and animal protein. It may be protective against bowel cancer by a mechanism consistent with several current hypotheses, for example, greater concentration of the hypothetical carcinogen in small stools, bacterial formation of carcinogens in a stagnant colon, or increased sensitivity to carcinogens that may be enhanced by a high-energy intake (Royal College of Physicians, 1980).

The basic hypothesis for the development of colonic neoplasia is that there is a dysplasia of the colonic mucosa, occurring most commonly in the sigmoid colon, half as frequently in the cecum and ascending colon, and uncommonly in the transverse colon. This change initially shows as a polypoid adenomatous lesion, the mucosa of which may become dysplastic and occasionally malignant and invasive. An unproven assumption is that a carcinogen and possibly a cocarcinogen are involved. The ability of the mucosa to resist evolution to dysplastic change may be important. The mechanisms of such resistance are not understood (Hill *et al.*, 1978).

Fiber may prevent the dysplasia → adenomatous polyp → carcinoma sequence from occurring by diluting a putative carcinogen so that the concentration never achieves oncogenic potential. Such a dilution can be achieved by wheat bran. Wheat bran increases stool weight in a predictable manner (Eastwood *et al.*, 1973). The water-holding capacity of wheat bran dictates its ability to increase stool weight, with the water-holding capacity paralleling the effect on colon function. Fine bran, with a low water-holding capacity, has less effect on colon function than coarse bran, which has a high water-holding capacity (Smith *et al.*, 1981). Wheat bran acts as a sponge along the gastrointestinal tract, holding water and ensuring a dilution of fecal contents. The fecal daily bile acid output does not change as a result of adding wheat bran to the diet. As stool weight increases, the concentration of bile acids decreases (Eastwood *et al.*, 1973). The Burkitt hypothesis suggests that a small stool and a prolonged colon transit time result in a concentrated stool remaining for an extended period in the rectosigmoid colon, so any carcinogenic propensity has a better chance of being realized.

It is well demonstrated that fiber content of the diet is the most important dictate of stool weight (Eastwood *et al.*, 1984a). The dry weight of feces is composed of two constituents—fiber and bacteria. Dietary fiber influences stool weight in one of two ways. The most important and predictable mechanism is through the water-holding capacity of the fiber. The other mechanism is through an increase in the colonic bacterial mass through the proliferation of bacteria that draw their energy from fermentable polysaccharides such as pectin. The fecal weight increase through this mechanism is small and not predictable (Stephen and Cummings, 1980).

When fermentable dietary fiber is broken down in the colon, important metabolites are produced, including the volatile fatty acids acetate, butyrate, and propionate (Cummings, 1985). A series of observations suggests that butyrate is an important intermediate in colonic mucosal metabolism (Roediger, 1980). Cell culture work suggests that butyrate has properties as a potent colonic cell-differentiating agent with potential for stabilizing colonic mucosal turnover (Cummings, 1985).

The place of dietary fiber in the etiology of colonic cancer may be twofold. Wheat bran is a most important dietary supplement with the ability to hold water and dilute the stool, thereby diluting potential carcinogens, e.g., fecal bile acids. A second possibility is that fiber, by being metabolized, nourishes the colonic mucosa and modulates mucosal cell turnover. This is achieved in part by suitable mucosal exposure to volatile fatty acids, especially butyrate. In order to have an increased concentration in the distal colon, it is necessary to have an increased production of butyrate. The most widely available sources of such fermentation substrates, e.g., cellulose, hemicellulose, and pectins, are fruit and vegetables.

A number of studies have looked at the proliferation and differentiation of cells within the colon. These have identified stages of abnormal cell development associated with an increased susceptibility to cancer. This is regardless of whether this is in the colon of individuals with genetic predisposition to cancer or susceptibility caused by dietary factors. Epithelial cell proliferation and differentiation are modified in diseases that predispose individuals to these conditions. It is still not clear whether the colon cancer susceptibility includes a transmissible agent. In the early stages of conversion to carcinoma of the colon, proliferative cell-cycle control mechanisms become modified, leading to continued DNA synthesis in maturing cells and delayed onset of normal terminal differentiation. This occurs in familial polyposis and in individuals who have had sporadic adenomas or previous familial and nonfamilial colon cancer and ulcerative colitis. A number of markers have been identified that suggest abnormal gastrointestinal cell proliferation.

In humans at increased risk for familial colon cancer, the markers of increased colonic epithelial cell proliferation have been studied before and after oral dietary supplementation with calcium (Lipkin and Neumark, 1985). Following calcium supplementation, epithelial cell proliferation in the group as a whole was significantly reduced, yielding an altered colonic crypt profile approaching that observed in individuals with a low risk for colonic cancer. The effect was greatest in individuals having pronounced hyperproliferation of colonic epithelial cells, and where the proliferation was close to normal there was little change. Calcium will render a number of substances insoluble and will have much the same effect as dietary fiber. It is clear, therefore, that physical conditions in the colon, whether they affect the transport of insoluble or soluble salts or render substances biologically inert, can have an effect on colonic mucosal stability. Fiber may be important in this complex physicochemical system.

REFERENCES

Au, J., Smith, A. N., and Eastwood, M. A., 1988, Diverticular disease in Scotland over 15 years, *Proc. R. Coll. Physicians Edinburgh* **18**:271–276.

Bradford Hill, A., 1977, *A Short Textbook of Medical Statistics,* 10th ed., Hodder & Stoughton, London.

Burkitt, D. P., 1971, Epidemiology of cancer of the colon and rectum, *Cancer* **28**:3–13.

Burkitt, D. P., and Trowell, H. C. (eds.), 1975, *Refined Carbohydrate Foods and Disease. Some Implications of Dietary Fibre,* Academic Press, London.

Card, W. I., Lucas, R. W., and Spiegelhalter, D. G., 1984, The logical description of a disease class as a Boolean function with special reference to the irritable bowel syndrome, *Clin. Sci.* **66**:307–315.

Chaudhary, N. A., and Truelove, S. C., 1962, The irritable bowel syndrome. A study of the clinical features, predisposing causes and prognosis in 130 cases, *Q. J. Med.* **31**:307–322.

Cummings, J. H., 1985, Cancer of the large bowel, in: *Dietary Fibre, Fibre-Depleted Foods and Disease* (H. C. Trowell, D. P. Burkitt, and K. W. Heaton, eds.), Academic Press, London, pp. 161–189.

Eastwood, M. A., Kirkpatrick, J. R., Mitchell, W. D., Bone, A., and Hamilton, T., 1973, Effects of dietary supplements of wheat bran and cellulose on faeces and bowel function, *Br. Med. J.* **4**:392–394.

Eastwood, M. A., Smith, A. N., Brydon, W. G., and Pritchard, J., 1978, A comparison of bran ispaghula and lactulose on colon function in diverticular disease, *Gut* **19**:1144–1147.

Eastwood, M. A., Brydon, W. G., Smith, D. M., and Smith, J. H., 1982, A study of diet serum lipids and faecal constituents in spouses, *Am. J. Clin. Nutr.* **31**:290–293.

Eastwood, M. A., Brydon, W. G., Baird, J. D., Elton, R. A., Smith, J. H., and Pritchard, J. L., 1984a, Fecal weight and composition, serum lipids, and diet among subjects aged 18–80 years not seeking health care, *Am. J. Clin. Nutr.* **40**:628–634.

Eastwood, M. A., Walton, B. A., Brydon, W. G., and Anderson, J. R., 1984b, Faecal weight, constituents, colonic motility and lactose tolerance in irritable bowel syndrome, *Digestion* **3**:7–12.

Eastwood, M. A., Eastwood, J., and Ford, M. J., 1987, The irritable bowel syndrome, a disease or a response? *J. R. Soc. Med.* **80**:219–221.

Fries, J. F., 1980, Aging, natural death and the compression of morbidity, *N. Engl. J. Med.* **303**:130–135.

Garman, C. M. B., and Kellett, J. M., 1979, Alcoholism in a general hospital, *Br. Med. J.* **2**:469–472.

Harvey, R. F., Salih, S. Y., and Reid, A. E., 1983, Organic and functional disorders in 2000 gastroenterology outpatients, *Lancet* **1**:632–634.

Harvey, R. F., Mauad, E. C., and Brown, A. M., 1987, Prognosis in the irritable bowel syndrome. A five year prospective study, *Lancet* **1**:963–965.

Henry, H. H., and Swash, M. (eds.), 1985, *Coloproctology and the Pelvic Floor. Pathophysiology and Management,* Butterworths, London.

Hill, M. J., Morson, B. C., and Bussey, A. J. R., 1978, Aetiology of adenoma carcinoma sequence in large bowel, *Lancet* **1**:245–247.

Ingram, P. W., Evans, G., and Oppenheim, A. N., 1965, Right iliac fossa pain in young women, *Br. Med. J.* **2**:149–151.

Jones, V. A., McLaughlan, P., Shorthouse, M., Workman, E., and Hunter, J. O., 1982, Food intolerance, a major factor in the pathogenesis of the irritable bowel syndrome, *Lancet* **2**:1115–1117.

Lee, J. A. H., 1961, Appendicitis in young women, *Lancet* **2**:815–817.

Le Fanu, J., 1987, *Eat Your Heart Out. The Fallacy of the Healthy Diet,* Macmillan, London.

Lipkin, M., and Neumark, H., 1985, Effect of adding dietary calcium on colonic epithelial cell proliferation in subjects of high risk for familial colon cancer, *N. Engl. J. Med.* **313**:1381–1384.

McCallum, A., Eastwood, M. A., Smith, A. N., and Fulton, P. M., 1988, Colonic diverticulosis in patients with colorectal cancer and in controls, *Scand. J. Gastroenterol.* **23**:284–286.

Ornstein, M. H., Littlewood, E. R., Baird, I. M., Fowler, J., North, W. R. S., and Cox, A. G., 1981, Are fibre supplements really necessary in diverticular disease of the colon? A controlled clinical trial, *Br. Med. J.* **282**:1353–1356.

Painter, N. S., 1975, *Diverticular Disease of the Colon,* Heinemann Medical Books, London.

Parks, T. G., 1968, Diverticular disease, *Proc. R. Soc. Med.* **61**:932.

Parks, T. G., 1969, Natural history of diverticular disease of the colon. A review of 521 cases, *Br. Med. J.* **4**:639–645.

Parry, D. A., Barnes, G. R., and Craig, A. S., 1978, A comparison of the size distribution of collagen fibrils in connective tissue as a function of age and possible relation to fibril size, distribution and mechanical properties, *Proc. R. Soc. Lond.* **203**:305–321.

Preston, D. M., and Lennard-Jones, J. E., 1985, Anismus in chronic constipation, *Dig. Dis. Sci.* **30:** 413–418.

Roediger, W. E. W., 1980, Role of anaerobic bacteria in the metabolic welfare of the colonic mucosa in man, *Gut* **21:**793–798.

Royal College of Physicians of London, 1980, *Medical Aspects of Dietary Fibre,* Pitman Medical, Tunbridge Wells, UK.

Smith, A. N., Drummond, E., and Eastwood, M. A., 1981, The effect of coarse and soft Canadian red spring and French soft wheat bran on colonic motility in patients with diverticular disease, *Am. J. Clin. Nutr.* **34:**2460–2463.

Stephen, A. M., and Cummings, J. H., 1980, Mechanism of action of dietary fibre in human colon, *Nature* **284:**283–284.

Thomson, H. J., Busuttil, A., Eastwood, M. A., Smith, A. N., and Elton, R. A., 1986, *J. Struct. Mol. Res.* **96:**22–30.

Thomson, H. J., Busuttil, A., Eastwood, M. A., Smith, A. N., and Elton, R. A., 1987, Submucosal collagen changes in the normal colon and in diverticular disease, *Int. J. Colorect. Dis.* **2:**208–213.

Varma, J. S., and Smith, A. N., 1988, Neurophysiological dysfunction in young women with intractable constipation, *Gut* **29:**963–968.

Vassiulas, C. A., 1988, Parasuicide and appendicectomy, *Br. J. Psychiatry* **152:**706–709.

Watters, D. A. T., Smith, A. N., Eastwood, M. A., Anderson, K. C., Elton, R. A., and Mugerwa, J. W., 1985, Mechanical properties of the colon: Comparison of the features of the African and European colon *in vitro, Gut* **26:**384–392.

Whiteway, J., and Morson, B. C., 1985, Elastosis in diverticular disease of the sigmoid colon, *Gut* **26:** 258–266.

20

Dietary Factors in the Etiology of Gallstones

KENNETH W. HEATON

1. THE NATURE OF THE DISEASE

In Western countries the vast majority of gallstones form within the gallbladder. They are asymptomatic in 70–80% of cases, causing trouble only when they pass (or get stuck trying to pass) through the cystic duct or the ampulla of Vater. The major component in 60–80% of gallstones is crystalline cholesterol monohydrate, but most stones contain calcium salts (especially carbonate, bilirubinate, phosphate, and palmitate), and these are sometimes the main or only component.

All the constituents of gallstones derive from the bile, and their presence in solid form implies that the bile in the gallbladder has become supersaturated with respect to that component. In the case of cholesterol, supersaturation means excess cholesterol in relation to its solubilizers, bile acids and phospholipids. This excess arises when there is a reduced pool of bile acids in the enterohepatic circulation and, more importantly, when the secretion of cholesterol into bile by the liver is increased.

Cholesterol precipitates only when the cholesterol–phospholipid vesicles that (together with mixed bile salt–phospholipid–cholesterol micelles) solubilize it in bile coalesce into larger, less stable particles, often with a liquid crystalline structure. This process is aided by a nucleating factor, probably in the form of a small glycoprotein. In the early stages of gallstone formation there is an increase in the mucus content of bile. Biliary sludge, which is the precursor of some if not most gallstones, consists of mucus in which are entrapped many cholesterol crystals (and/or bilirubinate granules). Another contributing factor is stasis or reduced emptying of the gallbladder, which allows time for sludge to accumulate and stones to grow.

In Western countries gallbladder calculi are present in around 10% of the popula-

KENNETH W. HEATON • University Department of Medicine, Bristol Royal Infirmary, Bristol BS2 8HW, United Kingdom.

tion, but precise prevalence data are only just becoming available from ultrasonographic surveys of random samples of the population. Prevalence in women increases from 1–2% in the 20s to 30–50% in old age; in men, prevalence is halved. For every person with gallstones there are probably two or three with the metabolic precursor state, namely, supersaturated gallbladder bile. Dietary factors are more likely to be involved in the production of supersaturated bile than in the later processes of nucleation and stone growth (for example, gallbladder emptying is impaired in pregnancy and probably in illness and with aging). This may confound attempts to identify dietary factors by case-control studies in which people with gallstones are compared with those without, since some of the controls will have the diet-induced metabolic abnormality, and to some extent the patients have become "cases" for nondietary reasons. On the other hand, most of the established risk factors for gallstones (age, female sex, estrogen usage, clofibrate treatment, obesity, and raised fasting plasma triglycerides) are associated with increased cholesterol saturation of bile rather than nucleating factors or other gallbladder factors, and these risk factors do emerge in case-control studies. Parity is a risk factor that probably reduces gallbladder emptying but may alter bile composition as well.

One way of looking for a dietary cause is to analyze bile cholesterol saturation before and after controlled manipulation of the diet.

2. DIETARY THEORIES

The belief that gallstones are a disease of civilization and especially of modern Western civilization is now backed by a reasonable body of evidence (Pixley, 1986; Heaton, 1988). There is no evidence relating gallstones to nondietary aspects of civilization such as smoking, physical inactivity, and chemical pollution, and drinking alcohol seems to be protective (Scragg, 1986). Dietary factors have long been considered paramount, and this is strongly supported by the finding of Pixley et al. (1985), using ultrasonography, that gallstones are only half as common in British vegetarians as in the rest of the population. There are, of course, several differences between vegetarian diets and normal diets. However, there is no epidemiologic or experimental evidence to incriminate meat or the type of protein in the diet as contributing to gallstones. Concerning the type of fat, the evidence is conflicting, but if there is a difference between saturated and unsaturated fats, it is the latter that favor gallstones (Scragg, 1986).

The dietary factors that carry any credence nowadays as causing gallstones are surplus energy intake, high intake of sugars (in the sense of added or refined sugars), low intake of dietary fiber, and possibly high intake of cholesterol. In addition, prolonged fasting can contribute in two ways. First, when the overnight fast is 15 hr rather than 10 hr the contents of the gallbladder are more saturated with cholesterol (Bloch et al., 1980). Secondly, failure to eat or drink for several days (as after gastrointestinal operations) leads to sludge formation in the gallbladder and in some cases to actual gallstones (Bolondi et al., 1985). However, there is only one small study to indicate that gallstone patients are particularly liable to miss breakfast (Capron et al., 1981), and the majority of patients with gallstones have not fasted for days on end (or had abdominal surgery).

3. SURPLUS ENERGY INTAKE

The association between gallstones and obesity is extensively documented (Bennion and Grundy, 1978; Scragg, 1986), and in Italian women, the relative risk of gallstones varies with body mass index across the whole range from 16 to 31 kg/m² (Angelico *et al.*, 1984a). The metabolic basis for the association is well understood. When bile is analyzed in a group of subjects there is a significant correlation between body mass index and the cholesterol saturation index (Haber and Heaton, 1979) and between body weight and the secretion of cholesterol into bile (Bennion and Grundy, 1975; Leiss *et al.*, 1987). Thus, it is likely that body fatness leads to gallstones by inducing the liver to secrete more cholesterol. The mechanism for this is uncertain. One possible intermediate factor is hyperinsulinemia; another is raised plasma estrogen levels—fat people tend to have had an early menarche (Garn *et al.*, 1986), and so do gallstone patients (Jørgensen, 1988).

However, obesity does not explain all gallstones. Some gallstone patients are slim, and cholesterol secretion is raised in them too. The association between gallstones and obesity may not hold in men and is even doubtful in older women (Scragg *et al.*, 1984).

In most but not all dietary case-control studies, high energy intake has been found to be a risk factor for gallstones. In the largest study (which, epidemiologically, is the only adequate study), there was a striking interaction between age and energy intake such that the increased risk associated with increased energy intake fell progressively with advancing age in both sexes (Scragg *et al.*, 1984) (Fig. 1).

The crucial factor is probably surplus energy intake rather than high energy intake, and it is not practicable to measure this on the large numbers of people necessary for epidemiologic studies. A possible explanation for the data shown in Fig. 1 is that those people who are susceptible to surplus energy intake are "picked off" young and are not represented in the older age groups, where other factors become dominant.

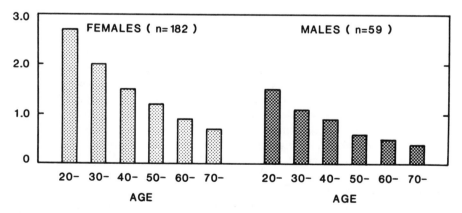

FIGURE 1. Relative risk of gallstones at different ages associated with an increase in energy intake of 500 kcal/day (2.09 MJ/day); the Adelaide case-control study (Scragg *et al.*, 1984).

If surplus energy intake is important in causing gallstones, there should be epidemiologic relationships between gallstones and other diseases blamed on overnutrition besides obesity. Among such diseases are cancers of the bowel, breast, uterus, and ovary, and it is interesting that women with such cancers have a 3.3-fold increased chance of having or having had gallstones (Lowenfels *et al.*, 1982). Furthermore, between countries there are good correlations between the autopsy prevalence of gallstones and the mortality from cancer of the uterus and large bowel (Lowenfels, 1981).

It would be interesting to see if the prevalence of gallstones in different communities correlates with the prevalence of diabetes. Certainly record prevalences of both diseases are found among the American Indians (Sampliner *et al.*, 1970).

If, as some believe, surplus energy intake is the main cause of cholesterol-rich gallstones, then the question of what dietary practices cause gallstones resolves itself into "What dietary practices promote surplus energy intake?" An adequate answer to this question must take into account cultural and psychological factors as well as the palatability, ease of consumption, and satiating value of foods and food components (and, of course, their energy value). It is certainly too simple to blame all surplus energy intake on overeating, that is, gluttony. The two dietary items that are most often blamed for inflating energy intake are refined carbohydrates, especially sugars, and fats. There is no evidence from epidemiology, case-control studies, or human experiments to incriminate dietary fat in causing gallstones. There is, however, considerable evidence to incriminate sugars. The role of highly processed starches has not yet been examined.

4. ADDED OR REFINED SUGARS

Direct evidence to incriminate added or refined sugars in the etiology of gallstones comes from case control studies. Three such studies have examined the role of sugars, and the two largest ones have shown a positive association between gallstones and sugar consumption (Alessandrini *et al.*, 1982; Scragg *et al.*, 1984; Pixley and Mann, 1988). The number of subjects in the Adelaide study was so large that the role of sugars could be examined separately in males and females and in subjects under and over the age of 50. In all four groups intake of sugars in drinks and sweets was higher in the gallstone patients than in the controls, significantly so in all except the older males (Table I).

Heaton and his colleagues in Bristol have done two studies in women with asymptomatic gallstones to determine the effect of dietary carbohydrate on bile composition (Thornton *et al.*, 1983a; Werner *et al.*, 1984). In both studies the subjects were randomized to eat two diets for 6 weeks each, the diets differing in sugars content by 100 g but otherwise being eaten *ad libitum*. In both studies the subjects gained weight and had a higher energy intake on the high-sugar diet, and in the study in which plasma lipids were measured, their fasting plasma triglycerides were higher and their plasma HDL-cholesterol was lower on this diet. On epidemiologic grounds, all these changes would be expected to increase the risk of gallstones (Scragg, 1986; Heaton, 1988). Furthermore, since bile cholesterol saturation tends to rise when plasma triglycerides are raised and when HDL-cholesterol is low (Thornton *et al.*, 1981, 1983b), one would expect bile to be more saturated on the higher sugar intake. However, this was found in only one of the two studies, and in this one there was a fall in the intake of dietary fiber at the same

TABLE I. Daily Intake of Sugars and Dietary Fiber in Patients with Gallstones and Matched Community Controls: The Adelaide Case-Control Study[a]

	Females				Males			
	Age < 50		Age > 50		Age < 50		Age > 50	
	Cases	Controls	Cases	Controls	Cases	Controls	Cases	Controls
Sugars: total (g)	147 ± 7[d]	113 ± 4	128 ± 8	126 ± 7	160 ± 15[b]	121 ± 10	135 ± 12	153 ± 10
Sugars: drinks and sweets (g)	53 ± 4[d]	28 ± 3	36 ± 4[c]	23 ± 2	58 ± 5[b]	40 ± 7	46 ± 6	40 ± 5
Dietary fiber (g)	19.2 ± 0.8	17.7 ± 0.7	18.1 ± 0.9	21.9 ± 1.2	17.4 ± 2.1	18.3 ± 1.6	18.5 ± 2.2	22.6 ± 1.9

[a]From Scragg et al. (1984).
[b]$P < 0.05$.
[c]$P < 0.01$.
[d]$P < 0.001$.

time as the rise in the intake of sugars, a change that could explain the rise in cholesterol saturation (see next section). Hence, the precise mechanism of the "lithogenic" effect of sugars, if there is one, is unclear. Possibilities are that the effect takes longer than 6 weeks to develop, that it depends on fiber intake being low, and that sugars promote gallstones by changing some parameter other than cholesterol saturation index.

5. DIETARY FIBER

With dietary fiber the situation is opposite to that with sugars. The experimental evidence that it is protective is quite good, namely, that a high intake of fiber in the form of wheat bran lowers bile cholesterol saturation, but the epidemiologic evidence is rather weak. In the second largest case-control study, patients with gallstones had a slightly but significantly reduced intake of fiber (Alessandrini *et al.*, 1982), but in the largest one dietary fiber emerged as protective only when the data were subjected to multiple regression analysis (Scragg *et al.*, 1984). In two studies of people with unsuspected gallstones detected on population screening, there was an inverse relationship between the frequency of vegetable consumption and the risk of gallstones, but this did not reach statistical significance (Attili *et al.*, 1984, 1987). However, the number of subjects was rather small in both studies.

Figure 2 summarizes the published data on the effect of wheat bran on the cholesterol saturation index of bile. Clearly, there is a beneficial effect that is consistent provided the initial saturation index is above 1.0, that is, the bile is supersaturated. The mode of action of bran is not proven, but the most likely explanation is via reduction in the circulating pool of the secondary bile acid deoxycholic acid (DCA). There is considerable evidence that an expanded DCA pool is a risk factor for supersaturated bile (Marcus and Heaton, 1988). For example, any regimen that lowers or raises the DCA content of bile also lowers or raises its cholesterol saturation. In all but one of the bran-feeding studies in which biliary DCA has been measured, it has been lowered by bran (Heaton, 1987). There are several mechanisms whereby bran might lower the DCA content of bile

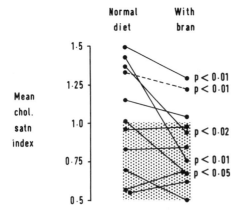

FIGURE 2. Mean cholesterol saturation index in groups of subjects before and after administration of wheat bran. Data extracted from papers by Pomare *et al.* (1976), Watts *et al.* (1978), McDougall *et al.* (1978), Wechsler *et al.* (1984), and Marcus and Heaton (1986a).

TABLE II. Possible Mechanisms whereby Bran Reduces
the DCA Content of Bile

Mechanism	By analogy with
Adsorption of DCA, preventing absorption	Lignin, other residues
Fermentation to short-chain fatty acids	Lactulose
leading to lower pH in colon and hence:	
Precipitation of DCA	
Reduced formation of DCA (inhibited	
dehydroxylation)	
Rapid transit, reducing absorption	Senna laxative

(Table II). Its laxative effect may be as important as any of its actions, since the DCA pool is sensitive to changes in intestinal transit time (Marcus and Heaton, 1986b).

Other forms of dietary fiber have not been examined in the same way. Purified wood cellulose lowers the DCA content of bile (Hillman *et al.*, 1986), but its effect has not been reported in subjects with supersaturated bile. Paradoxically, purified pectin and a diet rich in fruit and vegetable fiber increase the DCA content of bile but do not elevate the saturation index (Hillman *et al.*, 1986; Pomare, 1983).

6. TESTING THE FIBER–SUGAR HYPOTHESIS

The hypothesis that a diet rich in bran and low in added sugars would prevent the formation of gallstones was put to the test in the British/Belgian Gallstone Study Group's recently reported multicenter trial of postdissolution management (Hood *et al.*, 1988). In this study 31 patients whose gallstones had been completely dissolved by oral bile acids were randomly allocated to a dietary prescription that called for added sugars to be reduced as far as possible and for coarse wheat bran to be taken in a dose of at least 20 g/day. There were some problems with compliance, but all the same, total dietary fiber intake was estimated as about 35 g/day, compared with 20 g/day in placebo-treated patients, and intake of added sugars was substantially lowered. Despite these changes the recurrence rate of gallstones was not lowered at all (corrected cumulative recurrence 30% over 21 $\pm$ 2 months in diet and placebo groups). The most likely explanation for these disappointing findings is that the dietary changes failed to lower the cholesterol saturation index below 1.0 (below which crystallization of cholesterol is unlikely to occur), if only because most subjects were overweight and remained so. Reported changes in saturation index with bran or high-fiber diet are relatively modest and result in unsaturated bile in only a minority (Pomare *et al.*, 1976; Thornton *et al.*, 1983a). In support of this interpretation, recurrence of gallstones in the BBGSG Trial seemed to be determined by gallbladder factors rather than metabolic factors, since recurrence was prevented completely by concurrent administration of nonsteroidal antiinflammatory drugs, which are known to reduce mucus secretion by the gallbladder and to prevent gallstones in experimental animals.

The BBGSG Trial was probably too stringent a test for the hypothesis linking gallstone formation with dietary fiber and sugars. A different result might have been obtained if the diet-treated subjects had been persuaded to shed their excess body fat. There are anecdotal reports of fat patients' gallstones disappearing when they lose a great deal of weight, and, certainly reversal of obesity can result in drastic reduction of the cholesterol content of bile (Bennion and Grundy, 1975).

7. CHOLESTEROL INTAKE

The evidence for high cholesterol intake as a factor in human gallstone disease is weak and inconsistent. In the Adelaide case-control study, increased cholesterol intake was negatively associated with gallstones, at least in females (Scragg et al., 1984). Cholesterol-feeding studies have given conflicting results (Heaton, 1984). Animal models are misleading. When animals fed cholesterol develop gallstones, they also develop gross hypercholesterolemia. Hypercholesterolemia is not a feature of human gallstone disease. Indeed, in some studies, patients with gallstones have had reduced plasma cholesterol levels (van der Linden, 1961; Angelico et al., 1984b).

8. CONCLUSION

Overnutrition in general is the best established dietary factor in causing gallstones, but rapidly absorbed carbohydrate probably has a particular role. Dietary fiber may have some protective value through its effects on bile acid metabolism in the colon.

REFERENCES

Alessandrini, A., Fusco, M. A., Gatti, E., and Rossi, P. A., 1982, Dietary fibres and cholesterol gallstones: A case control study, Ital. J. Gastroenterol. **14:**156–158.

Angelico, F., and the GREPCO Group, 1984a, Factors associated with gallstone disease: Observations in the GREPCO study, in: Epidemiology and Prevention of Gallstone Disease (L. Capocaccia, G. Ricci, F. Angelico, M. Angelico, and A. F. Attili, eds.), MTP Press, Lancaster, pp. 185–192.

Angelico, M., and the GREPCO group, 1984b, Relationships between serum lipids and cholelithiasis; observations in the GREPCO study, in: Epidemiology and Prevention of Gallstone Disease (L. Capocaccia, G. Ricci, F. Angelico, M. Angelico, and A. F. Attili, eds.), MTP Press, Lancaster, pp. 77–84.

Attili, A. F., and the GREPCO group, 1984, Dietary habits and cholelithiasis, in: Epidemiology and Prevention of Gallstone Disease (L. Capocaccia, G. Ricci, F. Angelico, M. Angelico, and A. F. Attili, eds.), MTP Press, Lancaster, pp. 175–181.

Attili, A. F., and the Rome Group for the Epidemiology and Prevention of Cholelithiasis (GREPCO), 1987, Diet and gallstones: Result of an epidemiologic study performed in male civil servants, in: Nutrition in Gastrointestinal Disease (L. Barbara, G. Bianchi Porro, R. Cheli, and M. Lipkin, eds.), Raven Press, New York, pp. 225–231.

Bennion, L. J., and Grundy, S. M., 1975, Effects of obesity and caloric intake on biliary lipid metabolism in man, J. Clin. Invest. **56:**996–1011.

Bennion, L. J., and Grundy, S. M., 1978, Risk factors for the development of cholelithiasis in man, *N. Engl. J. Med.* **299:**1161–1167, 1221–1227.

Bloch, H. M., Thornton, J. R., and Heaton, K. W., 1980, Effects of fasting on the composition of gallbladder bile, *Gut* **21:**1087–1089.

Bolondi, L., Gaiani, S., Testa, S., and Labò, G., 1985, Gallbladder sludge formation during prolonged fasting after gastrointestinal tract surgery, *Gut* **26:**734–738.

Capron, J. P., Delamarre, J., Herve, M. A., Dupas, J. L., Poulain, P., and Descombes, P., 1981, Meal frequency and duration of overnight fast: A role in gallstone formation? *Br. Med. J.* **283:**1435.

Garn, S. M., LaVelle, M., Rosenberg, K. R., and Hawthorne, V. M., 1986, Maturational timing as a factor in female fatness and obesity, *Am. J. Clin. Nutr.* **43:**879–883.

Haber, G. B., and Heaton, K. W., 1979, Lipid composition of bile in diabetics and obesity-matched controls, *Gut* **20:**518–522.

Heaton, K. W., 1984, The role of diet in the aetiology of gallstones, *Rev. Clin. Nutr.* **54:**549–560.

Heaton, K. W., 1987, Effect of dietary fiber on biliary lipids, in: *Nutrition in Gastrointestinal Disease* (L. Barbara, G. Bianchi Porro, R. Cheli, and M. Lipkin, eds.), Raven Press, New York, pp. 213–222.

Heaton, K. W., 1988, Prevention of gallstones—clues from epidemiology, in: *Bile Acids in Health and Disease* (T. C. Northfield, R. Jazrawi, and P. Zentler-Munro, eds.), MTP Press, Lancaster, pp. 157–169.

Hillman, L. C., Peters, S. G., Fisher, C. A., and Pomare, E. W., 1986, The effects of the fibre components pectin, cellulose, and lignin on bile salt metabolism and biliary lipid composition in man, *Gut* **27:**29–36.

Hood, K., Gleeson, D., Ruppin, D., and Dowling, H., 1988, Can gallstone recurrence be prevented? The British/Belgian post-dissolution trial, *Gastroenterology* **94:**A548.

Jørgensen, T., 1988, Gallstones in a Danish population: Fertility period, pregnancies, and exogenous female sex hormones, *Gut* **29:**433–439.

Leiss, O., Becker, M., and von Bergmann, K., 1987, Effect of age and individual endogenous bile acids on biliary lipid secretion in humans, in: *Bile Acids and the Liver—Falk Symposium 45 (1986—Basel)* (G. Paumgartner, A. Stiehl, and W. Gerok, eds.), MTP Press, Lancaster, pp. 217–231.

Lowenfels, A. B., 1981, Gallstones and the risk of cancer, 1981, *Gut* **21:**1090–1092.

Lowenfels, A. B., Domellöf, L., Lindström, C. G., Bergman, F., Monk, M. A., and Sternby, N. H., 1982, Cholelithiasis, cholecystectomy, and cancer: A case-control study in Sweden, *Gastroenterology* **83:**672–676.

Marcus, S. N., and Heaton, K. W., 1986a, Effects of a new, concentrated wheat fibre preparation on intestinal transit, deoxycholic acid metabolism and the composition of bile, *Gut* **27:**893–900.

Marcus, S. N., and Heaton, K. W., 1986b, Intestinal transit, deoxycholic acid and the cholesterol saturation of bile—three inter-related factors, *Gut* **27:**550–558.

Marcus, S. N., and Heaton, K. W., 1988, Deoxycholic acid and the pathogenesis of gallstones, *Gut* **29:** 522–533.

McDougall, R. M., Yakymyshyn, L., Walker, K., and Thurston, O. G., 1978, The effect of wheat bran on serum lipoproteins and biliary lipids, *Can. J. Surg.* **21:**433–435.

Pixley, F., 1986, Epidemiology, in: *Gallstone Disease and Its Management* (M. C. Bateson, ed.), MTP Press, Lancaster, pp. 1–23.

Pixley, F., and Mann, J., 1988, Dietary factors in the aetiology of gallstones: A case control study, *Gut* **29:**1511–1515.

Pixley, F., Wilson, D., McPherson, K., and Mann, J., 1985, Effect of vegetarianism on development of gallstones in women, *Br. Med. J.* **291:**11–12.

Pomare, E. W., 1983, Fibre and bile acid metabolism, in: *Fibre in Human and Animal Nutrition* (G. Wallace and L. Bell, eds.), Royal Society of New Zealand, Wellington, pp. 179–182.

Pomare, E. W., Heaton, K. W., Low-Beer, T. S., and Espiner, H. J., 1976, The effect of wheat bran upon bile salt metabolism and upon the lipid composition of bile in gallstone patients, *Am. J. Dig. Dis.* **21:**521–526.

Sampliner, R. E., Bennett, P. H., Comess, L. J., Rose, F. A., and Burch, T. A., 1970, Gallbladder disease in Pima Indians. Demonstration of high prevalence and early onset by cholecystography, *N. Engl. J. Med.* **283**:1358–1364.

Scragg, R. K. R., 1986, Aetiology of cholesterol gallstones, in: *Gallstone Disease and Its Management* (M. C. Bateson, ed.), MTP Press, Lancaster, pp. 25–55.

Scragg, R. K. R., McMichael, A. J., and Baghurst, P. A., 1984, Diet, alcohol, and relative weight in gallstone disease: A case-control study, *Br. Med. J.* **288**:1113–1119.

Thornton, J. R., Heaton, K. W., and Macfarlane, D. G., 1981, A relation between high-density-lipoprotein cholesterol and bile cholesterol saturation, *Br. Med. J.* **283**:1352–1354.

Thornton, J. R., Emmett, P. M., and Heaton, K. W., 1983a, Diet and gallstones: Effects of refined and unrefined carbohydrate diets on bile cholesterol saturation and bile acid metabolism, *Gut* **24**:2–6.

Thornton, J., Symes, C., and Heaton, K., 1983b, Moderate alcohol intake reduces bile cholesterol saturation and raises HDL cholesterol, *Lancet* **2**:819–822.

van der Linden, W., 1961, Some biological traits in female gallstone-disease patients, *Acta Chirurg. Scand. (Suppl.)* **269**:7–94.

Watts, J. McK., Jablonski, P., and Toouli, J., 1978, The effect of added bran to the diet on the saturation of bile in people without gallstones, *Am. J. Surg.* **135**:321–324.

Wechsler, J. G., Swobodnik, W., Wenzel, H., Heuchemer, T., Nebelung, W., Hutt, V., and Ditschuneit, H., 1984, Ballaststoffe vom Typ Weizenkleie senken Lithogenität der Galle, *Deutsch. Med. Wochenschr.* **109**:1284–1288.

Werner, D., Emmett, P. M., and Heaton, K. W., 1984, The effects of dietary sucrose on factors influencing cholesterol gallstone formation, *Gut* **25**:269–274.

21

Fiber-Depleted Starch Foods and NIDDM Diabetes

HUGH TROWELL†

1. INTRODUCTION

The hypothesis that fiber-depleted starch foods are the main factor producing NIDDM diabetes (henceforth called diabetes) started in my mind in 1929 soon after I joined the Kenya Medical Service. It continued on and off in my mind until it culminated in two articles published in 1987, some 59 years later.

It is impossible to tell so long a story in one piece. I have therefore divided it into three sections: first, the early suggestions, 1929–1958, after which I left East Africa; second, the years of conflict and confusion, 1960–1986; third, the years of acceptance, 1987–1988.

2. EARLY SUGGESTIONS 1929–1958

I was posted to Machakos hospital in Kenya in 1929 and worked under Dr. Callanan; both of us were members of the Kenya Medical Service. Dr. Callanan had pioneered the British Medical Research Council investigation of the health and diseases of two African tribes (Orr and Gilks, 1931). One tribe was the pastoral Masai, eating much milk, meat, and blood, for they were a pastoral tribe. The other tribe was the Akikuyu, an agricultural Kenyan tribe. Their diet was derived from foods grown in their numerous small gardens, for they had no food shops; instead, they bartered foods in their markets. They ate mainly corn (maize) and millet as full-fiber whole-grain foods. These had been pounded in wooden mortars and not sieved to remove any bran. These foods provided a high-starch diet 80% of daily energy. It was also very low fat, only 10% of daily energy, and had a normal amount of protein, largely of plant origin.

---

†Deceased.

HUGH TROWELL • Makerere University, Uganda.

I was posted to Nairobi African Hospital, 200 beds, in 1930–1937. Nine-tenths of the patients were urban men, mostly Akikuyu. All of them looked rather thin, and obesity was very rare at that time. No male patient was reported to have diabetes during all these 8 years. One Akikuyu woman, however, had obvious NIDDM diabetes. She was a stout nursemaid employed by a British employer and fed by her mistress (Trowell, 1981). The customary diet of employed nursemaids was based on cheap white rice. In Kenya I wrote an elementary book for African nurses (Trowell, 1937). It discussed of necessity only the common diseases of Africans. Diabetes was therefore omitted in this book and in the second edition (1946).

Transferred to Uganda in 1937, I began to teach medicine in Mulago hospital; eventually this became the Makerere University Hospital. I wrote a more advanced textbook for Uganda nurses (Trowell, 1939). It stated that "foods that contain fibers, e.g., maize meal, beans, etc., as eaten by Africans cause them to pass bulky stools two or three daily." No association of fiber and diabetes occurred in either of these early textbooks.

3. CONFLICT AND CONFUSION 1960–1986

In my last year in Africa I wrote *Non-Infective Disease in Africa* (Trowell, 1960). It listed 39 diseases of unknown etiology which appeared to be "uncommon in many Africans, common in Europeans." Obesity and diabetes were both on this list. Diabetes had been often called a "disease of civilization." Civilizations occur, as the name suggests, in cities. Cities necessitate the storage of cereals and cereal flour; the latter usually contains less fiber.

This book summarized the knowledge of diabetes at that time, thus: "high-starch carbohydrate foods and low-fat diets are protective in diabetes but diets that contain processed starch foods and are also high fat predispose to diabetes." All this as early as 1960, 28 years ago.

This book was studied in African medical schools and has been cited in several medical articles. In Britain rather naturally this book on African epidemiology appeared irrelevant. I have traced only two references to this book in British medical journals, and no reference to the diabetes picture in Africans.

Eventually these suggestions about diabetes blossomed 14 years later in *The Lancet* (Trowell, 1974). This reported that diabetes death rates fell 55% in civilian males in England and Wales starting in 1942 and continued until 1954. These dates coincided exactly with all the years of the high-fiber, 85% extraction-rate National flour, which started during the war.

It was therefore "postulated that fiber-depleted starchy carbohydrate foods are a risk factor, and conversely that fiber-rich starchy foods are a protective factor, with respect to diabetes mellitus in susceptible genotypes." Unfortunately, only the protective factor, the high-fiber starchy foods, was investigated. No attention was paid to the possibility that fiber-depleted starch foods might be a risk factor. Soon after the 1974 publication, several articles reported that high-fiber starchy foods improved the treatment of diabetes. The Diabetes Association of the United States and of Britain recommended that diabetic patients should eat increased amounts of high-fiber starch foods and less fat. No doctor apparently looked at the other side of the coin; this suggested that fiber-depleted starch foods might cause diabetes.

Diabetes mortality rates had fallen 55% during the years of the high-fiber National flour. This must have been related to decreased mortality from associated coronary heart disease. In the late 1970s no article has been found that reported less angina or less macrovascular disease in diabetes patients eating a high-fiber diet. I am not aware that any medical article reported this anomaly, but the recommended diets allowed patients to eat white bread and white rice, both postulated to be diabetogenic foods.

During the years 1960–1984 many articles discussed low fiber intakes or fiber-deficiency diets and found no evidence to support a supposed hypothesis that these diets caused diabetes (Trowell, 1987b). Typical of this misunderstanding of the fiber-depleted starch hypothesis is a paragraph heading of a chapter on the etiology of NIDDM diabetes (Mann and Houston, 1983): they discussed "diets low in starchy foods and dietary fibre." They did not discuss fiber-depleted starch foods causing diabetes.

4. ACCEPTANCE OF FIBER-DEPLETED STARCH HYPOTHESIS 1985 ONWARD

In the second chapter of our book *Dietary Fibre, Fibre-Depleted Foods and Disease* (Trowell *et al.*, 1985), the three editors discussed and listed fiber-depleted starch foods. This forced contributors to consider a possible role of fiber-depleted starch foods in disease. Jim Mann of the Radcliffe Infirmary, Oxford, wrote a chapter on the etiology of NIDDM diabetes. He concluded his chapter by stating, "there is strong evidence that fibre-depleted starch foods increased the risk of developing diabetes." Open Sesame!

Ever since 1960 I had collected data that supported the hypothesis that fiber-depleted starch foods such as white bread (Trowell, 1987a) and white rice (Trowell, 1987b) were causative factors in NIDDM diabetes. These recent articles can be easily consulted and are not discussed here.

As soon as reprints of these two articles became available in March 1988, I began sending them to colleagues known personally by me. All were eminent members of the British medical profession. Sir Richard Doll wrote, "You make a good case for believing that fibre-depleted starch foods are the cause of non-insulin dependent diabetes." Sir Francis Avery Jones was more concise: "Fantastic! I do congratulate you once again. Greetings. Very best wishes." Dr. D. Pyke, former President of the British Diabetic Association, wrote, "Thank you for your letter and the interesting reprint. I greatly admire your energy and tenacity." Professor Barry Lewis delivered judgment: "Very many thanks for the two reprints. I found them eminently persuasive."

The position cannot rest there; other factors, many already known, increase the prevalence of diabetes; others aid remission of the disease. Obesity is a risk factor, and slimming aids recovery. High-fat intakes increase the risk of macrovascular complications and possibly microvascular complications. Prolonged periods of strenuous physical exercise protect. The consumption of large amounts of uncooked or lightly cooked raw plant foods protects. Salt intakes should be decreased.

5. CONCLUSIONS

Considerable evidence is presented that prolonged consumption of fiber-depleted starch foods, notably wheat white bread, white flour, and white and brown rice (Trowell, 1987b), produces NIDDM diabetes in susceptible phenotypes. Many other factors increase the complications and possibly the prevalence of diabetes. The avoidance of the causative foods is essential for the remission of diabetes; it may have to continue for life. Other factors increase the prevalence and the complications of diabetes. Overweight and obesity are risk factors; patients must be slimmed. High-fat diets increase overweight and increase the prevalence and severity of macrovascular and probably microvascular complications. Other factors aid the remission of diabetes: prolonged physical exercise, raw or lightly cooked plant foods, even low-salt diets all aid remission. Several of these suggestions are discussed in an article concerning the remission of NIDDM diabetes (Trowell, 1988).

ACKNOWLEDGMENTS. I thank Mrs. Jenny Allen, District Librarian, and her staff, Salisbury General Hospital, and Mrs. Natalie Adams for their assistance.

REFERENCES

Mann, J., 1985, Diabetes mellitus: Some aspects of aetiology and management of non-insulin dependent diabetes, in: *Dietary Fibre, Fibre-Depleted Foods and Disease* (H. Trowell, D. Burkitt, and K. Heaton, eds.), Academic Press, London, pp. 268–287.

Mann, J., and Houston, A. C., 1983, The aetiology of non-insulin dependent diabetes mellitus, in: *Diabetes in Epidemiological Perspective* (J. I. Mann, K. Pyörälä, and T. Teuscher, eds.), Churchill Livingstone, Edinburgh, pp. 122–164.

Orr, J. B., and Gilks, J. L., 1931, *The Physique and Health of Two African Tribes*, His Majesty's Stationery Office, London.

Trowell, H. C., 1937, *A Handbook for Dressers and Nurses in the Tropics*, Sheldon Press, London.

Trowell, H. C., 1939, *Diagnosis and Treatment of Disease in the Tropics*, Baillière, Tindall, and Cox, London, p. 163.

Trowell, H. C., 1960, *Non-Infective Disease in Africa*, Edward Arnold, London, pp. 306–311, 465–473.

Trowell, H., 1974, Diabetes mellitus death-rates in England and Wales, 1920–70, and food supplies, *Lancet* 2:998–1001.

Trowell, H. C., 1981, Diabetes mellitus emerges, in: *Western Diseases; Their Emergence and Prevention,* Harvard University Press, Cambridge, MA, pp. 22–26.

Trowell, H., 1987a, Dietary factors in the aetiology of diabetes mellitus, *Scand. J. Gastroenterol. [Suppl.]* **129**:142–144.

Trowell, H., 1987b, Diabetes mellitus and rice—a hypothesis, *Hum. Nutr.* **41F**:145–152.

Trowell, H., 1988, Remission of Diabetes (NIDDM). Editorial Article, *Nutrition.*

Trowell, H., Burkitt, D. P., and Heaton, K., 1985, *Dietary Fibre, Fibre-Depleted Foods and Disease,* Academic Press, London, pp. 21–30.

22

Dietary Fiber in the Management of Diabetes

THOMAS M. S. WOLEVER

1. INTRODUCTION

Over the past two decades there has been a radical change in dietary guidelines for diabetes, which now stress high carbohydrate and fiber intakes. The initial impetus for the increase in carbohydrate came from early studies of Himsworth (1935–36) and, later, Stone and Connor (1965), showing improved blood lipids and glucose tolerance on a high-carbohydrate diet. Further motivation for change originated in the early 1970s from Trowell's suggestion (1973) that the development of diabetes might be related to a lack of fiber in the diet. This spurred many new concepts about the dietary treatment of diabetes and lead to the classic studies of Anderson (Kiehm *et al.*, 1976; Anderson and Ward, 1979) and others (Simpson *et al.*, 1981; Rivellese *et al.*, 1980) showing the beneficial effects of high-carbohydrate, high-fiber diets.

However, there has been debate and vigorous disagreement about the role of high carbohydrate and dietary fiber in the diabetes diet (Reaven, 1980; National Institutes of Health, 1986; American Diabetes Association, 1987). Early on, Ahrens *et al.* (1961) and others (Bierman and Hamlin, 1961; Reaven, 1979) showed that high-carbohydrate diets could induce hyperlipidemia. The addition of fiber to a high-carbohydrate diet was shown to ameliorate the rise in blood glucose and triglycerides (Rivellese *et al.*, 1980; Anderson and Chen, 1979). Nevertheless, some investigators are still concerned about this effect, even if a high-carbohydrate, high-fiber diet is consumed (Hollenbeck *et al.*, 1985, 1986). At least some of the disagreement results from a lack of knowledge about the types, amounts, and physiological effects of fiber and carbohydrate in specific foods. It is to be hoped that as the analysis of dietary fiber improves, and as more data are

THOMAS M. S. WOLEVER • Department of Nutritional Sciences, Faculty of Medicine, and Division of Endocrinology and Metabolism, St. Michael's Hospital, University of Toronto, Toronto, Ontario M5S 1A8, Canada.

acquired, consensus can be reached. Since there are major interactions between the effects of fiber and carbohydrate in diabetes, I review both and attempt to provide some explanations for conflicting results.

2. FIBER HYPOTHESIS: DIABETES

The original rationale for the use of fiber in diabetes was that by forming a barrier to the ingestion, digestion, and absorption of nutrients, the rate and absolute amount of glucose absorbed would be reduced, leading to flatter blood glucose and insulin responses. Recent advances in our understanding of human gastrointestinal physiology have suggested a new rationale. It is now appreciated that much of the carbohydrate entering the colon is fermented with the production of short-chain fatty acids (SCFA) (Cummings and Branch, 1986). Increasing fiber intake increases the amount of carbohydrate (starch and fiber) entering the colon and hence increases SCFA production. The SCFA may be absorbed from the colon and have systemic effects on carbohydrate and lipid metabolism.

There are several ways to delay carbohydrate absorption, and most of these result in small increases in delivery of carbohydrate to the colon: slow consumption of carbohydrate (Jenkins et al., 1982a); the use of amylase inhibitors (Jenkins et al., 1981a); the consumption of slowly digested carbohydrate foods (Jenkins et al., 1982b); and the addition of viscous types of dietary fiber (Jenkins et al., 1978a). Test-meal studies have shown that all these maneuvers result in reduced postprandial blood glucose and insulin responses. Although small increases in carbohydrate malabsorption can be demonstrated, these are not nearly enough to account, directly, for the large differences in blood glucose response (Wolever et al., 1986; Jenkins et al., 1987a). With respect to purified fiber supplements and whole carbohydrate foods, the results of in vitro tests of viscosity (Jenkins et al., 1978a; Edwards et al., 1987), the rate of diffusion of glucose out of fiber solutions (Blackburn et al., 1984; Jenkins et al., 1986), and the rate of amylolitic digestion of starch (Jenkins et al., 1982b) relate well to the in vivo blood glucose and insulin responses. There has also been a direct demonstration of the ability of guar to slow the rate of glucose absorption in vivo in man (Blackburn et al., 1984).

3. PURIFIED DIETARY FIBERS

3.1. Test-Meal Studies

Insoluble fibers such as wheat bran, which do not increase the viscosity of water, have little effect on the postprandial blood glucose response, whereas soluble, viscous fibers such as guar reduce glycemic responses (Jenkins et al., 1978a). Early studies showed that the maximum effect of guar was achieved if it was mixed intimately with the ingested carbohydrate load (Jenkins et al., 1979). This has been confirmed recently (Fuessel et al., 1986).

3.2. Longer-Term Studies

The different effects of different types of fiber observed in test-meal studies are reflected in the results of longer-term studies. There have been 21 studies of the effect of purified viscous fiber (mostly guar) in diabetes (Jenkins *et al.*, 1977, 1978b, 1980a; Smith *et al.*, 1982; Kuhl *et al.*, 1983; Gatti *et al.*, 1984; Smith and Holm, 1982; Johansen,, 1981; Carroll *et al.*, 1981; Najemnik *et al.*, 1984; Aro *et al.*, 1981; Botha *et al.*, 1981; Cohen *et al.*, 1980; Vaaler *et al.*, 1983; Ray *et al.*, 1983; Fuessl *et al.*, 1987; Atkins *et al.*, 1987; Tagliaferro *et al.*, 1985; Osilesi *et al.*, 1985; Holman *et al.*, 1987; Paganus *et al.*, 1987). The average of the results of these studies indicates about twice the reduction of blood glucose compared to the 11 studies using insoluble fibers (Vaaler *et al.*, 1983; Mahalko *et al.*, 1984; Mayne *et al.*, 1982; Monnier *et al.*, 1981; Bosello *et al.*, 1980; Karlstrom *et al.*, 1984; Miranda and Horwitz, 1978), with twice as many studies showing statistically significant improvements (Table I). Significant reductions of serum triglycerides have been seen in only one of 11 studies where soluble fiber was added to the diet and in only two out of ten studies of insoluble fiber (Table I). However, serum cholesterol was significantly reduced in 14 (88%) of the 16 studies using guar but in only one (10%) of ten studies of various insoluble fibers ($P < 0.005$, Table I). In a recent double-blind, randomized trial of 15 g guar versus 15 g bran daily for 4 weeks in 18 type 2 diabetics, bran had no effect on blood glucose or cholesterol levels, but on guar fasting blood glucose, Hb_{Alc}, and serum total and LDL-cholesterol all fell by 10% ($P < 0.05$) (Fuessl *et al.*, 1987).

4. HIGH-FIBER FOODS

4.1. Test-Meal Studies

In keeping with the results of studies using purified fiber, the glycemic response of foods high in cereal fiber (e.g., whole-meal bread) is no different from that of their refined counterparts (e.g., white bread) (Jenkins *et al.*, 1981b, 1983, 1984). However, foods rich in soluble fiber, such as legumes and barley, are digested slowly and produce low blood glucose responses (Jenkins *et al.*, 1980c; Wolever *et al.*, 1987). A number of studies have shown a close relationship between the rate of digestion of foods *in vitro* and their blood glucose responses *in vivo* (Jenkins *et al.*, 1982b, 1984; O'Dea *et al.*, 1981; Brand *et al.*, 1985; Thorburn *et al.*, 1987; Wolever *et al.*, 1988a). Factors that affect the rate of digestion of foods are very complex and include dietary fiber, anti-nutrients, particle size, food form, amylose: amylopectin ratio, starch gelatinization, starch–protein and starch–lipid interactions, and the degree of cooking (Jenkins *et al.*, 1988a). Until all the factors that affect the rate of digestion of foods are understood, it will not be possible to predict the glycemic respose of foods from their chemical composition. Therefore, in order to classify foods according to their physiological effect, the glycemic index (GI) was developed.

TABLE I. Mean Percentage Changes of Fasting Blood Glucose (FBG), Urinary Glucose Excretion, Hb$_{Alc}$, and Serum Cholesterol (Chol) and Triglyceride (TG) in 21 Studies of Soluble Fiber (20 Guar Gum, One Xanthan Gum) and 11 Studies Using Various Insoluble Fibers in the Treatment of Diabetes

	Dose (g/day)	Duration (days)	FBG	Urine glucose	Hb$_{Alc}$	Serum Chol	Serum TG
Soluble fiber							
Mean ± S.E.M.	18 ± 1	59 ± 18[a]	−15 ± 4%	−53 ± 4%	−6.3 ± 1.3%	−13 ± 1.5%	−7 ± 5%
Range	(12–29)	(5–90)	(−39 to +13%)	(−70 to −30%)	(−10 to 0%)	(−30 to −8%)	(−25 to +11%)
No. (%) of studies with significant reductions			12/18 (67%)	11/13 (85%)	5/8 (63%)	14/16 (88%)	1/11 (9%)
Insoluble fiber							
Mean ± S.E.M.	30 ± 4	32 ± 6	−5.5 ± 1.5%	+25 ± 19%	−3.9 ± 1.7%	−0.4 ± 1.8%	+0.6 ± 2.5%
Range	(14–52)	(10–90)	(−16 to +2%)	(−37 to +119%)	(−13 to +3%)	(−9 to +10)	(−15 to +11)
No. (%) of studies with significant reductions			3/11 (27%)	2/8 (25%)	2/9 (22%)	1/10 (10%)	2/10 (20%)

[a]One study lasting for 1 year has not been included in the mean.

4.2. The Glycemic Index

The GI is defined as the incremental area under the glycemic response curve for a food expressed as a percentage of the glycemic response for the same amount of carbohydrate (50 g) from white bread taken by the same subject (Jenkins *et al.*, 1981c; Wolever and Jenkins, 1986). The GI may be a useful tool to assess the effects of high-fiber diets, since the addition of fat and protein to foods does not alter the relative differences between their glycemic effects (Wolever *et al.*, 1987, 1988a; Collier *et al.*, 1984; Bornet *et al.*, 1987). Apart from two studies where the GI of individual foods was not assessed before their incorporation into a meal (Coulston *et al.*, 1987; Laine *et al.*, 1987), in most studies the GI of individual foods predicts the relative glycemic responses of mixed meals (Wolever *et al.*, 1985, 1986; Wolever and Jenkins, 1985; Collier *et al.*, 1986; Golay *et al.*, 1986; Parillo *et al.*, 1985; Chew *et al.*, 1988) (Fig. 1).

5. HIGH-CARBOHYDRATE, HIGH-FIBER DIETS

High-carbohydrate diets may cause increases in blood glucose and triglyceride levels unless accompanied by increases in dietary fiber (Kiehm *et al.*, 1976; Anderson and Ward, 1979; Rivellese *et al.*, 1980; Coulston *et al.*, 1983; Perrotti *et al.*, 1984; Liu *et al.*, 1983; H. C. R. Simpson *et al.*, 1981, 1982; R. W. Simpson *et al.*, 1979a,b; Kinmonth *et al.*, 1982; Karlstrom *et al.*, 1987). On the other hand, high-fiber diets appear to have greater effects against a background of higher carbohydrate intakes (Rivellese *et al.*, 1980; Jenkins *et al.*, 1980b).

There have been several studies in which high-carbohydrate, high-fiber diets have

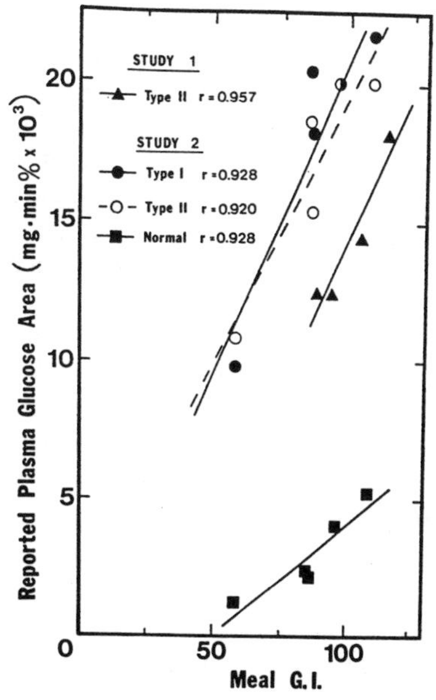

FIGURE 1. Regressions of blood glucose response areas on meal glycemic index (G.I.) for groups of normal and diabetic subjects who took different mixed test meals. Retrospective analysis by Wolever *et al.* (1985) of data from Nuttal *et al.* (1983) (study 1) and Bantle *et al.* (1983) (study 2). (From Wolever *et al.*, 1985; reprinted with permission.)

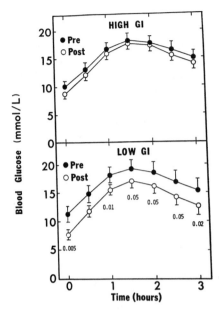

FIGURE 2. Mean ± S.E.M. blood glucose responses of eight patients with type 2 diabetes who took 50 g carbohydrate from white bread before (Pre) and after 2 weeks (Post) on high- and low-glycemic-index (GI) diets. The numbers under the graphs on the lower panel represent the significance of the differences between the blood glucose levels before and after the low-GI diet. (From Jenkins et al., 1988a; reprinted with permission.)

had no effect or deleterious effects in diabetes (Hollenbeck et al., 1985, 1986; Lindsay et al., 1984). The conflicts may be able to be explained by examination of the GI of the specific foods used in the diets. In general, in those studies that have shown beneficial effects, the carbohydrate and fiber intakes were increased by the use of slowly digested, low-glycemic-index foods such as beans and pasta. In those studies where there was no effect, carbohydrate and fiber intakes were increased with rapidly digested starchy cereal foods with a high glycemic index such as whole-meal bread, rice, and high-fiber breakfast cereals (Wolever and Collier, 1987). This suggests that in order to predict the effects of a diet in diabetes, both the chemical composition (i.e., carbohydrate and fiber content) and the physiological effect (i.e., glycemic index) of a diet must be defined.

In order to test this hypothesis, we have studied, in normal (Jenkins et al., 1987b), hyperlipidemic (Jenkins et al., 1985, 1987c), and diabetic subjects (Jenkins et al., 1988b), the effects of changes in the GI of the diet independent of changes in total carbohydrate and fiber content. Postprandial blood glucose excursions were reduced to the extent predicted by the reduction in diet GI. This resulted in falls in glycosylated serum proteins and, in type 2 diabetics, by a reduction in fasting blood glucose (Fig. 2). There was significantly less insulin secreted on the low-GI diet, as assessed by urinary C-peptide excretion. Serum cholesterol was reduced on the low-GI diet in normal and diabetic subjects. With a more modest reduction in diet GI, the lipid-lowering effect was only consistent, in hyperlipidemic subjects, in those with a high triglyceride level. Here, there were significant falls in both cholesterol and triglyceride, with no change in HDL-cholesterol. There was also increased colonic fermentation on the low-GI diet, as judged by increased breath hydrogen levels. Thus, the results of the glycemic index studies show the same range of metabolic effects as seen in studies using viscous fiber. This may be because of common mechanisms of action, namely, slow absorption and increased SCFA production.

6. MECHANISM OF ACTION

6.1. Slowing Carbohydrate Absorption

It has been suggested that slowing the absorption of carbohydrate *per se* has prolonged effects on carbohydrate metabolism. This has been demonstrated by the so-called "second-meal effect." Here, the blood glucose and insulin responses to a standard meal are affected by slowing the rate of carbohydrate absorption in the *previous* meal. This has been demonstrated for the 4-hr period between lunch and breakfast when carbohydrate absorption from the first test meal was reduced using guar (Jenkins *et al.*, 1980d), low-glycemic-index foods (Jenkins *et al.*, 1982a), or continuous glucose sipping (Ocana *et al.*, 1988), and for the 11-hr period between dinner and breakfast, when the rate of absorption of dinner carbohydrate was reduced using low-glycemic-index foods (Wolever *et al.*, 1988b). The hypothesis to explain the effect is that rapidly absorbed carbohydrate causes a rapid rise in the blood glucose level, which stimulates a large insulin secretion. The high insulin level stimulates rapid glucose uptake by tissues, resulting in a rapid fall of blood glucose with an undershoot. This, in turn, leads to a counterregulatory response that arrests the fall in blood glucose and results in relative resistance to the action of insulin at the time of the consumption of the second meal. Evidence for this includes increased free fatty acid and ketone body levels 4 hr after glucose alone compared to glucose plus guar (Jenkins *et al.*, 1980d) and increased urinary catecholamines and serum glucagon and growth hormone 4 hr after an oral glucose load taken as a bolus as compared to sipped continuously for 3.5 hr (Ocana *et al.*, 1988).

6.2. Effect of Short-Chain Fatty Acids

The fermentation of carbohydrate in the colon results in the production of acetic (AC), propionic (PR), and butyric (BU) acids. Butyric acid is believed to be taken up by the colonic mucosa, and PR by the liver, so that only AC reaches the peripheral circulation (Cummings and Branch, 1986). The short-chain fatty acids could have systemic effects that explain, at least in part, the blood glucose- and insulin-lowering effects of high-fiber diets. For example, AC may have blood glucose- and free-fatty-acid-lowering effects similar to those of acetoacetate (Jenkins, 1967). In experimental animals, AC and PR stimulate insulin secretion and reduce blood glucose levels (Asplund *et al.*, 1985; Brockman, 1982). Propionic acid is a gluconeogenic substrate in the horse (Ford and Simmons, 1985).

Since the effects of AC and PR in man are not known, I have begun to study the effects of rectal infusion of short-chain fatty acids in human subjects. Infusion of 180 mmole of AC plus 60 mmole PR over 30 min resulted in a peak of serum AC 1 hr after the start of the infusion. Propionic acid was undetectable in peripheral blood before or after the infusion. The blood glucose level was significantly higher after the AC and PR infusion than after the control saline infusion, with a tendency toward higher serum insulin levels and significantly reduced free fatty acid levels. This suggests that PR and/or AC may be gluconeogenic substrates in man. These effects may not reproduce those of the physiological situation where the fermentation of starch and fiber is likely to result in a slower and more prolonged absorption of AC and PR from the cecum rather than the rectum and descending colon. Nevertheless, they indicate that short-chain fatty acids may have effects on carbohydrate metabolism in humans.

7. CONCLUSIONS

Long-term benefits have been demonstrated using high-carbohydrate, high-fiber diets in both insulin- and non-insulin-dependent diabetes. These are associated with and may result from slowing the rate of carbohydrate digestion and absorption within the GI tract. Slow-release carbohydrate may be obtained in the diet either by the addition of certain types of dietary fiber or by the use of starchy low-glycemic-index foods. In addition, however, the consumption of low-glycemic-index foods and dietary fiber also increases the amount of fermentable carbohydrate that enters the colon. This will result in the absorption of short-chain fatty acids, which may also have effects on carbohydrate metabolism.

ACKNOWLEDGMENTS. The author was supported in part by a grant to the Department of Nutritional Sciences from the Bristol-Myers Company, New York.

REFERENCES

Ahrens, E. H., Hirsch, J., Oette, K., Farquhar, J. W., and Stein, Y., 1961, Carbohydrate induced and fat induced lipemia, *Trans. Assoc. Am. Physicians* **74:**134.

American Diabetes Association, 1987, Policy statement: Nutritional recommendations and principles for individuals with diabetes mellitus: 1986, *Diabetes Care* **10:**126–132.

Anderson, J. W., and Chen, W. L., 1979, Plant fiber: Carbohydrate and lipid metabolism, *Am. J. Clin. Nutr.* **32:**346–363.

Anderson, J. W., and Ward, K., 1979, High carbohydrate, high fiber diets for insulin treated men with diabetes mellitus, *Am. J. Clin. Nutr.* **32:**2312–2321.

Aro, A., Uusitupa, M., Vontilainen, E., Hersio, K., Korhonen, T., and Siitonen, O., 1981, Improved diabetic control and hypocholesterolemic effect induced by long-term dietary supplementation with guar gum in type 2 (insulin-independent) diabetes, *Diabetologia* **21:**29–33.

Asplund, J. M., Orskov, E. R., Hovell, F. D., and Macleod, N. A., 1985, The effect of intragastric infusion of glucose, lipids or acetate on fasting nitrogen excretion and blood metabolites in sheep, *Br. J. Nutr.* **54:**189–195.

Atkins, T. W., Al-Hussary, N. A. J., and Taylor, K. G., 1987, The treatment of poorly controlled non-insulin-dependent diabetic subjects with granulated guar gum, *Diabetes Res. Clin. Pract.* **3:**153–159.

Bantle, J. P., Laine, D. C., Castle, G. W., Thomas, J. W., Hoogwerf, B. J., and Goetz, F. C., 1983, Postprandial glucose and insulin responses to meals containing different carbohydrates in normal and diabetic subjects, *N. Engl. J. Med.* **309:**7–12.

Bierman, E., and Hamlin, J. I., 1961, Hyperlipidemic effect of a low fat, high carbohydrate diet in diabetes, *Diabetes* **10:**432–437.

Blackburn, N. A., Redfern, J. S., Jarjis, H., Holgate, A. M., Hanning, I., Scarpello, J. H. B., Johnson, I. T., and Read, N. W., 1984, The mechanism of action of guar gum in improving glucose tolerance in man, *Clin. Sci.* **66:**329–336.

Bornet, F. R. J., Costaglioloa, D., Blayo, A., Fontvielle, A., Haardt, M. J., Letanoux, M., Tchobroutsky, G., and Slama, G., 1987, Insulinogenic and glycemic indexes of six starch-rich foods taken alone and in a mixed meal by type 2 diabetics, *Am. J. Clin. Nutr.* **45:**588–595.

Bosello, O., Ostuzzi, R., Armellini, F., Micciolo, R. M., and Ludovico, A. S., 1980, Glucose tolerance and blood lipids in bran fed patients with impaired glucose tolerance, *Diabetes Care* **3:**46–49.

Botha, A. P. J., Steyn, A. F., Esterhuysen, A. J., and Slabbert, M., 1981, Glycosylated haemoglobin, blood glucose and serum cholesterol levels in diabetics treated with guar gum, *S. Afr. Med. J.* **59:** 333–334.

Brand, J. C., Nicholson, P. L., Thorburn, A. W., and Truswell, A. S., 1985, Food processing and the glycemic index, *Am. J. Clin. Nutr.* **42:**1192–1196.

Brockman, P. R., 1982, Insulin and glucagon responses in plasma to intraportal infusions of propionate and butyrate in sheep, *Comp. Biochem. Physiol.* **73A:**237–238.

Carroll, D. G., Dykes, V., and Hodgson, W., 1981, Guar gum is not a panacea in diabetes management, *N.Z. Med. J.* **93:**292–234.

Chew, I., Brand, J. C., Thorburn. A. W., and Truswell, A. S., 1988, Application of glycemic index to mixed meals, *Am. J. Clin. Nutr.* **47:**53–56.

Cohen, M., Leong, V. W., Salmon, E., and Martin, F. I. R., 1980, The role of guar and dietary fibre in the management of diabetes mellitus, *Med. J. Aust.* **1:**59–61.

Collier, G., McLean, A., and O'Dea, K., 1984, Effect of coingestion of fat on the metabolic responses to slowly and rapidly absorbed carbohydrates, *Diabetologia* **26:**50–54.

Collier, G. R., Wolever, T. M. S., Wong, G. S., and Josse, R. G., 1986, Prediction of glycemic response to mixed meals in non-insulin dependent diabetic subjects, *Am. J. Clin. Nutr.* **44:**349–352.

Coulston, A., Liu, G., and Reaven, G. M., 1983, Plasma glucose, insulin and lipid responses to high carbohydrate low fat diets in normal humans, *Metabolism* **32:**52–56.

Coulston, A. M., Hollenbeck, C. B., Swislocki, A. L. M., and Reaven, G. M., 1987, Effect of source of dietary carbohydrate on plasma glucose and insulin responses to mixed meals in subjects with NIDDM, *Diabetes Care* **10:**395–400.

Cummings, H., and Branch, W. J., 1986, Fermentation and the production of short-chain fatty acids in the human colon, in: *Dietary Fiber: Basic and Clinical Aspects* (G. V. Vahouny and D. Kritchevsky, eds.), Plenum Press, New York, pp. 131–149.

Edwards, C. A., Blackburn, N. A., Craigen, L., Davison, P., Tomlin, J., Sugden, K., Johnson, I. T., and Read, N. W., 1987, Viscosity of food gums determined *in vitro* related to their hypoglycemic actions, *Am. J. Clin. Nutr.* **46:**72–77.

Ford, E. J. H., and Simmons, H. A., 1985, Gluconeogenesis from caecal propionate in the horse, *Br. J. Nutr.* **53:**55–60.

Fuessl, S., Adrian, T. E., Bacarese-Hamilton, A. J., and Bloom, S. R., 1986, Guar in NIDD: Effect of different modes of administration on plasma glucose and insulin responses to a starch meal, *Pract. Diabetes* **3**(5):258–260.

Fuessl, H. S., Williams, G., Adrian, T. E., and Bloom, S. R., 1987, Guar sprinkled on food: Effect on glycaemic control, plasma lipids and gut hormones in non-insulin dependent diabetic patients, *Diabetic Med.* **4:**463–468.

Gatti, E., Catenazzo, G., Camisasca, E., Torri, A., Denegri, E., and Sirtori, D. R., 1984, Effects of guar-enriched pasta in the treatment of diabetes and hyperlipidemia, *Ann. Nutr. Metab.* **28:**1–10.

Golay, A., Coulston, A. M., Hollenbeck, C. B., Kaiser, L. L., Wursch, P., and Reaven, G. M., 1986, Comparison of metabolic effects of white beans processed into two different forms, *Diabetes Care* **9:**260–266.

Himsworth, H. P., 1935–36, The dietetic factor determining the glucose tolerance and sensitivity to insulin of healthy men, *Clin. Sci.* **2:**67–94.

Hollenbeck, C. B., Riddle, M. C., Conner, W. E., and Leklem, J. E., 1985, The effects of subject-selected high carbohydrate, low fat diets on glycemic control in insulin dependent diabetes mellitus, *Am. J. Clin. Nutr.* **41**:293–299.

Hollenbeck, C. B., Coulston, A. M., and Reaven, G. M., 1986, To what extent does increased dietary fiber improve glucose and lipid metabolism in patients with noninsulin-dependent diabetes mellitus (NIDDM)? *Am. J. Clin. Nutr.* **43**:16–24.

Holman, R. R., Steemson, J., Darling, P., and Turner, R. C., 1987, No glycemic benefit from guar administration in NIDDM, *Diabetes Care* **10**:68–71.

Jenkins, D. J. A., 1967, Ketone bodies and the inhibition of free-fatty acid release, *Lancet* **2**:338–340.

Jenkins, D. J. A., Wolever, T. M. S., Hockaday, T. D. R., Leeds, A. R., Haworth, R., Bacon, S., Apling, E. C., and Dilawari, J., 1977, Treatment of diabetes with guar gum, *Lancet* **2**:779–780.

Jenkins, D. J. A., Wolever, T. M. S., Leeds, A. R., Gassull, M. A., Dilawari, J. B., Goff, D. V., Metz, G. L., and Alberti, K. G. M. M., 1978a, Dietary fibres, fibre analogues and glucose tolerance: Importance of viscosity, *Br. Med. J.* **1**:1392–1394.

Jenkins, D. J. A., Wolever, T. M. S., Nineham, R., Taylor, R. H., Metz, G. L., Bacon, S., and Hockaday, T. D. R., 1978b, Guar crispbread in the diabetic diet, *Br. Med. J.* **2**:1744–1746.

Jenkins, D. J. A., Nineham, R., Craddock, C., Craig-McFeely, P., Donaldson, K., Leigh, T., and Snook, J., 1979, Fibre and diabetes, *Lancet* **1**:434–435.

Jenkins, D. J. A., Wolever, T. M. S., Taylor, R. H., Reynolds, D., Nineham, R., and Hockaday, T. D. R., 1980a, Diabetic glucose control, lipids, and trace elements on long term guar, *Br. Med. J.* **1**: 1353–1354.

Jenkins, D. J. A., Wolever, T. M. S., Bacon, S., Nineham, R., Leeds, A. R., Rowden, R., Love, M., and Hockaday, T. D. R., 1980b, Diabetic diets: High carbohydrate combined with high fiber, *Am. J. Clin. Nutr.* **33**:1729–1733.

Jenkins, D. J. A., Wolever, T. M. S., Taylor, R. H., Barker, H., and Fielden, H., 1980c, Exceptionally low blood glucose response to dried beans: Comparison with other carbohydrate foods, *Br. Med. J.* **2**:578–580.

Jenkins, D. J. A., Wolever, T. M. S., Nineham, R., Sarson, D. L., Bloom, S. R., Ahern, J., Alberti, K. G. M. M., and Hockaday, T. D. R., 1980d, Improved glucose tolerance four hours after taking guar with glucose, *Diabetologia* **19**:21–24.

Jenkins, D. J. A., Taylor, R. H., Goff, D. V., Fielden, H., Miseiwicz, J. J., Sarson, D. L., Bloom, S. R., and Alberti, K. G. M. M., 1981a, Scope and specificity of acarbose in slowing carbohydrate absorption in man, *Diabetes* **30**:951–954.

Jenkins, D. J. A., Wolever, T. M. S., Taylor, R. H., Barker, H. M., Fielden, H., and Gassull, M. A., 1981b, Lack of effect of refining on the glycemic response to cereals, *Diabetes Care* **4**:509–513.

Jenkins, D. J. A., Wolever, T. M. S., Taylor, R. H., Barker, H. M., Fielden, H., Baldwin, J. M., Bowling, A. C., Newman, H. C., Jenkins, A. L., and Goff, D. V., 1981c, Glycemic index of foods: A physiological basis for carbohydrate exchange, *Am. J. Clin. Nutr.* **34**:362–366.

Jenkins, D. J. A., Wolever, T. M. S., Taylor, R. H., Griffiths, C., Krzeminska, K., Lawrie, J. A., Bennett, C. M., Goff, D. V., Sarson, D. L., and Bloom, S. R., 1982a, Slow release carbohydrate improves second meal tolerance, *Am. J. Clin. Nutr.* **35**:1339–1346.

Jenkins, D. J. A., Ghafari, H., Wolever, T. M. S., Taylor, R. H., Barker, H. M., Fielden, H., Jenkins, A. L., and Bowling, A. C., 1982b, Relationship between the rate of digestion of foods and postprandial glycaemia, *Diabetologia* **22**:450–455.

Jenkins, D. J. A., Wolever, T. M. S., Jenkins, A. L., Lee, R., Wong, G. S., and Josse, R. G., 1983, Glycemic response to wheat products: Reduced response to pasta but no effect of fiber, *Diabetes Care* **6**:155–159.

Jenkins, D. J. A., Wolever, T. M. S., Thorne, M. J., Jenkins, A. L., Wong, G. S., Josse, R. G., and Csima, A., 1984, The relationship between glycemic response, digestibility, and factors influencing the dietary habits of diabetics, *Am. J. Clin. Nutr.* **40**:1175–1191.

Jenkins, D. J. A., Wolever, T. M. S., Kalmusky, J., Giudici, S., Giordano, C., Wong, G. S., Bird, J. H., Patten, R., Hall, M., Buckley, G. C., and Little, J. A., 1985, Low glycemic index foods in the management of hyperlipidemia, *Am. J. Clin. Nutr.* **42**:604–617.

Jenkins, D. J. A., Jenkins, M. A., Wolever, T. M. S., Taylor, R. H., and Ghafari, H., 1986, Slow release carbohydrate: Mechanism of action of viscous fibers, *J. Clin. Nutr. Gastroenterol.* **1**:237–241.

Jenkins, D. J. A., Cuff, D., Wolever, T. M. S., Knowland, D., Thompson, L. U., Cohen, Z., and Prokipchuk, E., 1987a, Digestibility of carbohydrate foods in an ileostomate: Relationship to dietary fiber, *in vitro* digestibility, and glycemic response, *Am. J. Gastroenterol.* **82**:709–717.

Jenkins, D. J. A., Wolever, T. M. S., Collier, G. R., Ocana, A., Rao, A. V., Buckley, G., Lam, K. Y., Meyer, A., and Thompson, L. U., 1987b, The metabolic effects of a low glycemic index diet, *Am. J. Clin. Nutr.* **46**:968–975.

Jenkins, D. J. A., Wolever, T. M. S., Kalmusky, J., Guidici, S., Giordano, C., Patten, R., Wong, G. S., Bird, J. N., Hall, M., Buckley, G., Csima, A., and Little, J. A., 1987c, Low-glycemic index diet in hyperlipidemia: Use of traditional starchy foods, *Am. J. Clin. Nutr.* **46**:66–71.

Jenkins, D. J. A., Wolever, T. M. S., Buckley, G., Lam, K. Y., Giudici, S., Kalmusky, J., Jenkins, A. L., Patten, R., Bird, J., Wong, G. S., and Josse, R. G., 1988a, Low glycemic index starchy foods in the diabetic diet, *Am. J. Clin. Nutr.* **48**:248–254.

Jenkins, D. J. A., Wolever, T. M. S., and Jenkins, A. L., 1988b, Starchy foods and the glycemic index, *Diabetes Care* **11**:149–159.

Johansen, K., 1981, Decreased urinary glucose excretion and plasma cholesterol level in non-insulin-dependent diabetic patients with guar, *Diabetes Metab.* **7**:87–90.

Karlstrom, B., Vessby, B., Asp, N.-G., Boberg, M., Gustafsson, I.-B., Lithell, H., and Werner, I., 1984, Effects of an increased content of cereal fibre in the diet of type II (non-insulin-dependent) diabetics, *Diabetologia* **26**:272–277.

Karlstrom, B., Vessby, B., and Asp, N.-G., 1987, Effects of leguminous seeds in a mixed diet in non-insulin-dependent diabetic patients, *Diabetes Res.* **5**:199–205.

Kiehm, T. G., Anderson, J. W., and Ward, K., 1976, Beneficial effects of a high carbohydrate high fiber diet in hyperglycemic men, *Am. J. Clin. Nutr.* **29**:895–899.

Kinmonth, A.-L., Angus, R. M., Jenkins, P. A., Smith, M. A., and Baum, D., 1982, Whole foods and increased dietary fibre improve blood glucose control in diabetic children, *Arch. Dis. Child.* **57**:187–194.

Kuhl, C., Molsted-Pedersen, L., and Hornnes, P. J., 1983, Guar gum and glycemic control of pregnant insulin-dependent diabetic patients, *Diabetes Care* **6**:152–154.

Laine, D. C., Thomas, W., Levitt, M. D., and Bantle, J. P., 1987, Comparison of predictive capabilities of diabetic exchange lists and glycemic index of foods, *Diabetes Care* **10**:387–394.

Lindsay, A. N., Hardy, S., Jarrett, L., and Rallinson, M. L., 1984, High-carbohydrate, high-fiber diet in children with type I diabetes mellitus, *Diabetes Care* **7**:63–67.

Liu, G. C., Coulston, A. M., and Reaven, G. M., 1983, Effect of high carbohydrate–low fat diets on plasma glucose, insulin and lipid responses in hypertriglyceridemic humans, *Metabolism* **32**:750–753.

Mahalko, J. R., Sandstead, H. H., Johnson, L. K., Inman, L. F., Milne, D. B., Warner, R. C., and Haunz, E. A., 1984, Effect of consuming fiber from corn bran, soy hulls, or apple powder on glucose tolerance and plasma lipids in type II diabetes, *Am. J. Clin. Nutr.* **39**:25–34.

Mayne, P. D., McGill, A. R., Gormley, T. R., Tomplin, G. H., Julian, T. R., and O'Moore, R. R., 1982, The effect of apple fibre on diabetic control and plasma lipids, *Ir. J. Med. Sci.* **151**:36–42.

Miranda, P. M., and Horwitz, D. L., 1978, HIgh fiber diets in the treatment of diabetes mellitus, *Ann. Intern. Med.* **88**:482–486.

Monnier, L. H., Blotman, M. J., Colette, C., Monnier, M. P., and Mirouze, J., 1981, Effects of dietary fibre supplementation in stable and labile insulin-independent diabetics, *Diabetologia* **20**:12–17.

Najemnik, C., Kritz, H., Irsigler, K., Laube, H., Knick, B., Klimm, H. D., Wahl, P., Vollmar, J., and Brauning, C., 1984, Guar and its effects on metabolic control in type II diabetic subjects, *Diabetes Care* **7**:215–220.

National Institutes of Health, 1986, *Diet and Exercise in Non-Insulin-Dependent Diabetes Mellitus*, National Institutes of Health Consensus Development Conference Statement, Vol. 6, No. 8, Dec 10, 1986, National Institutes of Health, Bethesda, MD.

Nuttal, F. Q., Mooradian, A. D., DeMarais, R., and Parker, S., 1983, The glycemic effect of different meals approximately isocaloric and similar in protein, carbohydrate, and fat content as calculated using the ADA exchange lists, *Diabetes Care* **6**:432–435.

Ocana, A. M., Jenkins, D. J. A., Wolever, T. M. S., Cunnane, S., Singer, W., and Bloom, S. R., 1988, "Nibbling versus gorging": Metabolic effects of prolonging the rate of glucose absorption, *Diabetes* **37**(Suppl. 1):109A.

O'Dea, K., Snow, P., and Nestel, P., 1981, Rate of starch hydrolysis *in vitro* as a predictor of metabolic responses to complex carbohydrate *in vivo*, *Am. J. Clin. Nutr.* **34**:1991–1993.

Osilesi, O., Trout, D. L., Glover, E. E., Harper, S. M., Koh, E. T., Behall, K. M., O'Dorisio, T. M., and Tartt, J., 1985, Use of xanthan gum in dietary management of diabetes mellitus, *Am. J. Clin. Nutr.* **42**:597–603.

Paganus, A., Maenpaa, J., Akerblom, H. K., Stenman, U.-H., Knip, M., and Simell, O., 1987, Beneficial effects of palatable guar and guar plus fructose diets in diabetic children, *Acta Paediatr. Scand.* **76**:76–81.

Parillo, M., Giacco, R., Riccardi, G., Pacioni, D., and Rivellese, A., 1985, Different glycaemic responses to pasta, bread, and potatoes in diabetic patients, *Diabetic Med.* **2**:374–377.

Perrotti, N., Santoro, D., Genovese, S., Giacco, A., Rivellese, A., and Riccardi, G., 1984, Effect of digestible carbohydrates on glucose control in insulin-dependent diabetic patients, *Diabetes Care* **7**: 354–359.

Ray, T. K., Mansell, K. M., Knight, L. C., Malmud, L. S., Owen, O. E., and Boden, G., 1983, Long-term effects of dietary fiber on glucose tolerance and gastric emptying in noninsulin-dependent diabetic patients, *Am. J. Clin. Nutr.* **37**:376–381.

Reaven, G. M., 1979, Effect of variations in carbohydrate intake on plasma glucose, insulin, and triglyceride responses in normal subjects and patients with chemical diabetes, in: *Treatment of Early Diabetes* (R. A. Camerini-Davalos and B. A. Hanover, eds.), Plenum Press, New York, pp. 253–262.

Reaven, G. M., 1980, How much carbohydrate? *Diabetologia* **19**:409–413.

Rivellese, A., Riccardi, G., Giacco, A., Pancioni, D., Genovese, S., Mattioli, P. L., and Mancini, M., 1980, Effect of dietary fibre on glucose control and serum lipoproteins in diabetic patients, *Lancet* **2**:447–450.

Simpson, H. C. R., Simpson, R. W., Lousley, S., Carter, R. D., Geekie, M., Hockaday, T. D. R., and Mann, J. I., 1981, A high carbohydrate leguminous fibre diet improves all aspects of diabetic control, *Lancet* **1**:1–5.

Simpson, H. C. R., Carter, R. D., Lousley, S., and Mann, J. I., 1982, Digestible carbohydrate—an independent effect on diabetic control in type II (non-insulin-dependent) diabetic patients? *Diabetologia* **23**:235–239.

Simpson, R. W., Mann, J. I., Eaton, J., Carter, R. D., and Hockaday, T. D. R., 1979a, High carbohydrate diets in insulin-dependent diabetes, *Br. Med. J.* **2**:523–525.

Simpson, R. W., Mann, J. I., Eaton, J., Moore, R. A., Carter, R., and Hockaday, T. D. R., 1979b, Improved glucose control in maturity onset diabetes treated with high carbohydrate-modified fat diet, *Br. Med. J.* **1**:1752–1756.

Smith, C. J., Roseman, M. S., Levitt, N. S., and Jackson, W. P. U., 1982, Guar biscuits in the diabetic diet, *S. Afr. Med. J.* **61**:196–198.

Smith, U., and Holm, G., 1982, Effect of a modified guar gum preparation on glucose and lipid levels in diabetics and healthy volunteers, *Atherosclerosis* **45**:1–10.

Stone, D. B., and Connor, W. E., 1965, Prolonged effects of a low cholesterol, high carbohydrate diet upon the serum lipids in diabetic patients, *Diabetes* **12**:127–132.

Tagliaferro, V., Cassader, M., Bozzo, C., Pisu, E., Bruno, A., Marena, S., Cavallo-Perin, P., Cravero, L., and Pagano, G., 1985, Moderate guar-gum addition to usual diet improves peripheral sensitivity to insulin and lipaemic profile in NIDDM, *Diabetes Metab.* **11**:380–385.

Thorburn, A. W., Brand, J. C., and Truswell, A. S., 1987, Slowly digested and absorbed carbohydrate in traditional bushfoods: A protective factor against diabetes? *Am. J. Clin. Nutr.* **45**:98–106.

Trowell, H. C., 1973, Dietary fibre, ischaemic heart disease and diabetes mellitus, *Proc. Nutr. Soc.* **32:** 151–157.

Vaaler, S., Hanssen, K. F., Dahl-Jorgensen, L., Frolich, W., Aaseth, J., Odengaard, B., and Aagenaes, O., 1983, Improvement in diabetes control after bread-enrichment with wheat bran and guar gum, in: *European Association for Study of Diabetes International Symposium on Diabetes and Nutrition: Crete.*

Wolever, T. M. S., and Collier, G. R., 1987, Dietary fiber and noninsulin-dependent diabetes mellitus, *Am. J. Clin. Nutr.* **46:**866–867.

Wolever, T. M. S., and Jenkins, D. J. A., 1985, Application of the glycemic index to mixed meals, *Lancet* **2:**944.

Wolever, T. M. S., and Jenkins, D. J. A., 1986, The use of the glycemic index in predicting the blood glucose response to mixed meals, *Am. J. Clin. Nutr.* **43:**167–174.

Wolever, T. M. S., Nuttall, F. Q., Lee, R., Wong, G. S., Josse, R. G., Csima, A., and Jenkins, D. J. A., 1985, Prediction of the relative blood glucose response of mixed meals using the white bread glycemic index, *Diabetes Care* **8:**418–428.

Wolever, T. M. S., Cohen, Z., Thompson, L. U., Thorne, M. J., Jenkins, M. J. A., Prokipchuk, E. J., and Jenkins, D. J. A., 1986, Ileal loss of available carbohydrate in man: Comparison of a breath hydrogen method with direct measurement using a human ileostomy model, *Am. J. Gastroenterol.* **81:**115–122.

Wolever, T. M. S., Jenkins, D. J. A., Josse, R. G., Wong, G. S., and Lee, R., 1987, The glycemic index: Similarity of values derived in insulin-dependent and non-insulin-dependent diabetic patients, *J. Am. Coll. Nutr.* **6:**296–305.

Wolever, T. M. S., Jenkins, D. J. A., Collier, G. R., Lee, R., Wong, G. S., and Josse, R. G., 1988a, Metabolic responses to test meals containing different carbohydrate foods: 1. Relationship between rate of digestion and plasma insulin response, *Nutr. Res.* **8:**573–581.

Wolever, T. M. S., Jenkins, D. J. A., Ocana, A. M., Rao, A. V., and Collier, A. V., 1988b, Second meal effect: Low glycemic index foods eaten at dinner improve subsequent breakfast glycemic response, *Am. J. Clin. Nutr.* **48:**1041–1047.

23

Production and Absorption of Short-Chain Fatty Acids

S. E. FLEMING and SUNGSHIN YEO

1. INTRODUCTION

Short-chain fatty acids (SCFAs) are produced when undigested dietary constituents are fermented by anaerobic bacteria in the intestine (Bancroft *et al.*, 1944). Dietary fiber has been suggested to be the main substrate for the production of SCFAs. The SCFAs are readily absorbed and play significant physiological roles in the host. For example, butyrate is proposed to be a preferable fuel to glucose, glutamate, or ketone bodies for colonocytes (Roediger, 1982), whereas acetate and propionate are mainly converted to fatty acids and glucose, respectively, in the liver (Demigne *et al.*, 1986). In addition, SCFAs are involved in the maintenance of secretory and absorptive functions in the large intestine by affecting movement of water and electrolytes (Argenzio *et al.*, 1975). Therefore, investigation of the production of SCFAs through the fermentation of dietary fiber and evaluation of the mechanisms involved in the absorption are of interest. In this chapter we focus on absorption of SCFAs with a brief overview of production.

2. DIETARY FIBER AND PRODUCTION OF SHORT-CHAIN FATTY ACIDS

2.1. Production of Short-Chain Fatty Acids

The SCFAs, mainly acetate, propionate, and butyrate, are produced in the cecum and the proximal colon of nonruminant animals including humans and are major end products of the fermentation process. There are no studies that provide information on the production of SCFAs *in vivo* using humans. Since the proximal part of the human

S. E. FLEMING and SUNGSHIN YEO • Department of Nutritional Sciences, University of California, Berkeley, California 94720.

large intestine is not readily accessible, most human studies have been conducted using the distal part of the colon or indirect techniques. However, concentrations of SCFAs in the portal blood, feces, or distal part of the colon such as the rectum have little significance in determining the extent to which SCFAs are produced in the lumen of the gut because of significant uptake by the colonic epithelium. Cummings et al. (1987) measured concentrations of SCFAs in the portal blood and intestinal contents of the terminal ileum and the large intestine of sudden death victims. In their studies, total SCFA concentration (mmole/kg) was low in the terminal ileum at 13 ± 6 but high in all regions of the colon ranging from 131 ± 9 in the cecum to 80 ± 11 in the descending colon. Total SCFA concentration (μmole/liter) was also measured in the portal blood and was 375 ± 70. These data indicate that there are concentration differences among the various parts of the large intestine, and the concentration of SCFAs in the portal blood was extremely low compared to the concentration of the intestinal contents. In another study, the total concentration of SCFAs in the feces was reported to be about 75 mmole/kg feces, making the SCFAs the main anions in the feces of healthy human subjects (Hoverstad et al., 1984). Acetate is the principal SCFA and represents 60% of the SCFAs in the feces (Hoverstad et al., 1984; Cummings et al., 1987). Acetate comprised 70% and 90% of total SCFAs in the portal plasma and peripheral plasma, respectively (Cummings et al., 1987).

These levels of SCFAs in the human are very similar to those found in animals, suggesting that a fermentation pattern in the human colon exists and that it is similar to the fermentation pattern in the rumen and the colon of herbivorous species (Cummings, 1984).

2.2. Effects of Dietary Fiber Consumption on SCFA Production

One of the mechanisms by which dietary fiber exerts its physiological roles may be the production of SCFAs in the intestine by the anaerobic process of fermentation. The major substrate of the fermentation process is dietary fiber, mainly plant cell wall polysaccharides. Since the cecum and the proximal colon of the human are not readily accessible for studies, most in vivo studies on production of SCFAs with dietary fiber have been conducted by measuring changes in the concentration of SCFAs after feeding dietary fiber to nonruminant animals including rats, rabbits, and pigs as well as ruminant animals. From these studies it appears that SCFA concentration is increased with consumption of pectin (Illman et al., 1982), oat bran (Storer et al., 1983), and some wheat fibers (Ehle et al., 1982; Sambrook, 1979) as compared to fiber-free meals.

It has also been suggested that the chemical and physical structure of dietary fiber also can significantly influence the profile of SCFAs produced in the large intestine. Thomsen et al. (1984) observed that adding pectin to a fiber-free experimental diet resulted in an increased concentration of acetate in the contents of the intestine, whereas the propionate and n-butyrate concentrations were not altered. Stanogias and Pearce (1985) also reported that the level as well as the source of dietary fiber significantly influenced the molar proportion of SCFAs.

In a previous series of experiments using miniature pigs as animal models, we have evaluated the effects of different types of dietary fiber on cecal concentrations of SCFAs

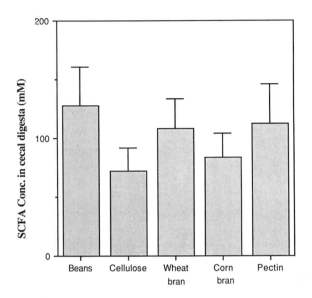

FIGURE 1. Concentrations of total SCFAs in the cecal digesta of pigs fed red-kidney-bean-, cellulose-, wheat-bran-, corn-bran-, or pectin-containing diets. Each data point is the mean of values for specimens collected at 2-hr intervals for 12 hr post-prandial in days 4, 7, and 11 of each metabolic period. Results are expressed as the mean ± S.D. of data.

and the pH of cecal digesta (Fleming *et al.*, 1988). Animals were fed diets containing red kidney beans, cellulose, wheat bran, corn bran, or pectin to provide 7% total dietary fiber. Specimens were collected through a surgically implanted cecal cannula to determine the concentration of SCFAs and pH of luminal contents. The mean daily total SCFA concentrations were higher for the bean diet than for the corn bran or cellulose diet (Fig. 1). The proportion of each individual SCFA was also different among diet groups; the bean-containing diet resulted in higher concentrations of acetate and lower concentrations of butyrate than did the bran-containing diet. An inverse relationship between the concentration of SCFAs and pH of luminal contents was observed (Fig. 2) when pigs were fed diets containing the five types of dietary fibers. Cummings *et al.* (1987) also reported that pH changed with region of the gut from 5.6 ± 0.2 in the cecum to 6.6 ± 0.1 in the descending colon, and these pH values were inversely related to total SCFA concentration in each segment of the colon.

Some have suggested that SCFAs stimulate cell proliferation (Sakata, 1987; Sakata and von Engelhardt, 1983) in the large intestine, whereas others have suggested that butyrate alone reduces cell proliferation (Kim *et al.*, 1982; Kruh, 1982) and stimulates differentiation of cells including colon carcinoma cells (Herz and Halwer, 1982; Whitehead *et al.*, 1986). Therefore, a better understanding of both production of total SCFAs and the concentration of these SCFAs that are most closely associated with having an influence on colonic cell biology would be beneficial in reducing the risk of colon cancer by dietary management.

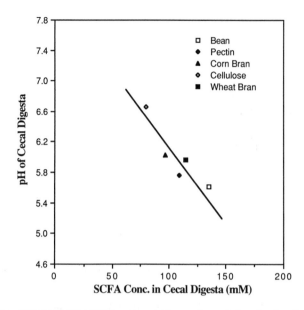

FIGURE 2. The pH versus total SCFA concentrations of cecal digesta of animals fed the red kidney bean, pectin, corn bran, cellulose, and wheat bran diets. Each data point is the mean of values for specimens collected at 8, 10, and 12 hr post-prandial on days 4, 7, and 11 of each metabolic period.

3. ABSORPTION OF SHORT-CHAIN FATTY ACIDS

3.1. Kinetic Characteristics of Absorption

The absorption of SCFAs has been subject to considerable study because SCFAs are the major anions in the large intestine and in feces of adult humans (Rubinstein *et al.*, 1969). Both passive and active transport systems have been suggested for the transport of SCFAs across the luminal membrane.

Evidence for absorption of SCFAs by passive diffusion includes the effects of lipid solubility of SCFAs and a positive relationship between the luminal concentration and absorption. Dawson *et al.* (1964) perfused the proximal jejunum of seven subjects with a solution containing a mixture of acetate, propionate, and butyrate. They found a very low absorption rate that was proportional to chain length. Since this probably reflected greater solubility in lipids, they suggested that the absorption of SCFAs takes place by simple passive diffusion. In several human studies, McNeil *et al.* (1978) have examined colonic absorption of SCFAs using a dialysis bag filled with test solutions containing a mixture of acetate, propionate, and butyrate in physiological concentration and inserted into the rectum. They found that the absorption was linearly concentration dependent and that the absorption of SCFA was accompanied by the absorption of Na, Cl, and water with accumulation of bicarbonate accompanied by increased luminal pH.

There has been rather strong evidence for absorption by both facilitated and active transport systems in order to explain the anomalously rapid absorption of SCFAs (McNeil *et al.*, 1978). For instance, Smyth and Taylor (1957) conducted an *in vitro* experiment to test the ability of the distal small intestine to transfer SCFAs against a concentration gradient, which is evidence of the presence of an active or facilitated transport system. In their experiment, glucose and SCFAs were present initially at equal concentrations in the mucosal and serosal fluids. The sacs of everted small intestine were shaken under aerobic conditions for 90 min. Considerable movement of fatty acids from the mucosal fluid to the serosal fluid was observed. This resulted in the serosal fluid acquiring a higher concentration of fatty acid than the mucosal fluid, and, hence, the transfer of fatty acid took place against the concentration gradient. The authors also demonstrated significant inhibiting effects: in anaerobic conditions there was almost no transfer of SCFAs; the absence of glucose from the luminal and serosal fluids caused a significant reduction in transfer of fatty acids; and metabolic inhibitors such as phlorizin and 2,4-dinitrophenol resulted in a considerable reduction in the transport of both water and SCFAs.

In addition to these *in vitro* studies, there have been some *in vivo* human studies suggesting that SCFAs are transported by facilitated or active transport systems. Schmitt *et al.* (1976) measured jejunal absorption of acetate, propionate, and butyrate using a perfusion technique in which a triple-lumen tube was inserted into the jejunum and test solutions were delivered by a perfusion pump. They observed that SCFAs were absorbed rapidly, and the rates rose in linear fashion up to 20 mM but reached saturation at higher concentrations (20–40 mM). They also observed saturation kinetics when they investigated the absorption of SCFAs from the human ileum by the same perfusion technique (Schmitt *et al.*, 1977). These workers concluded that an active transport system contributed significantly to the absorption of SCFAs from the intestine.

Conflicts among the conclusions that have been made regarding the kinetics of SCFA absorption may be caused by differences in experimental methodologies: for instance, most of the studies demonstrating that SCFA absorption occurs by an active transport system were conducted using *in vitro* tissue preparation (Smyth and Taylor, 1957) or *in vivo* perfusion of the small intestine (Schmitt *et al.*, 1976, 1977). The mechanisms involved in the movement of weak electrolytes in the small intestine may be different from those in the colon; thus, it may not be appropriate to speculate on transport of SCFAs in the colon from those studies conducted using the small intestine.

In our laboratory, we have conducted an *in vivo* experiment consisting of two studies by infusing SCFA-containing solutions into rat cecal preparations in an attempt to explain the inconsistencies among previously reported studies. The cecum was chosen because it is a major site for the production of SCFAs in the rat and is a discrete intestinal segment.

In the first part of these studies, test solutions containing 45, 90, and 120 mM acetate, propionate, or butyrate were infused into the surgically prepared rat cecum for 60 min in order to determine the presence of saturation kinetics with respect to SCFA absorption. The rate of absorption of SCFAs from the lumen was quantified by determining the rate of disappearance of the fatty acid from the luminal infusate. As shown in Fig. 3, the concentration of SCFAs in the luminal infusate decreased in a linear fashion with respect to time for each of the three SCFAs evaluated. The rate of absorption of acetate, propionate, and butyrate significantly increased with increases in the concentra-

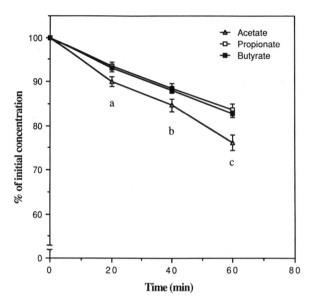

FIGURE 3. Changes in the concentrations of acetate, propionate, and butyrate in infusates as percentages of initial concentrations during the 60 min of infusion in *in vivo* rat cecal preparations. Each data point represents the mean ± S.E.M. (n = 51). Data sets for each of the three SCFAs were statistically analyzed individually. Superscripts refer to each of the three SCFAs. Values with different superscripts are significantly different at $P < 0.05$. (From Yeo, 1988.)

tion of SCFAs in the infusate (Fig. 4). This indicates that absorption of SCFAs takes place predominantly by simple passive diffusion. In addition, when absorption of acetate at low concentrations (1–45 mM) was evaluated, the relationship between the transport of acetate and concentration remained linear (Fig. 5). This is consistent with our observations at high concentrations (45–120 mM) of SCFAs and suggests that high concentrations of SCFAs did not mask the effects of an active transport system that would have been observed only at low concentrations. Therefore, we did not find saturation kinetics suggestive of a carrier-mediated transport system.

As mentioned earlier, saturation of SCFA transport has been observed by Schmitt *et al.* in the human jejunum (1976) and in the human ileum (1977). Rechkemmer and von Engelhardt (1982) suggested that in the guinea pig proximal colon SCFA anions may permeate by the paracellular pathway. In the cecum, dissociated forms of SCFAs also could be transported paracellularly, and this mechanism may play a significant role in transport of SCFAs. If so, it is likely to play a more important role in the large intestine than in the small intestine and may explain some inconsistencies between data derived from these sections of the gut. It was also suggested that some workers failed to note saturation because of the use of concentrations well below the reported K_m value of approximately 25 mM (Lucas, 1984). However, we did not observe saturation of SCFA transport at concentrations above the K_m value. Therefore, the inconsistencies among

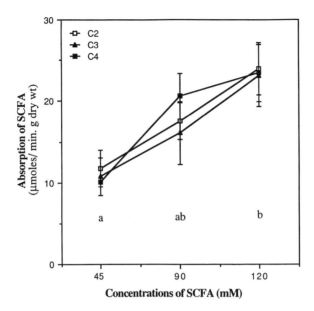

FIGURE 4. Absorption of acetate (C_2), propionate (C_3), and butyrate (C_4) from the luminal infusates with respect to concentration (45, 90, and 120 mM). Incubation was performed for 60 min at 37°C. Each data point represents the mean ± S.E.M. (n = 10, 7, and 8 for acetate, propionate, and butyrate, respectively). Data sets for each of the three SCFAs were individually statistically analyzed. Superscripts refer to each of the three SCFAs. Values with different superscripts are significantly different at $P < 0.05$. (From Yeo, 1988.)

studies could result from differences in the segments of the intestine (i.e., small intestine versus colon).

In the second part of the study, influence of chain length of SCFAs on absorption was evaluated by comparing the SCFA disappearance rate from test solutions containing 90 mM acetate, propionate, or butyrate. As seen in Fig. 6, the absorption of SCFAs was not significantly different among these three SCFA-containing solutions. Thus, absorption was not affected by chain length. Using long-chain fatty acids, Sallee and Dietschy (1973) reported that absorption rate increases with chain length. These workers calculated passive permeability coefficients (*P) to the lipid membrane by using the measured net movement and the initial concentration of a molecule in the *in vitro* preparation of the small intestine. Permeability to the lipid membrane is usually considered to be positively related to lipid solubility. Since Dawson *et al.* (1964) reported that lipid solubility of SCFAs increased with their chain length, we speculated that absorption as well as the calculated *P values would increase with chain length of SCFAs. However, we found that the *P values were not related to chain length, which is consistent with our *in vivo* data. Sallee and Dietschy (1973) also reported that *P values are not different for SCFAs with two to six carbon atoms but increase along with the chain length for long-chain fatty acids. This indicates that *P values of SCFAs are not associated with

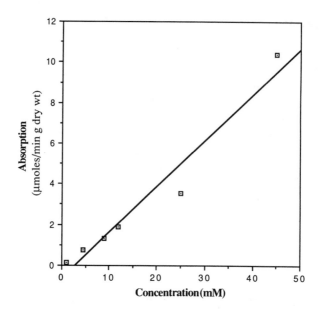

FIGURE 5. Absorption of acetate at low concentrations (1–45 mM). Incubation was performed for 60 min at 37°C. The plot of absorption versus concentration of data sets fits best to a linear relationship. (From Yeo, 1988.)

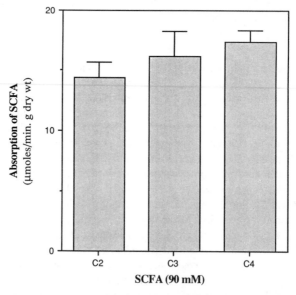

FIGURE 6. Effect of chain length on the absorption of SCFAs. Test solutions containing 90 mM acetate, propionate, or butyrate were infused into the rat cecum for 60 min. Each data point represents the mean ± S.E.M. (n = 8, 7, and 6 for acetate, propionate, and butyrate, respectively). There were no significant differences among the values at $P < 0.05$. (From Yeo, 1988.)

their lipid solubility. Therefore, SCFAs appear to be transported across the luminal membrane of the rat cecum by passive diffusion, and the relative rates of transport of SCFAs do not appear to be influenced by differences in their lipid solubility.

3.2. Effects of pH on SCFA Absorption, Total CO_2, and HCO_3^-

One of the physiological roles of SCFAs is their importance to the normal secretory and absorptive functions of the large intestine, since absorption of SCFAs is closely linked to the movement of water and electrolytes from the lumen (Argenzio et al., 1975). For example, in a number of studies both in vivo and in vitro using the intestine of horses (Argenzio et al., 1977), pigs (Argenzio and Southworth, 1975), goats (Argenzio et al., 1975), and humans (Ruppin et al., 1980), accumulation of HCO_3^-, luminal alkalinization, and a marked decrease in intraluminal Pco_2 was associated with absorption of SCFAs. The accumulation of HCO_3^- was independent of the chloride–bicarbonate exchange system, since it occurred in the absence of luminal chloride (McNeil et al., 1979) and was independent of chloride absorption (Argenzio and Whipp, 1979). The mechanism by which HCO_3^- enters or leaves the lumen was considered by an examination of changes in the luminal Pco_2 and HCO_3^- concentrations. According to the equilibrium $CO_2 + H_2O \leftrightarrow H_2CO_3 \leftrightarrow H^+ + HCO_3^-$, the net removal of H^+ or HCO_3^- ions should shift the reaction to the right and result in a decrease in the luminal Pco_2 (Hubel, 1974). In the presence of SCFAs, hydrogen ions could be used to protonate SCFA anions to facilitate their transport as undissociated acids. As a result, when SCFAs are absorbed, bicarbonate would accumulate in the lumen, pH would increase, and Pco_2 would decrease.

Since most of these experiments had been conducted in the distal colon or in in vitro preparations, we conducted an experiment using an in vivo rat cecal preparation to model the proximal colon. Interrelationships of luminal pH, total CO_2, HCO_3^- concentrations, and absorption of SCFAs were investigated. Test solutions containing 90 mM acetate, propionate, or butyrate of pH 5.4, 6.4, or 7.4 were infused to investigate the relationship between pH of the luminal contents and absorption of SCFAs. Differences in the rates at which SCFAs were absorbed were not observed when solutions were infused at low (5.4) and high (7.4) pH except that for butyrate significantly greater absorption occurred with the pH 5.4 solution than with the pH 6.4 or 7.4 solutions (Fig. 7). The reason that we did not observe the differences in absorption rates of SCFAs when pH of solutions were altered (5.4–7.4) may be that the pH of test solutions rose rapidly to 7.4 so that the ionization of SCFAs would become quickly equivalent for the three test solutions. For example, based on the pH measurements in solutions taken after 20 min of infusion and the pK_a values of SCFAs, ionization of SCFAs was greater than 90% for all three solutions (i.e., initial pH of 5.4, 6.4, and 7.4). In an in vivo experiment using the human rectum, McNeil et al. (1978) also observed no significant differences in absorption rates for SCFAs when solutions of pH 5.5 and 7.2 were compared.

Interrelationships among the luminal pH, total CO_2, and HCO_3^- were evaluated using our in vivo rat cecum model. Figure 8 presents changes in luminal pH, total CO_2, HCO_3^-, and concentrations of butyrate with respect to time during the infusion of a butyrate-containing solution (pH 6.8) and a saline solution (pH 6.8). The pH and concentrations of total CO_2 and HCO_3^- of luminal contents increased progressively

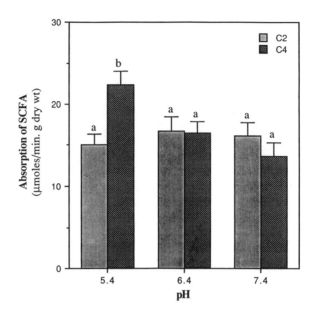

FIGURE 7. Absorption of acetate (C_2) and butyrate (C_4) at pH 5.4, 6.4, and 7.4. Test solutions containing 90 mM acetate or butyrate adjusted to pH 5.4, 6.4, or 7.4 were infused into the rat cecum for 60 min. Each data point represents the mean ± S.E.M. (n = 12, 10 for acetate and butyrate, respectively). Values with different superscripts are significantly different at $P < 0.05$. (From Yeo, 1988.)

during the 60 min following SCFA infusion, and the changes were significantly greater with infusion of butyrate than with infusion of saline. The quantity of total CO_2 in the lumen was closely associated with the concentration of HCO_3^- since, at the range of pH 6.8 to 7.6, the proportion of total CO_2 present in the form of HCO_3^- is 83% to 97%. Thus, within this range of changes in pH the HCO_3^- would constitute the major form of total CO_2. Our total CO_2 measurements using the rat cecum agreed with results of a previous *in vivo* human study using the rectum, which showed that SCFA transport was accompanied by a significant increase in total CO_2 and a decrease in hydrogen ion concentration in the perfusate (Ruppin *et al.*, 1980). Thus, the mechanisms by which SCFA absorption occur in the proximal colon appear to be similar to those of the distal colon. In addition, our experiments demonstrated that changes in the luminal pH and accumulation of HCO_3^- are closely related to absorption of SCFAs.

3.3. Proposed Mechanisms of SCFA Absorption

A number of investigators have proposed mechanisms to explain the transport of SCFAs across the luminal membrane of mucosal cells, and there have been two alternative mechanisms (Argenzio *et al.*, 1977) suggested to explain the luminal accumulation of HCO_3^- and luminal alkalinization observed during absorption of SCFAs.

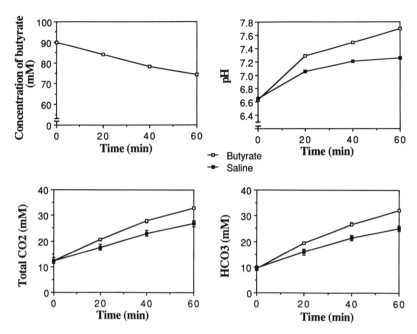

FIGURE 8. Changes in concentrations of butyrate, total CO_2, HCO_3^-, and pH of infusates as a function of time after infusing either saline or a butyrate-containing (90 mM) test solution into the rat cecal preparation. Each data point represents the mean $\pm$ S.E.M. ($n = 5$). The S.E.M. for several data sets were too small to be visualized on the figures. (From Yeo, 1988.)

One of the mechanisms is an ionic transport system and involves hydration of CO_2 inside the mucosal cell compartment and equilibration of the dissociation products of the reaction. This process provides HCO_3^-, which exchanges for SCFA anions at the luminal border of the cell to facilitate SCFA absorption. The driving force for this mechanism would be the rapid removal of SCFA anions from the cell by metabolism or by transport of the lipid-soluble protonated form of SCFAs across the serosal membrane into the blood.

The second mechanism is a nonionic transport system and includes hydration of CO_2 in the lumen, which would provide a continuous source of hydrogen ions to form the more lipid-soluble and readily diffusable undissociated fatty acids. The driving force for this reaction would be the rapid removal of hydrogen ions from the lumen, which would result in accumulation of bicarbonate ion within the lumen. The protonated SCFAs would be either metabolized in the cell or diffused into the blood. It is also possible that dissociation could take place inside the cell compartment so that ionic forms of SCFAs could cross the serosal membrane, leaving hydrogen ions inside the cell for recycling through the action of carbonic anhydrase. In the goat, pony, pig, and human large intestine, absorption of SCFAs stimulated Na absorption; therefore, it was postulated that a Na–H exchange system may be present on the surface of the luminal membrane that could generate a hydrogen ion gradient for protonating the ionic form of SCFAs in the lumen (von Engelhardt and Rechkemmer, 1983).

This second mechanism of nonionic absorption could be supported by a hypothesis that was proposed to elucidate the transport of weak electrolytes from the luminal to the serosal side of the cell; Hogben *et al.* (1959) suggested that transport of certain weak electrolytes including SCFAs across the mucosa of the small intestine could be attributed to a microlayer that is located adjacent to the brush border and maintained at a pH lower than that of the bulk lumen contents. The region of lower pH has been called the "acid microclimate" and is thought to be independent of the hydrogen ion concentration of the bulk lumen contents (Lucas *et al.*, 1980). The ionic form of SCFAs could be protonated in this region so that the nonionic form of the SCFAs could be transported more easily by passive diffusion through the luminal membrane of the cell. The hydration of CO_2 by carbonic anhydrase was proposed to generate hydrogen ions for the acidic microclimate and, thus, for the nonionic diffusion of SCFAs (Powell, 1987). In order to facilitate absorption of SCFAs predominantly by simple passive diffusion, the acid microclimate would play a significant role as a source of hydrogen ions, since more than 95% of SCFAs would be present as anions at the normal luminal pH (6.0–7.0).

Two mechanisms for absorption of SCFAs, an ionic and a nonionic transport system, have been postulated. However, whether SCFAs are transported predominantly by an ionic or a nonionic transport mechanism would not be clearly determined without further *in vitro* experiments to examine the intracellular events such as changes in intracellular pH or concentration of bicarbonate ions during SCFA absorption.

3.4. Influence of High-Fiber Diet on Absorption of SCFAs

The source as well as the amount of dietary fiber intake is closely related to the production of SCFAs in the large intestine of mammals including humans, and this appears to be one of the mechanisms by which dietary fiber plays its physiological roles. Most studies on SCFA absorption have been designed to use *in vitro* or *in vivo* models in which luminal contents were washed in order to isolate the effects of SCFA absorption from production. However, under the normal physiological conditions, absorption of SCFAs takes place as a result of interactions with other luminal factors such as viscosity, osmolality, motility, and luminal volume. These luminal factors could be altered by dietary fiber consumption. Thus, dietary fiber consumption may have significant effects on the absorption as well as the luminal concentration of SCFAs. It would be interesting to evaluate the influence of dietary fiber and changes in the associated luminal characteristics on the absorption of SCFAs.

4. CONCLUSIONS

Short-chain fatty acids are produced as an end product of anaerobic microbial breakdown of dietary fiber in the proximal large intestine of nonruminant animals. The total amount and the profile of individual SCFAs can be altered by consumption of different amounts and types of dietary fiber. In addition, the concentration of SCFAs appears to be inversely related to the pH of the luminal contents. Because there has been *in vitro* evidence that both pH and concentration of SCFAs in the luminal contents may be associated with the risk of colon cancer, further *in vivo* studies are necessary to relate the effects of dietary fiber consumption, luminal concentration of SCFAs, and luminal pH to cell proliferation and differentiation.

Based on our most recent studies and on a critical evaluation of the literature we conclude that absorption of SCFAs takes place predominantly by simple passive diffusion. We also conclude that chain length has little effect on their absorption. Luminal pH did not affect acetate absorption, but butyrate absorption was significantly higher at low pH than at neutral pH. Finally, during SCFA absorption there were increases in the concentrations of luminal bicarbonate and total CO_2 and in luminal pH. From these results, we speculate that SCFAs are transported primarily in the form of undissociated acids. The significance of this nonionic transport of SCFAs is that the apparent increase in luminal bicarbonate concentration that occurs in conjunction with SCFA absorption acts as a buffer. This suggests that luminal pH would not be dramatically reduced when diets containing large quantities of fermentable constituents are consumed unless the absorption processes are impaired. Thus, intraluminal pH would remain within the physiological range so that damage to the colonic epithelium would be minimized and net water and electrolyte secretion would be maintained. Thus, it appears that the potential positive effects of SCFAs on the host exceed the potential for negative effects. Studies that determine the effects of SCFAs on health of the intestine and energy contribution to the host should be emphasized.

ACKNOWLEDGMENTS. The authors wish to acknowledge Mark D. Fitch for technical assistance, Mark Hudes for statistical consultation, and Inese Hincenbergs for animal care. We also thank Professor T. Machen for scientific discussion. This work was supported by NIH grant R01 CA40845.

REFERENCES

Argenzio, R. A., Southworth, M., 1975, Sites of organic acid production and absorption in gastrointestinal tract of the pig, *Am. J. Physiol.* **228**:454–460.

Argenzio, R. A., and Whipp, S. C., 1979, Inter-relationship of sodium, chloride, bicarbonate and acetate transport by the colon of the pig, *J. Physiol. (Lond.)* **293**:365–382.

Argenzio, R. A., Miller, N., and von Engelhardt, W., 1975, Effect of volatile fatty acids on water and ion absorption from the goat colon, *Am. J. Physiol.* **229**:997–1002.

Argenzio, R. A., Southworth, M., Lowe, J. E., and Stevens, C. E., 1977, Interrelationship of Na, HCO_3, and volatile fatty acid transport by equine large intestine, *Am. J. Physiol.* **233**:E469–E478.

Bancroft, T., McNally, R. A., and Phillipson, A. T., 1944, Absorption of volatile fatty acids from the alimentary tract of the sheep and other animals, *J. Exp. Biol.* **20**:120–129.

Cummings, J. H., 1984, Fiber metabolism in large intestine, in: *Dietary Fiber in Health and Disease* (G. V. Vahouny and D. Kritchevsky, eds.), Plenum Press, New York, pp. 9–22.

Cummings, J. H., Pomare, E. W., Branch, W. J., Naylor, C. P. E., and MacFarlane, G. T., 1987, Short chain fatty acids in human large intestine, portal, hepatic and venous blood, *Gut* **28**:1221–1227.

Dawson, A. M., Holdsworth, C. D., and Webb, J., 1964, Absorption of short chain fatty acids in man, *Proc. Soc. Exp. Biol. Med.* **117**:97–100.

Demigne, C., Yacoub, C., and Remesy, C., 1986, Effects of absorption of large amounts of volatile fatty acids on rat liver metabolism, *J. Nutr.* **116**:77–86.

Ehle, F. R., Jeraci, J. L., Robertson, J. B., and Van Soest, P. J., 1982, The influence of dietary fiber on digestibility, rate of passage and gastrointestinal fermentation in pigs, *J. Anim. Sci.* **55**:1071–1081.

Fleming, S. E., Fitch, M. D., and Chansler, M. W., 1988, Influence of fiber consumption on pH and short-chain fatty acid concentrations in cecal contents of miniature swine, *FASEB J.* **4**:A862.

Herz, F., and Hawler, M., 1982, Synergistic induction of alkaline phosphatase on colonic carcinoma cells by sodium butyrate and hyperosmolality, *Biochim. Biophys. Acta* **718**:220–223.

Hogben, C. A. M., Tocco, D. J., Brodie, B. B., and Schankee, L. S., 1959, On the mechanism of intestinal absorption of drugs, *J. Pharmacol. Exp. Ther.* **125**:275–282.

Hoverstad, T., Fausa, O., Bjorneklett, A., and Bohmer, T., 1984, Short-chain fatty acids in the normal human feces, *Scand. J. Gastroenterol.* **19**:375–381.

Hubel, K. A., 1974, The mechanism of bicarbonate secretion in rabbit ileum exposed to choleragen, *J. Clin. Invest.* **53**:964–970.

Illman, R. J., Trimble, R. P., Snoswell, A. M., and Tapping, D. L., 1982, Daily variation in the concentrations of volatile fatty acids in the splanchnic blood vessels of rats fed diets high in pectin and bran, *Nutr. Rep. Int.* **26**:437–446.

Kim, Y. S., Tsao, D., Marita, A., and Bella, A., 1982, Effect of sodium butyrate on three human colorectal adenocarcinoma cell lines in culture, in: *Colonic Carcinogenesis* (R. A. Malt and R. C. N. Williamson, eds.), MTP Press, Lancaster, pp. 317–323.

Kruh, J., 1982, Effects of sodium butyrate, a new pharmacological agent, on cells in culture, *Mol. Cell Biochem.* **42**:65–82.

Lucas, M., 1984, Fatty acid absorption, in: *Pharmacology of Intestinal Permeation II* (T. Z. Csaky, ed.), Springer-Verlag, Berlin, pp. 145–148.

Lucas, M. L., Lei, F. H., and Blair, J. A., 1980, The influence of buffer pH, glucose and sodium ion concentration on the acid microclimate in rat proximal jejunum *in vitro, Pflugers Arch.* **385**:137–142.

McNeil, N. I., Cummings, J. H., and James, W. P. T., 1978, Short chain fatty acid absorption by the human large intestine, *Gut* **18**:819–822.

McNeil, N. I., Cummings, J. H., and James, W. P. T., 1979, Rectal absorption of short chain fatty acids in the absence of chloride, *Gut* **20**:400–403.

Powell, D., 1987, Intestinal water and electrolyte transport, in: *Physiology of the Gastrointestinal Tract,* Vol. 2 (L. R. Johnson, J. Christensen, M. J. Jackson, E. D. Jacobson, and J. H. Walsh, eds.), Raven Press, New York, pp. 1286–1287.

Rechkemmer, G., and von Engelhardt, W., 1982, Absorptive processes in different colonic segments of the guinea-pig and the effects of short-chain fatty acid, in: *Falk Symposium 32, Colon and Nutrition* (H. Kasper and H. Goebell, eds.), MTP Press, Lancaster, p. 61.

Roediger, W. E. W., 1982, Utilization of nutrients by isolated epithelial cells of the rat colon, *Gastroenterology* **83**:424–429.

Rubinstein, R., Howard, A. V., and Wrong, O. M., 1969, *In vivo* dialysis of faeces as a method of stool analysis. IV. The organic anion component, *Clin. Sci.* **37**:549–564.

Ruppin, H., Bar-Meir, S., Soergel, H., Wood, C. M., and Schmitt, M. G., 1980, Absorption of short chain fatty acids by the colon, *Gastroenterology* **78**:1500–1507.

Sakata, T., 1987, Stimulatory effect of short-chain fatty acids on epithelial cell proliferation in the rat intestine: A possible explanation for trophic effects of fermentable fibre, gut microbes and luminal trophic factors, *Br. J. Nutr.* **58**:95–103.

Sakata, T., and von Engelhardt, W., 1983, Stimulatory effect of short chain fatty acids on the epithelial cell proliferation in rat large intestine, *Comp. Biochem. Physiol.* **74A**:459–462.

Sallee, V. L., and Dietschy, J. M., 1973, Determinations of intestinal mucosal uptake of short- and medium-chain fatty acids and alcohols, *J. Lipid Res.* **14**:475–484.

Sambrook, I. E., 1979, Studies on digestion and absorption in the intestines of growing pigs. 8. Measurements on the flow of total lipid, acid-detergent fiber and volatile fatty acids, *Br. J. Nutr.* **42**:279–287.

Schmitt, M. G., Soergel, K. H., and Wood, C. M., 1976, Absorption of short chain fatty acids from the human jejunum, *Gastroenterology* **70:**211–215.

Schmitt, M. G., Soergel, K. H., and Wood, C. M., 1977, Absorption of short chain fatty acids from the human ileum, *Dig. Dis.* **22:**340–347.

Smyth, D. H., and Taylor, C. B., 1957, Intestinal transfer of short chain fatty acids *in vitro, J. Physiol. (Lond.)* **141:**73–80.

Stanogias, G., and Pearce, G. R., 1985, The digestion of fibre by pigs: 2. Volatile fatty acid concentrations in large intestine digesta, *Br. J. Nutr.* **53:**531–536.

Storer, G. B., Trimble, R. P., Illman, R. J., Snoswell, A. M., and Tapping, D. L., 1983, Effects of dietary oat bran and diabetes on plasma and cecal volatile fatty acids in the rat, *Nutr. Res.* **3:**519–526.

Thomsen, L. L., Robertson, A. M., Wong, J., Lee, S. P., and Tasman-Jones, C., 1984, Intra-cecal short chain fatty acids are altered by dietary pectin in the rat, *Digestion* **29:**129–137.

von Engelhardt, W., and Rechkemmer, G., 1983, The physiological effects of short chain fatty acids in the hind gut, in: *Fibre in Human and Animal Nutrition* (G. Wallace and L. Bell, eds.), Royal Society of New Zealand Press, Palmerston North, p. 149.

Whitehead, R. H., Young, G. P., and Bhathal, P. S., 1986, Effects of short chain fatty acids on a new human colon carcinoma cell line (LIM1215), *Gut* **27:**1456–1463.

Yeo, S., 1988, Effects of luminal pH on SCFA absorption, total CO_2 and HCO_3^-, in: *Absorption of Short Chain Fatty Acids in the Rat Cecum,* Doctoral Thesis, University of California, Berkeley.

24

Short-Chain Fatty Acids

Production, Absorption, Metabolism, and Intestinal Effects

JOHN L. ROMBEAU, SCOTT A. KRIPKE, and R. GREGG SETTLE

1. INTRODUCTION

The short-chain fatty acids (SCFA), also called the volatile fatty acids (VFA), are the C1–6 organic fatty acids. These are formed in the gastrointestinal tract of mammals by microbial fermentation of carbohydrates (Wolin, 1981; Wrong, 1981; Cummings and Branch, 1986). Acetate, propionate and butyrate account for 83% of SCFA so formed (Nyman and Aso, 1982; Demigné and Remesy, 1985) and are produced in a nearly constant molar ratio 60 : 25 : 15, respectively (Cummings and Branch, 1986). Among their various properties, SCFA are readily absorbed by intestinal mucosa (Cummings *et al.*, 1987), are relatively high in caloric content (Yang *et al.*, 1970), are readily metabolized by intestinal epithelium and liver (Cummings, 1981), stimulate sodium and water absorption in the colon (Roediger and Rae, 1982), and are trophic to intestinal mucosa (Sakata, 1987; Kripke *et al.*, 1988d).

This chapter reviews the production, absorption, and metabolism of SCFA in the mammalian intestine with particular attention to the nonruminant. In addition, we examine the recent evidence concerning the stimulatory effect of SCFA on intestinal mucosal growth and explore potential applications for these fatty acids in modern nutritional support.

JOHN L. ROMBEAU, SCOTT A. KRIPKE, and R. GREGG SETTLE • Department of Surgery and Otorhinolaryngology, University of Pennsylvania School of Medicine, and Medical Research Service, Veterans Administration Medical Center, Philadelphia, Pennsylvania 19104.

2. PRODUCTION

The SCFA are produced in the mammalian intestinal tract as a byproduct of anaerobic bacterial metabolism of carbohydrate. In the ruminant, fermentation occurs primarily in the forestomach as well as in the hindgut (Czerkawski, 1986), whereas in the nonruminant the principal fermentation chambers are the cecum and colon (Wrong, 1981). The straight-chain fatty acids acetate, propionate, and butyrate are the major endproducts of bacterial carbohydrate metabolism in the colon. Hydrogen gas, carbon dioxide, methane, and water are also produced (Wolin, 1975). Miller and Wolin (1979) have developed a stoichiometric relationship for carbohydrate fermentation in the colon in which 34.4 moles of monosaccharide yields 64 moles of SCFA, 23.75 moles of methane, 34.23 moles of carbon dioxide, and 10.5 moles water. Acetate, propionate, and butyrate account for 83% of the SCFA produced, and the remaining SCFA are distributed among isovaleric, isobutyric, valeric, lactic, formic, and succinic acids (Baldwin, 1970).

Carbohydrates reach the large bowel in at least three forms: (1) nonstarch polysaccharides (dietary fiber: plant cell wall polysaccharides that are resistant to the digestive enzymes of the upper gastrointestinal tract) including cellulose, pectins, and hemicelluloses, (2) other polysaccharides that resist digestion, such as resistant starch (Englyst *et al.*, 1987), and (3) simple carbohydrates that escape absorption in the small bowel (Wolin, 1981; Bond and Levitt, 1976; Bond *et al.*, 1980; Ravich *et al.*, 1983). In addition, sloughed cells and endogenous secretions provide a small amount of fermentable substrate. Conditions that cause malabsorption of carbohydrate result in increased delivery of fermentative substrate to the cecum and colon (Weser, 1979).

The microflora of both rumen and colon are composed of hundreds of different bacterial species. Although the interactions of various populations of bacteria are complex, certain basic patterns are evident. Of importance to this review, it has been shown that 99% of the colonic microflora are nonsporulating anaerobic rod-shaped organisms (Hill and Draser, 1975). *Bacteroides* species are the predominant bacteria in the fecal flora, accounting for 32% of all organisms isolated (Macy, 1979). *Bacteroides* spp., also dominant in the rumen, are capable of metabolizing a wide range of substrates including complex carbohydrates of plant origin, food gums, mucin glycoprotein, and sloughed goblet cells (Salyers *et al.*, 1977; Keys *et al.*, 1969). The pathways by which acetate, propionate, and butyrate are formed from carbohydrate have been reviewed in detail elsewhere (Macy, 1979; Miller and Wolin, 1979; Wolin and Miller, 1983). In general, these SCFA are generated by the metabolism of pyruvate, which is produced by oxidation of glucose via the glycolytic Embden–Meyerhof pathway (Miller and Wolin, 1979; Wolin and Miller, 1983). In addition to carbohydrate, other dietary constituents can serve as substrate for the production of SCFA. Microbial proteolysis followed by deamination of amino acids results in SCFA production (El-Shazly, 1952; Elsden and Hilton, 1978), and SCFA production from branched-chain amino acids has been demonstrated (Ruchim *et al.*, 1984).

Because SCFA are produced by endogenous bacteria, intestinal SCFA concentrations are highest where the bacterial population is most abundant. Therefore, in nonruminants, including pig (Clemens *et al.*, 1975), dog (Banta *et al.*, 1979), rat (Remesy and Demigné, 1976), and man (Hoverstad, 1986), SCFA concentrations are high in the distal small bowel and large intestine, very low in the proximal small bowel, and moderate in the mouth. In fact, in the large intestine of man, SCFA concentrations are roughly equivalent to those found in the rumen (Cummings, 1981; McNeill, 1984; Warner, 1964) (approximately 75 mM acetate, 30 mM propionate, and 20 mM butyrate), reflecting the similarity in bacterial flora between these two organs. Furthermore, the administration of antimicrobial agents active against colonic anaerobic flora will profoundly reduce the concentration of SCFA in feces (Hoverstad, 1986).

As mentioned above, dietary fiber and resistant starch are the principal substrates for hindgut SCFA production in nonruminants, and therefore colonic SCFA production can be altered by changing the dietary content of these substances. The cecal content of SCFA in the rat is reduced ten- to 12-fold after removal of fiber from a high-fiber diet (Demigné and Remesy, 1985). Removal of fiber, however, does not dramatically change the luminal concentrations of either acetate or propionate but does significantly reduce the cecal concentration of butyrate (Demigné and Remesy, 1985). Additionally, the removal of the fermentative chamber itself (cecum) in the rat reduces the hindgut generation of SCFA by 60–70% (Remesy and Demigné, 1976). Not surprisingly, resection of the cecum reduces the digestibility of dietary fiber in several species (Ambuhl *et al.*, 1979; Williams and Senior, 1982).

The generation of SCFA as free fatty acids in the cecal lumen produces hydrogen ions and thus lowers luminal pH. Compared to fiber-free diets, high-fiber diets increase the rate of production of SCFA and acidify the cecal contents (Demigné and Remesy, 1985; Lupton *et al.*, 1988). The pH of rat cecal contents is about 7.0 in animals consuming a fiber-free diet and decreases to about 6.0 in animals fed a high-fiber diet (Demigné and Remesy, 1985). As described below, luminal pH may affect blood flow, oxygen uptake, and mucosal growth in the colon and modulates the absorption of water and sodium (Kvietys and Granger, 1981; Thomson, 1982; Demigné and Remesy, 1985; Lupton, 1985).

Although carbohydrate fermentation is primarily responsible for the production of SCFA in the nonruminant mammal, a small amount of butyrate is generated by hydrolysis of bovine milk fat in the upper gastrointestinal tract (Hilditch and Williams, 1964; Bugaut, 1987). Butyrate is likely completely hydrolyzed in the stomach and proximal small bowel by gastric and pancreatic lipase and absorbed in free fatty acid form (Bugaut, 1987).

3. ABSORPTION

Colonic absorption rates for the three principal short-chain fatty acids in several nonruminant species are similar: human, 6.1–12.6 μmole/cm^2 per hr; horse, 8 μmole/cm^2 per hr; pig, 8–10 μmole/cm^2 per hr as shown in Fig. 1 (Cummings, 1981). The SCFA absorption by rumen mucosa occurs at 10.5 μmole/cm^2 per hr (Stevens *et al.*, 1980). Increasing the effective absorptive surface area and increasing the duration of exposure to the absorptive surface both increase the capacity to absorb SCFA. Therefore, daily absorption of SCFA *in vivo* per kilogram body weight varies according to species based on anatomic differences in the configuration and size of the cecocolon. For

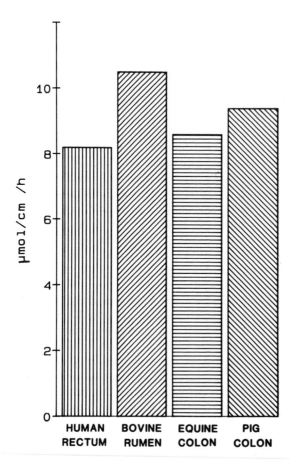

FIGURE 1. Similar colonic absorption rates for the three principal SCFA in several nonruminant species. (Adapted from Cummings, 1981.)

example, in the dog's relatively short colon, about 7.5 mmole/kg per day is absorbed, compared to 95 mmole in the pig (Stevens *et al.,* 1980). Of course, luminal SCFA production as well as absorption rates will differ according to the hindgut volume and motility pattern. Animal studies have demonstrated that prolonged retention of digesta is positively correlated with production of SCFA (Van Soest *et al.,* 1982). In the dog colon, production is low compared to the pig and rat, which possess voluminous cecums (Banta *et al.,* 1979). Clearly, decreased production leads to decreased absorption. Increased production, as occurs on feeding a very high-fiber diet, results in increased absorption as measured by increased flux across the rat cecum (Demigné and Remesy, 1985).

Small intestine as well as colon and rumen has the capacity to absorb SCFA. Although normal SCFA concentrations within the small bowel are low, small intestinal mucosa can absorb these acids as efficiently as colonic and rumen epithelium (Smyth and Taylor, 1958). Several investigators have further claimed that there is little difference between the relative rates of absorption of the three principal SCFA from the rumen (Weller *et al.,* 1967), the small intestine of human subjects (Schmitt *et al.,* 1976, 1977)

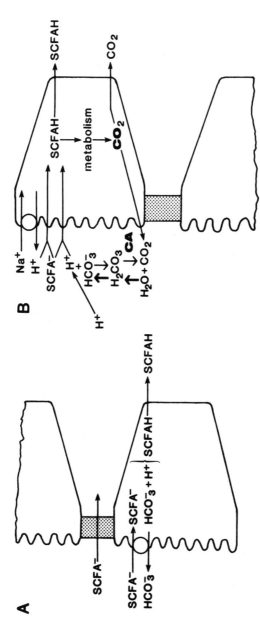

FIGURE 2. Proposed pathways for absorption of SCFA in the ionized (A) or unionized (B) form in the proximal and distal colon. (From Engelhardt and Rechkemmer, 1983a.)

or rats (Barry and Smyth, 1960), and the colon of rats or sheep (Umesaki et al., 1979; Meyers et al., 1967).

Whether in small intestine, colon, or rumen, SCFA absorption is generally considered to be passive and increases linearly with increasing concentration gradient (Ruppin et al., 1980; Englehardt and Rechkemmer, 1983a,b). The exact mechanism for the absorptive process is not known, but recent investigations have elucidated several important characteristics. First, absorption occurs in both the ionized and nonionized forms (Ruppin et al., 1980; Cummings, 1981; Englehardt and Rechkemmer, 1983a,b). Second, transport from the lumen is invariably associated with the luminal accumulation of bicarbonate ion (Umesaki et al., 1979; Argenzio and Whipp, 1979). Third, SCFA absorption stimulates sodium absorption (Argenzio et al., 1975; Umesaki et al., 1979; Rechkemmer and Engelhardt, 1982; Roediger and Moore, 1981). Fourth, a luminal source of hydrogen ions appears to be important for absorption of the protonated (nonionized) form (Engelhardt and Rechkemmer, 1983a,b). Fifth, bulk luminal pH is probably relatively unimportant in affecting absorption (McNeill et al., 1978; Engelhardt and Rechkemmer, 1983a,b; Fleming and Arce, 1986). And, finally, the SCFA themselves, via their metabolism in epithelial cells, may provide a fuel source for active sodium transport (Engelhardt and Rechkemmer, 1983a,b).

A detailed review of the current understanding of SCFA absorption is beyond the scope of this chapter and has been published elsewhere (Bugaut, 1987). However, the principal characteristics outlined above have been summarized in Fig. 2. Briefly stated, in the proximal colon some of the SCFA seem to be transported in the ionized form via a paracellular pathway or as a SCFA—bicarbonate exchange mechanism (Fig. 2A, proximal colon). The SCFA, however, are more easily absorbed in the unionized form. H^+ ions needed for this rapid absorption of SCFA may be available from (1) the Na–H exchange, (2) the bicarbonate gain in the lumen, and (3) the bulk solution in the lumen (Fig. 2B, proximal colon). In the distal colon SCFA are not absorbed in ionized form; an interrelationship between sodium transport and SCFA absorption has not been identified. The source of most of the H^+ ions needed for SCFA absorption in the distal colon is not known (Engelhardt and Rechkemmer, 1983a).

4. METABOLISM

The SCFA are readily metabolized by man. Fifty percent of [1-^{14}C]-labeled acetic and propionic acids and 63% of labeled butyric acid appear as [14]CO_2 in the breath within 6 hr after being instilled into the cecum and colon of man (Hoverstad et al., 1982). The appearance of [14]CO_2 in the breath 10 to 15 min after instillation suggested rapid uptake and metabolism of SCFA. Yang et al. (1970) have reported similar results in rats, with 38–55% of the radioactive CO_2 recovered in the breath after 2.5 hr.

The three major SCFA, once absorbed, are metabolized by the cecal, colonic, and rumen mucosa. In the ruminant, nearly all the butyrate generated in the rumen is metabolized by the epithelium itself, whereas 50% of the propionate and 30% of the acetate produced are also utilized by the mucosa (Bergman and Wolff, 1971; Bergman, 1975; Annison et al., 1957). In vitro studies have confirmed the rumen epithelial metabolism of the principal SCFA (Weekes and Webster, 1975; Pennington, 1952).

Acetate, propionate, and butyrate are activated in the rumen epithelial cell to their respective acyl-CoA derivatives by an acyl-CoA synthetase (Ash and Baird, 1973; Nocek et al., 1980). Formation of these CoA derivatives is the first step in epithelial SCFA metabolism and is modulated by the acids themselves. For instance, it has been shown that butyrate strongly inhibited propionyl-CoA synthetase activity in bovine (Ash and Baird, 1973) and ovine (Scaife and Tichivangana, 1980) rumen epithelium in vitro.

Hindgut mucosa in ruminants and nonruminants is similar to the rumen epithelium in its capacity to metabolize the SCFA. Marty and Vernay (1984) have shown that ^{14}C-labeled SCFA infused into loops of rabbit hindgut were metabolized, with butyrate the most extensively metabolized, followed by acetate and propionate, an order identical to that in the rumen (Ash and Baird, 1973). Others have demonstrated similar results with rat elevated cecal sacs (Mottaz and Worbe, 1977) and guinea pig isolated colonocytes (Engelhardt and Rechkemmer, 1983b).

The oxidation of butyrate by hindgut mucosa has been demonstrated by Roediger (1980, 1982) and Ardawi and Newsholme (1985) using suspensions of colonocytes from both rat and man. Roediger has shown that oxidation of butyrate in the rat colonocyte in the presence of glucose accounts for 86% of the overall oxygen consumption (Roediger, 1982). In addition, butyrate suppressed glucose oxidation, probably by inhibiting pyruvate dehydrogenase. Glycolysis is not suppressed, and therefore, in the presence of butyrate, glucose metabolism results in increased lactate accumulation. Rat colonocytes also converted butyrate to the ketone bodies acetoacetate and β-hydroxybutyrate, a finding similar to that in the rabbit colon (Henning and Hird, 1972).

In human colonocyte suspensions prepared from either proximal or distal colon mucosa, added butyrate suppressed glucose oxidation by about 50% (Roediger, 1980). About 75% of the oxygen consumed by the colonocytes was attributable to butyrate oxidation when butyrate was the only substrate available. Human colonocytes also produced ketone bodies (primarily acetoacetate) from n-butyrate (Roediger, 1980). Ketone body production in man is greater in the proximal than distal colon (Roediger, 1980), which parallels findings in other nonruminant mammals (rat and rabbit) demonstrating production of ketones from butyrate in the proximal large bowel but not in the distal colon (Remesy and Demigné, 1976; Henning and Hird, 1972). Less is known about the metabolism of acetate and propionate by colonocytes. Acetate appears to be largely unmetabolized by the proximal large intestine of the rabbit but is extensively metabolized by the distal colon (Marty et al., 1985). Acetate is primarily metabolized to glutamate and aspartate as measured by ^{14}C incorporation from [1-^{14}C]-labeled acetate (Marty et al., 1985).

These studies point up the similarities between rumen and colonic epithelial metabolism of SCFA, in particular butyrate. They also provide evidence that butyrate is preferentially oxidized by these epithelial cells when compared to glucose. Indeed, butyrate is preferentially oxidized by colonocytes in vitro when compared to several common fuels: the order of utilization of respiratory fuels by these cells is butyrate > acetoacetate > L-glutamine > D-glucose (Roediger, 1980, 1982; Ardawi and Newsholme, 1985). Butyrate and the other primary SCFA are therefore important respiratory fuels for colonic and rumen epithelial cells, and their production in these fermentative chambers by anaerobic bacteria may be important to the maintenance of a healthy local mucosa.

Those SCFA not metabolized by the mucosal epithelium are transported to the liver in the portal blood. Portal blood concentrations of the SCFA are four to ten times higher than systemic levels, indicating a substantial clearance function for the liver (Dankert et al., 1981; Hoverstad, 1986). Hepatic SCFA handling has been most extensively studied for acetate. The net uptake of acetate by the liver is directly related to its portal concentration in both rats and humans (Buckley and Williamson, 1977; Herrmann et al., 1985). Acetate may be cleared from portal blood by the liver as effectively as are butyrate and propionate, but the liver releases endogenous acetate, producing normal systemic acetate levels (Lazarus et al., 1988). The perfusion of isolated rat liver with [2-^{13}C]-labeled acetate resulted in the extensive labeling of the amino acids glutamate and glutamine and the ketone bodies acetoacetate and β-hydroxybutyrate in the liver as measured by NMR spectroscopy (Desmoulin et al., 1985). Similarly, the infusion of [1-^{13}C]-labeled butyrate leads to the labeling of acetoacetate, β-hydroxybutyrate, glutamate, and glutamine in the rat liver (Cross et al., 1984). Propionate may be used by the liver as a substrate for gluconeogenesis (Elliot, 1980; Grohn, 1985). Glucose was produced by isolated rat hepatocytes when propionate was provided as the sole substrate, but it was not produced when acetate or butyrate was provided (Anderson and Bridges, 1984). There is evidence that acetate and butyrate may be incorporated into longer-chain fatty acids (Remesy and Demigné, 1976; Hoverstad et al., 1982).

Interestingly, ketone bodies and glutamine produced in the colon mucosa and liver from the SCFA are the preferred oxidative fuels of the enterocyte. The preferential order of utilization of respiratory fuels studied in the small intestinal mucosa is L-glutamine > ketone bodies > D-glucose (Windmueller and Spaeth, 1978; Hanson and Parsons, 1978; Souba et al., 1985). Therefore, SCFA metabolism may indirectly provide preferred energy sources for the small bowel mucosa (Kripke et al., 1988d; Settle, 1988).

5. PHYSIOLOGICAL EFFECTS

One result of production, absorption, and metabolism of the SCFA is the provision of energy to the host (Settle, 1988). It is well known that bacterial production of SCFA in the ruminant foregut provides up to 70% of the energy requirement in cows (Carrol and Hungate, 1954) and sheep (Bergman et al., 1965). Recent investigations have shown that production and absorption of SCFA in the nonruminant hindgut occur at rates capable of providing 5–30% of basal metabolic requirements (Yang et al., 1970; Marty and Vernay, 1984; Rerat et al., 1985; Stevens et al., 1980). The potential contribution of SCFA to the energy requirement of the host has been variously derived (1) from the production rates of SCFA in colonic contents incubated in vitro (Stevens et al., 1980), (2) from in vivo experiments in which SCFA production within the colon was measured by isotopic dilution (Yang et al., 1970), and (3) by measuring arteriovenous differences across the colonic wall (Demigné and Remesy, 1985). Based on these data it has been estimated that the absorption of SCFA from fiber fermentation may provide from 5% to 30% of the daily energy requirements (Wrong, 1981). Colonic luminal perfusion studies estimate that the human colon has the ability to absorb up to 540 kcal/day in the form of SCFA (Ruppin et al., 1980). Other studies have indicated that in man colonic absorption of SCFA may normally supply 5–10% of daily energy requirements, dependent on the quantity of fiber and resistant starch in the diet (Hoverstad, 1986; Cummings, 1984; McNeill, 1984).

In addition to the simple provision of a caloric source to the host, SCFA production in the cecum and colon of nonruminants may strongly influence normal gastrointestinal function by colonic blood flow, pancreatic secretion, sodium and water absorption in the colon, and intestinal mucosal growth.

5.1. Colonic Blood Flow

Kvietys and Granger (1981) have demonstrated that the SCFA, infused into the autoperfused dog colon in physiological concentrations, stimulated a 24% increase in blood flow. This was comparable to the colonic hyperemia induced by a meal (Kvietys et al., 1980). Acetate proved to be the primary factor responsible for the increase in colonic blood flow. In the rat as well, cecal blood flow increased when the cecal content or concentration of SCFA was increased by diet (Demigné and Remesy, 1985). Stimulation of local blood flow may be mediated by a direct vasodilatory effect of the SCFA. Acetate has been shown to increase renal and forelimb blood flow in the dog (Frohlich, 1965) and stimulated an increase in canine mesenteric flow when infused intraarterially (Bing, 1979).

5.2. Pancreatic Secretion

The intravenous infusion of SCFA, specifically butyrate, has been shown to increase pancreatic enzyme secretion (Harada and Kato, 1983), whereas standard TPN produces pancreatic atrophy (Hughes and Dowling, 1980). In anesthetized sheep, intravenous butyrate infusion induced a 13-fold rise in pancreatic juice flow, a 26-fold increase in protein output, and a 37-fold increment in amylase output above basal levels within 5 min. This secretory response was comparable to that obtained with pancreozymin infusion. In an isolated pancreatic lobule preparation, amylase release increased in response to butyrate in a concentration-dependent manner (Harada and Kato, 1983). A similar effect has not been demonstrated in the rat (Harado and Kato, 1983).

5.3. Sodium and Water Absorption

A linear relationship between absorption of fatty acid (acetate and propionate) and sodium absorption has been demonstrated in the goat colon (Argenzio et al., 1975). The presence of 70 mM acetate in the goat colon doubled the rate of sodium absorption from the colonic lumen compared to control. Parsons and Paterson (1965) have shown that n-butyrate increases sodium movement by 82% across mucosal sacs of rat colon. In man, a relationship between propionate and sodium absorption has been reported (Ruppin et al., 1980). Roediger and Moore (1981), using the perfused isolated human colon model, demonstrated a sixfold increase in sodium absorption from the lumen with the infusion of 20 mM n-butyrate and proposed that the presence of bacteria and their metabolic byproducts (SCFA) may be important factors in the maintenance of efficient colonic

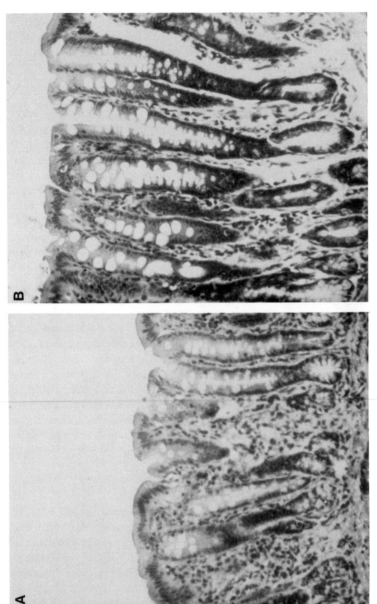

FIGURE 3. Light micrographs (×50) of representative cross sections of proximal colons from rats receiving intracolonic saline infusion (left) and 20 mM butyrate infusion (right). Crypt depth per specimen (in micrometers) was determined by measuring ten well-oriented crypts per sample with an eyepiece micrometer and calculating the mean value. Compared to the control groups, the 20 and 40 mM butyrate and SCFA infusions significantly increased crypt depth in the proximal colon: 20 mM, 236 (5); 40 mM 242 (14); 150 mM, 222 (15); SCFA, 240 (8); PSS, 190 (16); NO INF, 183 (8); and TRANSX, 179 (5). (From Kripke, 1988d.)

absorption of sodium. Furthermore, these investigators suggested that reductions in the SCFA content of the colon may contribute to the pathogenesis of diarrhea in certain clinical situations (following the use of poorly absorbed antibiotics, for example).

5.4. Mucosal Growth

It is well known that animals maintained on a fiber-free diet develop intestinal mucosal atrophy that is most pronounced in the distal small intestine and colon (Ryan *et al.*, 1979; Morin *et al.*, 1980, Ecknauer *et al.*, 1981) and is reversible by the addition of fiber to the diet (Ecknauer *et al.*, 1981; Gordon *et al.*, 1983; Jacobs and Schneeman, 1981; Jacobs and White, 1983; Jacobs and Lupton, 1984). Although most types of fiber exert a strong trophic effect on distal intestinal mucosa, only the most fermentable types of dietary fiber produce any trophic effect on the proximal jejunal mucosa (Vahouny and Cassidy, 1986; Koruda *et al.*, 1987). It has therefore been recently proposed that the SCFA produced from fiber during its fermentation by anaerobic bacteria in the colon mediate the stimulatory effect of fiber on intestinal mucosal growth.

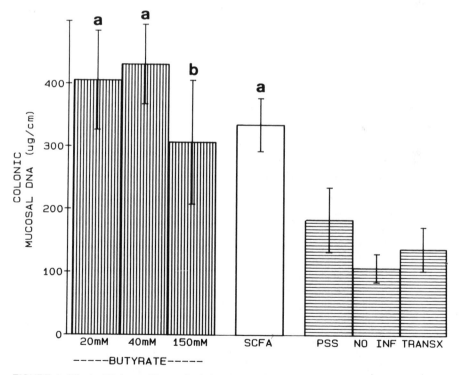

FIGURE 4. Effect of 7 days of intracolonic butyrate or SCFA infusion on colonic mucosal DNA (μg/cm): 20 mM, 40 mM, 150 mM are butyrate infusions; SCFA, infusion of 70 mM acetate plus 35 mM propionate plus 20 mM butyrate; PSS, saline infusion; NO INF, no infusion; TRANSX, transection and anastomosis of proximal colon without cecectomy or infusion. a, $P < 0.005$ versus PSS, NO INF, TRANSX; b, $P < 0.05$ versus NO INF, TRANSX. (From Kripke, 1988d.)

Short-chain fatty acids have been shown to increase intestinal mucosal growth under various experimental conditions. Tutton and Barkla (1982) showed increased rat jejunal crypt mitotic rate after intraperitoneal injection of SCFA. Sakata and Engelhardt (1983) instilled an SCFA solution (75 mM acetate, 35 mM propionate, 20 mM butyrate) or a control electrolyte solution into pouches formed from rat colon *in vivo* and after 60 min were able to demonstrate an increased mitotic index in the colonic epithelium exposed to the SCFA and in colonic mucosa not directly exposed to the SCFA. Further studies revealed that the twice-daily intracecal infusion of 3 ml of a SCFA solution significantly increased the crypt cell production rate (CCPR) as measured by the metaphase arrest technique in jejunum, cecum, and proximal and distal colon (Sakata, 1987). Butyrate appeared to be the most effective stimulator of CCPR when compared to acetate and propionate.

Kripke *et al.* (1988d) showed that in a rat model of colonic mucosal atrophy the chronic intracolonic infusion of butyrate in physiological concentrations significantly stimulated colonic mucosal growth in the rat (Fig. 3) and that of the three principal SCFA, butyrate was the primary trophic factor in the colon (Fig. 4). These investigators further demonstrated that the hindgut infusion of a SCFA solution comprising acetate, propionate, and butyrate at normal intracecal levels stimulated mucosal growth in the

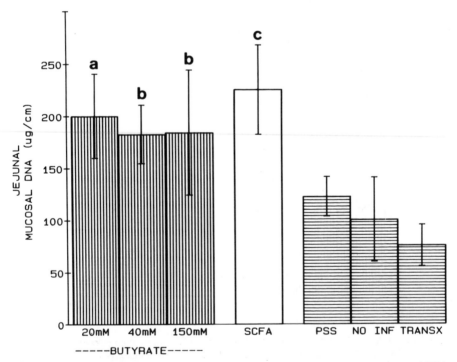

FIGURE 5. Effect of 7 days of intracolonic butyrate or SCFA infusion on jejunal mucosal DNA (μg/cm). a, $P < 0.05$ versus NO INF, TRANSX; b, $P < 0.05$ versus TRANSX; $P < 0.05$ versus PSS, NO INF, TRANSX (see Fig. 4 for group definitions). (From Kripke, 1988d.)

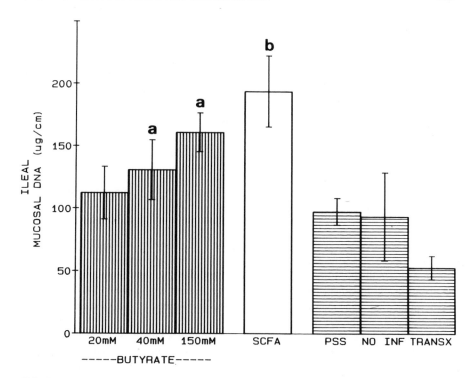

FIGURE 6. Effect of 7 days of intracolonic butyrate or SCFA infusion on ileal mucosal DNA (μg/cm). a, $P < 0.05$ versus TRANS; b, $P < 0.05$ versus PSS, NO INF, TRANSX, 20 mM (see Fig. 4 for group definitions). (From Kripke, 1988d.)

ileum and increased mucosal DNA in the jejunum (Fig. 5) and ileum (Fig. 6) in a manner that closely approximated the effect of fiber in the intestinal tract (Kripke *et al.*, 1988d).

Therefore, intracolonic SCFA have been shown to replicate the trophic effects of dietary fiber in the absence of mechanical bulk effects and may therefore be the principal mediators of the intestinal effects of fiber in the diet. The mechanisms by which SCFA exert their trophic effect in the colon have not been entirely elucidated. Provision of preferred oxidative fuels into a colon denied its normal intraluminal nutrient load may account for the effects described above. The presence of luminal nutrients in general is a major stimulus to intestinal mucosal growth (Weser *et al.*, 1982, 1986; Spector *et al.*, 1981; Jacobs *et al.*, 1987), and fuels preferred by the intestine are especially effective in this regard (Fox *et al.*, 1987; Jacobs *et al.*, 1987).

The manner in which SCFA modulate small intestinal mucosal proliferation is unknown. Many peptide hormones influence mucosal growth, most prominent among these is enteroglucagon (Bloom, 1980). Immunohistochemical studies have located the enteroglucagon-producing cells in the distal ileum, cecum, and proximal colon (Miazza *et al.*, 1985), and intracecal infusion of carbohydrate or fat has been shown to stimulate enteroglucagon secretion (Bloom, 1980; Hoverstad, 1986). The SCFA, normally generated in the cecum, may be important physiological stimulators of enteroglucagon release, and this hypothesis warrants further study.

Host metabolism of the SCFA provides another possible mechanism for indirect small bowel effects of colonic SCFA (Kripke *et al.*, 1988d; Settle, 1988). As mentioned previously hepatic metabolism of butyrate and acetate results in the production of glutamine and the ketone bodies acetoacetate and β-hydroxybutyrate, which are the preferred oxidative fuels of enterocytes (Windmueller and Spaeth, 1978). The enteral or parenteral provision of glutamine and acetoacetate has been shown to be trophic to both small and large intestinal mucosa (Fox *et al.*, 1987; Kripke *et al.*, 1988a).

6. POTENTIAL NUTRITIONAL AND SURGICAL APPLICATIONS OF SHORT-CHAIN FATTY ACIDS

There are several potential applications of SCFA in patients with intestinal dysfunction or in those requiring nutritional support (parenteral and enteral feeding) (Settle, 1988). The stimulatory effects of SCFA on the intestinal mucosa suggest that they may be useful in the treatment of various forms of intestinal dysfunction. The SCFA also represent an alternative energy source for clinical conditions in which the absorption or utilization of common energy sources is compromised.

Patients receive enteral and parenteral nutrition for a variety of reasons, including the inability or unwillingness to consume adequate nutrition, the inability to digest or absorb nutrients, or as a therapeutic modality, such as for bowel rest. Most enteral feeding formulas contain readily digestible polysaccharides and lack fiber. Fiber-free enteral diets induce atrophy in the colon and distal small bowel (Janne *et al.*, 1977; Ryan *et al.*, 1979; Morin *et al.*, 1980; Ecknauer *et al.*, 1981; Sircar *et al.*, 1983), and total parenteral nutrition (TPN) induces atrophy throughout the intestinal tract (Levine *et al.*, 1974; Johnson *et al.*, 1975; Hughes and Dowling, 1980; Goldstein *et al.*, 1985.

As a therapeutic modality, bowel rest, either intravenously (TPN) or enterally (fiber-free elemental diet), is commonly prescribed for patients with disease, injury, or loss of intestine. Although the efficacy of bowel rest cannot be questioned in specific disease entities, one effect of bowel rest is to deprive the intestinal mucosa of its preferred oxidative fuels. Although luminal nutrients are most trophic to the intestinal mucosa (Levine *et al.*, 1974; Spector *et al.*, 1981; Weser *et al.*, 1982, 1986), the provision of enteral nutrients may not always be feasible. Providing the intestinal mucosa with its preferred oxidative fuels intravenously may counteract the lack of luminal nutrients and the atrophy associated with TPN. Moreover, the provision of the preferred oxidative fuels of the intestinal mucosa (SCFAs, ketone bodies, or glutamine) in TPN formulations inhibits the intestinal mucosal atrophy associated with TPN in animals (Hwang *et al.*, 1986; Koruda *et al.*, 1988; Kripke *et al.*, 1988a,b).

We have found that the intracolonic infusion of SCFA in animals fed a fiber-free enteral diet enhanced the healing of a colonic anastomosis as measured by a significant increase in the bursting strength of the anastomosis (Rolandelli *et al.*, 1986). We have also shown that, following massive small bowel resection, TPN supplemented with SCFA significantly reduced the mucosal atrophy associated with TPN and thus facilitated adaptation to small bowel resection (Koruda *et al.*, 1988). A recent study in patients with diversion colitis showed a significant reduction in inflammation of the defunctionalized colon when SCFA were routinely instilled into the diverted colon (Harig and Soergel, 1989). Thus, either the enteral or parenteral provision of SCFA may provide a useful adjuvant therapy in patients with intestinal disease, injury, or loss.

An increasingly recognized source of sepsis in critically ill patients is purported to be disruption of the gut barrier with ensuing transmigration of endotoxin and translocation of enteric bacteria into the portal venous and lymphatic systems (Carrico et al., 1986). Most critically ill patients cannot be fed into the gut and are frequently given TPN. Since SCFA are preferred fuels for the gut, an attractive hypothesis for the maintenance of gut barrier function is to provide SCFA intravenously as a supplement to TPN to prevent the mucosal atrophy associated with TPN. The SCFA might provide a further benefit in septic patients if carnitine deficiencies exist, since the transport of SCFA is not dependent on the carnitine system.

Short-chain fatty acids may also be useful in disorders of lipid digestion, absorption, and transport. One such situation is in short bowel syndrome. We have recently demonstrated that short-chain triglycerides (a 50 : 50 mixture of triacetin and tributyrin), when substituted for the medium-chain triglycerides contained in a commercially available enteral diet, resulted in significantly better intestinal adaptation, greater weight gain, and more positive nitrogen balance in an animal model of short bowel syndrome (Kripke et al., 1988c).

In a broader context SCFA might be used in almost all situations for which medium-chain fatty acids (MCFA) have been proposed (Settle, 1988). A potential advantage to the use of SCFA over MCFA is that they are water soluble and can thus be given as the free acid or as a salt within the physiological limits of acid and salt tolerance. Similar to MCFA, SCFA can be provided as water-soluble monoglycerides (Birkhahn and Border, 1981). In animals there have been no adverse effects reported when SCFA are given in the form of a monoglyceride (monobutyrin), providing up to 70% of nonprotein calories (Kripke et al., 1988a; Birkhahn et al., 1977; Birkhahn and Border, 1978; Kirvela and Takala, 1986). The SCFA, like the MCFA, can also be provided in triglyceride form. Uniquely, triacetin (glycerol triacetate) is a water-soluble triglyceride (Windholz, 1987). It has been infused intravenously without apparent adverse effects on calcium or phosphate metabolism and without apparent toxicity even at high doses (Bailey et al., 1987, 1988).

Further research is needed to identify promising applications of SCFA and to define more clearly their potential role in nutritional support.

ACKNOWLEDGMENTS. The author is very grateful for the secretarial support of Mrs. Phyllis Schultz.

REFERENCES

Ambuhl, S., Williams, V. J., and Senior, W., 1979, Effect of cecectomy in the young adult female rat on the digestibility of food offered *ad-libitum* and in restricted amounts, *Aust. J. Biol. Sci.* **32**:205–214.

Anderson, J. W., and Bridges, S. R., 1984, Short chain fatty acid fermentation products of plant fiber affect glucose metabolism of isolated rat hepatocytes, *Proc. Soc. Exp. Biol. Med.* **177**:372–376.

Annison, E. F., Hill, K. J., and Lewis, D., 1957, Studies on the portal blood of sheep. 2. Absorption of volatile fatty acids from the rumen of sheep, *Biochem. J.* **66:**592–599.

Ardawi, M. S. M., and Newsholme, E. A., 1985, Fuel utilization in coloncytes of the rat, *Biochem. J.* **231:**713.

Argenzio, R. A., and Whipp, S. C., 1979, Inter-relationship of sodium, chloride, bicarbonate and acetate transport by the colon of the pig, *J. Physiol. (Lond.)* **295:**365–381.

Argenzio, R. A., Miller, N., and von Engelhardt, W., 1975, Effect of volatile fatty acids on water and ion absorption from the goat colon, *Am. J. Physiol.* **229:**997–1002.

Ash, R., and Baird, G. D., 1973, Activation of volatile fatty acids in bovine liver and rumen epithelium. Evidence for control by autoregulation, *Biochem. J.* **136:**311–319.

Bailey, J., Rodriquez, N., Marsh, H., Haymond, M., and Miles, J., 1987, Metabolic effects of an intravenous short chain triglyceride infusion in dogs, *J. Parent. Entr. Nutr.* **11**(Suppl. 1):6S.

Bailey, J., Marsh, H., Heath, H., and Miles, J., 1988, Effects of intravenous short chain triglycerides on mineral metabolism and energy expenditure in dogs, *J. Parent. Ent. Nutr.* **12**(Suppl. 1):14S.

Baldwin, R. L., 1970, Energy metabolism in anaerobes, *Am. J. Clin. Nutr.* **23:**1508.

Banta, C. A., Clemens, E. T., Krinsky, M. M., and Sheffy, B. E., 1979, Sites of organic acid production and patterns of digesta movement in the gastrointestinal tract of dogs, *J. Nutr.* **109:**1592–1600.

Barry, R. J. C., and Smyth, D. H., 1960, Transfer of short chain fatty acids by the intestine, *J. Physiol. (Lond.)* **152:**48–66.

Bergman, E. N., 1975, Production and utilization of metabolites by the alimentary tract as measured in portal and hepatic blood, in: *Digestion and Metabolism in the Ruminant* (I. W. McDonald and A. C. I. Warner, eds.), University of New England Publishing Unit, Armidale, Australia, pp. 292–305.

Bergman, E. N., and Wolff, J. E., 1971, Metabolism of volatile fatty acids by liver and portal-drained viscera in sheep, *Am. J. Physiol.* **221:**586–592.

Bergman, E. N., Reid, R. S., Murray, M. G., Brockway, J. M., and Whitelaw, F. G., 1965, Interconversions and production of volatile fatty acids in the sheep rumen, *Biochem. J.* **97:**53–58.

Bing, F. C., 1979, Dietary fiber in historical perspective, in: *Contemporary Nutrition Controversies* (T. P. Labuza and A. E. Sloan, eds.), West Publishing, St. Paul, pp. 170–183.

Birkhahn, R. H., and Border, J. R., 1978, Intravenous feeding of the rat with short chain fatty acid esters, II. Monoacetoacetin, *Am. J. Clin. Nutr.* **31:**436–444.

Birkhahn, R. H., and Border, J. R., 1981, Alternate or supplemental energy sources, *J. Parent. Ent. Nutr.* **5:**24–31.

Birkhahn, R. H., McMenamy, R. H., and Border, J. R., 1977, Intravenous feeding of the rat with short chain fatty acid esters, I. Glycerol monobutyrate, *Am. J. Clin. Nutr.* **30:**2078–2082.

Bloom, S. R., 1980, Gut and brain–endocrine connections (The Gaulstonian Lecture, 1979), *J. R. Coll. Physicians* **14:**51–57.

Bond, J. H., and Levitt, M. D., 1976, Fate of soluble carbohydrates in the colon of rats and man, *J. Clin. Invest.* **57:**1158–1164.

Bond, J. H., Currier, B. E., Buchwald, H., and Levitt, M. D., 1980, Colonic conservation of carbohydrate, *Gastroenterology* **78:**444–447.

Buckley, B. M., and Williamson, D. H., 1977, Origins of blood acetate in the rat, *Biochem. J.* **166:**539–545.

Bugaut, M., 1987, Occurrence, absorption and metabolism of short chain fatty acids in the digestive tract of mammals, *Comp. Biochem. Physiol.* **86B:**439–472.

Carrico, C. J., Meakins, J. L., Marshall, J. C., Fry, D., and Maier, R. V., 1986, Multiple-organ-failure syndrome, *Arch. Surg.* **121:**196–208.

Carroll, E. J., and Hungate, R. E., 1954, The magnitude of the microbial fermentation in the bovine rumen, *Appl. Microbiol.* **2:**205–214.

Clemens, E. T., Stevens, C. E., and Southworth, M., 1975, Sites of organic acid production and pattern of digesta movement in the gastrointestinal tract of swine, *J. Nutr.* **105:**759–768.

Cross, T. A., Paul, C., and Oberhansli, R., 1984, Ketogenesis in the living rat followed by ^{13}C-NMR spectroscopy, *Biochemistry* **23:**6398–6402.

Cummings, J. H., 1981, Short chain fatty acids in the human colon, *Gut* **22:**763–779.

Cummings, J. H., 1984, Colonic absorption: The importance of short chain fatty acids in man, *Scand. J. Gastroenterol.* **20:**88–99.

Cummings, J. H., and Branch, W. J., 1986, Fermentation and the production of short chain fatty acids in the human large intestine, in: *Dietary Fiber: Basic and Clinical Aspects* (G. B. Vahouny and D. Kritchevsky, eds.), Plenum Press, New York, pp. 131–152.

Cummings, J. H., Pomare, E. W., Branch, W. J., Naylor, C. P. E., and McFarlane, G. T., 1987, Short chain fatty acids in human large intestine, portal, hepatic and venous blood, *Gut* **28:**1221–1227.

Czerkawski, J. W., 1986, Energetics of rumen fermentation, in: *An Introduction to Rumen Studies,* Pergamon Press, Oxford, pp. 85–101.

Dankert, J., Zijlstra, J. B., and Wolthers, B. G., 1981, Volatile fatty acids in human peripheral and portal blood: Quantitative determination by vacuum distillation and gas chromatography, *Clin. Chim. Acta* **110:**301–307.

Demigné, C., and Remesy, C., 1985, Stimulation of absorption of volatile fatty acids and minerals in the cecum of rats adapted to a very high fiber diet, *J. Nutr.* **115:**53–60.

Desmoulin, F., Canioni, P., and Cozzone, P. J., 1985, Glutamate–glutamine metabolism in the perfused rat liver: ^{13}C-NMR study using (2-^{13}C-enriched acetate), *FEBS Lett.* **184:**29–32.

Ecknauer, R., Sircar, B., and Johnson, L. R., 1981, Effect of dietary bulk on small intestinal morphology and cell renewal in the rat, *Gastroenterology* **81:**781–786.

Elliot, J. M., 1980, Propionate metabolism and vitamin B-12, in: *Digestive Physiology and Metabolism in Ruminants* (Y. Ruckebush and P. Thivend, Eds.), Lancaster Medical Publisher, Lancaster, U.K.

Elsden, S. R., and Hilton, M. D., 1978, Volatile acid production from threonine, valine, leucine and isoleucine by clostridia, *Arch. Microbiol.* **117:**165–172.

El-Shazly, K., 1952, Degradation of protein in the rumen of sheep. 2. The action of rumen microorganisms on amino acids, *Biochem. J.* **51:**647–653.

Engelhardt, W., and Rechkemmer, G., 1983a, Absorption of inorganic ions and short chain fatty acids in the colon of mammals, in: *Intestinal Transport: Fundamental and Comparative Aspects* (M. Gilles-Baillein and R. Gilles, eds.), Springer-Verlag, Berlin, pp. 26–45.

Engelhardt, W., and Rechkemmer, G., 1983b, The physiological effects of short chain fatty acids in the hind gut, *Bull. R. Soc. N.Z.* **20:**149–155.

Englyst, H. N., Trowell, H., Southgate, D. A. T., and Cummings, J. H., 1987, Dietary fiber and resistant starch, *Am. J. Clin. Nutr.* **46:**873–874.

Fleming, S. E., and Arce, D. S., 1986, Volatile fatty acids: Their production, absorption, utilization and roles in human health, *Clin. Gastroenterol.* **15:**787–814.

Fox, A. D., Kripke, S. A., Berman, J. M., Settle, R. G., and Rombeau, J. L., 1987, Reduction of the severity of enterocolitis by glutamine-supplemented enteral diets, *Surg. Forum* **38:**43–44.

Frohlich, E. D., 1965, Vascular effects of the Krebs intermediate metabolites, *Am. J. Physiol.* **208:**149–153.

Goldstein, R. M., Hebiguchi, T., Luk, G. D., Tagi, F., Gularte, T. R., Franklin, F. A., Niemiee, P. W., and Dudgear, D. H., 1985, The effects of total parenteral nutrition on gastrointestinal growth and development, *J. Pediatr. Surg.* **20:**785–791.

Gordon, D. T., Besch-Williford, C., and Ellersieck, M. R., 1983, The action of cellulose on the intestinal mucosa and elemental absorption by the rat, *J. Nutr.* **113:**2545–2556.

Grohn, Y., 1985, Propionate loading test for liver function during experimental liver necrosis in sheep, *Am. J. Vet. Res.* **46:**952–958.

Hanson, P. J., and Parsons, D. S., 1978, Factors affecting the utilization of ketone bodies and other substrates by rat jejunum: Effects of fasting and of diabetes, *J. Physiol. (Lond.)* **278:**55–67.

Harada, E., and Kato, S., 1983, Effect of short-chain fatty acids on the secretory response of the ovine exocrine pancreas, *Am. J. Physiol.* **244:**G284–290.

Harig, J. M., and Soergel, K. H., 1989, Treatment of diversion colitis with short chain fatty acid (SCFA) irrigation, *N. Engl. J. Med.* **320:**23–28.

Henning, S. J., and Hird, F. J. R., 1972, Transport of acetate and butyrate in the hind-gut of rabbits, *Biochem. J.* **130:**791–796.

Herrmann, D. B. J., Herz, R., and Frolich, J., 1985, Role of gastrointestinal tract and liver in acetate metabolism in rat and man, *Eur. J. Clin. Invest.* **15**:221–226.

Hilditch, T. P., and Williams, P. N. (eds.), 1964, *The Chemical Constitution of the Natural Fats,* 4th ed., John Wiley & Sons, New York.

Hill, M. J., and Drasar, B. S., 1975, The normal colonic bacterial flora, *Gut* **16**:318–323.

Hoverstad, T., 1986, Studies of short chain fatty acid absorption in man, *Scand. J. Gastroenterol.* **21**: 257–260.

Hoverstad, T., Bohmer, T., and Fausa, O., 1982, Absorption of short chain fatty acids from the human colon measured by the $^{14}CO_2$ breath test, *Scand. J. Gastroenterol.* **17**:373–378.

Hughes, C. A., and Dowling, R. H., 1980, Speed of onset of adaptive mucosal hypoplasia and hypofunction in the intestine of parenterally fed rats, *Clin. Sci.* **59**:317–327.

Hwang, T. L., O'Dwyer, S. T., Smith, R. J., and Wilmore, D. W., 1986, Preservation of small bowel mucosa using glutamine-enriched parenteral nutrition, *Surg. Forum* **37**:56–58.

Jacobs, D. O., Evans, D. A., and O'Dwyer, S. T., 1987, Disparate effects of 5-fluorouracil on the ileum and colon of enterally fed rats with protection by dietary glutamine, *Surg. Forum* **38**:45–47.

Jacobs, L. R., and Lupton, J. R., 1984, Effect of dietary fibers on rat large bowel mucosal growth and cell proliferation, *Am. J. Physiol.* **246**:G378–385.

Jacobs, L. R., and Schneeman, B. O., 1981, Effects of dietary wheat bran on rat colonic structure and mucosal cell growth, *J. Nutr.* **111**:798–803.

Jacobs, L. R., and White, F. A., 1983, Modulation of mucosal cell proliferation in the intestine of rats fed a wheat bran diet, *Am. J. Clin. Nutr.* **37**:945–953.

Janne, P., Carpentier, Y., and Willems, G., 1977, Colonic mucosal atrophy induced by a liquid elemental diet in rats, *Am. J. Dig. Dis.* **22**:808–812.

Johnson, R., Copeland, E. M., Dudrick, S. J., Lichtenberger, L. M., and Castro, G. A., 1975, Structural and hormonal alterations in the gastrointestinal tract of parenterally fed rats, *Gastroenterology* **68**:1177–1183.

Keys, J. E., Jr., Van Soest, P. J., and Young, E. P., 1969, Comparative study of the digestibility of forage cellulose and hemicellulose in ruminants and nonruminants, *J. Anim. Sci.* **29**:11–15.

Kirvela, O. K., and Takala, J. A., 1986, Comparison of monoglyceryl acetoacetate and glucose as parenteral energy substrate after experimental trauma, *Eur. J. Surg. Res.* **18**(2):80–85.

Koruda, M. J., Rolandelli, R. H., Settle, R. G., Zimmaro, D. M., Hastings, J., and Rombeau, J. L., 1987, The effect of short chain fatty acids on the small bowel mucosa, *J. Parent. Ent. Nutr.* (*Suppl.*) **11**:8S.

Koruda, M. J., Rolandelli, R. H., Settle, R. G., Zimmaro, D. M., and Rombeau, J. L., 1988, Effect of parenteral nutrition supplemented with short chain fatty acids on adaptation to massive small bowel resection, *Gastroenterology* **95**:715–720.

Kripke, S. A., Fox, A. D., Berman, J. M., DePaula, J. A., Settle, R. G., Birkhahn, R. H., and Rombeau, J. L., 1988a, Inhibition of TPN-associate intestinal mucosal atrophy with mono-acetoacetin, *J. Surg. Res.* **44**:436–444.

Kripke, S. A., Fox, A. D., Berman, J. M., DePaula, J. A., Rombeau, J. L., and Settle, R. G., 1988b, Inhibition of TPN-associated colonic atrophy with beta-hydroxy-butyrate, *Surg. Forum* **39**:48–50.

Kripke, S. A., De Paula, J. A., Berman, J. M., Fox, A. D., Palacio, J. C., Settle, R. G., and Rombeau, J. L., 1988c, Short chain triglycerides improve mucosal adaptation in short bowel syndrome, *Am. J. Clin. Nutr.* (in press).

Kripke, S. A., Fox, A. D., Berman, J. M., Settle, R. G., and Rombeau, J. L., 1988d, Stimulation of intestinal mucosal growth with intracolonic infusion of short chain fatty acids, *J. Parent. Ent. Nutr.* **13**:109–116.

Kvietys, P. R., and Granger, N. D., 1981, Effect of volatile fatty acids on blood flow and oxygen uptake by the dog colon, *Gastroenterology* **80**:962–969.

Kvietys, P. R., Gallavan, R. H., and Chou, C. C., 1980, Contribution of bile to postprandial intestinal hyperemia, *Am. J. Physiol.* **238**:G284–288.

Lazarus, D. D., Zimmaro, D. M., Rolandelli, R. R., Settle, R. G., and Stein, T. P., 1988, Non-gut origin of plasma acetate in humans and rats, *FASEB J.* **2**:A444.

Levine, G. M., Deren, J. J., Steiger, E., and Zinno, R., 1974, Role of oral intake in maintenance of gut mass and disaccharide activity, *Gastroenterology* **67**:975–982.

Lupton, J. R., 1985, Influence of luminal pH on rat large bowel epithelial cell cycle, *Am. J. Physiol.* **249**:G382–388.

Lupton, J. R., Coder, D. M., and Jacobs, L. R., 1988, Long-term effects of fermentable fibers on rat colonic pH and epithelial cell cycle, *J. Nutr.* **118**:840–845.

Macy, J. M., 1979, The biology of gastrointestinal *Bacteroides, Annu. Rev. Microbiol.* **33**:561–594.

Marty, J. F., and Vernay, M., 1984, Absorption and metabolism of the volatile fatty acids in the hind-gut of the rabbit, *Br. J. Nutr.* **51**:265–277.

Marty, J. F., Vernay, M. Y., and Abravanel, G. M., 1985, Acetate absorption and metabolism in the rabbit hind-gut, *Gut* **26**:562–569.

McNeill, N. I., 1984, The contribution of the large intestine to energy supplies in man, *Am. J. Clin. Nutr.* **39**:338–342.

McNeill, N. I., Cummings, J. H., and James, W. P. T., 1978, Short chain fatty acid absorption by the human large intestine, *Gut* **19**:819–822.

Meyers, L. L., Jackson, J. D., and Packett, L. V., 1967, Absorption of volatile fatty acids from the cecum of sheep, *J. Anim. Sci.* **26**:1450–1458.

Miazza, B. M., Al-Mukhtar, M. Y. T., Salmeron, M., Miazza, B. M., Al-Mokhter, M. Y., Salmeran, M., Ghatei, M. A., Felce-Dadez, M., Filali, A., Villet, R., Wright, N. A., Bloom, S. R., and Cranbaud, J. C., 1985, Hyperenteroglucagonemia and small intestinal mucosal growth after colonic perfusion of glucose in rats, *Gut* **26**:518–524.

Miller, T. L., and Wolin, M. J., 1979, Fermentations by saccharolytic intestinal bacteria, *Am. J. Clin. Nutr.* **32**:164–172.

Morin, C. L., Ling, V., and Bourassa, D., 1980, Small intestinal and colonic changes induced by a chemically defined diet, *Dig. Dis. Sci.* **25**:123–128.

Mottaz, P., and Worbe, J. F., 1977, Transfer of volatile fatty acids within the isolated cecum of the rat, *C. R. Soc. Biol.* **171**:375–380.

Nocek, J. E., Herbein, J. H., and Polan, C. E., 1980, Influence of ration physical form, ruminal degradable nitrogen and age on rumen epithelial propionate and acetate transport and some enzymatic activities, *J. Nutr.* **110**:2355–2364.

Nyman, M., and Aso, N. G., 1982, Fermentation of dietary fiber components in rat intestinal tract, *Br. J. Nutr.* **47**:357–366.

Parsons, D. S., and Paterson, C. R., 1965, Fluid and solute transport across rat colonic mucosa, *Q. J. Exp. Physiol.* **50**:220–231.

Pennington, R. J., 1952, The metabolism of short chain fatty acids in the sheep, 1. Fatty acid utilization and ketone body production by rumen epithelium and other tissues, *Biochem J.* **56**:410–416.

Ravich, W. J., Bayless, T. M., and Thomas, M., 1983, Fructose: Incomplete intestinal absorption in humans, *Gastroenterology* **84**:26–29.

Rechkemmer, G., and Engelhardt, W., 1982, Absorptive processes in different colonic segments of the guinea pig and the effects of short chain fatty acids, in: *Colon and Nutrition* (H. Kasper and H. Goebell, eds.), MTP Press, Lancaster, pp. 61–67.

Remesy, C., and Demigné, C., 1976, Partition and absorption of volatile fatty acids in the alimentary canal of the rat, *Ann. Rech. Vet.* **7**:39–55.

Rerat, A., Fiszlewicz, M., Herpin, P., Vangelade, P., and Durand, M., 1985, Mesure de l'apparition dans la veine porte des acides gras volatile fomés au cours de la digestion chey le porc eveille, *C.R. Acad. Sci. [Paris]* **300**:467–470.

Roediger, W. E. W., 1980, Role of anaerobic bacteria in the metabolic welfare of the colonic mucosa in man, *Gut* **21**:793–798.

Roediger, W. E. W., 1982, Utilization of nutrients by isolated epithelial cells of the rat colon, *Gastroenterology* **83**:424–429.

Roediger, W. E. W., and Moore, A., 1981, Effect of short chain fatty acids on sodium absorption in isolated human colon perfused through the vascular bed, *Dig. Dis. Sci.* **26**:100–106.

Roediger, W. E. W., and Rae, D. A., 1982, Trophic effect of short chain fatty acids on mucosal handling of ions by the defunctionalized colon, *Br. J. Surg.* **69**:23–25.

Rolandelli, R. H., Koruda, M. J., Settle, R. G., and Rombeau, J. L., 1986, Effects of intraluminal infusion of short chain fatty acids on the healing of colonic anastomosis in the rat, *Surgery* **100**:198–203.

Ruchim, M. A., Makino, D., and Zarling, F. J., 1984, Volatile fatty acid production from amino acid degradation by human fecal suspensions, *Gastroenterology* **86**:1225.

Ruppin, H., Bar-Meir, S., Soergel, K. H., Wood, C. M., and Schmitt, M. D., Jr., 1980, Absorption of short-chain fatty acids by the colon, *Gastroenterology* **78**:1500–1507.

Ryan, G. P., Dudrick, S. J., Copeland, E. M., and Johnson, L. R., 1979, Effects of various diets on colonic growth in rats, *Gastroenterology* **77**:658–663.

Sakata, T., 1987, Stimulatory effect of short chain fatty acids on epithelial cell proliferation in the rat intestine: A possible explanation for the trophic effects of fermentable fiber, gut microbes and luminal trophic effects, *Br. J. Nutr.* **58**:95–103.

Sakata, T., and Engelhardt, W. V., 1983, Stimulatory effect of short chain fatty acids on the epithelial cell proliferation in the rat large intestine, *Comp. Biochem. Physiol.* **74A**:459–462.

Salyers, A. A., Vercellotti, J. R., and West, S. E. H., and Wilkins, T. D., 1977, Fermentation of mucin and plant polysaccharides by strains of bacteroides from the human colon, *Appl. Environ. Microbiol.* **33**:319–322.

Scaife, J. R., and Tichivangana, J. Z., 1980, Short chain acyl-Co A synthetases in ovine rumen epithelium, *Biochim. Biophys. Acta* **619**:445–450.

Schmitt, M. G., Jr., Soergel, K. H., and Wood, C. M., 1976, Absorption of short chain fatty acids from the human jejunum, *Gastroenterology* **70**:211–215.

Schmitt, M. G., Jr., Soergel, K. H., and Wood, C. M., 1977, Absorption of short chain fatty acids from the human ileum, *Dig. Dis. Sci.* **22**:340–347.

Settle, R. G., 1988, Short chain fatty acids and their potential role in nutritional support, *J. Parent. Ent. Nutr.* **12**:104S–107S.

Sircar, B., Johnson, L. R., and Lichtenberger, L. M., 1983, Effect of synthetic diets on gastrointestinal mucosal DNA synthesis in rats, *Am. J. Physiol.* **244**:G327–335.

Smyth, D. H., and Taylor, C. B., 1958, Intestinal transfer of short chain fatty acids *in-vitro*, *J. Physiol. (Lond.)* **141**:73–80.

Souba, W. W., Scott, T. E., and Wilmore, D. W., 1985, Intestinal consumption of intravenously administered fuels, *J. Parent. Ent. Nutr.* **9**:18–22.

Spector, M. H., Traylor, J., Young, E. A., and Weser, E., 1981, Stimulation of mucosal growth by gastric and ileal infusion of single amino acids in parenterally nourished rats, *Digestion* **21**:33–40.

Stevens, C. E., Argenzio, R. A., and Clemens, E. T., 1980, Microbial digestion: Rumen versus large intestine, in: *Digestive Physiology and Metabolism in Ruminants* (Y. Ruckebusch and P. Thivend, eds.), Lancaster Medical Publisher, Lancaster, United Kingdom, pp. 685–706.

Thomson, A. B. R., 1982, Influence of dietary modifications on uptake of cholesterol, glucose, fatty acids, and alcohols into rabbit intestine, *Am. J. Clin. Nutr.* **35**:556–565.

Tutton, P. J. M., and Barkla, D. H., 1982, Further studies on the effect of adenosine cyclic monophosphate derivatives on cell proliferation in jejunal crypts of rat, *Clin. Exp. Pharmacol. Physiol.* **9**:671–674.

Umesaki, Y., Yajima, T., Yokokura, T., and Mutai, M., 1979, Effect of organic acid absorption on bicarbonate transport in rat colon, *Pflugers Arch.* **379**:43–47.

Vahouny, G. V., and Cassidy, M. M., 1986, Dietary fiber and intestinal adaptation, in: *Dietary Fiber: Basic and Clinical Aspects* (G. B. Vahouny and D. Kritchevsky, eds.), Plenum Press, New York, pp. 181–209.

Van Soest, P. J., Jeraci, J., and Foose, T., 1982, Comparative fermentation of fibre in man and other animals, in: *Fiber in Human and Animal Nutrition,* Royal Society of New Zealand, Auckland, pp. 75–80.

Warner, A. C. I., 1964, Production of volatile fatty acids in the rumen: Methods of measurement, *Nutr. Abstr. Rev.* **34**:339–352.

Weekes, T. E. C., and Webster, A. J. F., 1975, Metabolism of propionate in the tissues of the sheep gut, *Br. J. Nutr.* **33**:425–438.

Weller, R. A., Gray, F. V., Pilgrim, A. F., and Jones, G. B., 1967, The rates of production of volatile fatty acids in the rumen. 4. Individual and total volatile fatty acids, *Aust. J. Agri. Res.* **18**:107–118.

Weser, E., 1979, Short bowel syndrome, *Gastroenterology* **77**:572.

Weser, E., Vandeventer, A., and Tawil, T., 1982, Stimulation of small bowel mucosal growth by midgut infusion of different sugars in rats maintained by total parenteral nutrition, *J. Pediatr. Gastroenterol. Nutr.* **1**:411–416.

Weser, E., Babbitt, J., and Hoban, M., 1986, Intestinal adaptation: Different growth responses to disaccharides compared with monosaccharides in rat small bowel, *Gastroenterology* **91**:1521–1527.

Williams, V. J., and Senior, W., 1982, Effects of cecectomy on the digestibility of food and rate of passage of digesta in the rat, *Aust. J. Biol. Sci.* **35**:373–380.

Windholz, M. (ed.), 1987, *The Merck Index,* Merck & Co., Rahway, NJ.

Windmueller, H. G., and Spaeth, A. E., 1978, Identification of ketone bodies and glutamine as the major respiratory fuels *in-vivo* for postabsorptive rat small intestine, *J. Biol. Chem.* **253**:69.

Wolin, M. J., 1975, Interactions between the bacterial species of the rumen, in: *Digestion and Metabolism in the Ruminant* (I. W. McDonald and A. C. Warner, eds.), University of New England Publishers, Armidale, Australia, pp. 134–148.

Wolin, M. J., 1981, Fermentation in the rumen and human large intestine, *Science* **213**:1463–1468.

Wolin, M. J., and Miller, T. L., 1983, Interactions of microbial populations in cellulose fermentations, *Fed. Proc.* **42**:109–113.

Wrong, O. M., 1981, Carbohydrates, in: *The Large Intestine: Its Role in Mammalian Nutrition and Homeostasis* (O. M. Wrong, C. J. Edmonds, and V. S. Chadwick, eds.), Halsted Press, New York, pp. 107–112.

Yang, M. G., Manoharen, K., and Mickelsen, O., 1970, Nutritional contributions of volatile fatty acids from the cecum of rats, *J. Nutr.* **100**:545–550.

25

Soluble Fiber

Hypocholesterolemic Effects and Proposed Mechanisms

JAMES W. ANDERSON, DEE A. DEAKINS, and SUSAN R. BRIDGES

1. INTRODUCTION

Water-soluble fibers significantly lower serum cholesterol in humans and animals, whereas water-insoluble fibers do not (Anderson and Tietyen-Clark, 1986). Soluble fibers, the dietary fibers extracted from plant cell walls by hot water, include pectins, gums, and some hemicelluloses. Insoluble fibers include other hemicelluloses, cellulose, and lignin (Southgate, 1977). Soluble fibers are important components of most plant foods and comprise, on average, the following percentages of total dietary fiber for these food groups: vegetables 32%, cereals 32%, beans 25%, and fruits 38% (Anderson and Bridges, 1988). Soluble fibers constitute about 30% of the total dietary fiber intake for a representative Western diet (Anderson et al., 1989).

The mechanisms responsible for the hypocholesterolemic effects of soluble fibers are not well delineated. In this chapter, we compare the reported effects of various soluble fibers on blood cholesterol from both human and experimental animal studies. We also review possible mechanisms and data to support these proposed effects.

2. CLINICAL STUDIES IN HUMANS

Several reviews document the hypocholesterolemic effects of soluble fibers in humans (Kay and Truswell, 1980; Chen and Anderson, 1986; Anderson and Tietyen-Clark, 1986; Reiser, 1987) and the lack of effect of insoluble fiber (Kay and Truswell,

JAMES W. ANDERSON, DEE A. DEAKINS, and SUSAN R. BRIDGES • Metabolic Research Group, VA Medical Center, University of Kentucky, Lexington, Kentucky 40511.

1980; Anderson and Tietyen-Clark, 1986). Since comparisons of study results are difficult because of variation in study designs, a set of inclusion and exclusion criteria
previously developed (Anderson and Tietyen-Clark, 1986) was used for reviewing
human clinical studies for this chapter. Studies selected had a minimum of eight subjects
to insure statistical validity, lasted at least 3 week for inpatient and 6–18 weeks for
outpatient studies, included a minimum of 10 g of dietary fiber daily, and were controlled for fat and cholesterol. Table I summarizes studies that met these criteria.

2.1. Gums

Guar gum, a galactomannan extracted from the endosperm of the Indian cluster
bean, has been used extensively in clinical studies. Table I summarizes studies meeting
the inclusion criteria outlined above. In six clinical studies the administration of guar
supplements in doses from 7.5 to 21 g/day produced decreases in serum cholesterol

TABLE I. Soluble Fiber Effects on Serum Lipids in Humans[a]

References	Type of fiber	Amount of dietary fiber (g/day)	Serum lipids (% change)		Study design[b]
			Cholesterol	Triglycerides	
Aro et al. (1981)	Guar	21	−14*[c]	+11	DB, CO, RA
Kyllastinen and Lahikainen (1981)	Guar	16	−13*	−3	SC
Simons et al. (1982)	Guar	18	−15*	−4	PC, SB
Aro et al. (1984)	Guar	15	−8*	−4	DB, CO, RA
Bosello et al. (1984)	Guar	16	−12*	−22*	SC
Uusitupa et al. (1984)	Guar	7.5–21	−11*	NS	DB, CO, RA
Superko et al. (1988)	Guar	15	−6*	+12	SB, CO, RA
Osilesi et al. (1985)	Xanthan gum	12	−7*	−0	SC, CO
Zavoral et al. (1983)	Locust bean gum	8–30	−5 to −19*	−10	CO, RA
Kay and Truswell (1977)	Pectin	15	−15*	+1	SC
Judd and Truswell (1982)	Pectin	15	−17*	+13	CO, RA
Burton and Manninen (1982)	Psyllium	<25	−16*	0	SC
Anderson et al. (1988a)	Psyllium	10	−15*	−13	DB, P, PC
Lo et al. (1986)	Soy fiber	18	−8*	−7	SB, CO
Anderson et al. (1984a)	Beans	19	−19*	−4	SC, P, RA
Kirby et al. (1981)	Oat bran	15	−13*	−5	SC, P, RA
Anderson et al. (1984a)	Oat bran	15	−19*	−19*	SC, P, RA

[a] Adapted from Anderson and Tietyen-Clark (1986).
[b] SC, self-controlled; PC, placebo-controlled; P, parallel; DB, double blind; SB, single blind; CO, crossover; RA, random allocation.
[c] Asterisks indicate values significantly lower than control values, $P < 0.05$.

ranging from 6% to 15% with an average reduction 12.0% and a median reduction of 12.0%. In all six studies these reductions were statistically significant. The response of serum triglycerides was variable and reached statistical significance in only one study.

Osilesi and associates (1985) studied the effects of xanthan gum, a biosynthetic edible gum produced under highly controlled conditions by the microorganism *Xanthomonas campestris,* in healthy subjects. After 6 weeks of use of this supplement, serum cholesterol was significantly reduced by 7%.

Zavarol and associates (1983) studied the effects of locust bean gum (LBG), a gum that closely resembles guar gum and is extracted from the endosperm of the locust bean. In a crossover design, 29 normolipidemic or hyperlipidemic children or adults were given LBG supplements of 8–30 g/day for 8 weeks and identical food supplements without LBG for 8 weeks. The LBG supplements were accompanied by significant reductions in serum cholesterol ranging from 5% to 19%.

Although a variety of gums have significant cholesterol-lowering effects, most are not very palatable. Many investigators report a moderate to high frequency of side effects such as nausea, vomiting, abdominal discomfort, and excess gas production. To overcome these problems several investigators have tested the effects of various gums in food products such as breads (Jenkins *et al.,* 1980b), bars (McIvor *et al.,* 1986), pasta (Gatti *et al.,* 1984), and other food products (Zavarol *et al.,* 1983; Ellis *et al.,* 1988; Superko *et al.,* 1988).

2.2. Pectins

Pectins, polygalacturonic acid polymers found predominantly in fruits and vegetables, have significant hypocholesterolemic effects (Reiser, 1987). The early studies of Keys and colleagues (1961) have been confirmed by many subsequent clinical studies. Table I includes the studies that meet previously described criteria.

When 15 g of pectin was given daily, serum cholesterol decreased significantly by 15–17% while serum triglycerides did not change significantly. Other clinical studies were not included in Table I because they used combined therapy (Ginter *et al,* 1979) or did not meet the inclusion criteria (Jenkins *et al.,* 1975, 1979; Raymond *et al.,* 1977; Miettinen and Tarpilla, 1977; Nakamura *et al.,* 1982; Challen *et al.,* 1983; Sable-Amplis *et al.,* 1983; Hillman *et al.,* 1985; Vargo *et al.,* 1985). Although pectin has important hypocholesterolemic effects, it shares many properties with the gums and is not palatable unless presented in a modified form to enhance acceptability.

2.3. Psyllium

Psyllium hydrophilic mucilloid, a highly branched acidic arabinoxylan extracted from the husk of *Plantago ovata* seeds (Kennedy *et al.,* 1979), has been widely used for management of large bowel disorders for more than 50 years. Garvin and colleagues (1965) demonstrated that psyllium administration had hypocholesterolemic effects, and subsequent studies as summarized by Anderson *et al.* (1988a) confirmed these observations. Table I summarizes the studies meeting the inclusion criteria outlined above. Other studies, summarized elsewhere (Anderson *et al.,* 1988a; Abraham and Mehta,

1988), reported that administration of psyllium in doses of 3.6 to 24.2 g/day decreased serum cholesterol by 5–20%.

Burton and Manninen (1982), in a self-controlled study, noted a 16% reduction in serum cholesterol for psyllium-treated subjects. Using a double-blind, placebo-controlled, parallel design, we studied 26 mildly or moderately hypercholesterolemic subjects (Anderson et al., 1988a). Subjects used 3.4 g of psyllium or placebo three times daily and maintained their usual diets. After 8 weeks, total serum cholesterol was 14.8% lower than control ($P < 0.01$), and low-density lipoprotein cholesterol was 20.2% lower than control values ($P < 0.01$) for the psyllium group while values were not significantly different from control for the placebo group. Since psyllium was well tolerated by these subjects, it may have a practical role in the long-term management of certain individuals with hypercholesterolemia and may deserve consideration before using more potent hypocholesterolemic pharmacological agents, which have substantial side effects.

2.4. Soybean Polysaccharide

Although water-soluble gums such as guar and locust bean gum extracted from beans have well-documented hypocholesterolemic effects, recent studies (Shorey et al., 1985; Lo et al., 1986) indicate that a largely insoluble soybean polysaccharide also has significant hypocholesterolemic effects. Soy polysaccharide, isolated from the soybean cotyledon, contains approximately 16% cellulose, 50% hemicellulose (largely arabinogalactans), and only 9% soluble fiber. Shorey and colleagues (1985), using a placebo-controlled, blinded crossover design, noted that the administration of 25 g/day of soy polysaccharide produced a significant decrease in serum cholesterol of 5–11% over an 8-week period. Lo and colleagues (1986) also noted that administration of 25 g daily of soy polysaccharide significantly lowered serum cholesterol by 8%. Other studies (Schweizer et al., 1983; Tsai et al., 1983; Sasaki et al., 1985) did not meet the criteria for inclusion in Table I. Studies with soy polysaccharide, because of its palatability and acceptability, suggest that this product may have a role in the management of hypercholesterolemia.

2.5. Beans

Dried beans as well as their fiber and gum extracts have significant cholesterol-lowering effects. Groen et al., (1962) noted that Benedictine monks, who consumed typical Western diets, had significantly higher serum cholesterols than Trappist monks, who consumed frugal lactovegetarian diets including 100–150 g of dried beans per day. To examine the hypocholesterolemic effect of beans, Luyken and associates (1962) fed 100 g of dried beans to healthy subjects and noted an 11–12 mg/dl reduction in serum cholesterol.

Mathur and colleagues (1968) studied the effects of Bengal gram (chickpea), a staple in the diet of North India; after raising serum cholesterol by using a high-fat diet, they noted that substituting Bengal gram for flour and cereal products lowered serum cholesterol by 22%. Jenkins and colleagues (1983) noted that incorporating 140 g of dried beans into the diet of hyperlipidemic subjects produced a significant reduction in serum triglycerides and a nonsignificant reduction in serum cholesterol.

Table I summarizes our study indicating that 100 g of dried beans daily has the potential to lower serum cholesterol by 19% without affecting serum triglycerides. As summarized recently (Anderson and Gustafson, 1988), dried beans as part of a prudent diet can substantially enhance the hypocholesterolemic effects of a low-fat and low-cholesterol diet.

2.6. Oat Products

Oatmeal and oat bran are rich in the water-soluble β-glucan or oat gum, which has significant hypocholesterolemic effects. DeGroot and colleagues (1963) reported that the intake of 140 g of rolled oats daily reduced serum cholesterol by 12%. More recent metabolic ward and ambulatory studies indicate that even without alterations in fat and cholesterol intake, the regular daily use of 50–100 g of oat bran lowers serum cholesterol by 12–19% (Kirby et al., 1981; Anderson et al., 1984a; Storch et al., 1984). Studies meeting the inclusion criteria are summarized in Table I. When fat and cholesterol intake were consistent on both control and oat bran diets on a metabolic ward, a daily intake of 100 g of oat bran decreased serum cholesterol by 13% in 10 days and by 19% in 21 days. Subsequent studies have confirmed the hypocholesterolemic effects of oat bran and oatmeal (Table II).

Oat products have the characteristic of selectively decreasing serum LDL-cholesterol while preserving or actually increasing serum HDL-cholesterol. In our first study, we incorporated 100 g/day of oat bran into a high-carbohydrate (70% of energy), low-fat (12% of energy) diet for diabetic individuals. Over a 3-week period the high-carbohydrate, low-fat diet including oat bran decreased serum cholesterol by 38%, decreased serum LDL-cholesterol by 58%, and *increased* serum HDL-cholesterol by 82%. As summarized in Table II, oat bran feeding lowers LDL-cholesterol by 8.3–28.8% (average decrease of 18.1%) while changing HDL-cholesterol by −5.6% to +8.3% (average increase of 0.6%). Oatmeal may have an even more specific effect in lowering LDL-

TABLE II. Effect of Oat Products on Lipids and Lipoproteins in Humans (% Change)

	Grams/day	Cholesterol	LDL-Chol.	HDL-Chol.	LDL/HDL
Oat bran					
Kirby et al. (1981)	100	−13.0	−13.6	−2.0	−11.8
Anderson et al. (1984a)	100	−19.3	−23.0	−5.6	N/A
Anderson et al. (1984b)	100	−23.1	−22.7	−19.9	−3.6[b]
Van Horn et al. (1986)[a]	39	−2.7	N/A	N/A	N/A
Oatmeal					
Judd and Truswell (1981)	125	−8.0	N/A	+4.7	N/A
Van Horn et al. (1986)[a]	35	−3.3	N/A	N/A	N/A
Turnbull and Leeds (1987)	150	−5.0	−13.9	+16.6	−26.3[b]
Van Horn et al. (1988)[a]	56	−3.1[b]	N/A	N/A	N/A

[a] Subjects followed a fat- and cholesterol-restricted diet (American Heart Association) for 4–6 weeks prior to receiving oat product.
[b] Estimated.

cholesterol while preserving HDL-cholesterol: while the LDL-cholesterol reduction averaged 11.6%, the HDL-cholesterol increase averaged 7.9%, leading to significant and favorable reductions in the LDL:HDL-cholesterol ratio.

Kinosian and Eisenberg (1988) compared the cost effectiveness of treating hyper-cholesterolemic individuals with either bile-acid-binding resins or oat bran and noted that oat bran was much more cost effective than cholestyramine or colestipol. Thus, in addition to a low-fat and low-cholesterol diet, regular use of oat products may have an important role in reducing risk for coronary heart disease (Anderson and Gustafson, 1988).

2.7. Time Response

The serum cholesterol response to soluble fiber is very rapid, as depicted in Fig. 1. As previously described (Anderson et al., 1984a), the control and oat bran diets pro-vided nearly identical quantities of fat, cholesterol, and energy, with the principal difference being the high fiber content of the oat bran. Incorporating oat bran into a representative Western diet lowered serum cholesterol by 22.6% in an average of 11 days and sustained reductions of 19.5% after 21 days on the diet.

In the free-living or ambulatory setting, the time response to soluble fiber is slower, as depicted in Fig. 2. Hypercholesterolemic men were encouraged to maintain the same diets throughout the study (Anderson et al., 1988a). Three base-line measurements were averaged for the initial (0-week) value. Subjects used 3.4 g of psyllium three times daily and had the following serum cholesterol values expressed as the percentage below control values: 2 weeks, 8.6%; 4 weeks, 11.0%; 6 weeks, 10.6%, 7 weeks, 11.8%; and 8 weeks, 14.8%. The LDL-cholesterol followed the same time response with a 20.2% reduction at 8 weeks. In contrast to the metabolic ward study using oat bran, where the greatest reduction in serum cholesterol was seen in less than 2 weeks (Fig. 1), the

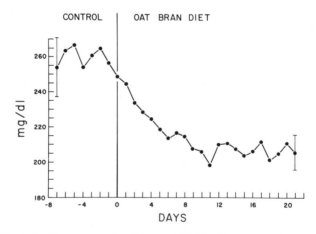

FIGURE 1. Effect of oat bran on serum cholesterol. Eight subjects were given 100 g/day of oat bran for 21 days on a metabolic research ward.

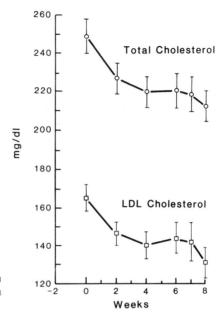

FIGURE 2. Effect of psyllium on total serum cholesterol and LDL-cholesterol. Thirteen men received 3.4 g of psyllium t.i.d. for 8 weeks.

greatest reduction in serum cholesterol in the ambulatory study using psyllium was at 8 weeks. These different response times may be related to the constant nature of the diet on the metabolic ward compared to the day-to-day variation in diet in free-living people.

2.8. Dose Responses to Soluble Fiber

A graded response of serum cholesterol to different doses of dietary fiber (a dose–response curve) is not established. Figure 3 illustrates the data from Table I plotted as hypocholesterolemic effect versus grams of fiber per day. Oat bran providing 7.2 g of dietary fiber produced a 19% reduction in serum cholesterol and was as effective as doses of guar and psyllium up to 25 g/day (Anderson and Tietyen-Clark, 1986). As summarized by Superko and colleagues (1988), a dose–response curve for guar is not established.

Reiser (1987), Kay and Truswell (1977), and Judd and Truswell (1985a) quote the studies of Palmer and Dixon (1966) to support a dose—response curve for pectin. Palmer and Dixon (1966) conducted a placebo-controlled, crossover study comparing the effects of five doses of pectin (2–10 g/day) administered in a random order. The reductions in serum cholesterol were small compared to those reported in Table I, with reductions ranging from 2.0% for 2 g/day to 5.5% for 10 g/day. None of the pectin doses produced significantly greater reductions than the cellulose placebo, which was accompanied by a 4.2% reduction. Although these investigators noted a graded response, the magnitude of changes was so small and statistically insignificant that this study, in our assessment, does not document a dose–response curve for serum cholesterol to pectin.

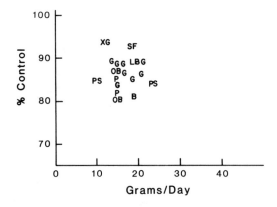

FIGURE 3. Dose effect of soluble fiber on serum cholesterol of humans. B, beans; G, guar; OB, oat bran; LB, locust bean gum; PE, pectin; PS, psyllium; SF, soy fiber; XG, xanthan gum.

3. ANIMAL STUDIES

Wells and Ershoff (1961) reported that dietary fiber decreases cholesterol concentrations in the serum and liver of rats. Subsequently, many studies have documented the cholesterol-lowering effects of different dietary fibers (see Table III). In general, water-soluble gums, pectins, and oat products containing oat gum exert significant cholesterol-lowering effects, whereas fiber sources providing predominantly water-insoluble fibers do not have significant cholesterol-lowering effects.

Table III compares dietary feeding studies using rats in which cellulose was included as a control fiber and cholesterol was fed to increase serum cholesterol. Test fibers and cellulose were both fed at 5–10% of diet. Most studies included cholesterol at 0.5% to 1.0% of diet except Mathe et al. (1977), who used cholesterol at only 0.02% to 0.05% of diet. About half of the experimental protocols (Kiriyama et al., 1969; Chen and Anderson, 1979a,b; Elliott et al., 1981; Jennings et al., 1988; Lopez-Guisa et al., 1988; Shinnick et al., 1988) also added cholic acid at 0.13% to 0.25% of diet to further increase serum and liver cholesterol and to provide a more sensitive protocol for examining the cholesterol-lowering effects of fiber.

Wheat bran feeding does not significantly affect serum or liver cholesterol, whereas oat bran feeding significantly decreases both serum and liver cholesterol. Oat hulls, rich in cellulose, do not significantly affect serum or liver cholesterol. Oat gum, the β-glucan extracted from oat bran, effectively lowers serum and liver cholesterol and appears as potent as any fiber that has been carefully tested. Agar and alginate do not have significant cholesterol-lowering effects. In this summary compiled from studies of several investigators, guar gum and pectin appear similar in potency to oat gum and significantly lower serum and liver cholesterol.

Figure 4 compares the serum and liver cholesterol responses to different fibers studied in our laboratory using the same protocol for cholesterol-fed rats. All diets were prepared by ICN Nutritional Biochemicals, Cleveland, Ohio, and provided approximately 46.5% sucrose, 16% starch, 15% casein, 6% cottonseed oil, 1% cholesterol, 0.2% cholic acid, 10% cellulose or test fiber, and other components as outlined elsewhere (Chen et al., 1981).

TABLE III. Influence of Plant Fibers on Serum and Liver Cholesterol in Rats

Source of fiber	No. of studies	Percentage change		References
		Serum chol.	Liver chol.	
Wheat bran	4	110.9	100.0	Elliott et al. (1981)
				Mathe et al. (1977)
				Chen and Anderson (1979b)
Oat bran	3	77.3[a]	51.7[a]	Chen and Anderson (1979a)
				Chen et al. (1981)
				Shinnick et al. (1988)
Oat hull	1	84.6	92.7	Lopez-Guisa et al. (1988)
Oat gum	2	68.4[a]	36.0[a]	Chen et al. (1981)
				Jennings et al. (1988)
Agar	2	97.8	136.6	Wells and Ershoff (1961)
				Tsai et al. (1976)
Alginate	2	87.7	78.2	Wells and Ershoff (1961)
				Kiriyama et al. (1969)
Guar gum	2	75.1[a]	46.6[a]	Chen and Anderson (1979a)
				Chen and Anderson (1979b)
Pectin	10	68.7[a]	36.2[a]	Wells and Ershoff (1961)
				Leveille and Sauberlich (1966)
				Kiriyama et al. (1969)
				Chang and Johnson (1976)
				Tsai et al. (1976)
				Mathe et al. (1977)
				Chen and Anderson (1979a)
				Chen et al. (1981)
				Elliott et al. (1981)
				Judd and Truswell (1985b)

[a]Values significantly lower than for cellulose control.

Over a 10-year period, we fed 140 rats the cellulose diet for 3 weeks in 18 different experiments; serum cholesterols were remarkably consistent, having mean values of 154 mg/dl ($\pm$28 mg/dl as standard deviation). Serum cholesterol values were similar with cellulose intakes of 4%, 6%, and 10% of diet (Fig. 4a). Wheat bran increased serum cholesterol slightly (Chen and Anderson, 1979b), corn bran did not significantly alter serum values (J. W. Anderson, unpublished data), and oat bran had significant hypocholesterolemic effects (Chen and Anderson, 1979a). Oat gum produced a dose-related decrease in serum cholesterol with concentrations ranging from 3% to 9% of diet (J. W. Anderson, unpublished results). Guar (Chen and Anderson, 1979a) and oat bran (Chen et al., 1981) had similar hypocholesterolemic effects. Chitosan and cholestyramine (Jennings et al., 1988) had significantly greater hypocholesterolemic effects than oat gum. Pectin, psyllium, guar, and oat gum had similar hypocholesterolemic effects in the cholesterol-fed rat model (Chen and Anderson, 1979a; Chen et al., 1981; J. W. Anderson, unpublished data).

Figure 4b compares the effects of these same fibers on liver cholesterol concentra-

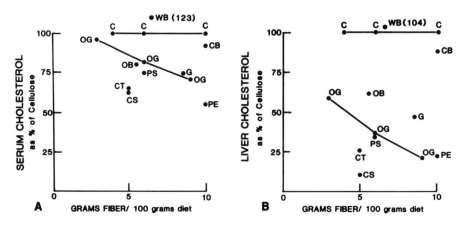

FIGURE 4. Comparison of dietary fibers and other products on serum (a) and liver (b) cholesterol concentrations of rats. Data summarized from Chen and Anderson (1979a,b), Chen et al. (1981), Jennings et al. (1988), and J. W. Anderson (unpublished data). C, cellulose; CT, chitosan; CS, cholestyramine; CB, corn bran; PE, pectin; G, guar; OG, oat gum; WB, wheat bran; PS, psyllium.

tions. For the same 140 rats fed cellulose diets, liver cholesterol averaged 42.6 mg/g (standard deviation ±7 mg/g). Liver cholesterols were similar with cellulose intakes of 4%, 6%, and 10% of diet. Wheat bran (Chen and Anderson, 1979b) and corn bran (J. W. Anderson, unpublished data) did not significantly alter liver cholesterol. Oat gum produced a dose-related decrease in liver cholesterol with intakes ranging from 3% to 9% of diet (J. W. Anderson, unpublished results). Oat bran and guar (Chen and Anderson, 1979a) lowered liver cholesterol less than oat gum (Chen et al., 1981), whereas pectin (Chen et al., 1981) and psyllium (J. W. Anderson, unpublished data) had similar effects. Chitosan and cholestyramine had more potent cholesterol-lowering effects than the soluble fibers studied. The cholesterol-lowering effects of pectin, psyllium, guar, and oat gum are consistently greater for liver than for serum in this cholesterol-fed rat model.

4. PHYSICOCHEMICAL QUALITIES AND HYPOCHOLESTEROLEMIC EFFECTS

Many of the physicochemical characteristics of fibers affect their hypocholesterolemic properties. Some of the important physicochemical attributes of dietary fibers are the following: particle size, preparation, and surface area; water-holding capacity; solubility features including viscosity and gelling attributes; binding properties including ion exchange and absorption of organic molecules; and degradability including fermentation by bacteria (Dreher, 1987).

Smaller particles may have greater bile-acid-binding capacity but increase fecal bulk less than larger particles (Dreher, 1987). Water-holding capacity may relate to viscosity and gelling characteristics but does not have a defined role in the hypocholesterolemic effects of dietary fiber.

The viscosity and gelling properties of soluble fibers may have important effects on the hydrolysis and absorption of lipids and the absorption of bile acids. Fibers that increase the viscosity of intestinal contents or that produce gels may decrease intestinal motility, thereby decreasing the mixing of nutrients, digestive enzymes, and other intestinal components required to interact for optimal micelle formation and absorption (Schneeman, 1987). Decreases in intestinal motility may decrease bile acid pool size (Hardison *et al.*, 1979).

The physicochemical properties of pectins are better characterized than most other fibers and correlate with their hypocholesterolemic potential. In general, the larger-molecular-weight pectins with more extensive esterification of uronic acid groups with methyl groups decrease serum cholesterol more than lower-molecular-weight, lower-methoxylated pectins (Judd and Truswell, 1985b).

The binding properties of dietary fibers may affect the absorption of cholesterol, fatty acids, and bile acids. Wheat bran and cabbage, for example, have high cation-exchange capacities, whereas psyllium and cellulose have low capacities (Dreher, 1987). Whereas dietary fibers with ion-exchange properties have cation-exchange properties, the therapeutic bile-acid-binding resins such as cholestyramine and colestipol are strong anion-exchange resins. Thus, it seems unlikely that the ion-exchange properties of dietary fibers have an important effect on bile acid binding and excretion. It seems more likely that the organic-binding characteristics and viscosity–gelling properties of fibers are important in altering the absorption and metabolism of bile acids (Selvandran, 1987).

Most dietary fibers are partially fermented by bacteria in the human colon, and some, such as pectin, are almost completely fermented. Fermentation products include hydrogen, carbon dioxide, methane, water, and short-chain fatty acids (SCFA) such as acetate, propionate, and butyrate. These SCFA are almost completely absorbed from the colon and may affect hepatic cholesterol synthesis (Chen *et al.*, 1984). In addition, the presentation of more dietary fiber to the colon stimulates the growth of more bacteria, leading to further conversion of primary to secondary bile acids.

5. GASTROINTESTINAL EVENTS AND HYPOCHOLESTEROLEMIC EFFECTS

Dietary fibers have major effects on gastrointestinal function, resulting in alterations in absorption of nutrients, altered hormonal responses, and production of SCFA. These events could act individually or collectively to decrease serum cholesterol. The effects of dietary fiber on different segments of the gastrointestinal tract can be examined to explore potential mechanisms by which dietary fiber may decrease serum cholesterol.

5.1. Mouth

High-fiber foods are less energy dense and require more chewing and longer to eat than low-fiber foods (Duncan *et al.*, 1983). Although not extensively studied, high-fiber foods appear to produce greater satiety between meals than low-fiber foods (Blundell and Burley, 1987). These events may lead to lower energy intake and slight reductions in body weight, which could produce small reductions in serum cholesterol.

5.2. Stomach

Soluble fibers tend to delay gastric emptying (Meyer and Doty, 1988), which may contribute to earlier fullness and satiation leading to decreased energy intake.

5.3. Small intestine

Through their viscosity and gelling properties, soluble fibers may delay or even decrease the digestion and absorption of nutrients (Edwards et al., 1987). These effects could be mediated by decreasing the mixing of intestinal contents, decreasing intestinal motility, and increasing the unstirred water layer, presenting a physical barrier to nutrient absorption. Certain fibers increase small intestinal transit, whereas others may slow small intestinal transit, providing less or more time for nutrient absorption. Dietary fibers may affect small intestinal cell function through morphological and cell proliferation effects. These various effects could independently affect cholesterol homeostasis or work in combination with other effects.

5.4. Large Intestine

Dietary fibers have well-documented effects on fecal weight, stool frequency, transit time, and intraluminal pressures. These effects may not be important in cholesterol metabolism, since wheat bran, one of the most active effectors of these events, does not lower serum cholesterol. More important, it seems that the proliferation of colonic bacteria and their metabolic processes and products such as conversion of primary to secondary bile acids and short-chain fatty acid production contribute substantially to the hypocholesterolemic effect of soluble fiber.

6. PROPOSED MECHANISMS FOR HYPOCHOLESTEROLEMIC EFFECTS

Although numerous investigators have documented the cholesterol-lowering effects of dietary fiber (Tables I–III), the mechanisms responsible for these effects remain unclear. Several hypotheses have emerged regarding the cholesterol-lowering mechanisms, including these: (1) fiber may alter bile acid absorption and metabolism; (2) fiber may modify lipid absorption and metabolism; (3) short-chain fatty acids (SCFA) resulting from fiber fermentation may affect cholesterol or lipoprotein metabolism; and (4) fiber may change insulin or other hormone concentrations or tissue sensitivity to hormones. Evidence related to these proposed mechanisms is here reviewed.

6.1. Altered Bile Acid Absorption and Metabolism

One of the earlier hypotheses concerning the hypocholesterolemic effects of dietary fiber suggested that fiber bound bile acids and increased their fecal excretion. Thus, the need increased for cholesterol flux to bile acid synthetic pathways, and there was less cholesterol available for lipoprotein synthetic pathways (Story and Lord, 1987). Eastwood and Hamilton (1968) noted that bile acids were strongly bound to vegetable fibers; subsequent investigators have confirmed and extended these observations (Kritchevsky and Story, 1974; Eastwood and Mowbray, 1976; Story and Kritchevsky, 1976; Vahouny et al., 1980). There is no consistent relationship between the capacity of certain fibers to bind bile acids and their hypocholesterolemic effects. Lignin is one of the most potent bile-acid-binding fibers (Story, 1985) but does not have hypocholesterolemic effects (Hillman et al., 1985). Pectin has weak in vitro bile acid binding but is one of the most potent hypocholesterolemic fibers (Tables I and III). However, Vahouny (1985) concluded that there was, in general, a reasonably good relationship between the in vitro bile acid binding by various fibers and their hypocholesterolemic effects.

Dietary fibers increase fecal bile acid excretion by several mechanisms: binding bile acids, forming gels or highly viscous solutions in the intestine, and interfering with micelle formation. Pectins and gums may affect bile acid absorption in the small intestine through their viscosity and gelling properties, leading to decreased micelle formation and entrapment of bile acids and lipids to a greater extent than through their physical binding properties (Selvendran, 1987).

The effects of different fibers on fecal bile acid excretion have been extensively studied (Pilch, 1987). In general, fibers that have significant hypocholesterolemic effects in humans also significantly increase fecal bile acid excretion; fibers that do not have hypocholesterolemic effects in humans usually do not increase fecal bile acid excretion significantly. However, a number of exceptions to these generalizations are noted (Ross et al., 1983; Anderson et al., 1984a). Table IV compares the effects of three fiber sources on serum cholesterol and fecal excretion of bile acids and neutral sterols. Oat bran significantly lowers serum cholesterol and significantly increases bile acid excretion. Beans significantly decrease serum cholesterol and slightly decrease bile acid excretion. Wheat bran affects neither serum cholesterol nor fecal bile acid excretion significantly.

TABLE IV. Comparison of Oat Bran, Beans, and Wheat Bran on Fecal Sterols in Humans (% Change)

	Serum cholesterol	Fecal bile acids	Fecal neutral sterols	References
Oat bran	−19.3	+59.5	+15.3	Kirby et al. (1981)
				Anderson et al. (1984a)
Beans	−18.8	−29.9	+1.8	Anderson et al. (1984a)
Wheat bran	0	+6.8	+7.7	Eastwood et al. (1973)
				Walters et al. (1975)
				Jenkins et al. (1975)
				Cummings et al. (1976)
				Kay and Truswell (1977)
				Tarpila et al. (1978)

Although dietary fibers with hypocholesterolemic effects have actions that resemble those of bile-acid-binding resins, the resins increase bile acid excretion to a much greater extent (Stanley *et al.*, 1973; Miettinen, 1978). Some fiber sources such as beans (Table IV) have cholesterol-lowering effects without increasing fecal bile acid excretion while others produce only small increases in fecal bile acid excretion. These different observations suggest that other mechanisms may contribute to the hypocholesterolemic effects of dietary fibers (Kay and Truswell, 1980).

Alterations in the ratio of primary to secondary bile acids may also contribute to the effects of dietary fibers on serum cholesterol (Story, 1985). In rats, several dietary fibers increase the fecal excretion of chenodeoxycholic acid and its derivatives and decrease the fecal excretion of cholic acid and its derivatives (Story, 1985). In humans, oat bran also increases the fecal excretion of chenodeoxycholic acid (Kirby *et al.*, 1981). Since chenodeoxycholic acid and its derivatives are less well absorbed than cholic acid and its derivatives, and since chenodeoxycholic acid may alter hepatic cholesterol synthesis, changes in the ratio of chenodeoxycholic to cholic acid could affect hepatic cholesterol metabolism and serum cholesterol (LaRusso *et al.*, 1977; Ahlberg *et al.*, 1977; Story, 1985).

Quantitative measures of the effects of fibers on bile acid secretion rates are not available. Hillman and colleagues (1986) compared the effects of pectin, lignin, and cellulose supplements on biliary excretion of [^{14}C]deoxycholic acid in humans and found that pectin intake increased the biliary secretion of [^{14}C]deoxycholate metabolites of taurocholic acid, whereas cellulose produced smaller changes, and lignin had no effect.

Thus, most investigators feel that increased fecal bile acid excretion and alterations in bile acid metabolism may lead to the hypocholesterolemic effects of some dietary fibers such as oat gum and pectin, but other mechanisms probably contribute as well. Most investigators feel that further studies are required to quantitate the contribution that altered bile acid excretion and metabolism make to hypocholesterolemic effects.

6.2. Modified Lipid Absorption and Metabolism

6.2.1. Proposed Mechanisms

Dietary fibers may affect the absorption of lipids as well as bile acids from the small intestine by these and other mechanisms: (1) altered gastric function; (2) decreased availability of bile acids; (3) interference with effective micelle formation; (4) altered digestive enzyme availability or activity; (5) changes in intestinal motility and mixing of intestinal contents; (6) alterations in the unstirred water layer or other physiological changes in the absorptive surface; (7) alterations in the morphology and function of intestinal absorptive cells; and (8) changes in flow through capillaries or lymphatics.

6.2.2. Animal Studies

In the rat soluble fibers delay gastric emptying and slow the presentation of nutrients to the digestive enzymes and absorptive surfaces of the small intestine (Brown *et al.*, 1988). Viscous fibers such as guar decrease the grinding and/or digestive action of the stomach, allowing large, poorly digested pieces of meat to appear in the midintestine of the dog and leading to marked slowing of lipid digestion (Meyer and Doty, 1988).

As outlined above and elsewhere, many different dietary fibers bind bile acids and decrease their availability for optimal fat digestion and absorption (Story, 1985). Vahouny and colleagues extensively studied the effects of different fibers on micelle formation *in vitro* and correlated this with fat absorption (Vahouny *et al.*, 1980, 1981). These studies indicate that viscous fibers sequester lipid and bile acid components of intestinal contents and decrease lipid and cholesterol digestion in the upper small intestine.

Dietary fibers may decrease the activities of digestive enzymes by several mechanisms such as viscosity effects, pH effects, absorption of enzymes, and other fiber–enzyme interactions (Isaksson *et al.*, 1982; Schneeman and Gallaher, 1985). These mechanisms could also contribute to decreased digestion of lipids in the upper small intestine (Gagne and Acton, 1983; Schneeman and Gallaher, 1985; Meyer and Doty, 1988).

Recent studies indicate that decreased motility of the small intestine may make a major contribution to decreased digestion and absorption of nutrients (Brown *et al.*, 1988). Whether the decreased motility or changes in the thickness of the unstirred water layer (Elsenhans *et al.*, 1980) are more important in decreasing nutrient absorption is unclear. Morphological (Cassidy *et al.*, 1982) and functional changes in absorptive mucosal cells also could contribute to decreased absorption, but these changes are not well characterized. The effects of dietary fiber on flow through intestinal capillaries and lymphatics have not been explored.

These animal studies suggest that dietary fiber, acting through several different mechanisms in the small intestine and the colon, increases fecal excretion of bile acids and slows lipid digestion with absorption at more distal than usual sites in the small intestine (Imaizumi *et al.*, 1982; Meyer and Doty, 1988). Most of these mechanisms appear to operate in humans but are less well characterized.

6.2.3. Human Studies

Early clinical studies documented that dietary fiber intake increased fecal fat excretion by about 2–4 g/day (Baird *et al.*, 1977; Kay and Truswell, 1977; Cummings *et al.*, 1979). It seems unlikely that fat losses of this small magnitude would decrease serum cholesterol by 15–20%. More likely, fibers affect sites and rates of absorption of cholesterol, resulting, perhaps, in gut secretion of lipoprotein particles of different sizes and different compositions (Bosello *et al.*, 1984; Judd and Truswell, 1985a; Redard *et al.*, 1988).

Soluble fibers significantly slow gastric emptying in humans (Holt *et al.*, 1979; Ray *et al.*, 1983; Blackburn *et al.*, 1984) and either increase the unstirred water layer or produce hypomotility to slow the digestion and absorption of nutrients (Blackburn *et al.*, 1984).

Preliminary studies suggest that soluble-fiber-rich meals containing guar alter the absorption rate of fats and lead to a delayed chylomicron triglyceride peak when compared to control meals (Judd and Truswell, 1985a). Much further work is required in human subjects before these hypotheses can be validated.

6.3. Effects of SCFA on Lipid Metabolism

In the human colon, polysaccharides such as dietary fibers are fermented by a rich and variegated bacterial population. Short-chain fatty acids (SCFA) including acetate, propionate, and butyrate are major end products of this process (Cummings *et al.,* 1987). We proposed that these SCFA might affect hepatic cholesterol metabolism leading to hypocholesterolemic effects (Anderson and Chen, 1979). Preliminary studies in rats support this hypothesis, but much more data are required.

6.3.1. Animal Studies

In rats, oat bran feeding, which has potent hypocholesterolemic effects, is accompanied by significantly higher portal vein concentrations of SCFA than is cellulose feeding in rats (Illman and Topping, 1985; Chen and Anderson, 1986). Propionate fed to rats leads to significant reductions in serum cholesterol (Chen *et al.,* 1984; Illman *et al.,* 1988). *In vitro* studies using isolated rat liver cells (Anderson and Bridges, 1984) suggest that physiological concentrations of propionate may significantly attenuate hepatic cholesterol synthesis.

Although pharmacological doses of propionate and its derivatives inhibit important hepatic lipid regulatory enzymes both *in vitro* and *in vivo* (Bush and Milligan, 1971; Hall *et al.,* 1984), effects of physiological doses are not well documented. Since propionate levels as high as 0.1 mM in the portal vein of oat-bran-fed rats are reported (Illman and Topping, 1985; Chen and Anderson, 1986), we investigated the effects of doses in the range 0.1–25 mM on cholesterol metabolism in the isolated rat hepatocyte. Cholesterol synthesis was estimated by measuring incorporation of tracer amounts of [^{14}C]acetate (Calandra *et al.,* 1979) or [^{3}H]water (Lakshmanan *et al.,* 1977).

Figure 5 illustrates the effects of increasing concentrations of propionate on cholesterol synthesis from acetate using tracer amounts of [^{14}C] acetate. Propionate concentrations as low as 0.25 mM significantly inhibited cholesterol synthesis using the [^{14}C]acetate tracer. Propionate concentrations of 1.0 mM produced a 16% reduction in

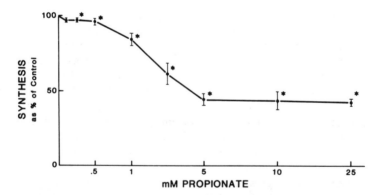

FIGURE 5. Effect of propionate on *de novo* cholesterol synthesis from [^{14}C]acetate (5 mM) in isolated rat hepatocytes. Values are the mean ± S.E.M. of 3–9 rats. *Significantly differs from control ($P < 0.05$).

cholesterol synthesis. When [³H]water was used as tracer, propionate concentrations of 0.5 mM and above produced significant inhibition.

Illman and colleagues have conducted a series of experiments and concluded that propionate does not make a physiological contribution to the hypocholesterolemic effects of oat bran or pectin (Illman *et al.*, 1982, 1988; Illman and Topping, 1985). Ahrens *et al.* (1986) reported that oral pectin feeding significantly lowered serum cholesterol in minipigs and that intracecal administration of pectin did not affect serum cholesterol, suggesting that the small intestinal effects predominated and that cecal SCFA production, without the small intestinal effects, did not lower serum cholesterol. Komai and Kimura (1987) noted that feeding larger doses of pectin (20% of diet) decreased serum cholesterol in both conventional and germ-free mice, suggesting that pectin fermentation is not required for the hypocholesterolemic effects of fiber.

Further studies are required to resolve the controversy regarding the role of propionate in the regulation of serum cholesterol. The peak levels of propionate that are achieved in the portal vein of oat-bran-fed animals must be determined. Differences between our observations of propionate effects on cholesterol metabolism with isolated hepatocytes and those of Illman *et al.* (1988) using perfused rat livers need to be carefully studied. Additional studies are required to uncover mechanisms and potential sites of enzyme inhibition associated with propionate use.

6.3.2. Human Studies

Many studies document that increased dietary fiber intake leads to increased fecal excretion of SCFA (Williams and Olmsted, 1936; Cummings *et al.*, 1976; Spiller *et al*, 1980; Ehle *et al.*, 1982; Fleming and Rodriguez, 1983). Presumably increased intake of dietary fiber also increases portal vein concentrations of SCFA, but this has not been documented. Pomare and colleagues (1985) studied the changes in peripheral venous acetate concentrations in response to lactulose and pectin and reported the time curves. They found fasting venous blood acetate levels to be 53.8 ± 4.4 μmole/liter. After a 20-g dose of lactulose, blood acetate levels increased to 181.3 ± 23.9 μmole/liter in 2–4 hr. Following a 20-g dose of pectin, blood acetate levels started to rise after 6 hr and remained elevated at about twice fasting levels for 18 hr.

6.4. Changed Hormone Concentrations

Increased intake of dietary fiber affects serum insulin values and levels of several other important regulatory hormones, contributing, perhaps, to the regulation of lipid metabolism. Insulin is a key regulatory hormone both for lipid synthesis in many body tissues and for lipid mobilization from adipose tissue and liver. Glucagon antagonizes many of these effects on lipid metabolism. Several studies suggest that serum glucose, insulin, and glucagon concentrations affect rates of hepatic cholesterol and lipoprotein synthesis, particularly the secretion of very-low-density lipoproteins (Bhathena *et al.*, 1974; Reaven and Bernstein, 1978).

High-carbohydrate, high-fiber diets compared to conventional diets lower insulin requirements of diabetic individuals (Anderson and Ward, 1979) and decrease serum insulin concentrations of normal subjects (Fukagawa *et al.*, 1984). These high-fiber diets also increase insulin sensitivity of diabetic individuals (Tagliaferro et al., 1985; Anderson et al., 1988b) and normal subjects (Munoz et al., 1979; Fukagawa et al., 1984). These reductions in ambient insulin concentrations would be expected to reduce hepatic cholesterol and fatty acid synthesis (Jenkins et al., 1983).

Dietary fiber intake also decreases serum glucagon concentrations in diabetic individuals (Miranda and Horwitz, 1978) and normal individuals (Munoz et al., 1979). The effects of fiber intake on other pancreatic and gastrointestinal hormones are summarized elsewhere (Anderson and Sieling, 1985; Anderson, 1982).

Thus, dietary fiber produces significant changes in serum levels of insulin, glucagon, and other pancreatic and gastrointestinal hormones as well as in peripheral insulin sensitivity. Because of the important role these hormones play in regulation of cholesterol and fatty acid metabolism, some of the fiber effects may be mediated by these hormonal alterations.

7. CONCLUSIONS

Soluble fibers, constituting about 30% of a typical Western diet, have important hypocholesterolemic effects in humans and experimental animals. Serum cholesterol decreases 10–20% when increased amounts of soluble fibers are ingested from either soluble-fiber-rich foods, such as oat bran and beans, or concentrated forms of soluble fibers, such as guar gum, pectins, and psyllium. Oat products have the capability of decreasing serum LDL-cholesterol without affecting or even slightly increasing HDL-cholesterol in humans. Soluble fibers act rapidly to lower serum cholesterol, with maximal responses seen in about 11 days in metabolic ward studies and near-maximal responses seen in about 4 weeks in ambulatory studies. Although the response of serum cholesterol to doses of different fibers is easily demonstrable in a cholesterol-fed rat model, no dose response between soluble fiber and serum cholesterol is established in humans.

Proposed mechanisms by which soluble fibers lower serum cholesterol are still under intense study. Multiple mechanisms may interact to produce the cholesterol-lowering effects, and each fiber may work through various effectors differently. Available data support, in part, all of the currently proposed mechanisms: (1) most soluble fibers increase fecal excretion of bile acids and alter the percentages of various primary and secondary bile acids excreted by the liver; (2) most soluble fibers decrease the absorption of lipids and cholesterol in the proximal intestine and increase their absorption in the midintestine, leading, presumably, to alterations in the size and composition of lipoproteins secreted by the intestine; (3) most soluble fibers are fermented in the colon to yield short-chain fatty acids, which are carried to the liver and have the potential to affect cholesterol and lipoprotein synthesis and secretion; and (4) dietary fibers tend to decrease serum insulin concentrations and increase peripheral insulin sensitivity while altering other pancreatic and gastrointestinal hormone levels, possibly affecting lipid synthesis and metabolism.

Although much more study is required to elucidate the mechanisms by which dietary fibers lower serum cholesterol, the clinical effect is an important one. Because of the high prevalence of hypercholesterolemia among Western people, and because of the relationship between hypercholesterolemia and risk for coronary heart disease, soluble fibers may have practical utility in the management of hypercholesterolemia for selected individuals.

REFERENCES

Abraham, Z. D., and Mehta, T., 1988, Three-week psyllium-husk supplementation: Effects on plasma cholesterol concentrations, fecal steroid excretion and carbohydrate absorption in men, *Am. J. Clin. Nutr.* **47**:67–74.

Ahlberg, J., Angelin, B., Einarsson, K., Hellstrom, K., and Leijd, B. C., 1977, Influence of deoxycholic acid on biliary lipids in man, *Clin. Sci. Mol. Med.* **53**:249–256.

Ahrens, F., Hagemeister, H., Pfeuffer, M., and Barth, C. A., 1986, Effects of oral and intracecal pectin administration on blood lipids in minipigs, *J. Nutr.* **116**:70–76.

Anderson, J. W., 1982, Dietary fiber, in: *Dietary Fiber in Health and Disease* (G. V. Vahouny and D. Kritchevsky, eds.), Plenum Press, New York, pp. 151–167.

Anderson, J. W., and Bridges, S. R., 1984, Short-chain fatty acid fermentation products of plant fiber affect glucose metabolism of isolated rat hepatocytes, *Proc. Soc. Exp. Biol. Med.* **177**:372–376.

Anderson, J. W., and Bridges, S. R., 1988, Dietary fiber content of selected foods, *Am. J. Clin. Nutr.* **47**:440–447.

Anderson, J. W., and Chen, W. L., 1979, Plant fiber: Carbohydrate and lipid metabolism, *Am. J. Clin. Nutr.* **32**:346–363.

Anderson, J. W., and Gustafson, N., 1988, Hypocholesterolemic effects of oat and bean products, *Am. J. Clin. Nutr.* **48**:749–753.

Anderson, J. W., and Sieling, B., 1985, Nutrition and diabetes, *Nutr. Update* **2**:49–69.

Anderson, J. W., and Tietyen-Clark, J., 1986, Dietary fiber: Hyperlipidemia, hypertension, and coronary heart disease, *Am. J. Gastroenterol.* **81**:907–919.

Anderson, J. W., and Ward, K., 1979, High-carbohydrate, high-fiber diets for insulin-treated men with diabetes mellitus, *Am. J. Clin. Nutr.* **32**:2312–2321.

Anderson, J. W., Story, L., Sieling, B., Chen, W.-J. L., Petro, M. S., and Story, J., 1984a, Hypocholesterolemic effects of oat-bran and bean intake for hypercholesterolemic men, *Am. J. Clin. Nutr.* **40**:1146–1155.

Anderson, J. W., Story, L., Sieling, B., and Chen, W.-J. L., 1984b, Hypocholesterolemic effects of high-fibre diets rich in water-soluble plant fibres, *J. Can. Dietet. Assoc.* **45**:140–149.

Anderson, J. W.,, Zettwoch, N., Feldman, T., Tietyen-Clark, J., Oeltgen, P., and Bishop, C. W., 1988a, Cholesterol-lowering effects of psyllium hydrophilic mucilloid for hypercholesterolemic men, *Arch. Intern. Med.* **148**:292–296.

Anderson, J. W., Zeigler, J., and Deakins, D., 1988b, High carbohydrate, high fiber diets increase insulin sensitivity and carbohydrate disposal in type I diabetic individuals, *Diabetes* **37**(Suppl. 1): 11A.

Anderson, J. W., Bridges, S. R., Tietyen, J., and Gustafson, N., 1989, Dietary fiber content of a simulated American diet and selected research diets, *Am. J. Clin. Nutr.* **49**:353–357.

Aro, A., Uusitupa, M., Voutilainen, E., Hersio, K., Korhonen, T., and Siitonen, O., 1981, Improved diabetic control and hypocholesterolaemic effect induced by long-term dietary supplementation with guar gum in type 2 (insulin-dependent) diabetes, *Diabetologia* **21**:29–33.

Aro, A., Uusitupa, M., Voutilainen, E., and Korhonen, T., 1984, Effects of guar gum in male subjects with hypercholesterolemia, *Am. J. Clin. Nutr.* **39**:911–916.

Baird, I. M., Walters, R. L., Davies, P. S., Hill, M. J., Drasar, B. S., and Southgate, D. A. T., 1977, The effects of two dietary fiber supplements on gastrointestinal transit, stool weight, and frequency and bacterial flora, and fecal bile acids in normal subjects, *Metabolism* **26**:117–128.

Bhathena, S. J., Avigan, J., and Schreiner, M. E., 1974, Effect of insulin on sterol and fatty acid synthesis and hydroxmethylglutaryl CoA reductase activity in mammalian cells grown in culture, *Proc. Natl. Acad. Sci. U.S.A.* **71:**2174–2178.

Blackburn, N. A., Redfern, J. S., Jarjis, H., Holgate, A. M., Hanning, I., Scarpello, J. H. B., Johnson, I. T., and Read, N. W., 1984, The mechanism of action of guar gum in improving glucose tolerance in man, *Clin. Sci.* **66:**329–336.

Blundell, J. E., and Burley, V. J., 1987, Satiation, satiety, and the action of fibre on food intake, *Int. J. Obesity* **11**(Suppl. 1):9–25.

Bosello, O., Cominacini, L., Zocca, I., Garbin, U., Ferrari, F., and Davoli, A., 1984, Effects of guar gum on plasma lipoproteins and apolipoproteins C-II and C-III in patients affected by familial combined hyperlipoproteinemia, *Am. J. Clin. Nutr.* **40:**1165–1174.

Brown, N. J., Worlding, J., Rumsey, R. D. E., and Read, N. W., 1988, The effect of guar gum on the distribution of a radiolabelled meal in the gastrointestinal tract of the rat, *Br. J. Nutr.* **59:**223–231.

Burton, R., and Manninen, V., 1982, Influence of a psyllium-based fibre preparation on faecal and serum parameters, *Acta Med. Scand. [Suppl.]* **668:**91–94.

Bush, R. S., and Milligan, L. P., 1971, Study of inhibition of ketogenesis by propionate in bovine liver, *J. Anim. Sci.* **51:**121–127.

Calandra, S., Tarugi, P., Battistini, N., and Ferrari, R., 1979, Cholesterol synthesis in isolated rat hepatocytes: Effect of homologous and heterologous serum lipoproteins, *Metabolism* **28:**843–850.

Cassidy, M. M., Lightfoot, F. G., and Vahouny, G. V., 1982, Dietary fiber, bile acids, and intestinal morphology, in: *Dietary Fiber in Health and Disease* (G. V. Vahouny and D. Kritchevsky, eds.), Plenum Press, New York, pp. 53–71.

Challen, A. D., Branch, W. J., and Cummings, J. H., 1983, The effect of pectin and wheat bran on platelet function and haemostasis in man, *Hum. Nutr. Clin. Nutr.* **37C:**209–217.

Chang, M. W., and Johnson, M. A., 1976, Influence of fat level and type of carbohydrate on the capacity of pectin in lowering serum and liver lipids of young rats, *J. Nutr.* **106:**1562–1568.

Chen, W.-J. L., and Anderson, J. W., 1979a, Effects of plant fiber in decreasing plasma total cholesterol and increasing high-density lipoprotein cholesterol, *Proc. Soc. Exp. Biol. Med.* **162:**310–313.

Chen, W. L., and Anderson, J. W., 1979b, Effects of guar gum and wheat bran on lipid metabolism of rats, *J. Nutr.* **109:**1028–1934.

Chen, W.-J. L., and Anderson, J. W., 1986, Hypocholesterolemic effects of soluble fiber, in: *Dietary Fiber: Basic and Clinical Aspects* (G. V. Vahouny and D. Kritchevsky, eds.), Plenum Press, New York, pp. 275–286.

Chen, W.-J. L., Anderson, J. W., and Gould, M. R., 1981, Effects of oat bran, oat gum, and pectin on lipid metabolism of cholesterol fed rats, *Nutr. Rep. Int.* **24:**1093–1098.

Chen, W.-J. L., Anderson, J. W., and Jennings, D., 1984, Propianate may mediate the hypocholesterolemic effects of certain soluble plant fiber in cholesterol-fed rats, *Proc. Soc. Exp. Biol. Med.* **175:**215–218.

Cummings, J. H., Hill, M. J., Jenkins, D. J. A., Pearson, J. R., and Wiggins, H. S., 1976, Changes in fecal composition and colonic function due to cereal fiber, *Am. J. Clin. Nutr.* **29:**1468–1473.

Cummings, J. H., Southgate, D. A. T., Branch, W. J., Wiggins, H. S., Houston, H., Jenkins, D. J. A., Jivraj, T., and Hill, M. J., 1979, The digestion of pectin in the human gut and its effect on calcium absorption and large bowel function, *Br. J. Nutr.* **41:**477–485.

Cummings, J. H., Pomare, E. W., Branch, W. J., Naaylor, C. P. E., and MacFarlane, G. T., 1987, Short-chain fatty acids in human large intestine, portal, hepatic, and venous blood, *Gut* **28:**1221–1227.

DeGroot, A. P., Luyken, R., and Pikaar, N. A., 1963, Cholesterol-lowering effect of rolled oats, *Lancet* **2:**303–304.

Dreher, M. L., 1987, *Handbook of Dietary Fiber,* Marcel Dekker, New York.

Duncan, K. H., Bacon, J. A., and Weinsier, R. L., 1983, The effects of high and low energy density diets on satiety, energy intake, and eating time of obese and nonobese subjects, *Am. J. Clin. Nutr.* **37:**763–767.

Eastwood, M. A., and Hamilton, D., 1968, Studies of the adsorption of bile salts to non-absorbed components of diet, *Biochim. Biophys. Acta* **152:**165–173.

Eastwood, M. A., and Mowbray, L., 1976, The binding of the components of mixed micelles to dietary fiber, *Am. J. Clin. Nutr.* **29**:1461–1467.

Eastwood, M. A., Kirkpatrick, J. R., Mitchell, W. D., Bone, A., and Hamilton, T., 1973, Effects of dietary supplements of wheat bran and cellulose on faeces and bowel function, *Br. Med. J.* **4**:392–394.

Edwards, C. A., Blackburn, N. A., Craigen, L., Davison, P., Tomlin, J., Sugden, K., Johnson, I. T., and Read, N. W., 1987, Viscosity of food gums determined *in vitro* related to their hypoglycemic actions, *Am. J. Clin. Nutr.* **46**:72–77.

Ehle, F. R., Robertson, J. B., and Van Soest, P. J., 1982, Influence of dietary fibers on fermentation in the human large intestine, *J. Nutr.* **112**:158–166.

Elliott, J., Mulvihill, E., Duncan, C., Forsythe, R., and Kritchevsky, D., 1981, Effects of tomato pomace and mixed-vegetable pomace on serum and liver cholesterol in rats, *J. Nutr.* **111**:2203–2211.

Ellis, P. R., Kamalanathan, T., Dawoud, F. M., Strange, R. N., and Coultate, T. P., 1988, Evaluation of guar biscuits for use in the management of diabetes: Tests of physiological effects and palatability in non-diabetic volunteers, *Eur. J. Clin. Nutr.* **42**:425–435.

Elsenhans, B., Sufke, U., Blume, R., and Caspary, W. F., 1980, The influence of carbohydrate gelling agents on rat intestinal transport of monosaccharides and neutral amino acids *in vitro,* Clin Sci. **59**:373–380.

Fleming, S. E., and Rodriguez, M. A., 1983, Influence of dietary fiber on fecal excretion of volatile fatty acids by human adults, *J. Nutr.* **113**:1613–1625.

Fukagawa, N. K., Minaker, K. L., Hageman, G., Young, V. R., and Anderson, J. W., 1984, High-carbohydrate, high-fiber diets increase peripheral insulin sensitivity of healthy young men, *Clin Res.* **32**:796A.

Gagne, C. M., and Acton, J. C., 1983, Fiber constituents and fibrous food residues effects on the *in vitro* enzymatic digestion of protein, *J. Food Sci.* **48**:734–738.

Garvin, J. E., Forman, D. T., Eiseman, W. R., and Phillips, C. R., 1965, Lowering of human serum cholesterol by an oral hydrophilic colloid, *Proc. Soc. Exp. Biol. Med.* **120**:744–746.

Gatti, E., Catenazzo, G., Camisasca, E., Torri, A., Denegri, E., and Sirtori, C. R., 1984, Effects of guar-enriched pasta in the treatment of diabetes and hyperlipidemia, *Ann. Nutr. Metab.* **28**:1–10.

Ginter, E., Kubec, F. J., Vozar, J., and Bobek, P., 1979, Natural hypocholesterolemic agent: Pectin plus ascorbic acid, *Int. J. Vitam. Nutr. Res.* **49**:406–412.

Groen, J. J., Tijong, K. B., Koster, M., Willebrands, A. F., Verdonck, G., and Pierloot, M., 1962, The influence of nutrition and ways of life on blood cholesterol and the prevalence of hypertension and coronary heart disease among Trappist and Benedictine monks, *Am. J. Clin. Nutr.* **10**:456–470.

Hall, I. H., Chapman, J. M., Voorstad, P. J., and Cocolas, G. H., 1984, Hypolipidemic activity of 3-N-(naphthalimido) propionic acid in rodents, *J. Pharm. Sci.* **73**:956–961.

Hardison, G. M., Tomaszewski, N., and Grundy, S. M., 1979, Effect of acute alterations in small bowel transit time upon the biliary excretion rate of bile acids, *Gastroenterology* **76**:568–574.

Hillman, L. C., Peters, S. G., Fisher, C. A., and Pomare, E. W., 1985, The effects of the fiber components pectin, cellulose, and lignin on serum cholesterol levels, *Am. J. Clin. Nutr.* **42**:207–213.

Hillman, L. C., Peters, S. G., Fisher, C. A., and Pomare, E. W., 1986, Effects of the fibre components pectin, cellulose, and lignin on bile salt metabolism and biliary lipid composition in man, *Gut* **27**:29–36.

Holt, S., Heading, R. C., Carter, D. C., Prescott, L. F., and Tothill, P., 1979, Effect of gel fibre on gastric emptying and absorption of glucose and paracetamol, *Lancet* **1**:636–639.

Illman, R. J., and Topping, D. L., 1985, Effects of dietary oat bran on faecal steroid excretion, plasma volatile fatty acids and lipid synthesis in rats, *Nutr. Res.* **5**:839–846.

Illman, R. J., Trimble, R. P., Snoswell, A. M., and Topping, D. L., 1982, Daily variations in the concentrations of volatile fatty acids in the splanchnic blood vessels of rats fed diets high in pectin and bran, *Nutr. Rep. Int.* **26**:439–447.

Illman, R. J., Topping, D. L., McIntosh, G. H., Trimble, R. P., Storer, G. B., Taylor, M. N., and Cheng

B.-Q., 1988, Hypocholesterolaemic effects of dietary propionate: Studies in whole animals and perfused rat liver, *Ann. Nutr. Metab.* **32**:97–107.

Imaizumi, K., Tominaga, A., Mawatari, K., and Sugano, M., 1982, Effect of cellulose and guar gum on the secretion of mesenteric lymph chylomicrons in meal fed rats, *Nutr. Rep. Int.* **26**:263–269.

Isaksson, G., Lundquist, I., and Ihse, I., 1982, Effect of dietary fiber on pancreatic enzyme activity *in vitro*, *Gastroenterology* **82**:918–924.

Jenkins, D. J. A., Leeds, A. R., Newton, C., and Cummings, J. H., 1975, Effect of pectin, guar gum, and wheat fibre on serum cholesterol, *Lancet* **1**:1116–1117.

Jenkins, D. J. A., Reynolds, D., Leeds, A. R., Waller, A. R., and Cummings, J. H., 1979, Hypocholesterolemic action of dietary fiber unrelated to fecal bulking effect, *Am. J. Clin. Nutr.* **32**:2430–2435.

Jenkins, D. J. A., Wolever, T. M. S., Taylor, R. H., Reynolds, R., and Hockaday, T. D. R., 1980a, Diabetic glucose control, lipids, and trace elements on long term guar, *Br. Med. J.* **280**:1353–1354.

Jenkins, D. J. A., Reynolds, D., Slavin, B., Leeds, A. R., Jenkins, A. L., and Jepson, E. M., 1980b, Dietary fiber and blood lipids: Treatment of hypercholesterolemia with guar crispbread, *Am. J. Clin. Nutr.* **33**:575–581.

Jenkins, D. J. A., Wong, G. S., Patten, R., Bird, J., Hall, M., Buckley, G. C., McGuire, V., Reichert, R., and Little, J. A., 1983, Leguminous seeds in the dietary management of hyperlipidemia, *Am. J. Clin. Nutr.* **38**:567–573.

Jennings, C. D., Boleyn, K., Bridges, S. R., Wood, P., and Anderson, J. W., 1988, A comparison of the lipid-lowering and intestinal morphological effects of cholestyramine, chitosan, and oat gum in rats, *Proc. Soc. Exp. Biol. Med.* **189**:13–20.

Judd, P. A., and Truswell, A. S., 1981, The effect of rolled oats on blood lipids and fecal steroid excretion in man, *Am. J. Clin. Nutr.* **34**:2061–2067.

Judd, P. A., and Truswell, A. S., 1982, Comparison of the effects of high- and low-methoxyl pectins on blood and faecal lipids in man, *Br. J. Nutr.* **48**:451–458.

Judd, P. A., and Truswell, A. S., 1985a, Dietary fibre and blood lipids in man, in: *Dietary Fibre Perspectives: Reviews and Bibliography* (A. R. Leeds, ed.), John Libbey, London, pp. 23–39.

Judd, P. A., and Truswell, A. S., 1985b, The hypocholesterolaemic effects of pectins in rats, *Br. J. Nutr.* **53**:409–442.

Kay, R. M., and Truswell, A. S., 1977, Effect of citrus pectin on blood lipids and fecal steroid excretion in man, *Am. J. Clin. Nutr.* **30**:171–175.

Kay, R. M., and Truswell, A. S., 1980, Dietary fiber: Effects on plasma and biliary lipids in man, in: *Medical Aspects of Dietary Fiber* (G. A. Spillar, ed.), Plenum Press, New York, pp. 153–173.

Kennedy, J. F., Sandhu, J. S., and Southgate, D. A. T., 1979, Structural data for the carbohydrate of ispaghula husk, *Carbohyr. Res.* **75**:265–274.

Keys, A., Grande, F., and Anderson, J. T., 1961, Fiber and pectin in the diet and serum cholesterol concentrations in man, *Proc. Soc. Exp. Biol. Med.* **106**:555–558.

Kinosian, B. P., and Eisenberg, J. M., 1988, Cutting into cholesterol: Cost-effective alternatives for treating hypercholesterolemia, *J.A.M.A.* **259**:2249–2254.

Kirby, R. W., Anderson, J. W., Sieling, B., Rees, E. D., Chen, W. L., Miller, R. E., and Kay, R. M., 1981, Oat bran intake selectively lowers serum low density lipoprotein cholesterol concentrations in hypercholesterolemic men, *Am. J. Clin. Nutr.* **34**:824–829.

Kiriyama, S., Okazaki, Y., and Yoshida, A., 1969, Hypocholesterolemic effect of polysaccharide and polysaccharide-rich foodstuffs in cholesterol-fed rats, *J. Nutr.* **97**:382–388.

Komai, M., and Kimura, S., 1987, Effect of dietary fiber on fecal steroid profiles in germfree and conventional mice, *Nutr. Rep. Int.* **36**:365–375.

Kritchevsky, D., and Story, J. A., 1974, Binding of bile salts in vitro by nonnutritive fiber, *J. Nutr.* **104**:458–462.

Kyllastinen, M., and Lahikainen, T., 1981, Long term dietary supplementation with a fiber product (guar gum) in elderly diabetics, *Curr. Ther. Res.* **30**:872–879.

Lakshmanan, M. R., Muesing, R. A., Cook, G. A., and Veech, R. L., 1977, Regulation of lipogenesis in isolated hepatocytes by triglyceride-rich lipoproteins, *J. Biol. Chem.* **252**:6581–6584.

LaRusso, N. F., Szczepanik, P. A., and Hofmann, A. F., 1977, Effect of deoxycholic acid ingestion on bile acid metabolism and biliary lipid secretion in normal subjects, *Gastroenterology* **72:**132–140.

Leveille, G. A., and Sauberlich, H. E., 1966, Mechanisms of the cholesterol-depressing effect of pectin in the cholesterol-fed rat, *J. Nutr.* **88:**209–214.

Lo, G. S., Goldberg, A. P., Lim, A., Grundhauser, J. J., Anderson, C., and Schonfeld, G., 1986, Soy fiber improves lipid and carbohydrate metabolism in primary hyperlipidemic subjects, *Atherosclerosis* **62:**239–248.

Lopez-Guisa, J. M., Harned, M. C., Dubielzig, R., Rao, S. C., and Marlett, J. A., 1988, Processed oat hulls as potential dietary fiber sources in rats, *J. Nutr.* **118:**953–962.

Luyken, R., Pikaar, N. A., Polman, H., and Schippers, F. A., 1962, The influence of legumes on the serum cholesterol level, *Voeding* **23:**447–453.

Mathe, D., Lutton, C., Rautureau, J., Coste, T., Gouffier, E., Sulpire, J. C., and Chevallier, F., 1977, Effects of dietary fiber and salt mixtures on the cholesterol metabolism of rats, *J. Nutr.* **107:**466–474.

Mathur, K. S., Khan, M. A., and Sharma, R. D., 1968, Hypocholesterolaemic effect of Bengal gram: A long-term study in man, *Br. Med. J.* **1:**30–31.

McIvor, M. E., Cummings, C. C., Van Duyn, M. A., Leo, T. A., Margolis, S., Behall, K. M., Michnowski, J. E., and Mendeloff, A. I., 1986, Long-term effects of guar gum on blood lipids, *Atherosclerosis* **60:**7–13.

Meyer, J. H., and Doty, J. E., 1988, GI transit and absorption of solid food: Multiple effects of guar, *Am. J. Clin. Nutr.* **48:**267–273.

Miettinen, T. A., 1978, Effects of dietary fibers and ion-exchange resins on cholesterol metabolism in man, in: *International Conference on Atherosclerosis* (L. A. Carlson, R. Paoletti, C. R. Sirtori, and G. Weber, eds.), Raven Press, New York, pp. 193–198.

Miettinen, T. A., and Tarpila, S., 1977, Effect of pectin on serum cholesterol, fecal bile acids, and biliary lipids in normolipidemic and hyperlipidemic individuals, *Clin. Chim. Acta* **79:**471–477.

Miranda, P. M., and Horwitz, D. L., 1978, High-fiber diets in the treatment of diabetes mellitus, *Ann. Intern. Med.* **88:**482–486.

Munoz, J. M., Sandstead, H. H., and Jacob, R. A., 1979, Effects of dietary fiber on glucose tolerance of normal man, *Diabetes* **28:**496–502.

Nakamura, H., Ishikawa, T., Tada, N., Kagami, A., Kondo, K., Miyazima, E., and Takeyama, S., 1982, Effect of several kinds of dietary fibres on serum and lipoprotein lipids, *Nutr. Rep. Int.* **26:**215–221.

Osilesi, O., Trout, D. L., Glover, E. E., Harper, S. M., Koh, E. T., Behall, K. M., O'Dorisio, T. M., and Tartt, J., 1985, Use of xanthan gum in dietary management of diabetes mellitus, *Am. J. Clin. Nutr.* **42:**597–603.

Palmer, G. H., and Dixon, D. G., 1966, Effects of pectin dose on serum cholesterol levels, *Am. J. Clin. Nutr.* **18:**437–442.

Pilch, S. M., 1987, *Physiologic Effects and Health Consequences of Dietary Fiber,* Federation of American Societies for Experimental Biology, Life Sciences Research Office, Bethesda, MD, pp. 36–39.

Pomare, E. W., Branch, W. J., and Cummings, J. H., 1985, Carbohydrate fermentation in the human colon and its relation to acetate concentrations in venous blood, *J. Clin. Invest.* **75:**1448–1454.

Ray, T. K., Mansell, K. M., Knight, L. C., Malmud, L. S., Owen, O. E., and Boden, G., 1983, Long-term effects of dietary fiber on glucose tolerance and gastric emptying in noninsulin dependent diabetic patients, *Am. J. Clin. Nutr.* **37:**376–381.

Raymond, T. L., Connor, W. E., Lin, D. S., Warner, S., Fry, M. M., and Connor, S. L., 1977, The interaction of dietary fibers and cholesterol upon the plasma lipids and lipoproteins, sterol balance, and bowel function in human subjects, *J. Clin. Invest.* **60:**1429–1437.

Reaven, G. M., and Bernstein, R. M., 1978, Effect of obesity on the relationship between very low density lipoprotein production rate and plasma triglyceride concentration in normal and hypertriglyceridemic subjects, *Metabolism* **27:**1047–1054.

Redard, C. L., Schneeman, B. O., and Davis, P. A., 1988, Differences in postprandial lipemia due to gender and dietary fiber, *FASEB J.* **2:**A1418.

Reiser, S., 1987, Metabolic effects of dietary pectins related to human health, *Food Tech.* **41:**91–99.

Ross, A. H. M., Eastwood, M. A., Anderson, J. R., and Anderson, D. M. W., 1983, A study of the effects of dietary gum arabic in humans, *Am. J. Clin. Nutr.* **37:**368–375.

Sable-Amplis, R., Sicart, R., and Agid, R., 1983, Further studies on the cholesterol-lowering effect of apple in human: Biochemical mechanisms involved, *Nutr. Res.* **3:**325–328.

Sasaki, J., Funakoshi, M., and Arakawa, K., 1985, Effect of soybean crude fiber on the concentrations of serum lipids and apolipoproteins in hyperlipemic subjects, *Ann. Nutr. Metab.* **29:**274–278.

Schneeman, B. O., 1987, Soluble vs insoluble fiber—different physiological responses, *Food Tech.* **41:** 81–82.

Schneeman, B. O., and Gallaher, D., 1985, Effects of dietary fiber on digestive enzyme activity and bile acids in the small intestine, *Proc. Soc. Exp. Biol. Med.* **180:**409–414.

Schweizer, T. F., Bekhechi, A. R., Koellreutter, B., Reimann, S., Pometta, D., and Bron, B. A., 1983, Metabolic effects of dietary fiber from dehulled soybeans in humans, *Am. J. Clin. Nutr.* **38:**1–11.

Selvendran, R. R., 1987, Chemistry of plant cell walls and dietary fibre, *Scand. J. Gastroenterol.* **22:** (Suppl. 129):33–41.

Shinnick, F. L., Longacre, M. J., Ink, S. L., and Marlett, J. A., 1988, Oat fiber: Composition versus physiological function in rats, *J. Nutr.* **118:**144–151.

Shorey, R. L., Day, P. J., Willis, R. A., Lo, G. S., and Steinke, F. H., 1985, Effects of soybean polysaccharide on plasma lipids, *J. Am. Diet. Assoc.* **85:**1461–1465.

Simons, L. A., Gayst, S., Balasubramaniam, S., and Ruys, J., 1982, Long-term treatment of hyper-cholesterolaemia with a new palatable forumulation of guar gum, 1982, *Athrosclerosis* **45:**101–108.

Southgate, D. A. T., 1977, The definition and analysis of dietary fibre, *Nutr. Rev.* **35:**31–37.

Spiller, G. A., Chernoff, M. C., Hill, R. A., Gates, J. E., Nassar, J. J., and Shipley, E. A., 1980, Effect of purified cellulose, pectin, and a low-residue diet on fecal volatile fatty acids, transit time, and fecal weight in humans, *Am. J. Clin. Nutr.* **33:**754–759.

Stanley, M. M., Paul, D., Gacke, D., and Murphy, J., 1973, Effects of cholestyramine, metamucil and cellulose on fecal bile salt secretion in man, *Gastroenterology* **65:**889–894.

Storch, K., Anderson, J. W., and Young, V. R., 1984, Oat-bran muffins lower serum cholesterol of healthy young people, *Clin. Res.* **34:**740A.

Story, J. A., 1985, Dietary fiber and lipid metabolism, *Proc. Soc. Exp. Biol. Med.* **180:**447–452.

Story, J. A., and Kritchevsky, D., 1976, Comparison of the binding of various bile acids and bile salts *in vitro* by several types of fiber, *J. Nutr.* **106:**1292–1294.

Story, J. A., and Lord, S. L., 1987, Bile salts: *In vitro* studies with fibre components, *Scand. J. Gastroenterol.* [Suppl.] **129:**174–180.

Superko, H. R., Haskell, W. L., Sawrey-Kubicek, L., and Farquhar, J. W., 1988, Effects of solid and liquid guar gum on plasma cholesterol and triglycerides concentrations in moderate hyper-cholesterolemia, *Am. J. Cardiol.* **62:**51–55.

Tagliaferro, V., Cassader, M., Bozzo, C., Pisu, E., Bruno, A., Marena, S., Cavallo-Perin, P., Cravero, L., and Pagano, G., 1985, Moderate guar-gum addition to usual diet improves peripheral sensitivity to insulin and lipaemic profile in NIDDM, *Diabetes Metab.* **11:**380–385.

Tarpila, S., Miettinen, T. A., and Metsaranta, L., 1978, Effects of bran on serum cholesterol, faecal mass, fat bile acids, and neutral sterols and biliary lipids in patients with diverticular disease of the colon, *Gut* **19:**137–145.

Tsai, A. C., Elias, J., Kelley, J. J., Lin, R. C., and Robson, J. R., 1976, Influence of certain dietary fibers on serum and tissue cholesterol levels in rats, *J. Nutr.* **106:**118–123.

Tsai, A. C., Mott, E. L., Owen, G. M., Bennick, M. R., Lo, G. S., and Steinke, F. H., 1983, Effects of soy polysaccharide on gastrointestinal functions, nutrient balance, steriod excretions, glucose tolerance, serum lipids and other parameters in humans, *Am. J. Clin. Nutr.* **38:**504–511.

Turnbull, W. H., and Leeds, A. R., 1987, Reduction of total and LDL-cholesterol in plasma by rolled oats, *J. Clin. Nutr. Gastroenterol.* **2:**177–180.

Uusitupa, M., Tuomilehto, J., Karttunen, P., and Wolf, E., 1984, Long term effects of guar gum on metabolic control, serum cholesterol and blood pressure levels in type 2 (non-insulin-dependent) diabetic patients with high blood pressure, *Ann. Clin. Res.* [Suppl.] **43**:126–131.

Vahouny, G. V., 1985, Dietary fibers: Aspects of nutrition, pharmacology, and pathology, in: *Nutritional Pathology: Pathobiochemistry of Dietary Imbalances* (H. Sidransky, ed.), Marcel Dekker, New York, pp. 207–277.

Vahouny, G. V., Tombes, R., Cassidy, M. M., Kritchevsky, D., and Gallo, L. L., 1980, Dietary fibers: V. Binding of bile salts, phospholipids and cholesterol from mixed micelles by bile acid sequestrants and dietary fibers, *Lipids* **15**:1012–1018.

Vahouny, G. V., Tombes, R., Cassidy, M. M., Kritchevsky, D., and Gallo, L. L., 1981, Dietary fibers: VI. Binding of fatty acids and monolein from mixed micelles containing bile salts and lecithin, *Proc. Soc. Exp. Biol. Med.* **166**:12–16.

Van Horn, L. V., Liu, K., Parker, D., Emidy, L., Laio, Y. L., Pan, W. H., Guimetti, D., Hewitt, J., and Stamler, J., 1986, Serum lipid response to oat product intake with a fat modified diet, *J. Am. Diet. Assoc.* **86**:759–764.

Van Horn, L. V., Emidy, L. A., Liu, K., Liao, Y., Ballew, C., King, J., and Stamler, J., 1988, Serum lipid response to a fat-modified, oatmeal-enhanced diet, *Prev. Med.* **17**:377–386.

Vargo, D., Doyle, R., and Floch, M. H., 1985, Colonic bacterial flora and serum cholesterol: Alterations induced by dietary citrus pectin, *Am. J. Gastroenterol.* **80**:361–364.

Walters, R. L., Baird, I. M., Davies, P. S., Hill, M. J., Drasar, B. S., Southgate, D. A. T., Green, M., and Morgan, B., 1975, Effects of two types of dietary fibre on faecal steriod and lipid excretion, *Br. Med. J.* **2**:536–538.

Wells, A. F., and Ershoff, B. H., 1961, Beneficial effects of pectin in prevention of hypercholesterolemia and increase in liver cholesterol in cholesterol-fed rats, *J. Nutr.* **74**:87–92.

Williams, R. D., and Olmsted, W. H., 1936, The effect of cellulose, hemicellulose and lignin on the weight of the stool: A contribution to the study of laxation in man, *J. Nutr.* **11**:433–449.

Zavoral, J. H., Hannan, P., Fields, D. J., Hanson, M. N., Frantz, I. D., Kuba, K., Elmer, P., and Jacobs, D. R., 1983, The hypolipidemic effect of locust bean gum food products in familial hypercholesterolemic adults and children, *Am. J. Clin. Nutr.* **38**:285–294.

26

Dietary Fiber and Bile Acid Metabolism

JON A. STORY and EMILY J. FURUMOTO

1. INTRODUCTION

Bile acids are the major route of steroid excretion, a phenomenon that has been exploited as a mechanism for lowering serum cholesterol and coronary heart disease risk (Lipid Research Clinics Program, 1984). Alteration of bile acid metabolism by dietary fiber has been of interest as a result of the potential involvement of changes in bile acid metabolism in the etiology of a variety of important diseases of Western populations (Story, 1979). Dietary fiber also increases fecal bulk and, as a result, dilutes the contents of the colon, including bile acids. Bile acid concentrations have been linked with colon cancer susceptibility; thus, alterations in bile acid concentrations by dietary fiber have been suggested as a mechanism for reduced cancer risk, a second epidemiologic observation.

We have summarized our experience with alteration of bile acid metabolism by dietary fiber in earlier reviews (Story and Thomas, 1982; Story, 1986). Our early studies indicated that adsorption of bile acids by dietary fiber resulted in increased excretion and possibly increased concentrations of bile acids in feces (Kritchevsky and Story, 1974; Story and Kritchevsky, 1976). On closer examination we also observed that some sources of dietary fiber also altered the spectrum of bile acids excreted. These observations implied that dietary fiber might reduce cholesterol levels by increasing bile acid excretion and by increasing the proportion of bile acids derived from chenodeoxycholic acid (mechanism discussed in Section 4). This may also involve, in some cases, increased concentrations of bile acids, some of which have been shown to potentiate mutagenicity (Wilpart *et al.*, 1983), in the colon. As a result, we have continued to examine the role of a variety of sources of dietary fiber in bile acid metabolism with attention to total daily excretion and concentration as well as the relative amounts of the various bile acids excreted. We have also started examining possible mechanisms by which these changes are taking place.

JOHN A. STORY and EMILY J. FURUMOTO • Department of Foods and Nutrition, Purdue University, West Lafayette, Indiana 47907.

2. CHANGES IN BILE ACID EXCRETION AND CHOLESTEROL ACCUMULATION

2.1. Cholesterol-Fed Animals

We have examined the effects of several sources of dietary fiber on cholesterol accumulation and steroid excretion in rats using a semipurified diet with and without added dietary cholesterol (0.25%). This diet results in inconsistent effects on serum cholesterol but a consistent increase in liver cholesterol, and we have used it in the past to measure the effects of dietary fiber sources on cholesterol accumulation. First we compared white wheat bran with corn bran and barley bran, with cellulose as a control. Wheat bran and corn bran have been reported to have little effect on cholesterol levels but effectively lower fecal bile acid concentrations. Barley bran is a byproduct of the brewing industry and, as such, contains few of the $\beta(D)$-glucan gums present in whole barley. Thus, we expected these three materials to alter concentrations of bile acids but to have little effect on liver cholesterol accumulation.

As can be seen in Table I, serum cholesterol was highest in animals fed wheat bran, being significantly higher in comparison to those fed cellulose and the other sources when cholesterol was not present in the diet. Liver cholesterol levels were significantly increased by cholesterol feeding, but few alterations in this accumulation resulted from the various sources of dietary fiber. Animals fed corn bran accumulated significantly more liver cholesterol than animals fed wheat or barley bran.

As was mentioned above, our primary concerns in measuring steroid excretion are total daily steroid excretion as it relates to cholesterol balance and concentration of fecal bile acids as a measure of colon cancer susceptibility. Total amounts of fecal steroids excreted per day (Table II) were not altered dramatically by any of the dietary fiber sources in comparison to the amount observed in response to cellulose without added

TABLE I. Changes in Serum and Liver
Cholesterol in Response to Wheat Bran,
Corn Bran, and Barley Bran

Fiber source	Dietary cholesterol	Cholesterol[a]	
		Serum	Liver
Cellulose	−	100[b]	100[c]
	+	104[bd]	361[bd]
Wheat bran	−	119[d]	88[c]
	+	108[d]	312[b]
Corn bran	−	100[b]	98[c]
	+	102[b]	420[d]
Barley bran	−	102[b]	93[c]
	+	108[bd]	349[b]

[a]Relative values given based on cellulose/without cholesterol group as 100. Means with different superscripts are significantly different ($P < 0.05$).

TABLE II. Effects of Wheat Bran, Corn Bran, and Barley Bran
on Fecal Steroid Excretion[a]

Fiber source	Dietary cholesterol	Neutral steroids		Bile acids		Total steroids	
		Conc.	Daily	Conc.	Daily	Conc.	Daily
Cellulose	−	100[f]	100	100[b]	100[cd]	100[g]	100[e]
	+	308[cd]	304	137[bh]	149[bcdh]	224[cd]	241[bh]
Wheat bran	−	229[de]	102	183[bh]	85[d]	199[de]	95[e]
	+	553[h]	246	193[bh]	93[d]	380[h]	184[d]
Corn bran	−	157[ef]	133	103[b]	94[cd]	127[fg]	117[e]
	+	359[bc]	346	164[bh]	165[bh]	263[c]	273[h]
Barley bran	−	214[e]	138	144[bh]	104[bcd]	170[ef]	124[e]
	+	422[b]	284	214[h]	157[bch]	317[b]	232[bc]

[a]Relative values given based on cellulose/without cholesterol fed group as 100. Means with different superscripts are significantly different ($P < 0.05$).

dietary cholesterol. When cholesterol was added to the diet, total daily steroid excretion was significantly increased in all dietary groups, but animals fed wheat bran excreted significantly less than those given any of the other sources of dietary fiber. No substantial differences in bile acid concentrations were observed, suggesting that all these sources of dietary fiber had similar effects on the amount of bile acid delivered to the large intestine for excretion and on the various mechanisms of dilution of colonic contents. The similar physical nature of these sources of dietary fiber, in light of the loss of the majority of β(D)-glucan from barley bran, would support the concept that these largely water-insoluble materials act as diluents with little interaction with bile acids.

As is usually reported, cholesterol feeding increased the amount of 6β-hydroxylated bile acids in all dietary fiber groups (data not presented). A higher percentage of deoxycholic acid in cellulose-fed animals was replaced primarily by hyodeoxycholic in all dietary fiber groups.

We also examined three fruits and vegetables often suggested as sources of dietary fiber for human diets, two of which had previously been shown to be involved in reduction of serum cholesterol levels. These foods were added to diets at 15% by weight after lyphilization and grinding to a consistent particle size. Although dietary cholesterol increased liver cholesterol in all cases, few differences in effects of these foods were observed (Table III). Animals fed the potato diet had a significantly higher level of liver cholesterol than any of the other groups. Serum cholesterol levels were not significantly increased in response to dietary cholesterol in any of the groups except those fed potato. This group had the lowest level of serum cholesterol among groups without dietary cholesterol and one of the highest when cholesterol was added to the diet.

Some interesting differences in excretion of fecal steroid were observed (Table IV). Increases in concentrations of both neutral and acidic steroids were observed in comparison to the brans employed in the previous experiment. This may result from the higher digestibility of the fruits and vegetables compared to the brans. Among foods tested, concentrations of bile acids both with and without dietary cholesterol were lowest

TABLE III. Effects of Apple, Carrot, or
Potato on Serum and Liver Cholesterol
Levels in Rats

Fiber source	Dietary cholesterol	Cholesterol	
		Serum	Liver
Cellulose	−	100^{bcd}	100^c
	+	111^d	387^b
Apple	−	100^{bcd}	103^c
	+	104^{bd}	369^b
Carrot	−	92^{bc}	98^c
	+	98^{bcd}	487^{bd}
Potato	−	81^c	118^c
	+	107^{bd}	629^d

[a]Relative values given based on cellulose/without cholesterol group as 100. Means with different superscripts are significantly different ($P < 0.05$).

for potato among foods tested, but these concentrations were higher than those observed for cellulose-fed animals, again lowest among the groups of this and the previous experiment. Total steroid excretion was similar for animals fed apple and cellulose and much lower but similar for groups fed potato and carrot.

Most notable among changes in composition of fecal bile acids was the increase in the muricholic acids in potato-fed animals (data not shown). Muricholic acids accounted for about 7% of total bile acids in cellulose-fed animals without dietary cholesterol and 15% when cholesterol is added to the diet. Substitution of potato for cellulose resulted in a muricholic acid proportion of 37% of total bile acids without dietary cholesterol and 51% with dietary cholesterol. These levels of muricholic acids are far greater than any we have observed for any other source of dietary fiber. Synthesis of these bile acids

TABLE IV. Alterations in Fecal Steroid Excretion in Response to Apple, Carrot, and Potato in Rats[a]

Fiber source	Dietary cholesterol	Neutral steroids		Bile acids		Total steroids	
		Conc.	Daily	Conc.	Daily	Conc.	Daily
Cellulose	−	100^e	100^d	100^b	100^{bc}	100^f	100^c
	+	378^d	381^g	172^b	174^{bc}	289^{def}	292^g
Apple	−	382^d	83^d	493^b	106^{bc}	432^{cd}	93^c
	+	1389^g	316^b	1486^g	361^g	1438^g	336^g
Carrot	−	304^d	65^d	480^b	103^{bc}	382^{de}	81^c
	+	879^b	181^c	1037^g	215^b	951^b	196^b
Potato	−	145^e	64^d	161^b	69^c	152^{ef}	66^c
	+	740^c	238^c	475^b	143^{bc}	627^e	197^b

[a]Relative values given based on cellulose/without cholesterol group as 100. Means with different superscripts are significantly different ($P < 0.05$).

appears to be a response to increased levels of hepatic esterified cholesterol, and their synthesis may constitute a mechanism for maintaining cholesterol homeostasis (Thomas *et al.*, 1984). We have suggested that increases in excretion of these bile acids was one of the mechanisms by which some sources of dietary fiber reduced cholesterol accumulation (Story and Thomas, 1982; Story, 1986), but our data with animals fed potato now suggest that this increase in muricholic acid excretion may be a symptom of increased hepatic cholesterol rather than a mechanism for preventing this cholesterol accumulation. Observations of regulatory enzymes in these systems in animals fed cholesterol-free diets will be needed to clarify the importance and root causes of changes in synthesis of these unique bile acids.

2.2. Animals Fed Diets without Cholesterol

With diets similar to those described above but without added cholesterol, we examined the effects of a variety of dietary fiber sources with different physical properties and effects on cholesterol and bile acid metabolism. Few changes in liver cholesterol accumulation were observed in the absence of dietary cholesterol (Table V). However, as we had observed previously, serum cholesterol levels were significantly higher in groups of animals fed either white or hard red wheat bran.

As would be expected, removal of fiber from the diet caused an increase in both neutral and acidic steroid concentrations and reductions in daily excretion in feces (Table VI). As we had reported in our earlier work, concentrations of neutral steroids and bile acids were higher for oat bran than for any of the other dietary fiber sources (Story, 1986). This is probably a result of greater digestibility of oat bran by intestinal microflora, resulting in reduced fecal bulk. This phenomenon has not been observed in humans, where fecal weight is greater when oat bran is fed, and steroid concentrations are not appreciably altered (Kirby *et al.*, 1981; Anderson *et al.*, 1984). Corn bran and wheat bran of both types caused a 1.5- to twofold increase in steroid concentrations over controls fed cellulose. Daily steroid excretion was lower in both wheat bran groups,

TABLE V. Modification of Serum and Liver
Cholesterol Levels in Rats by Oat, Wheat,
and Corn Brans[a]

Fiber source	Cholesterol	
	Serum	Liver
No fiber	96[b]	103
Cellulose	100[b]	100
Oat bran	106[b]	86
Corn bran I	107[b]	92
Corn bran II	111[b]	87
White wheat bran	129[c]	81
Red wheat bran	132[c]	81

[a]Relative values given based on cellulose/without cholesterol group as 100. Means with different superscripts are significantly different ($P < 0.05$).

TABLE VI. Fecal Steroid Excretion in Response to Wheat, Oat, and Corn Brans

Fiber source	Neutral steroids		Bile acids		Total steroids	
	Conc.	Daily	Conc.	Daily	Conc.	Daily
No fiber	582^f	72^d	492^f	60^c	546^f	67^e
Cellulose	100^d	100^b	100^d	100^{bf}	100^e	100^{bc}
Oat bran	331^b	74^d	326^b	72^b	329^b	74^{de}
Corn bran I	152^{cd}	116^f	136^{cd}	103^f	146^d	111^{bf}
Corn bran II	208^c	121^f	191^c	109^f	201^c	116^f
White wheat bran	189^c	83^{cd}	201^c	88^{bf}	194^c	85^{cd}
Red wheat bran	203^c	97^{bc}	153^{cd}	72^b	183^{cd}	87^{cd}

[a]Relative values given based on cellulose/without cholesterol group as 100. Means with different superscripts are significantly different ($P < 0.05$).

possibly accounting for the observed increase in serum cholesterol levels. However, oat bran caused the greatest reduction in steroid excretion and had no effect on serum cholesterol.

As can be seen in Table VI, cellulose substantially reduces concentrations of bile acids in comparison to no fiber. Comparison of the effects of other sources of dietary fiber with those of cellulose suggests that corn bran and barley bran had similar effects. Other materials used in these experiments resulted in higher concentrations of bile acids than cellulose. Most notably, apple and carrot resulted in concentrations about five times those of cellulose. These observations are of interest in light of the relationship between bile acid concentration and colon cancer susceptibility and provide an excellent means of further understanding this relationship using the rat model and foods that are common sources of dietary fiber in humans.

3. CHANGES IN BILE ACID POOL SIZE AND TURNOVER

In an effort to examine further the effects of sources of dietary fiber that reduce cholesterol accumulation on bile acid metabolism, experiments were designed to measure changes in bile acid turnover and pool size in response to feeding oat bran and corn bran using cellulose as a control. Diets similar to those used above were employed, without added cholesterol. Bile acids were allowed to wash out by complete bile drainage via a bile duct cannula, and total and individual bile acid pool sizes and synthetic and secretion rates were measured. Total bile acid pool and synthetic rates are substantially lower in response to these semipurified diets than has been reported in the literature for animals fed commercial diets (Portman and Murphy, 1958). In these and other experiments, Portman (1960) determined grain to be the ingredient in commercial diets that was responsible for this difference.

Data from our experiments indicate some interesting differences in response to these three sources of dietary fiber. Oat bran resulted in about an 80% reduction in total bile acid pool size in comparison to cellulose, whereas corn bran caused a more than fourfold increase in pool size (Table VII). Daily secretion rate was similarly changed, as

TABLE VII. Effects of Corn Bran, Oat Bran,
and Cellulose on Bile Acid Synthesis,
Secretion, and Pool Size in Rats[a]

	Pool size	Basal synthetic rate	Daily secretion rate
Cellulose	100[d]	100[d]	100[d]
Oat bran	22[b]	60[d]	22[d]
Corn bran	481[c]	595[b]	409[c]

[a]Relative values given based on cellulose/without cholesterol group as 100. Means with different superscripts are significantly different ($P < 0.05$).

would be expected. Bile acid synthetic rate was 40% lower in response to oat bran feeding (compared to cellulose), although the difference was not statistically significant, and was six times higher in response to corn bran. Our earlier observation of increased excretion of bile acids in animals fed corn bran seems to account for these effects on bile acid secretion and synthesis and is similar to the observation made by Portman (1960). Oat bran, on the other hand, reduces bile acid synthesis and thus reduces pool size and rate of secretion, explaining earlier reported effects on bile acid excretion in response to oat bran in rats (Table VI; Story, 1985, 1986).

4. POTENTIAL MECHANISMS

Several mechanisms that attempt to explain the above observations have been suggested, but none seem to answer completely all questions posed by these data. Our initial hypothesis suggested that sources of dietary fiber that reduced cholesterol levels did so by increasing steroid excretion as a result of adsorption of bile acids onto dietary fiber in the small and large intestine. Clearly, this hypothesis does not explain the observations we have just outlined.

Corn bran, which does not prevent cholesterol accumulation in the animal model employed, increases steroid excretion and pool size, possibly through increased bile acid adsorption. Oat bran, which consistently prevents cholesterol accumulation, displays a quite different set of effects on these variables, i.e., reduced pool size, synthetic rate, and excretion, indicating that some other mechanism is playing an important part in its effect on cholesterol metabolism. One possibility is that an effector is inhibiting bile acid synthesis and reducing pool size of bile acids. The identity of this effector is unknown to us. In related studies we have observed an effect on the rate-limiting step of cholesterol synthesis, HMG-CoA reductase, in less than an hour in response to the first feeding of oat bran (Kelley and Story, 1987). This effect appeared similar to that observed with cholesterol feeding, suggesting that some quickly absorbed component of oat bran may be responsible for changes in cholesterol synthesis. Others have reported similar effects for compounds isolated from related plant materials (Qureshi et al., 1986).

Changes in synthetic rates for the bile acids may not be uniform for all bile acids and, as a result, may cause a change in the relative size of the various bile acid pool sizes. Differences in bile acid pool sizes have been shown to alter several phases of cholesterol metabolism. Of interest is the effect of an increase in the chenodeoxycholic acid (CDCA) pool size. As the CDCA pool increases, reductions in cholesterol absorption (Ponz de Leon *et al.*, 1979), bile acid reabsorption (Beher *et al.*, 1967), and cholesterol synthesis (Coyne *et al.*, 1976) have been observed. In rats, feeding cholesterol increases synthesis of CDCA and its derivatives as a mechanism for maintenance of cholesterol homeostasis. We have observed an increase in pool size of CDCA and its derivatives in response to oat bran in rats but have not seen a similar response in humans (Story, 1986).

Changes in the relative sizes of the bile acid pools may also influence colon cancer susceptibility. Several lines of evidence suggest that deoxycholic (DCA) and lithocholic (LCA) acids may potentially increase susceptibility to colon cancer. Wilpart *et al.* (1983) have observed an increased *in vitro* mutagenesis to dimethylhydrazine when DCA or LCA is included. Narisawa *et al.* (1974) reported increased carcinogenesis when DCA or LCA was intrarectally instilled with a direct-acting carcinogen in rats. Thorough examination of the important of changes in the relative amounts of bile acids excreted in response to dietary fiber have not been carried out.

Observations in humans would suggest that a combination of these mechanisms may be involved. Bile acid excretion increases in response to oat bran (Kirby *et al.*, 1981; Anderson *et al.*, 1984), but the magnitude of the increase in bile acid excretion has never been reported as adequate to explain changes in cholesterol levels (Kirby *et al.*, 1981; Anderson *et al.*, 1984), suggesting some additional mechanism. Further observations of effects of specific sources of dietary fiber on bile acid kinetics in conjunction with measurement of changes in bile acid excretion and cholesterol synthesis are needed in order to understand how these mechanisms fit together.

Mechanistic studies now need to establish which effectors are responsible for these changes in bile acid kinetics. Greater effectiveness of water-soluble dietary fiber sources in lowering serum cholesterol levels has prompted speculation that production of short-chain fatty acids in the cecum and colon and their subsequent absorption may be an effector of the observed changes (Chen and Anderson, 1986). Several laboratories have shown that diets with oat bran and some other sources of dietary fiber result in higher portal blood concentrations of short-chain fatty acids in rats compared to diets containing cellulose (Chen and Anderson, 1986). In addition, propionate has been shown to reduce incorporation of acetate into cholesterol in hepatocytes (Chen *et al.*, 1984). Obviously a great deal of work needs to be done to establish the viability of this hypothesis.

5. CONCLUSIONS

Our hypothesis concerning the ability of some sources of dietary fiber to lower cholesterol levels suggested that increased excretion of steroids in feces resulted in negative steroid balance. In our animal model for cholesterol accumulation, this hypothesis does not seem to be supported by the observations. In comparison to cellulose, cholesterol accumulation was reduced in response to diets with oat bran (Story, 1985, 1986) and increased when potato was included in the diet (Table IV). Excretion of steroids was reduced by oat bran feeding in comparison to cellulose and was unchanged by potato diets. Steroid excretion is not the apparent mechanism by which these two sources of dietary fiber alter cholesterol levels in rats. Earlier work showed that carrot and apple reduced serum cholesterol in humans (Gormley *et al.*, 1977; Robertson *et al.*, 1979), but these two foods did not have a similar effect in rats fed the diet employed in these experiments. Other sources of dietary fiber examined in this series of experiments have not previously been shown to alter cholesterol accumulation and, in our animal model, did not display an ability to reduce cholesterol accumulation and had little effect on total daily steroid excretion.

More recent evidence suggests that some effector is altering bile acid and/or cholesterol synthesis and, as a result, reducing the size of the bile acid pool. This reduction seems to be an important part of the cholesterol-lowering effect of dietary fiber, but little is known of the identity of these effectors. In addition, changes in the relative sizes of the bile acid pools may result, leading to possible changes in cholesterol metabolism as well as potentially altering susceptibility to colon cancer.

Reductions in fecal bile acid concentrations in response to many of the less digestible sources of dietary fiber appear to result from dilution of colonic contents by the fiber and water held by the fiber. Reductions in susceptibility to experimentally induced colon cancer in rats in relation to these reductions in bile acid concentrations are being examined.

In both cases these effects are being examined with the intention of improving our ability to make dietary recommendations to the public concerning selection of diets that will reduce risk for disease. We are far from being able to make recommendations for dietary fiber intake beyond a general suggestion for increased intake of dietary fiber from a wide variety of sources. Results reported here strongly support this general recommendation.

ACKNOWLEDGMENTS. this work was supported, in part, by the Indiana Agricultural Experiment Station (paper #12187), the Quaker Oats Company, Miller Brewing Company, Corn Products Corporation International, and the American Institute for Cancer Research.

REFERENCES

Anderson, J. W., Story, L., Sieling, B., Chen, W.-J. L., Petro, M. S., and Story, J. A., 1984, Hypercholesterolemic effects of oat bran or bean intake for hypercholesterolemic men, *Am. J. Clin. Nutr.* **40**:1146–1155.

Beher, W. T., Beher, M. E., and Rao, B., 1967, Turnover of cholic and chenodeoxycholic acids in normal and hypothsectomized rats, *Life Sci.* **6**:866–868.

Chen, W.-J. L., and Anderson, J. W., 1986, Hypocholesterolemic effects of soluble fibers, in: *Dietary Fiber: Basic and Clinical Aspects* (G. V. Vahouny and D. Kritchevsky, eds.), Plenum Press, New York, pp. 275–286.

Chen, W.-J. L., Anderson, J. W., and Jennings, D., 1984, Propionate may mediate the hypocholesterolemic effects of certain soluble plant fibers in cholesterol-fed rats, *Proc. Soc. Exp. Biol. Med.* **175**:215–218.

Coyne, M. J., Bonouris, G. G., Goldstein, L. I., and Schoenfield, L. J., 1976, Effect of chenodeoxycholic acid and phenobarbital on the rate limiting enzymes of hepatic cholesterol and bile acid synthesis in patients with gallstones, *J. Lab. Clin. Med.* **87**:281–291.

Gormley, T. R., Kevany, J., Egan, J. P., and McFarlane, R., 1977, Effect of apples on serum cholesterol levels in humans, *Ir. J. Food Sci. Technol.* **3**:101–109.

Kelley, M. J., and Story, J. A., 1987, Short-term changes in hepatic HMG-CoA reductase in rats fed diets containing cholesterol or oat bran, *Lipids* **22**:1057–1059.

Kirby, R. W., Anderson, J. W., Sieling, B., Rees, E. D., Chen, W.-J. L., Miller, R. E., and Kay, R. M., 1981, Oat bran intake selectively lowers serum low density lipoprotein concentrations: Studies of hypercholesterolemic men, *Am. J. Clin. Nutr.* **34**:824–829.

Kritchevsky, D., and Story, J. A., 1974, Binding of bile salts *in vitro* by nonnutritive fiber, *J. Nutr.* **104**:458–462.

Lipid Research Clinics Program, 1984, The lipid research clinics coronary primary prevention trial results. II. The relationship of reduction of incidence of coronary heart disease to cholesterol lowering, *J.A.M.A.* **251**:365–374.

Narisawa, T., Magadia, N. E., and Wynder, E. L., 1974, Promoting effect of bile acids on colon carcinogenesis after intrarectal instillation of N-methyl-N-nitroguanidine in rats, *Cancer Res.* **53**:1093–1097.

Ponz de Leon, M., Carulli, N., Loria, P., and Zirone, F., 1979, The effect of chenodeoxycholic acid (CDCA) on cholesterol absorption, *Gastroenterology* **77**:223–230.

Portman, O. W., 1960, Nutritional influences on the metabolism of bile acids, *Am. J. Clin. Nutr.* **8**:462–470.

Portman, O. W., and Murphy, P., 1958, Excretion of bile acids and β-hydroxysterols by rats, *Arch. Biochem. Biophys.* **76**:367–371.

Qureshi, A. A., Burger, W. C., Peterson, D. M., and Elson, C. E., 1986, The structure of an inhibitor of cholesterol biosynthesis isolated from barley, *J. Biol. Chem.* **261**:10544–10550.

Robertson, J. A., Brydon, W. G., Tadesse, K., Wenham, P., Walls, A., and Eastwood, M. A., 1979, The effect of raw carrot on serum lipids and colon function, *Am. J. Clin. Nutr.* **32**:1889–1892.

Story, J. A., 1979, Dietary fiber: Its role in diverticular disease, colon cancer and coronary heart disease, in: *Biochemistry of Nutrition*, Vol. 1 (A. Neuberger and T. H. Jukes, eds.), University Park Press, Baltimore, pp. 189–205.

Story, J. A., 1985, Dietary fiber and lipid metabolism, *Proc. Soc. Exp. Biol. Med.* **180**:447–452.

Story, J. A., 1986, Modification of steroid excretion in response to dietary fiber, in: *Dietary Fiber: Basic and Clinical Aspects* (G. V. Vahouny and D. Kritchevsky, eds.), Plenum Press, New York, pp. 253–264.

Story, J. A., and Kritchevsky, D., 1976, Comparison of the binding of various bile acids and bile salts by several types of fiber, *J. Nutr.* **106**:1292–1294.

Story, J. A., and Thomas, J. N., 1982, Modification of bile acid spectrum by dietary fiber, in: *Dietary Fiber in Health and Disease* (G. V. Vahouny and D. Kritchevsky, eds.), Plenum Press, New York, pp. 193–201.

Thomas, J. N., Kelley, M. J., and Story, J. A., 1984, Alteration of regression of cholesterol accumulation in rats by dietary pectin, *Br. J. Nutr.* **51**:339–345.

Wilpart, M., Mainguet, P., Maskens, A., and Roberfroid, M., 1983, Structure–activity relationship amongst biliary acids showing co-mutagenic activity towards dimethylhydrazine, *Carcinogenesis* **4**:1239–1241.

27

Antitoxic Effects of Dietary Fiber

DAVID KRITCHEVSKY

In 1943, Woolley and Krampitz reported that when glucoascorbic acid, an antagonist of ascorbic acid, was fed to mice as 10% of a purified diet, the animals exhibited deficiency symptoms and severe weight loss. When cerophyl (a dehydrated grass preparation) was added as 10% of the semipurified diet, the mice exhibited normal growth. Addition of glucoascorbic acid to a ration containing alfalfa meal, corn meal, and linseed oil meal did not elicit an adverse response. Ershoff (1954, 1957) confirmed the findings in mice and extended the work to rats. Rats fed 4% glucoascorbic acid showed poor survival (75%) and low weight gain as well as severe diarrhea. Addition to the diet of 5% or 10% alfalfa meal led to 100% survival and improved weight gain, but diarrhea persisted, although it was not as severe. When the diet contained 20% alfalfa meal or 20% rye, wheat, or oat grass, weight gain was optimal, and no diarrhea was observed.

Chow *et al.* (1953) investigated the effects of polyoxyethylene sorbitan monostearate (Tween 60) and other surface-active agents on the growth of rats fed semipurified diets. When the diet contained 22% casein, 69.5% sucrose, 4.5% corn oil, and 5% salt mix, addition of 5% Tween 60 inhibited growth by 25–50%; when 5% Tween 60 was added to a diet containing 62% soybean meal, 29.5% sucrose, 4.5% corn oil, and 4% salt mix weight gain was reduced by only 4–6%. The two basal diets were different in the level of sucrose and type of protein, but it is possible that the soybean meal exerted a protective effect. Ershoff (1960) found alfalfa to reverse the toxicity of other nonionic surface-active agents such as polyoxyethylene sorbitan monolaurate (Tween 20), sorbitan monolaurate (Span 20), and polyoxyethylene stearate (Myrj 45 and Myrj 52). In the studies with Tween 20, Ershoff (1960) found that feeding alfalfa meal (10% or 20%) or alfalfa residue (15%) resulted in 100% survival and optimal weight gain, whereas rats fed alfalfa juice (5%) and 15% Tween 20 showed growth inhibition and toxicity equivalent to those seen in rats fed the detergent alone. Cellulose (10%) added to the Tween-20-containing diet gave 100% survival but significantly lower weight gain than seen in the alfalfa-fed animals. Ershoff and Hernandez (1959) found that alfalfa also reversed the toxic effects of Tween 60. Ershoff and Marshall (1975) found that psyllium

DAVID KRITCHEVSKY • The Wistar Institute of Anatomy and Biology, Philadelphia, Pennsylvania 19104.

seed (5% or 10%) completely reversed the toxic effects of Tween 60. Wheat bran (10%) gave 100% survival, but weight gain after 14 days was 28% below that of the controls. When cellulose was fed at the 5% level, survival was 67% (compared to 33% on the fiber-free diets), and weight gain was 69% below control levels; when 10% cellulose was fed, survival rose to 100%, but weight gain was still reduced by 63%. The basal diet contained 66% sucrose, 24% casein, 5% cottonseed oil, and 5% mineral mix; all additions were made at the expense of the sucrose.

Kimura et al. (1980, 1982) investigated the mechanisms of action of a number of detergents including Tween 20, Tween 60, Span 20, sodium taurocholate, sodium deoxycholate, sodium dodecyl sulfate, and sodium laurylbenzene sulfonate. They found the detergents promoted exfoliation of the intestinal brush border membrane and inhibited intestinal disaccharidase activities. Fiber prepared from the root of the edible burdock prevented these adverse effect. The edible burdock (Arctium lappa) is a common Japanese foodstuff called Gobo.

The antioxidant 2,5-di-t-butylhydroquinone (DBH), when added to a semipurified diet at the level of 0.1–0.2%, inhibits weight gain in rats. Ershoff (1963) showed that this effect was not observed when DBH was added to a stock diet; DBH (0.2%) added to a semipurified diet led to 33% toxicity and reduced (by 78%) weight gain in male Holtzman rats and to 38% toxicity and 71% lower weight gain in Long–Evans rats. When 0.2% DBH was added to stock diet, there was 100% survival in both rat strains, but there still was lessened weight gain (15% in Holtzman rats and 24% in Long–Evans rats).

Ershoff and his colleagues assessed the toxicity of a number of food colors in rats by measuring survival and weight gain over a 3-week period. The basal diet was similar to that used in the detergent studies. Amaranth (FD & C red no. 2) is very toxic when fed as 5% of the diet. The effect is negated by feeding 10% cellulose, pectin, or alfalfa (Ershoff and Thurston, 1974) (Table I). The toxic effects of other dyes such as tartrazine (FD & C yellow no. 5), sunset yellow (FD & C yellow no. 6), brilliant blue FCF (FD & C blue no. 1), and FD & C red no. 40 are also reversed by 10% alfalfa leaf meal, carrot root powder, blond psyllium seed powder, or wheat bran (Ershoff, 1977). Kimura et al.

TABLE I. Influence of Fiber on Toxicity of Amaranth
(FD & C Red No. 2) in Rats[a]

Diet	Average weight gain (g) (21 days)	Survival
Basal (B)[b]	89.8	6/6
B + 5% Amaranth (BA)	—	0/6
BA plus Cottonseed oil (5%)	25.5	2/6
Casein (10%)	42.0	1/6
Pectin (10%)	70.8	6/6
Cellulose (10%)	86.9	6/6
Alfalfa (10%)	92.6	6/6

[a] After Ershoff and Thurston (1974).
[b] Sucrose, 66; casein, 24; cottonseed oil, 5; mineral mix, 5.

TABLE II. Effect of Graded Levels of Sodium
Cyclamate (C) on Weight Gain of
Immature Male Rats[a]

| Diet | Average weight gain (21 days) | |
	Sprague–Dawley	Long–Evans
Basal (B)[b]	120	120
B + 2.5% C	82	118
B + 5.0% C	35	75
Stock (S)	119	130
S + 2.5% C	118	126
S + 5.0% C	114	116
S + 10.0% C	86	ND[c]

[a]After Ershoff (1972).
[b]See footnote, Table I.
[c]ND, not done.

(1980) reported that Amaranth reduced intestinal sucrase and alkaline phosphatase activity.

Ershoff (1976) also reported that cholestyramine, a bile-acid-binding resin, reduced Amaranth toxicity. It is possible that bile acids enhance Amaranth toxicity by synergistic effects on intestinal brush border enzymes (Kimura *et al.*, 1982) or perhaps by enhancing their absorption in a lipid milieu. Dietary fibers have been shown to bind bile acids and bile salts (Kritchevsky and Story, 1974; Story and Kritchevsky, 1976).

Sodium cyclamate added to purified rations is progressively toxic for immature rats. As Table II shows, immature male Sprague–Dawley rats fed 2.5% or 5.0% sodium cyclamate for 21 days show reductions in weight gain of 31% and 71%, respectively. When added to stock diet, 2.5% or 5.0% sodium cyclamate does not affect weight gain in immature Sprague–Dawley rats; at 10% of the diet sodium cyclamate reduces weight gain by 28%. Sodium cyclamate (2.5%) does not inhibit weight gain in immature male Long–Evans rats, and at 5% it reduces weight gain by 38% (Ershoff, 1972). In a later study, Ershoff and Marshall (1975) investigated the effects of 26 different additions on cyclamate toxicity, with results summarized in Table III.

There have been a few studies of fiber effects on heavy metal toxicity. Wilson and De Eds (1950) reported that cadmium chloride was less toxic for rats fed 6% fiber than it was for those fed 3% fiber. The accumulation of strontium-89, calcium-45, or calcium-47 in blood and bone is reduced markedly by dietary calcium alginate (Paul *et al.*, 1966) or cellulose (Momcilovic and Gruden, 1981). Rowland *et al.* (1986) tested the effects of dietary fiber on tissue deposition and turnover time of methylmercury in BALB/c mice. Deposition of mercury in blood, brain, and intestine was reduced by 30% wheat bran, and half-time of mercury elimination was reduced by about 43% by 15% or 30% wheat bran (Table IV).

In early studies, Rusch (1944), Wilson and De Eds (1950), and Engel and Copeland (1952) showed that diets high in fiber-rich grains inhibited the carcinogenicity of azo dyes or 2-acetylaminofluorene. In the last 10–15 years there has been a proliferation of

TABLE III. Influence of Dietary Additions
on Toxicity of Sodium Cyclamate
in Immature Male Rats[a]

Diet	Weight gain (g)
Basal diet (B)[b]	78 ± 5
B + 5% Na cyclamate (BC)	19 ± 4[c]
BC + 10% Apple powder	25 ± 3
BC + 10% Potato meal	37 ± 4
BC + 10% Wheat bran	36 ± 4
BC + 10% Locust bean gum	38 ± 5
BC + 10% Guar gum	53 ± 3
BC + 10% Agar	53 ± 5
BC + 10% Irish moss powder	54 ± 5
BC + 10% Na alginate	62 ± 5
BC + 10% Gum tragacanth	65 ± 3
BC + 10% Gum karaya	71 ± 3
BC + 10% Unhulled sesame seed	28 ± 10[d]
BC + 10% Sunflower seed	22 ± 2
BC + 10% Alfalfa seed	46 ± 3
BC + 10% Flaxseed	56 ± 5
BC + 10% Blond psyllium seed	83 ± 4
BC + 10% Rice straw	42 ± 5
BC + 10% Sugar cane bagasse	48 ± 3
BC + 10% Parsley powder	54 ± 4
BC + 10% Watercress powder	56 ± 4
BC + 10% Cabbage powder	58 ± 3
BC + 10% Celery powder	61 ± 3
BC + 10% Dried orange peel	51 ± 5
BC + 10% Dried lemon peel	66 ± 3
BC + 10% Blond psyllium husk	77 ± 3
BC + 10% Alfalfa meal	55 ± 5
BC + 20% Alfalfa meal	72 ± 6
BC + 10% Carrot root powder	58 ± 3
BC + 20% Carrot root powder	80 ± 2

[a] After Ershoff and Marshall (1975).
[b] See footnote, Table I.
[c] 80% mortality.
[d] 50% mortality.

studies on the effects of dietary fibers on experimental carcinogenesis. The data have been reviewed (Kritchevsky, 1983, 1985, 1986). There is no unanimity or congruence among the studies because of marked differences in experimental design. The rats used are of different genders and strains, the diets commercial or semipurified, and the different carcinogens are administered by various routes. The fibers used are fed at levels ranging from 3% to 20%. As one example (Watanabe *et al.*, 1979), bran and pectin inhibit the carcinogenicity of azoxymethane (AOM) administered subcutaneously but are ineffective against methylnitrosourea (MNU) administered intrarectally, whereas alfalfa

TABLE IV. Concentration (μg/g) of Mercury in Tissues of BALB/c Mice Fed
Methylmercury[a]

| Fiber | Tissue[b] | | | | $T_{\frac{1}{2}}{}^c$ (days) |
	Blood	Kidney	Brain	Small intestine	
None	9.2	77.1	4.0	10.1	41.3
Pectin (5%)	9.6	79.8	5.0	8.9	38.3
Cellulose (5%)	10.0	90.2[d]	5.3	11.3	50.8
Bran (5%)	9.6	89.9[d]	4.6	10.1	35.3
Bran (15%)	8.4	86.2	4.4	9.0	23.9
Bran (30%)	7.3[d]	82.4	3.8[d]	8.5	23.3

[a] After Rowland et al. (1986).
[b] No difference in liver concentration: 14.0 in control; 14.0 $\pm$ 0.46 in all treatments.
[c] Half-time of mercury elimination.
[d] Significantly different from control.

is ineffective against AOM but enhances the carcinogenicity of MNU. Various aspects of fiber and experimental carcinogenesis are reviewed elsewhere in this volume.

As with so many other perceived panaceas, fiber is not an unmixed blessing. In cases where toxicity may be influenced by the metabolic activity of cecal microflora, the enhancing effect of fiber metabolic products on bacterial growth may enhance toxicity. Wise et al. (1982) tested dietary effects on the in vitro reduction of Amaranth, p-nitrobenzoic acid, and nitrate by cecal bacteria. The diets tested were stock diet, semipurified diet, and semipurified diet plus 5% pectin. There were no differences among the diets when reduction of Amaranth or p-nitrobenzoic acid was studied. Reduction of nitrate was similar for bacteria from rats fed the stock or the semipurified diets, but the 5% pectin diet resulted in a significant increase in nitrite production, which was paralleled by methemoglobinemia in rats fed nitrate and the pectin-containing diet. Fibers are metabolized to short-chain fatty acids (acetic, propionic, butyric), whose individual effects on bacterial metabolism require elucidation. deBethizy and Goldstein (1985) have summarized their data, which show that soluble (fermentable) fiber increases the number of anaerobic bacteria in the rat cecum and thus the capacity for nitroreduction and azoreduction. The increased metabolism of nitrobenzene and dinitrotoluene is correlated with their increased toxicity. Addition of 5% pectin to a semipurified diet increased microfloral β-glucuronidase activity by 198% and microfloral nitroreductase activity by 250%. deBethizy and Goldstein (1985) also found that metabolism of Amaranth was significantly enhanced in rats fed pectin but not in those fed cellulose, lignin, or Metamucil.

The effects of dietary fiber on xenobiotics are affected by the individual chemical and physical properties of the individual fibers. This area of research is relatively new. Early studies such as those of Ershoff concentrated on survival and weight gain, but it is evident that effects that depend on fiber metabolites or on microfloral growth and metabolism deserve further careful study.

ACKNOWLEDGMENTS. Supported in part by a Research Career Award (HL-00734) from the National Institutes of Health and by funds from the Commonwealth of Pennsylvania.

REFERENCES

Chow, B. F., Burnett, J. M., Ling, C. T., and Barrows, L., 1953, Effect of basal diets on the response of rats to certain dietary non-ionic surface-active agents, *J. Nutr.* **49:**563–577.

deBethizy, J. D., and Goldstein, R. S., 1985, The influence of fermentable dietary fiber on the disposition and toxicity of xenobiotics, in: *Xenobiotic Metabolism: Nutritional Effects* (J. W. Finley and D. E. Schwass, eds.), American Chemical Society, Washington, DC, pp. 37–50.

Engel, R. W., and Copeland, D. H., 1952, Protective action of stock diets against the cancer-inducing action of 2-acetyl-aminofluorene in rats, *Cancer Res.* **12:**211–215.

Ershoff, B. H., 1954, Protective effects of alfalfa in immature mice fed toxic doses of glucoascorbic acid, *Proc. Soc. Exp. Biol. Med.* **87:**134–136.

Ershoff, B. H., 1957, Beneficial effects of alfalfa and other succulent plants on glucoascorbic acid toxicity in the rat, *Proc. Soc. Exp. Biol. Med.* **95:**656–659.

Ershoff, B. H., 1960, Beneficial effects of alfalfa meal and other bulk containing or bulk-forming materials on the toxicity of non-ionic surface-active agents in the rat, *J. Nutr.* **70:**484–490.

Ershoff, B. H., 1963, Comparative effects of a purified and stock diet on DBH (2,5-di-*tert*-butylhydroquionone) toxicity in the rat, *Proc. Soc. Exp. Biol. Med.* **112:**362–365.

Ershoff, B. H., 1972, Comparative effects of a purified diet and stock ration on sodium cyclamate toxicity in rats, *Proc. Soc. Exp. Biol. Med.* **141:**857–862.

Ershoff, B. H., 1976, Protective effects of cholestyramine in rats fed a low-fiber diet containing toxic doses of sodium cyclamate or Amaranth, *Proc. Soc. Exp. Biol. Med.* **152:**253–256.

Ershoff, B. H., 1977, Effects of diet on growth and survival of rats fed toxic levels of tartrazine (FD & C yellow no. 5) and sunset yellow (FD & C yellow no. 6), *Proc. Soc. Exp. Biol. Med.* **107:**822–828.

Ershoff, B. H., and Hernandez, H. J., 1959, Beneficial effects of alfalfa meal and other bulk-containing or bulk-forming materials on symptoms of Tween 60 toxicity in the immature mouse, *J. Nutr.* **69:**172–178.

Ershoff, B. H., and Marshall, W. E., 1975, Protective effects of dietary fiber in rats fed toxic doses of sodium cyclamate and polyoxyethylene sorbitan monostearate (Tween 60), *J. Food Sci.* **40:**357–361.

Ershoff, B. H., and Thurston, E. W., 1974, Effects of diet on Amaranth (FD & C red no. 2) toxicity in the rat, *J. Nutr.* **104:**937–942.

Kimura, T., Furuta, H., Matsumoto, Y., and Yoshida, A., 1980, Ameliorating effects of dietary fiber on toxicities of chemicals added to a diet in the rat, *J. Nutr.* **110:**513–521.

Kimura, T., Imamura, H., Hasegawa, K., and Yoshida, A., 1982, Mechanisms of toxicities of some detergents added to a diet and of the ameliorating effect of dietary fiber in the rat, *J. Nutr. Sci. Vitaminol.* **28:**483–489.

Kritchevsky, D., 1983, Fiber, steroids and cancer, *Cancer Res.* **43:**2491S–2495S.

Kritchevsky, D., 1985, Dietary fiber and cancer, *Nutr. Cancer* **6:**213–219.

Kritchevsky, D., 1986, Fiber and cancer, in: *Dietary Fiber: Basic and Clinical Aspects* (G. V. Vahouny, and D. Kritchevsky, eds.), Plenum Press, New York, pp. 427–432.

Kritchevsky, D., and Story, J. A., 1974, Binding of bile salts *in vitro* by nonnutritive fiber, *J. Nutr.* **104:**458–462.

Momcilovic, B., and Gruden, N., 1981, The effect of dietary fibre on [85]Sr and [47]Ca absorption in infant rats, *Experientia* **37:**498–499.

Paul, T. M., Skoryna, S. C., and Waldron-Edward, D., 1966, Studies on the inhibition of absorption of radioactive strontium. V. The effect of administration of calcium alginate, *Can. Med. Assoc. J.* **95:**957–960.

Rowland, I. R., Mallett, A. K., Flynn, J., and Hargreaves, R. J., 1986, The effect of various dietary fibres on tissue concentration and chemical form of mercury after methylmercury exposure in mice, *Arch. Toxicol.* **59:**94–98.

Rusch, H. P., 1944, Extrinsic factors that influence carcinogenesis, *Physiol. Rev.* **24:**177–204.

Story, J. A., and Kritchevsky, D., 1976, Comparison of the binding of various bile acids and bile salts *in vitro* by several types of fiber, *J. Nutr.* **106:**1292–1294.

Watanabe, K., Reddy, B. S., Weisburger, J. H., and Kritchevsky, D., 1979, Effect of dietary alfalfa, pectin and wheat bran on azoxymethane or methylnitrosourea-induced colon carcinogenesis in F344 rats, *J. Natl. Cancer Inst.* **63:**141–145.

Wilson, R. H., and De Eds, F., 1950, Importance of diet in studies of chronic toxicity, *Arch. Ind. Hyg. Occup. Med.* **1:**73–80.

Wise, A., Mallett, A. K., and Rowland, I. R., 1982, Dietary fibre, bacterial metabolism and toxicity of nitrate in the rat, *Xenobiotica* **12:**111–118.

Woolley, D. W., and Krampitz, L. D., 1943, Production of a scurvy-like condition by feeding of a compound structurally related to ascorbic acid, *J. Exp. Med.* **78:**333–345.

28

National Cancer Institute Satellite Symposium on Fiber and Colon Cancer

ELAINE LANZA

1. INTRODUCTION

The vast majority of research studies on dietary fiber and cancer have focused on colon cancer. In the United States today the incidence and mortality rates for colon cancer are second only to those for lung cancer (Silverberg and Lubera, 1988). The mortality from this malignancy has not changed materially in the past 50 years, highlighting the lack of effective therapy. As a consequence, prevention appears to offer the best opportunity to control this disease.

Although the incidence of colon cancer varies considerably across the world, cancer rates are extremely high in most other westernized countries, including England, Australia, and most of Europe. The epidemiology of colon cancer, particularly time trends, and studies of migrants and religious groups indicate that environmental rather than genetic factors are of overwhelming importance in determining the incidence of this disease in a given population. When considering the possible environmental factors of importance in cancer of the colon, available data concentrate on diet. Most animal and epidemiologic studies have shown a positive association with high intake of fat, calories, and meat and a low intake of dietary fiber and vegetables.

The main purpose of this chapter is to review the scientific evidence that a diet high in dietary fiber can reduce the risk for colon cancer and discuss approaches that will lead to definitive clinical data either supporting or refuting this hypothesis. To begin I would like to summarize briefly some of the difficulties frequently encountered in diet and cancer research, problems that need to be solved in order to bridge the gap from basic animal and descriptive epidemiologic studies to human intervention research. There are

ELAINE LANZA • Cancer Prevention Research Program, Division of Cancer Prevention and Control, National Cancer Institute, National Institutes of Health, Bethesda, Maryland 20892.

four categories of research studies in diet and cancer: (1) animal studies, (2) descriptive epidemiologic studies, (3) analytical epidemiological studies, and (4) intervention trials. Each of these areas offers unique opportunities for advancing our knowledge about dietary fiber and colon cancer.

2. ANIMAL STUDIES

One question that is always raised when discussing animal models of carcinogenesis is the concern about the relevance of animal models to human cancer. Yet these studies could be made more relevant. Often the dose level of fiber is so high that the effect is pharmacological rather than nutritional. For example, in rodent diets the usual 5% dietary fiber diet is equivalent to 25 g of dietary fiber in a human diet, whereas a high-dietary-fiber diet (20% dietary fiber/day weight) is equivalent to a 100-g dietary fiber diet in humans. Another problem is that dose–response studies are rarely done for dietary components. The relationship between the dose of a dietary or nutritional factor and the magnitude of its impact on carcinogenesis, generally expressed as a relative risk, is clearly important but almost never done in dietary carcinogenesis studies.

The complex interrelationships and interactions among nutrients are also not appropriately considered in most animal studies. Are combinations of nutrients more effective than a single nutrient in preventing or causing cancer? Can a low level of one nutrient in the diet be compensated by higher levels of other nutrients? In analyzing highly correlative variables, can attribution of risk be separated for one factor from the other?

Another problem in fiber colon cancer animal model studies is the use of fiber-free controls. This is especially important in colon cancer if the mechanism involves bacteria metabolism. If the only control is without fiber, then extrapolation of data from animal to the human situation becomes even more difficult.

3. DESCRIPTIVE EPIDEMIOLOGY STUDIES

Table I summarizes the results from 48 epidemiologic studies published during the period 1970–1988. Thirty-eight of these studies identified an inverse or protective association between dietary fiber, fiber-rich diets, or other measures of fiber consumption and the risk of colon cancer. Seven studies showed no association or were considered equivocal, and three reported a direct or promotive association with colon cancer. A detailed critical review of most of these studies was published in 1987 (Greenwald *et al.*, 1987).

Six of seven large-scale international correlation studies report lower cancer rates in populations eating a high-fiber diet. Six of seven within-country studies comparing population subgroups found a protective effect. Other evidence supporting a protective role for fiber in colon cancer comes from metabolic epidemiology studies. These studies involve assessments of both eating patterns and metabolic data. Six of seven metabolic epidemiology studies show an inverse or protective effect for high fiber. Finally, three time-trend studies report protective effects for fiber.

The above four types of studies are all classified as ecological studies. These

TABLE I. Summary of Epidemiologic Studies on Dietary Fiber and Colon Cancer

Type of study	Number of studies	Associations[a]		
		Inverse	None	Direct
1. International correlation	7	6	1	0
2. Within-country correlation	7	6	1	0
3. Metabolic epidemiologic	7	6	1	0
4. Time trend	3	3	0	0
5. Case-control	23	16	4	3
6. Cohort	1	1	0	0
Total	48	38	7	3

[a] Associations between fiber intake or fiber-rich diets and colon cancer.

ecological or correlation studies have the ability to examine the influence of dietary changes over long periods of time, a critical issue in dietary exposure and cancer development. In ecological studies (international correlation studies, within-country correlation studies, and metabolic correlation studies), cancer rates of various populations may be correlated with data on food disappearance, dietary surveys, or blood chemistries. Ecological studies frequently rely on per-capita consumption data, not individual intake, and tend to focus on a mortality end point rather than cancer incidence. It is important to be cognizant of these details when comparing different studies and study types. For example, inconsistencies in the literature might be attributed to differences in the association of fiber with cancer incidence compared to cancer death or mortality rates.

Ecological studies are sometimes criticized because of a concept known as the "ecological fallacy." Two populations may differ in many factors (e.g., genetics, pollutant exposure, employment, weather) in addition to the factor under study. It may be that one or more of these factors actually account for observed differences in disease outcomes. Ecological study data are generally from a specific point in time, a potential limitation in cancer etiology research. Although ecological studies are sometimes considered less powerful and useful than case-control and cohort studies, the data allow multiple comparisons that permit relatively inexpensive, rapid testing and generation of hypotheses. Consistency among correlations from many different studies strengthens the argument for a cause-and-effect relationship (United States Department of Health and Human Services, 1982).

4. ANALYTICAL EPIDEMIOLOGY STUDIES

There are 23 case-control studies in the literature that assess the role of dietary fiber or diets rich in fiber in reducing the risk of colon cancer. Sixteen of these show an inverse or protective association, four show no association, and three show a positive or promotive association (Table I). In case-control studies, individual cases of the disease of interest are selected. They are compared to persons without the disease, the controls. The cases and controls are compared with respect to present or past exposure to hypothesized risk factors. Exposure can refer to specific dietary components, history of other disease, genetic constitution, or specific biochemical indices (e.g., hypercholesterolemia). If cases are found to have greater numbers of individuals with the hypothesized risk factor, the hypothesis is strengthened. The risk is often expressed as a relative odds or odds ratio that reflects the proportion of cases with the risk factor compared to the controls.

There are three major problems associated with diet-and-cancer case-control studies: (1) imprecision of dietary assessment, (2) highly correlated variables, and (3) narrow range of intake (Byar and Freedman, 1988). The measure of dietary intake, such as the dietary questionnaire, is not sufficiently sensitive to assess diet quality accurately, especially if the time period in question is 10 to 20 years in the past. Jain and her colleagues (1980) compared the data obtained from a food frequency questionnaire to a 30-day food record and found that fiber gave the lowest correlation of any nutrient, $r = 0.24$. It is a well-known fact that the variable with the larger measurement error will be more misclassified and will bias your estimate toward the null (Macquart-Moulin et al., 1986). Thus, the sample size needed to assess the effect of fiber in this study was larger than for other nutrients.

The problem of highly correlated variables makes it difficult to untangle the effects of individual dietary components. In case-control studies the question is often raised as to whether the reduced risk is from dietary fiber or from vegetables. In the United States 45% of the dietary fiber intake comes from vegetable sources (Block and Lanza, 1987), and thus the correlation between these variables is quite high. When variables are highly correlated, the statistical power of the study is significantly reduced, thus again necessitating a larger than expected sample size.

It is difficult to establish a true association in case-control studies if the range of exposure (e.g., food groups or nutrients) is narrow, as is often true in defined populations. For example, if the cases and controls are drawn from a population that is homogeneous in food or nutrient intake, differences between cases and controls may not be large enough to be seen. The amount of potentially beneficial food constituent ingested by most of the population may, therefore, be insufficient to result in an observable benefit.

5. INTERVENTION TRIALS

Intervention research, or clinical trials, is a newer approach in cancer prevention. Clinical trials provide the strongest research design to test whether or not an intervention will have the hypothesized effect. A specificity not possible in epidemiologic studies derives from the prospective nature and use of randomized controls in clinical trials. These trials may show whether or not data from animal studies are applicable to people and whether people will adhere to the intervention for the duration of the trial (Greenwald et al., 1986). Although randomized controlled intervention trials are useful in demonstrating a causal relationship, the cost and time involved in conducting a dietary intervention in which cancer is the end point is enormous. It has been estimated that a clinical trial with colon cancer as the end point would require a sample size of 100,000 and take from 5 to 10 years (Bruce et al., 1981). A number of scientists now believe that a great proportion of colon cancers develop from preexisting adenomatous polyps. Even though only a small fraction of adenomas may actually develop into polyps, the use of polyp recurrence as an outcome measure could considerably reduce the sample size needed to study colon cancer prevention.

The following chapters summarize the progress being made in these various areas. I hope that after reflecting on the data presented we will be better able to develop innovative study designs and new research tools that will help provide definitive answers to the dietary fiber–colon cancer hypothesis.

REFERENCES

Block, G., and Lanza, E., 1987, Dietary fiber sources in the United States by demographic group, *J. Natl. Cancer Inst.* **79**(1):83–91.

Bruce, W. R., McKeown-Eyssen, G., Ciampi, A., Dion, P. W., and Boyd, N., 1981, Strategies for dietary intervention studies in colon cancer, *Cancer* **47**:1121–1127.

Byar, D. P., and Freedman, L. S., 1989, Clinical trials in diet and cancer, *Prev. Med.* **18**:203–219.

Greenwald, P., Sondik, E., and Lynch, B., 1986, Diet and chemoprevention in NCI's research strategy to achieve national cancer control objectives, *Annu. Rev. Publ. Health* **7**:267–291.

Greenwald, P., Lanza, E., and Eddy, G., 1987, Dietary fiber in the reduction of colon cancer risk, *J. Am. Dietet. Assoc.* **87**(9):1178–1188.

Jain, M., Howe, G. R., Johnson, K. C., and Miller, A. B., 1980, Evaluation of a diet history questionnaire for epidemiologic studies, *Am. J. Epidemiol.* **111**(2):212–219.

Macquart-Moulin, G., Riboli, E., Cornee, J., Charnay, B., Berthezene, P., and Nicholas, D., 1986, Case-control study on colorectal cancer and diet in Marseilles, *Int. J. Cancer* **38**:183–191.

Silverberg, E., and Lubera, J., 1988, Cancer statistics, 1988, *CA* **38**:5–22.

U.S. Department of Health and Human Services, 1982, *The Health Consequences of Smoking: Cancer. A Report of the Surgeon General*, DHHS publication no. (PHS) 82-50179, United States Public Health Service, Rockville, MD.

29

Influence of Soluble Fibers on Experimental Colon Carcinogenesis

LUCIEN R. JACOBS

1. INTRODUCTION

The original fiber hypothesis of Burkitt (1971) proposed that a high consumption of fiber-containing foods is associated with a lower frequency of large bowel cancer. Although this association still seems valid, the mechanisms by which fiber-containing foods produce such an effect remain unclear. Furthermore, definitive clinical trials in humans to demonstrate the cancer-prevention properties of dietary fiber have yet to be successfully completed. In the meanwhile we have only the results from retrospective, correlation, and case-control studies in humans and controlled prospective studies in animals. In a recent review of the literature (Jacobs, 1988), I found that 62% of correlational and 48% of case-control studies showed evidence of a protective effect of dietary fiber. Moreover, out of the 11 case-control studies showing a protective effect, eight reports found vegetables to be the protective fiber-containing food. This suggests that not all fiber-containing foods are equally protective and that vegetables may be protective because of some ingredient they contain other than fiber. A further possible explanation is that not all forms of dietary fiber have a similar effect on colon carcinogenesis. Recent advances in the chemical analysis of dietary fibers and in our knowledge of the effects of fibers on gastrointestinal physiology have permitted a reanalysis of the fiber and cancer literature according to the physiochemical properties of individual fibers.

LUCIEN R. JACOBS • Department of Medicine, University of California, Los Angeles, and Section of Nutrition, Division of Gastroenterology, Cedars-Sinai Medical Center, Los Angeles, California 90048.

2. SOLUBLE AND FERMENTABLE FIBERS

Dietary fibers can be categorized as being insoluble or soluble. A currently used definition of soluble fiber (Pilch, 1987) is a fiber that is soluble in water, depending on the method of extraction. This includes gums, pectins, mucilages, and some hemicelluloses. Soluble fibers tend to be thought of as producing their greatest effects in the upper digestive system, slowing gastric emptying and delaying the site and rate of nutrient assimilation in the small intestine. The insoluble fibers, in general, produce a greater effect on the large bowel, increasing fecal bulk and speeding up the rate of colonic transit. However, such a simple classification overlooks the fate of soluble fibers when they reach the large intestine. In general most soluble fibers are also very fermentable, meaning that they are capable of being used as a substrate by microorganisms in anaerobic catabolism (Pilch, 1987). Anderson and Bridges (1988) analyzed a large number of foods and estimated their content of soluble fiber as a percentage of total fiber present. The average content in cereal products was 30.7%, vegetables 32.3%, legumes 24.5%, and fruits 38.9%. This information suggests that the average Western diet will normally contain a relatively low percent of soluble fiber as compared with the insoluble fraction. This is important to consider when interpreting animal data in which investigators have usually fed a single source of a soluble fiber.

3. SOLUBLE FIBERS AND EXPERIMENTAL COLON CANCER

The levels of fiber fed vary widely from study to study. Similarly, other dietary constituents such as the amount and type of fat may also be confounding variables. The majority of investigators have used the rat model together with the carcinogen 1,2-dimethylhydrazine (DMH) or its more potent metabolite azoxymethane (AOM). These carcinogens are usually administered systematically but can be given orally, following which they are metabolized to a DNA-methylating agent. Administration of DMH or AOM to rodents produces both benign adenomas and malignant carcinomas of the large bowel. Methylnitrosourea (MNU) is a direct-acting carcinogen and needs to be administered by repeated intrarectal installations. Pectin has been the most widely studied of all the soluble fiber supplements.

3.1. Pectin and Experimental Colon Cancer

These results are summarized in Table I. Fifty percent of these eight studies actually demonstrated tumor enhancement while three found no effect, and one found evidence of protection. The amount of fat in the diet did not appear to influence the effect of pectin on tumor development. Freeman et al. (1980) introduced the pectin into the diet only when carcinogen administration was complete. In contrast, Bauer et al. (1979, 1981) fed the pectin only during the period of carcinogen exposure and then removed the pectin during the promotional or postcarcinogen stage. Jacobs and Lupton (1986) and Watanabe et al. (1979) fed the pectin throughout the entire experiment. Although not entirely consistent, these data suggest that pectin stimulates tumor development by

TABLE I. Dietary Pectin and Chemically Induced Rat Colon Cancer[a]

Diet (% by wt.)						
Pectin	Fat	Carcinogen	Strain	Sex	Effect	Reference
4.5	8	DMH	Wistar	M	N	Freeman et al. (1980)
9.0	8	DMH	Wistar	M	N	Freeman et al. (1980)
5.0 LM	20	DMH	Sprague–Dawley	M	E	Bauer et al. (1981)
5.0 HM	20	DMH	Sprague–Dawley	M	E	Bauer et al. (1981)
6.5	20	DMH	Sprague–Dawley	M	E	Bauer et al. (1979)
10	8	DMH	Sprague–Dawley	M	E	Jacobs and Lupton (1986)
15	20	AOM	Fischer 344	F	P	Watanabe et al. (1979)
15	20	MNU	Fischer 344	F	N	Watanabe et al. (1979)

[a]DMH, 1,2-dimethylhydrazine; AOM, azoxymethane; MNU, methylnitrosourea; N, no effect; E, enhancement; P, protective; LM, low methoxylated; HM, high methoxylated.

somehow sensitizing the colonic epithelium to carcinogenic exposure. The results do not support the idea that pectin acts during the promotional or postcarcinogen stage of tumor development.

3.2. Guar Gum, Carrageen, and Agar

The results from these studies are summarized in Table II. Five percent guar gum with 20% fat had no effect, whereas 10% guar with 8% fat enhanced tumor development. Although only two dose levels were examined, these results do support the idea that tumor enhancement occurs only with the higher levels of soluble fibers, independent of the amount of dietary fat. This suggestion of a dose–response effect is further supported by the carrageenan data, where a 6% carrageenan diet was in fact protective when given with 6% fat, whereas 15% carrageenan enhanced tumorigenesis in animals consuming 20% fat. Agar at either 7% or 9% enhanced tumor development in mice fed either low- or high-fat diets.

3.3. Psyllium and Colon Cancer

The results of feeding different psyllium preparations on the development of DMH-induced colon cancer are summarized in Table III. Four out of the seven experiments (57%) found evidence of protection, two showed no effect, and one had evidence of tumor enhancement (Toth, 1984). It is of note that tumor enhancement occurred in male but not female mice and that despite a 50% reduction in tumor yield compared with controls, this was greater than the expected number of tumors and therefore interpreted as tumor enhancement. However, as with data in Table II, this result raises the question of whether such relatively high levels of soluble fiber increase tumor development in contrast to lower levels, which may be protective. Wilpart and Roberfroid (1987) randomized their animals to various experimental diets only after carcinogen administration was completed, in contrast to Roberts-Andersen et al. (1987) and Toth (1984) who both fed the psyllium throughout all stages of the carcinogenic process.

TABLE II. Soluble Fibers and Colon Cancer[a]

Diet (% by wt.)							
Fiber	Fat	Carcinogen	Animal	Strain	Sex	Effect	Reference
5% Guar gum	20% Peanut oil	DMH	Rat	Sprague–Dawley	M	N	Bauer et al. (1981)
10% Guar gum	8% Corn oil	DMH	Rat	Sprague–Dawley	M	E	Jacobs and Lupton (1986)
6% Carrageenan	6% Corn oil	DMH	Rat	Fischer 344	M	P	Arakawa et al. (1986)
15% Carrageenan	20% Corn oil	AOM	Rat	Fischer 344	F	E	Watanabe et al. (1978)
15% Carrageenan	20% Corn oil	MNU	Rat	Fischer 344	F	E	Watanabe et al. (1978)
7% Agar	2% Safflower oil	DMH	Mouse	CF$_1$	M	E	Glauert et al. (1981)
9% Agar	20% Beef tallow +2% Safflower oil	DMH	Mouse	CF$_1$	M	E	Glauert et al. (1981)

[a]DMH, 1,2-dimethylhydrazine; AOM, azoxymethane; MNU, methylnitrosourea; E, enhancement; P, protective; N, no effect.

TABLE III. Psyllium Fiber and Dimethylhydrazine-Induced Colon Cancer

Fiber	Fat	Animal	Strain	Sex	Effect	Reference
Diet (% by wt.)						
5% Fybogel[a]	5% palm/corn oil	Rat	Wistar	M	N	Wilpart and Roberfroid (1987)
5% Fybogel[a]	20% palm/corn oil	Rat	Wistar	M	P	Wilpart and Roberfroid (1987)
15% Fybogel[a]	5% palm/corn oil	Rat	Wistar	M	P	Wilpart and Roberfroid (1987)
15% Fybogel[a]	20% palm/corn oil	Rat	Wistar	M	P	Wilpart and Roberfroid (1987)
10% Sat-Isabgol[b,d]	20% lard	Rat	Sprague–Dawley	M	P	Roberts-Andersen et al. (1987)
20% Metamucil[c,d]	NM	Mouse	Swiss albino	M	E	Toth (1984)
20% Metamucil[c,d]	NM	Mouse	Swiss albino	F	N	Toth (1984)

[a]Mucillage from seed of Ispaghula ex Plantago ovata Forsk.
[b]Psyllium husk.
[c]Psyllium hydrophilic mucilloid (derived from husk).
[d]From Plantago ovata seeds.
[e]NM, not measured (Wayne Lab Blox powdered diet); N, no effect; P, protective; E, enhancement.

4. FERMENTATION AND COLON CANCER

In monogastric animals, dietary fiber remains more or less intact during its passage through the upper gastrointestinal tract. However, when fiber passes into the large intestine it becomes available for fermentation by anaerobic microorganisms, which degrade dietary fibers to varying degrees, producing short-chain fatty acids (SCFA), gases, and a more acidic pH (Cummings, 1983; Nyman and Asp, 1982). In general, soluble fibers are more fermentable than insoluble fibers. The earlier version of the fiber hypothesis (Burkitt, 1971) suggested that a high-fiber diet inhibited tumor development primarily by increasing the bulk of intestinal contents, diluting any carcinogens and tumor promoters, and reducing their contact time with the colonic mucosa by speeding up intestinal transit. In addition, it was suggested that a fiber-deficient diet would alter colonic bacterial activity and the degradation of bile acids and other substrates that have been implicated in malignant transformation. At that time it was not well recognized that soluble fibers are extensively metabolized in the large bowel and therefore have much less effect on colonic motility and fecal bulk than insoluble fibers. Thus, if bulking and transit are the key factors in reducing neoplasia, then the insoluble fibers should be the most effective inhibitors of colon cancer. On the other hand, if changes in colonic bacterial metabolism are the key factors, then fermentable fibers should be more effective tumor inhibitors.

So far the summary analysis supports the idea that the more soluble fibers such as pectin, guar, agar, and carrageenan are the least effective at tumor prevention and can even enhance tumor formation. Psyllium, which is the least fermentable of the group, seems to be the most consistently effective at tumor inhibition. However, the question that needs to be considered is: How do soluble fibers enhance carcinogenesis, and what is the relevance of these animal data to human colon cancer risk? The major end products of bacterial fermentation are the SCFA, of which butyrate in particular appears to exhibit antineoplastic activity (Kruh, 1982). The increased formation of SCFA also produces acidification of colonic contents, which in itself is believed to inhibit colon carcinogenesis (Thornton, 1981; Bruce, 1987). Human populations with a lower fecal pH have lower rates of colorectal cancer (Bruce, 1987; Thornton, 1981; van Dokkum *et al.*, 1983; Walker *et al.*, 1986). As pH drops, the solubility of free bile acids and free fatty acids is diminished, thereby decreasing their potential tumor promoter activity. Primary bile acids are degraded in the colon to secondary bile acids by bacterial 7α-dehydroxylase, the activity of which is markedly inhibited below pH 6.0–6.5 (Thornton, 1981). Therefore, a fall in pH will decrease the production of secondary bile acids within the colon and reduce the level and activity of these postulated tumor promoters. In addition, acidification of colonic contents increases the availability of luminal calcium for binding to free bile and fatty acids and will thereby further inhibit their mitogenic and cocarcinogenic effects (Wargovich *et al.*, 1984). The increase in luminal acidity also reduces the concentration of ammonia, a bacterial metabolite of protein and possible tumor promoter (Thornton, 1981). A number of studies have been performed to test the validity of the pH hypothesis and specifically to see whether fecal acidification inhibits colon carcinogenesis.

These results are summarized in Table IV. Three experiments fed lactulose, a synthetic disaccharide of galactose and fructose that is not hydrolyzed in the small

TABLE IV. Effect of Fecal Acidification on Dimethylhydrazine-Induced
Rat Colon Cancer[a]

Acidifying agent	Level	Route of administration	Acid pH	Effect	Reference
Lactulose	3.3%	Drinking water	NM	63% Incr (NS)	Ingram and Castleden (1980)
Lactulose	7.5%	Chow	Yes (NS)	31% Decr	Samelson et al. (1985)
Lactulose	10%	Defined diet	Yes	3.3× Incr	Jacobs (1986a)
Sodium sulfate	0.25%	Chow	Yes (NS)	34% Decr	Samelson et al. (1985)
Sorbitol	10%	Defined diet	Yes	82% Incr	Jacobs (1986b)

[a]Incr, increase; Decr, decrease; NS, not significant; all studies used male Sprague–Dawley rats except for the first which used male Wistar rats.

intestine but is fermented in the large intestine. Administration of 3.3% lactulose in drinking water and as 10% of a defined diet enhanced tumor development (statistically significant only at the 10% level), whereas a 7.5% level added to rat chow decreased tumorigenesis. Two of these reports documented effective fecal acidification. Other methods of fecal acidification using sodium sulfate and sorbitol, a hexahydric alcohol have demonstrated tumor inhibition and enhancement, respectively. Apart from the many differences in experimental design, it is not clear why these investigators report such contradictory data. One possible explanation is that acidification of colonic contents by using fermentable substrates is not ideal since feeding butyrate alone enhances colon carcinogenesis (Freeman, 1986). Thus, acidification of colonic contents may still be an effective way to inhibit colon cancer, but these recent data suggest that this might best be accomplished by using a nonfermentable substrate such as sodium sulfate (Samelson et al., 1985), which does not increase SCFA production within the colon.

TABLE V. Soluble Fibers and Cecal Short-Chain Fatty Acids

Diet	Changes relative to control[c]				Reference
	Total SCFA	Acetate	Propionate	Butyrate	
Starvation	D	D	D	D	Illman et al. (1986)
Wheat bran	N	D	N	I	Walter et al. (1986)
Chow[a]	I	I	I	I	Illman et al. (1982)
Pectin	I	I	N	I	Thomsen et al. (1984)
Pectin[a]	I	I	I	N	Illman et al. (1982)
Guar gum	I	I	I	N	Tulung et al. (1987)
Gum arabic[b]	I	I	I	N	Storer et al. (1984)
Oat bran	I	I	I	I	Illman and Topping (1985)

[a]Portal venous blood.
[b]Portal blood and cecal contents.
[c]D, decrease; N, no effect; I, increase.

Further evidence that SCFA do not inhibit the development of experimental colon cancer is presented in Table V. Starvation decreases SCFA levels uniformly. Wheat bran selectively increases butyrate without increasing total SCFA levels. Wheat bran and cellulose are the fibers that most consistently inhibit experimental colon cancer (Jacobs, 1988). As stated earlier, pectin and guar gum can enhance colon carcinogenesis, and yet both fibers increase total SCFA levels, with pectin increasing butyrate. Oat bran, which has been reported to enhance experimental colon cancer (Jacobs and Lupton, 1986), also increases total SCFA levels and specifically butyrate levels. Thus, the data show that soluble fibers are effective substrates for SCFA production and acidify colonic contents (Jacobs and Lupton, 1986), yet despite this, the soluble fibers have been shown to increase tumor development. Although indirect, these results do not support a role for SCFA as inhibitors of colon carcinogenesis. In fact, the data suggest that if anything SCFA enhance experimental colon cancer (Freeman, 1986; Jacobs and Lupton, 1986).

5. MICROBIAL METABOLIC ACTIVITY AND COLON CANCER

As discussed above, the intestinal microflora play a central role in the metabolism of dietary fiber. In addition, the large bowel microflora can metabolize many naturally occurring and xenobiotic compounds that gain access to the colon, producing products capable of interacting with host tissues. Gut microbial enzymes appear to play an important role in the development of chemically induced (DMH) colon cancer. Previous studies indicate that bacterial deconjugation enzymes increase the genotoxicity of DMH and its related compounds (Lacquer et al., 1981; Takada et al., 1982). Dietary components such as fiber that are resistant to digestion and absorption in the upper digestive tract can influence carcinogen activation by altering the environment of the large bowel or by providing a nutrient source to the microflora and thereby modifying bacterial biotransformation processes. Among the bacterial enzymes that have been more frequently studied are β-glucosidase, which converts glucosides to toxic aglycones, and β-glucuronidase, which activates the carcinogen AOM by hydrolyzing the conjugate of methylazoxymethanol in the intestinal lumen (Takada et al., 1982). Nitrate reductase reduces nitrates to nitroso products, while nitroreductase reduces aromatic nitro compounds to aromatic amines, which can then be converted to carcinogenic N-hydroxy compounds by a tissue N-hydroxylase. Azoreductase reduces azo compounds to aromatic amines and in the process produces highly reactive intermediates that react with proteins and nucleic acids.

The effects of fibers on microbial metabolic activity are summarized in Table VI. Cellulose decreased all enzyme activities, consistent with its usual protective effect, whereas wheat bran decreased or had no effect on all enzymes measured except for β-glucosidase, which it increased. Pectin had little effect except on nitrate reductase, which it increased. Guar gum increased all enzyme activities except for nitrate reductase. Carrageenan decreased all enzymes, whereas agar decreased or had no effect on enzyme activities. Overall, there appears to be little relationship between enzyme activities and tumor modulation. Even though dietary fibers significantly modify microbial enzyme activities and tumor frequency, the enzymes that have so far been measured do not correlate with tumor enhancement or inhibition. It therefore seems as if either other

TABLE VI. Fibers and Metabolic Activity of Cecal Microflora[a]

Fiber	β-Glucosidase	β-Glucuronidase	Nitrate reductase	Nitroreductase	Azoreductase	Reference
Cellulose	D	D	D	D	D	Mallett et al. (1983)
Wheat bran	I	N	D	D	—	Mallett et al. (1986)
Pectin	N	N/D	I	N	N	Mallett et al. (1984)
Guar gum	I	I	N	I	I	Mallett et al. (1984)
Carrageenan	D	D	D	D	D	Mallett et al. (1984)
Agar	D	D	N	N	N	Mallett et al. (1984)

[a]D, decrease; I, increase; N, no effect.

enzymes are more relevant or microbial enzymes are not a critical factor in this model system.

6. FECAL BULK AND BILE ACIDS

All of the fibers discussed produce some increase in bulking, but this is minimal with the highly degradable fibers such as pectin and guar and marked with fibers such as psyllium, carrageenan, and agar. This fact tends to argue against the idea that increased fecal bulk inhibits tumor production, since carrageenan, agar, and oat bran all were associated with tumor enhancement. The influence of fibers on bile acid metabolism are discussed in detail elsewhere in this volume. Guar gum (Vahouny *et al.*, 1987) and agar (Glauert *et al.*, 1981) decrease total fecal bile acid concentration, whereas pectin and carrageenan increase secondary bile acid concentration (Reddy *et al.*, 1980). Psyllium did not alter fecal bile acid concentration (Vahouny *et al.*, 1982) in contrast to oat bran, which increases total and secondary bile acid concentrations (Illman and Topping, 1985). Thus, pectin, carrageenan, and oat bran could promote colon carcinogenesis by increasing the colonic concentration of bile acids. However, for guar gum and agar, which reduce bile levels, another explanation must be looked for.

7. CELL PROLIFERATION AND COLON CANCER

Some dietary fibers have been found to stimulate epithelial cell growth and modify colonic mucosal morphology. We have found that wheat bran, pectin, and guar gum consistently stimulate colonic epithelial cell proliferation (Jacobs, 1984; Jacobs and Lupton, 1984, 1986; Jacobs and White, 1983). On the other hand, oat bran, psyllium, and carrageenan produce little in the way of growth stimulation, whereas agar produces hyperplasia (Glauert *et al.*, 1981). It is generally well recognized that many tumor promoters are also growth factors and that dietary stimulation of cell growth during carcinogen exposure enhances tumor development (Jacobs, 1983, 1984). The mechanism whereby fibers stimulate colonic cell proliferation is probably related to the microbial fermentation of fiber and production of SCFA, which, when infused into the colon, stimulate cell proliferation (Sakata and Yajima, 1984). However, other factors such as bile and fatty acids also stimulate crypt cell production and can damage the surface epithelium (Cohen *et al.*, 1980; Craven *et al.*, 1987). Furthermore, acidification of the extracellular environment may also be a stimulus to the cell cycle (Lupton *et al.*, 1985, 1988). In addition, changes in mineral bioavailability appear to be important, and in particular alterations in calcium, which, when present at high concentration in the extracellular environment, inhibits cell proliferation.

8. SUMMARY

Controlled prospective experiments in which soluble fibers have been fed to rodents treated with chemical carcinogens that induce colon cancer have demonstrated that a number of soluble fibers are associated with tumor enhancement. A number of these studies suggest that this tumor enhancement may be the result of feeding excessively high levels of these products. Other factors such as increased microbial metabolic activation of precarcinogens to carcinogens, higher levels of secondary bile acids in the colonic lumen, a lower pH, increased SCFA production, and stimulation of colonic cell proliferation may all play a role. In fact, it seems likely that the tumor enhancement observed in these studies is multifactorial in origin. The relevance of these studies to human colon cancer is of great concern, particularly in light of current recommendations to increase the consumption of soluble fibers for the purpose or reducing serum cholesterol and in the management of diabetes. The animal experiments indicate the need for further evaluation in humans of the effects of soluble fibers on colonic metabolism and cell biology. Until such studies have been performed, we can only speculate about whether supplements of soluble fibers will increase human colon cancer risk.

REFERENCES

Anderson, J. W., and Bridges, S. R., 1988, Dietary fiber content of selected foods, *Am. J. Clin. Nutr.* **47**:440–447.

Arakawa, S., Okumura, M., Yamada, S., Ito, M., and Tejima, S., 1986, Enhancing effect of carageenan on the induction of rat colonic tumors by 1,2-dimethylhydrazine and its relation to β-glucuronidase activities in feces and other tissues, *J. Nutr. Sci. Vitaminol.* **32**:481–485.

Bauer, H. G., Asp. N.-G., Oste, R., Dahlqvist, A., and Fredlund, P. E., 1979, Effect of dietary fiber on the induction of colorectal tumors and fecal β-glucuronidase activity in the rat, *Cancer Res.* **39**: 3752–3756.

Bauer, H. G., Asp. N.-G., Dahlqvist, A., Fredlund, P. E., Nyman, M., and Oste, R., 1981, Effect of two kinds of pectin and guar gum on 1,2-dimethlhydrazine initiation of colon tumors and on fecal β-glucuronidase activity in the rat, *Cancer Res.* **41**:2518–2523.

Bruce, W. R., 1987, Recent hypotheses for the origin of colon cancer, *Cancer Res.* **47**:4237–4242.

Burkitt, D. P., 1971, Epidemiology of cancer of the colon and rectum, *Cancer* **28**:3–13.

Cohen, B. I., Raicht, R. F., Deschner, E. E., Takahashi, M., Sarwal, A. N., and Fazzini, E., 1980, Effect of cholic acid feeding on *m*-methyl-N-nitrosourea-induced colon tumors and cell kinetics in rats, *J. Natl. Cancer Inst.* **64**:573–578.

Craven, P. A., Pfanstiel, J., and DeRubertis, F. R., 1987, Role of activation of protein kinase C in the stimulation of colonic epithelial proliferation and reactive oxygen formation by bile acids, *J. Clin. Invest.* **79**:532–541.

Cummings, J. H., 1983, Fermentation in the human large intestine: Evidence and implications for health, *Lancet* **1**:1206–1209.

Freeman, H. J., 1986, Effect of differing concentrations of sodium butyrate on 1,2-dimethylhydrazine-induced rat intestinal neoplasia, *Gastroenterology* **91**:596–602.

Freeman, H. J., Spiller, G. A., and Kim, Y. S., 1980, A double-blind study on the effects of differing purified cellulose and pectin fiber diets on 1,2-dimethylhydrazine-induced rat colonic neoplasia, *Cancer Res.* **40**:2661–2665.

Glauert, H. P., Bennink, M. R., and Sander, S. H., 1981, Enhancement of 1,2-dimethylhydrazine-induced colon carcinogenesis in mice by dietary agar, *Food Cosmet. Toxicol.* **19**:281–286.

Illman, R. J., and Topping, D. L., 1985, Effects of dietary oat bran on faecal steroid excretion, plasma volatile fatty acids and lipid synthesis in rats, *Nutr. Res.* **5**:839–846.

Illman, R. J., Trimble, R. P., Snoswell, A. M., and Topping, D. L., 1982, Daily variations in the concentrations of volatile fatty acids in the splanchnic blood vessels of rats fed diets high in pectin and bran, *Nutr. Rep. Int.* **26**:439–446.

Illman, R. J., Topping, D. L., and Trimble, R. P., 1986, Effects of food restriction and starvation-refeeding on volatile fatty acid concentrations in the rat, *J. Nutr.* **116:**1694–1700.

Ingram, D. M., and Castleden, W. M., 1980, The effect of dietary lactulose on experimental large bowel cancer, *Carcinogenesis* **1:**893–895.

Jacobs, L. R., 1983, Enhancement of rat colon carcinogenesis by wheat bran consumption during the stage of 1,2-dimethylhydrazine administration, *Cancer Res.* **43:**4057–4061.

Jacobs, L. R., 1984, Stimulation of rat colonic crypt cell proliferative activity by wheat bran consumption during the stage of 1,2-dimethylhydrazine administration, *Cancer Res.* **44:**2458–2463.

Jacobs, L. R., 1986a, Enhancement of experimental rat colon cancer with dietary lactulose, *Gastroenterology* **90:**1473.

Jacobs, L. R., 1986b, Enhancement of experimental colon cancer and production of colitis in rat fed sorbitol, *Clin. Res.* **35:**441A.

Jacobs, L. R., 1989, Dietary fiber, fiber-containing foods, and colon cancer risk, in: *Colorectal Cancer: From Pathogenesis to Treatment* (H. K. Seitz, U. A. Simanowski, and N. A. Wright, eds.), Springer-Verlag, Berlin, Heidelberg, pp. 139–159.

Jacobs, L. R., and Lupton, J. R., 1986, Relationship between colonic luminal pH, cell proliferation, and colon carcinogenesis in 1,2-dimethylhydrazine treated rats fed high fiber diets, *Cancer Res.* **46:** 1727–1734.

Jacobs, L. R., and White, F. A., 1983, Modulation of mucosal cell proliferation in the intestine of rats fed a wheat bran diet, *Am. J. Clin. Nutr.* **37:**945–953.

Kruh, J., 1982, Effect of sodium butyrate, a new pharmacological agent, on cells in culture, *Mol. Cell. Biochem.* **42:**65–82.

Laqueur, G. L., Matsumoto, H., and Yamamoto, R. S., 1981, Comparison of the carcinogenicity of methylazoxymethanol-β-D-glucosiduronic acid inconventional and germ free Sprague Dawley rats, *J. Natl. Cancer Inst.* **67:**1053–1055.

Lupton, J. R., Coder, D. M., and Jacobs, L. R., 1985, Influence of luminal pH on rat large bowel epithelial cell cycle, *Am. J. Physiol.* **249**(12):G382–G388.

Lupton, J. R., Coder, D. M., and Jacobs, L. R., 1988, Long-term effects of fermentable fibers on rat colonic pH and epithelial cell cycle, *J. Nutr.* **118:**840–845.

Mallett, A. K., Wise, A., and Rowland, I. R., 1983, Effect of dietary cellulose on the metabolic activity of the rat cecal microflora, *Arch. Toxicol.* **52:**311–317.

Mallett, A. K., Wise, A., and Rowland, I. R., 1984, Hydrocolloid food additives rat caecal microbial enzyme activities, *Food Chem. Toxicol.* **22:**415–418.

Mallett, A. K., Rowland, I. R., and Bearne, C. A., 1986, Influence of wheat bran on some reductive and hydrolytic activities of the rat cecal flora, *Nutr. Cancer* **8:**125–131.

Nyman, M., and Asp, N.-G., 1982, Fermentation of dietary fiber components in the rat intestinal tract, *Br. J. Nutr.* **47:**357–366.

Pilch, S. M. (ed.), 1987, *Physiological Effects and Health Consequences of Dietary Fiber*, Life Sciences Research Office, Federation of American Societies for Experimental Biology, Bethesda, MD.

Reddy, B. S., Watanabe, K., and Sheinfil, A., 1980, Effect of dietary wheat bran, alfalfa, pectin and carrageenan on plasma cholesterol and fecal bile acid and neutral sterol excretion in rats, *J. Nutr.* **110:**1247–1254.

Roberts-Andersen, J., Mehta, T., and Wilson, R. B., 1987, Reduction of DMH-induced colon tumors in rats fed psyllium husk or cellulose, *Nutr. Cancer* **10:**129–136.

Sakata, T., and Yajima, T., 1984, Influence of short chain fatty acids on the epithelial cell division of digestive tract, *Q. J. Exp. Physiol.* **69:**639–648.

Samelson, S. L., Nelson, R. L., and Nyhus, L. M., 1985, Protective role of fecal pH in experimental colon carcinogenesis, *J. R. Soc. Med.* **78:**230–233.

Storer, G. B., Illman, R. J., Trimble, R. P., Snoswell, A. M., and Topping, D. L., 1984, Plasma and cecal volatile fatty acids in male and female rats: Effects of dietary gum arabic and cellulose, *Nutr. Res.* **4:**701–707.

Takada, H., Hirooka, T., Hiramatsu, Y., and Yamamoto, M., 1982, Effect of β-glucuronidase inhibitor on azoxymethane-induced colonic carcinogenesis in rats, *Cancer Res.* **42:**331–334.

Thomsen, L. L., Roberton, A. M., Wong, J., Lee, S. P., and Tasman-Jones, C., 1984, Intra-cecal short chain fatty acids are altered by dietary pectin in the rat, *Digestion* **29:**129–137.

Thornton, J. R., 1981, High colonic pH promotes colorectal cancer, *Lancet* **1:**1081–1083.

Toth, B., 1984, Effect of Metamucil on tumour formation by 1,2-dimethylhydrazine dihydrocholoride in mice, *Food Chem. Toxicol.* **22:**573–578.

Tulung, B., Remesy, C., and Demigne, C., 1987, Specific effects of guar gum or gum arabic on adaptation of cecal digestion to high fiber diets in the rat, *J. Nutr.* **117:**1556–1561.

Vahouny, G. V., Khalafi, R., Satchithanandram, S., Watkins, D. W., Story, J. A., Cassidy, M. M., and Kritchevsky, D., 1987, Dietary fiber supplementation and fecal bile acids, neutral steroids and divalent cations in rats, *J. Nutr.* **117:**2009–2015.

van Dokkum, W., de Boer, B. C. J., van Faasen, A., Pikaar, N. A., and Hermus, R. J. J., 1983, Diet, faecal pH and colorectal cancer, *Br. J. Cancer* **48:**109–110.

Walker, A. R. P.,. Walker, B. F., and Walker, A. J., 1986, Fecal pH, dietary fibre inake, and proneness to colon cancer in four South African populations, *Br. J. Cancer* **53:**489–495.

Walter, D. J., Eastwood, M. A., Brydon, W. G., and Elton, R. A., 1986, An experimental design to study colonic fibre fermentation in the rat: The duration of feeding, *Br. J. Nutr.* **55:**465–479.

Wargovich, M. J., Eng, V. W. S., and Newmark, H. L., 1984, Calcium inhibits the damaging and compensatory proliferative effects of fatty acids on mouse colon epithelium, *Cancer Lett.* **23:**253–258.

Watanabe, K., Reddy, B. S., Wong, C. Q., and Weisburger, J. H., 1978, Effect of dietary undegraded carageenan on colon carcinogenesis in F344 rats treated with azoxymethane or methylnitrosurea, *Cancer Res.* **38:**4427–4430.

Watanabe, K., Reddy, B. S., Weisburger, J. H., and Kritchevsky, D., 1979, Effect of dietary alfalfa, pectin and wheat bran on azoxymethane- or methylnitrosourea-induced colon carcinogenesis in F344 rats, *J. Natl. Cancer Inst.* **63:**141–145.

Wilpart, M., and Roberfroid, 1987, Intestinal carcinogenesis and dietary fibers: The influence of cellulose on Fybogel chronically given after exposure to DMH, *Nutr. Cancer* **10:**39–51.

30

Insoluble Dietary Fiber and Experimental Colon Cancer
Are We Asking the Proper Questions?

DAVID M. KLURFELD

The effects of insoluble dietary fiber on colonic carcinogenesis in animal models can be summarized by stating that this type of fiber inhibits, enhances, and has no effect on development of intestinal tumors. This vague conclusion is predicated on the fact that any assessment of the role of dietary fiber in studies of colonic carcinogenesis can only be made with an appreciation of the many sources of variability that contribute to the confusion surrounding the effect of fiber on the tumorigenic process.

The first obvious source of potential variability is related to the type of animal used; not only does this include differences in species (rat and mouse being the only animals used to date in fiber studies), but there are published differences probably attributable to the strain and gender of rodents used.

Another source of conflicting data is the carcinogen used to induce tumors. 1,2-Dimethylhydrazine (DMH) is the most commonly used carcinogen to induce colonic tumors since it is the most potent and specific (LaMont and O'Gorman, 1978); it and azoxymethane (AOM) need to be metabolized in the liver to the alkylating agent, methyldiazonium, which in turn decomposes spontaneously to a carbonium ion. Other colonic carcinogens that have been used in studies of dietary fiber include N-methyl-N'-nitro-N-nitrosoguanidine (MNNG), 2',3-dimethyl-4-aminobiphenyl (DMAB), and methylnitrosourea (MNU). X irradiation has also been used in one study that resulted in a relatively low incidence of colonic neoplasms (Donham et al., 1984).

The route of administration of the carcinogen may also affect the outcome of tumor studies. Carcinogens have been injected subcutaneously, intraperitoneally, and intramuscularly. The same carcinogens have also been administered orally by gavage. Direct-acting carcinogens such as MNU or MNNG have been instilled intrarectally; this route

DAVID M. KLURFELD • The Wistar Institute of Anatomy and Biology, Philadelphia, Pennsylvania 19104.

of administration often results in leakage of variable amounts of the solution containing the carcinogen. Highly concentrated solutions of carcinogens are also potential sources of variability when injection volume varies slightly.

The dosage of carcinogen is another aspect of variability in the experiments. With DMH as an example, reported studies in which insoluble dietary fiber was fed used two to 26 doses in a range of 15 to 135 mg/kg body weight. Too low an effective dose can result in a tumor response small enough to preclude seeing a protective effect, whereas too high a dose may be overwhelming and prevent reduction of tumor response by dietary manipulation. Since DMH and most other carcinogens are also highly toxic, the immediate side effects need to be accounted for. The highest reported doses caused death of more than half the test animals, and the more commonly used doses (in the range of 30 mg/kg of the dihydrochloride salt) cause temporary significant reductions of food intake and weight gain (Kritchevsky et al., 1987).

Another source of variability with regard to carcinogens is the timing of administration relative to feeding the test fiber. Some investigators feed a nonpurified diet during carcinogen administration and then switch to the semipurified test diets. This is only accomplishable when there is a relatively short dosing regimen (6 weeks or less). Most studies reported having given the carcinogen at the same times as the various test diets. Since differences in dietary fiber could cause altered metabolism and enterohepatic circulation of a carcinogen and its metabolites, this is a potentially significant source of variability in that initiation will be affected as well as promotion. Switching from diets containing widely different amounts of dietary fiber just prior to or after administration of carcinogen could result in differences in colonic mucosal cytokinetics and subsequent initiation of colonic tumors that do not accurately reflect the tumor-promoting or -inhibiting effects of that fiber source.

The experimental diets used in studies of dietary fiber and colonic carcinogenesis are a major source of variability in making comparisons among the published studies. The basal diet is most often a semipurified form, but a number of papers have reported using nonpurified, grain-based diets to which a particular fiber has been added. Even among semipurified diets, there is an unresolved question of the various ingredients individually having effects on colonic tumorigenesis. With increased interest in unabsorbed carbohydrate, this dietary component may receive more attention in the future. In rodent diets, carbohydrate can also determine palatability and subsequent weight gain of the animals. Since fat intake can modify colon tumor response, this variable needs to be accounted for in comparisons of tumor data, since the range of dietary fat has been from 2% to 35% of the diet. Additionally, many types of fat have been used, including corn oil, safflower oil, peanut oil, beef tallow, lard, and blends of palm and corn oil. It is known from other tumor systems that there is a requirement for linoleic acid, which affects tumor yield (Klurfeld and Kritchevsky, 1986). Another way in which dietary fat may act as a tumor promoter for colon cancer is through provision of excessive calories. Both caloric intake and body weight can be influenced significantly by type and amount of dietary fiber included in a diet. Since energy intake and body weight are strongly correlated with colon tumor promotion (Klurfeld et al., 1987), these sources of variability need to be accounted for in interstudy comparisons.

The amount of dietary fiber incorporated into test diets has varied from 4.5% to 40% by weight. This wide range could affect interpretation of results if there are irritating effects relative to tumor promotion above a certain threshold level, which is quite possible with nonfermentable insoluble fibers. The choice of fiber level to be used is often derived arbitrarily or through short-term pilot studies in which maximally tolerated doses are determined. Few studies have examined dose–response patterns to different levels of dietary fiber. Both the physical and chemical characteristics of a single source of dietary fiber may vary enough to make a difference in tumorigenesis studies. The particle size of most dietary fibers can vary over a wide range and is not given in most rodent studies. Cellulose is available in many particle sizes and in amorphous powdered form as well as microcrystalline preparations. Additionally, some cellulose preparations that have been fed to rodents are of pharmaceutical grade, which is recommended for making tablets. Wheat bran can be derived from soft white or hard red varieties, which have been shown to have differing effects on lipid metabolism and may exert different effects in this tumor system.

When fiber is removed from its native source and concentrated, a number of physical and structural changes can occur. Although it is considered appropriate to feed high levels of single nutrients, including fiber, this is not the way any organism eats, so the relevance of this type of dietary regimen really must be questioned. It is likely that this question would not arise if the majority of animal experiments agreed with the hypothesis that increased dietary fiber would decrease risk of colon cancer; since there is considerable uncertainty of how the animal experiments should be interpreted, one needs to address the issue of whether the correct questions are being asked and if the current techniques will provide answers.

The final source of variability to be considered under diet is the choice of an appropriate control group. Many investigators have used a fiber-free diet, but this may be inappropriate for animals with a large cecum. This type of diet results in production of very little feces, significantly lengthened mouth-to-anus transit time, and a slowdown of colonic mucosal cytokinetics. It also changes the intestinal microflora, which leads to alterations in the fecal bile acid profile and in production of vitamins in the intestinal tract. Whether a low-fiber diet in the range of perhaps 2% is a more realistic control diet for rodent studies remains to be determined.

The endpoint of tumor experiments is not always the same. Although investigators are measuring tumor formation, almost all studies are terminated at a predetermined time point when a certain percentage of animals is expected to have tumors. In studies of insoluble dietary fiber, the duration of feeding has varied from 18 to 52 weeks. Although this seems quite short in the scheme of tumorigenesis as is understood from human cancer biology, this time frame represents a period of up to half of the normal life expectancy of a rat or mouse; however, this is a threefold range in time. A few studies have continued until each animal died or exhibited objective criteria for euthanasia.

The last source of variability encountered in animal experiments of dietary fiber and colon cancer is that of data analysis. The choice of number of animals per treatment is critical to whether a result will be interpretable. Epidemiologists and most clinical trials utilize power calculations to determine sample size prior to commencing a study; no published study on fiber and cancer in animals alludes to having done this. The range of animals per treatment used is ten to 200. From a practical view, depending on the magnitude of change expected, 20 to 30 animals per treatment are usually adequate. With this number of animals, a change in tumor incidence of approximately 25% is

TABLE I. Wheat Bran and Colon Cancer Studies[a]

Reference	Animals	Carcinogen	Percentage bran	Control	Percentage fat	Statistics	Tumor results[b]
Wilson et al. (1977)	M S–D 11/gp	DMH, 30 mg/kg, 4 or 8 doses	20	0% fiber	20	OK for incidence, not for T/TBR	69% controls 37% bran
Barbolt and Abraham (1978)	F S–D 10/gp	DMH, 30 mg/kg, 10 doses	20	Nonpurified diet		OK for T/TBR, not for incidence	T/TBR 6.4–1.8 Inc 100%–67%
Chen et al. (1978)	F CF1 200/gp	DMH, 20 mg/kg, 26 doses	40	0% fiber	6	?	Major path changes 39%–19%
Cruse et al. (1978)	F Wis 20/gp	DMH, 40 mg/kg, 20 doses	20	Nonpurified diet		OK	No diff. in mortality
Fleiszer et al. (1978)	M CB 25/gp	DMH, 20 mg/kg, 20 doses	28	Nonpurified diet		?	Inc 71%–6% 50% ↓ in weight gain
Bauer et al. (1979)	F S–D 40/gp	DMH, 15 mg/kg, 12 doses	20	0% fiber	20	Not acceptable	No differences
Nigro et al. (1979)	M S–D 25/gp	AOM, 8 mg/kg, 21 doses	10	0% fiber	35	Not acceptable	?
Watanabe et al. (1979)	F F344 30/gp	AOM, 8 mg/kg, 10 doses MNU, 2 mg, 6 doses	20,30 15	0% fiber 0% fiber	5–7 20	Not acceptable OK	? Inc 57%–33% T/TBR no diff No diff

Reference	Animals	Carcinogen		Diet		Statistics	Results[b]
Abraham et al. (1980)	F S–D 10/gp	DMH, 30 mg/kg, 10 doses	20	Nonpurified diet, 0.5% bile salts	20	Not acceptable	T/TBR no diff ? ↓ in % carcinoma
Barbolt and Abraham (1980)	M,F S–D 10/gp	DMH, 15,30 mg/kg, 10 doses	20	Nonpurified diet	20	Not acceptable	Inc low in all F, ↓ in M only on low dose
Reddy and Mori (1981)	M F344 50/gp	DMAB, 50 mg/kg, 20 doses	15	5% cellulose	5	OK for inc, not for T/TBR	Inc 46%–26%
Reddy et al. (1981)	M F344 51/gp	AOM, 8 mg/kg, 10 doses	15	5% cellulose	5	OK for inc, not for T/TBR	Adenomas 86%–47% Carcinoma 63%–39%
Barnes et al. (1983)	M F344 25/gp	DMH, 135 mg/kg, 2 doses	20	0% fiber	20	?	Inc 93%–75%
Jacobs (1983)	M S–D 12/gp	DMH, 20 mg/kg, 13 doses	20	0% fiber	8	OK	T/TBR 1.1–2.7
Clapp et al. (1984)	M Balb/c 50–70/gp	DMH, 20 mg/kg, 10 doses	20	0% fiber	5	No stats	Inc 11%–48% soft wheat 58% hard wheat
Calvert et al. (1987)	M F344 30/gp	DMH, 30 mg/kg, 5 doses	10	0% fiber, 0.02% bile salts	5	OK	Inc 86%–67%

[a] Abbreviations: M, male; F, female; S–D, Sprague–Dawley rat; CB, Chester Beatty rat; Wis, Wistar rat; CF1, CF1 mouse; F344, Fischer 344 rat; Hol, Holtzman rat; Balb/c, Balb/c mouse; T/TBR, tumors per tumor-bearing rat; inc, tumor incidence.

[b] Under results, first value is for control group, second for bran-fed group.

necessary to achieve a statistically significant result. The smaller the group, the greater is the difference necessary to derive statistical significance, so with ten animals per treatment a difference of about 50% is required. Unfortunately, only about 40% of the publications in this area have used correct statistical analysis. This is not unique to this aspect of scientific research and is similar to the percentage of papers using correct statistics in the New England Journal of Medicine (Godfrey, 1985). The major reason for using appropriate analyses is that proper interpretation (i.e., a greater degree of certainty) of the data is essential in drawing conclusions in which multiple groups are compared and several endpoints are examined.

The endpoints in animal studies of colon cancer and insoluble dietary fiber includes those relative to tumorigenesis and putative biological intermediates. The measures of tumor formation include incidence (percentage of animals that develop tumors), tumor multiplicity (tumors per tumor-bearing rat), metastases, mortality, and microscopic analysis. Regarding the last point, some investigators do not report microscopic assessment at all, and some that do give their own categorizations. There is no generally accepted scheme for categorization of colonic tumors, but a classification such as that by Pozharisski (1973) seems to include all the relevant forms of intestinal tumors found in rodents.

Many biological intermediates have been investigated as potential leads to tumor-mediating effects of insoluble dietary fiber on the colon. These include fecal steroids, fecal enzymes, short-chain fatty acids, fecal bulk, fecal water, intestinal microflora, transit time, cytokinetics, mucins, epithelial morphology, mutagenicity (Dolara et al., 1986), pH (Jacobs and Lupton, 1982), gut hormones (Goodlad et al., 1987), and trace compounds found associated with high-fiber foods such as lignans, phytate, and phytosterols. Although several of these show some promise as indicators of risk, there is no good predictor of subsequent tumor development. If one or more of these factors could be validated as a indicator of potential tumor growth, it would be a significant step toward understanding biological factors in tumor growth and how diet might intervene in this process.

The definition used in this review of an individual study that examined effects of insoluble dietary fiber on colonic carcinogenesis is a comparison of one type of fiber with a control group. By this definition there are 38 studies, but these have appeared in fewer papers since several papers have reported on more than one type of fiber. The most commonly used insoluble fiber source in animal experiments on colon tumorigenesis is wheat bran; studies using this type of dietary fiber are summarized in Table I. There are 16 studies; five used a nonpurified cereal-based diet as the base for adding wheat bran. This seems somewhat inappropriate for studying effects of fiber, since the fiber content of nonpurified rodent diets is relatively high and can vary from batch to batch since they are usually of the closed-formula variety. This simply means that the specifications for certain nutrients are met but the ingredients used to achieve those levels vary depending on actual analysis of the raw materials available seasonally. The fat content of the basal diets ranged from 5% to 35%, and the bran constituted 10% to 40% (ten studies fed 20% wheat bran). Four different carcinogens were used in these studies, with most investigators using DMH. However, the dose of DMH varied from 15 to 135 mg/kg, and the animals were given from two to 26 weekly doses. Despite these methodological differences among studies, comparison of the 16 reports reveals that only five found decreases in tumor yield independently attributable to wheat bran, four found no change, two showed increased tumorigenesis, four were equivocal in outcome, and one was uninterpretable.

An interesting study from a mechanistic view is that of Calvert et al. (1987) in which wheat bran was compared with a fiber-free diet. Enough bile salts were added to the wheat bran diet to increase the concentration of fecal bile acids in the wheat bran group to that of the fiber-free group. This amounted to about 0.02% of the diet, which is many times less than the pharmacological doses of bile acids found to be tumor promoters. Wheat bran caused a significant reduction in tumor yield despite equalization of total fecal bile acid concentrations. Although it is possible that specific bile acids are more important in this model than total concentration, the results are suggestive that some other aspect of protection is operative.

The uninterpretable study (Nigro et al., 1979) reported only tumors per rat and used multiple t-tests for comparing six groups. The differences were such that use of appropriate statistical analyses probably would not have revealed significant differences. The four equivocal studies reported differences that do not lend themselves to comparison with other studies or have internal inconsistencies. Chen et al. (1978) found that "major-invasive pathologic changes" were reduced from 39% to 19% with feeding of 40% wheat bran, but tumor incidence was identical in both groups. Watanabe et al. (1979) found no differences in tumor yields with NMU but reduced incidence (57% to 33%) and no change in tumor multiplicity with AOM. Abraham et al. (1980), using only ten rats per group fed a nonpurified diet with 0.5% bile salts and added 20% wheat bran to the diet of the test group. They found no difference in incidence or tumor multiplicity and used incorrect statistics to claim decreased percentage of adenocarcinomas. Finally, Barbolt and Abraham (1980) compared the response of male and female rats to two doses of DMH. A protective effect was seen only in the males treated with the lower dose. Although this is suggestive of a sex difference in response to the carcinogen and dietary fiber, this study also suffers the same design flaws of the previously cited study from the same group of researchers. The use of statistics by investigators in this group of studies is summarized as follows: no statistics, three; correct analyses, four; incorrect analyses, eight; and cannot determine from the paper which tests were used, one.

The next most commonly studied form of insoluble dietary fiber for its influence on colonic tumorigenesis is cellulose. Whereas wheat bran is a true food ingredient, cellulose is physically and chemically extracted (usually from wood pulp) and used only as an additive. However, it is the standard fiber component in semipurified animal diets and is added to foods for humans at up to 10%. The critical unanswered question for proper interpretation of results obtained with cellulose is whether processed cellulose behaves like cellulose in its native state. Thirteen studies have been reported in which cellulose was used to modify colonic carcinogenesis (Table II). Seven studies found no change in tumor yield, and three reported a decrease. A single study found increased tumors, and two studies were classified as equivocal. The same design limitations that applied to the studies of wheat bran apply to those using cellulose. That is, 4.5% to 40% cellulose was fed in diets containing 2% to 35% fat. Three experiments used nonpurified basal diets; 10 to 200 rats of six strains were used; follow-up ranged from 18 to 50 weeks; three carcinogens were used at many doses.

Two interesting studies by Donham and co-workers (1980, 1984) compared 10%

TABLE II. Cellulose and Colon Cancer Studies[a]

Reference	Animals	Carcinogen	Percentage fiber	Control	Percentage fat	Statistics	Tumor results
Ward et al. (1973)	M F344 11/gp	AOM, 15 mg/kg, ? doses	20,40	0% fiber	?	?	No effect
Freeman et al. (1978)	M Wis 20/gp	DMH, 25 mg/kg, 16 doses	4.5	0% fiber	8	OK	Inc 70-30% T/TBR 1.6-0.6
Nigro et al. (1979)	M S–D 25/gp	AOM, 8 mg/kg, 21 doses	10	0% fiber	35	Not acceptable	No effect
			20,30	0% fiber	5–7	Not acceptable	? decrease w/30%, 25% ↓ body weight
Donham et al. (1980)	M,F F344 115–200/gp	None	10	10% asbestos	?	OK	No difference
Freeman et al. (1980)	M Wis 36/gp	DMH, 35 mg/kg, 18 doses	4.5, 9.0	0% fiber	8	Not acceptable	? decrease No dose response
Trudel et al. (1983)	M S–D 22/gp	DMH, 20 mg/kg, 20 doses	20	Nonpurified	20	OK for histology	Inc 80%–60% 90%–42%
Donham et al. (1984)	M F344, M Hol 40/gp	X irradiation	10	10% asbestos	?	?	No difference
Klurfeld et al. (1986)	M F344 24/gp	DMH, 30 mg/kg, 6 doses	10,20,30	0% fiber	5	OK	Inc 71,75,92,100% T/TBR 2.4,2.8,5.1,6.5
Jacobs and Lupton (1986)	M S–D 24/gp	DMH, 20 mg/kg, 12 doses	10	0% fiber	8	OK	No difference
Galloway (1986)	M Swiss 58/gp	AOM, 10 mg/kg, 12 doses	27	2% cellulose	2,25	OK for overall, not for pairs	Inc 22%–4% low fat 64%–11% high fat
Prizont (1987)	M Wis 10/gp	DMH, 25 mg/kg, 10 doses	15	5% cellulose	5	OK	Only tumors/gp given, 7 vs. 0 tumors/gp
Roberts-Anderson et al. (1987)	M S–D 10/gp	DMH, 30 mg/kg, 8 doses	9.7	0% fiber	20	OK for inc, not for T/TBR	Inc 100%–60% T/TBR 1.9–1.3
Wilpart and Roberffroid (1987)	M Wis 30/gp	DMH, 30 mg/kg, 15 doses	5,15	Interaction	5,20	OK	No difference

[a] Abbreviations: see footnote a to Table I.

cellulose with 10% asbestos added to a nonpurified rodent diet. The first study used no carcinogen, and the animals were fed for their entire lives. Only 1–2% of rats in each group developed colon tumors. In the second study, localized x irradiation of the colon was used as the initiating agent, and the rats were fed for 50 weeks. Eight percent of the animals fed asbestos developed colon cancer, whereas 12% of those fed cellulose had colon tumors.

One of the two equivocal studies is by Freeman et al. (1980). This study used 4.5% and 9% microcrystalline cellulose of a pharmaceutical grade. No dose response was seen, and protection against tumorigenesis was observed only at an early point 2 to 4 weeks after the last dose of DMH. There was no effect when cellulose was fed only during the promotion phase. The other equivocal study (Roberts-Anderson, 1987) fed 9.7% cellulose to only ten rats per group and found decreased incidence but no change in tumor multiplicity. The only study to report increased tumorigenicity (Klurfeld, 1988) differs in a number of respects from most of the other studies. Fewer doses of DMH were used, and the test diets were fed only after the last dose of carcinogen was administered. Three dose levels of fiber were fed, and two physically different types of cellulose were compared. Powdered cellulose was a tumor promoter in this system, whereas microcrystalline cellulose gave an equivocal response. The use of statistics in these papers totaled six correct, five incorrect, and two in which one cannot tell what tests were used.

One additional problem in comparing these studies of cellulose feeding is that six identifiably different preparations were used. There are about 20 cellulose products for food use on the market, differing primarily in physical properties such as size and chemical derivitization. Of six companies surveyed that formulate semipurified diets for animal experimentation, two would not divulge their sources, and the specifications offered did not match those of any of the manufacturers.

A variety of other insoluble fiber sources have been tested for effects on colon cancer in animals. Eight types of fiber added to diets are summarized in Table III. Comparison of four studies that added corn bran reveals that three studies fed 20% bran and reported increased tumor yield, whereas addition of 4.5% corn bran led to protection at an early time point after DMH but not at a later time. Most of the other sources of dietary fiber, soybean bran, rice bran, citrus peel, carrot fiber, alfalfa, and lignin, show no consistent effect on colonic tumor formation.

Table IV summarizes the results of the 38 studies described in this chapter. The most common result is no change in colonic tumor yield with feeding of insoluble dietary fiber.

Only two studies were reassigned from claimed protective effects by the authors to equivocal or uninterpretable categories; the remainder that used incorrect statistics probably came to correct conclusions, but improver use of these tests precludes determining the degree of certainty of the conclusions. Most of the improper statistical comparisons involved use of multiple t-tests and the use of tests that require parametrically distributed data when there was not a normal distribution of results. Although one can argue that statistics should merely confirm the obvious, in studies of the nature described above reliance on correct analyses is critical to proper interpretation of the results. Of course, it is equally important not to be overly rigid in requirement for statistical significance. After all, there is little fundamental difference between a pair of results that provides P values of 0.04 and 0.06 except that the former is accepted by convention as being

TABLE III. "Miscellaneous" Insoluble Dietary Fiber and Colon Cancer Studies

Reference	Animals	Carcinogen	Percentage fiber	Control	Percentage fat	Statistics	Tumor results
Bauer et al. (1979)	M S–D 40/gp	DMH, 15 mg/kg, 12 doses	20 carrot	0% fiber	20	Not acceptable	No effect
Watanabe et al. (1979)	M F344 30/gp	AOM, 8 mg/kg, 10 doses MNU, 2 mg, 6 doses	15 alfalfa	5% cell	20	OK	Inc AOM 69%–83% Inc MNU 57%–53%
Reddy and Mori (1981)	M F344 50/gp	DMAB, 50 mg/kg, 20 doses	15 citrus peel	5% cell	5	OK for inc, not for T/TBR	No effect
Reddy et al. (1981)	M F344 50/gp	AOM, 8mg/kg, 10 doses	15 citrus peel	5% cell	5	OK for inc, not for T/TBR	Inc 90%–63% T/TBR 3.5–1.8
Barnes et al. (1983)	M F344 25/gp	DMH, 135 mg/kg, 2 doses	20 rice bran, soy bran, corn bran	0% fiber	20	?	Inc, no effect Corn ↑ total tumors
Reddy et al. (1983)	M F344 20/gp	DMAB, 50 mg/kg, 20 doses	15 corn bran 7.5 lignin	5% cell	5	OK	Control 30% Corn bran 67% Lignin 26%
Freeman et al. (1984)	M Wis 36/gp	DMH, 35 mg/kg, 18 doses	4.5 corn bran	0% fiber	8	OK	Early 67%–33% Late 76%–62%
Clapp et al. (1984)	Balb/c 50–70/gp	DMH, 20 mg/kg, 10 doses	20 soy bran, corn bran	0% fiber	5	None	Control 11% Soy bran 44% Corn bran 72%
Klurfeld (1988)	M F344 24/gp	DMH, 30 mg/kg, 6 doses	5,10,20 cutin	0% fiber		OK	Inc 65,67,75,83% T/TBR 1.9,1.6,2.5,3.1

a Abbreviations: see footnote a to Table I.

TABLE IV. Insoluble Dietary Fiber
and Colon Cancer Studies

Results		Statistical analysis	
No change	14 (37%)	None	4 (11%)
Decrease	9 (24%)	Correct	14 (37%)
Increase	6 (16%)	Incorrect	16 (42%)
Equivocal	7 (18%)	Cannot tell	4 (11%)
Uninterpretable	2 (5%)		

significant whereas the latter is not. This does not mean that the latter shows no change with treatment, but it is not as certain of showing consistent effects as the former.

Although the hypothesis articulated by Burkitt, Trowell, and Walker that dietary fiber will reduce the incidence of colon cancer is highly attractive to researchers, there is little objective evidence from animal studies using insoluble fiber to support it. This does not mean that the hypothesis is invalid but only that further research is required in order to determine if appropriate methods are being used to investigate this hypothesis. This is probably true for all tumor models, but the experimental data with, for example, the rodent mammary tumor system fit fairly well with the epidemiologic data and the hypotheses relating diet to cancer, so the relevance of this model to human breast cancer is relatively well accepted. Not only do the issues regarding the validity of the rodent model for colon cancer need to be resolved, but the question of whether feeding chemically or physically isolated dietary fiber is similar in effects to actual consumption of fiber-rich foods must be addressed.

REFERENCES

Abraham, R., Barbolt, T. A., and Rodgers, J. B., 1980, Inhibition by bran of the colonic cocarcinogenicity of bile salts in rats given dimethylhydrazine, *Exp. Mol. Pathol.* **33**:133–143.

Barbolt, T. A., and Abraham, R., 1978, The effect of bran on dimethylhydrazine-induced colon carcinogenesis in the rat, *Proc. Soc. Exp. Biol. Med.* **157**:656–659.

Barbolt, T. A., and Abraham, R., 1980, Dose-response, sex difference, and the effect of bran in dimethylhydrazine-induced intestinal tumorigenesis, *Toxicol. Appl. Pharmacol.* **55**:417–422.

Barnes, D. S., Clapp, N. K., Scott, D. A., Oberst, D. L., and Berry, S. G., 1983, Effects of wheat, rice, corn and soybean bran on 1,2-dimethylhydrazine-induced large bowel tumorigenesis in F344 rats, *Nutr. Cancer* **5**:1–9.

Bauer, H. G., Asp, N.-G., Oste, R., Dahlqvist, A., and Fredlund, P. E., 1979, Effect of dietary fiber on the induction of colorectal tumors and fecal β-glucuronidase activity in the rat, *Cancer Res.* **39**: 3752–3756.

Calvert, R. J., Klurfeld, D. M., Subramaniam, S., Vahouny, G. V., and Kritchevsky, D., 1987, Reduction of colonic carcinogenesis by wheat bran independent of fecal bile acid concentration, *J. Natl. Cancer. Inst.* **79**:875–880.

Chen, V. F., Patchefsky, A. S., and Goldsmith, H. S., 1978, Colonic protection from dimethylhydrazine by a high fiber diet, *Surg. Gynecol. Obstet.* **147**:503–506.

Clapp, N. K., Henke, M. A., London, J. F., and Shock, T. L., 1984, Enhancement of 1,2-dimethylhydrazine-induced large bowel tumorogenesis in Balb/c mice by corn, soybean, and wheat brans, *Nutr. Cancer* **6**:77–85.

Cruse, J. P., Lewin, M. R., and Clark, C. G., 1978, Failure of bran to protect against experimental colon cancer in rats, *Lancet* **2:**1278–1280.

Dolara, P., Caderni, F., Bianchini, F., and Tanganelli, E., 1986, Nuclear damage of colon epithelial cell by the food carcinogen 2-amino-3-methylimidazo[4,5-f]quinoline (IQ) is modulated by dietary lipids, *Nutr. Res.* **175:**255–258.

Donham, K. J., Berg, J. W., Will, L. A., and Leininger, J. R., 1980, The effects of long-term ingestion of asbestos on the colon of F344 rats, *Cancer* **45:**1073–1084.

Donham, K. J., Will, L. A., Denman, D., and Leininger, J. R., 1984, The combined effects of asbestos ingestion and localized x-irradiation of the colon in rats, *J. Environ. Pathol. Toxicol. Oncol.* **5:**299–308.

Fleiszer, D., Murray, D., MacFarlane, J., and Brown, R. A., 1978, Protective effect of dietary fiber against chemically induced bowel tumours in rats, *Lancet* 2:552–553.

Freeman, H. J., Spiller, G. A., and Kim, Y. S., 1978, A double-blind study on the effect of purified cellulose dietary fiber on 1,2-dimethylhydrazine-induced rat colonic nepolasia, *Cancer Res.* **38:**2912–2917.

Freeman, H. J., Spiller, G. A., and Kim, Y. S., 1980, A double-blind study on the effects of differing purified cellulose and pectin fiber diets on 1,2-dimethylhydrazine-induced rat colonic neoplasia, *Cancer Res.* **40:**2661–2665.

Freeman, H. J., Spiller, G. A., and Kim, Y. S., 1984, Effect of high hemicellulose corn bran in 1,2-dimehylhydrazine-induced rat intestinal neoplasia, *Carcinogenesis* 5:261–264.

Galloway, D. J., Owen, R. W., Jarrett, F., Boyle, P., Hill, M. J., and George, W. D., 1986, Experimental colorectal cancer: The relationship of diet and faecal bile acid concentration to tumour induction, *Br. J. Surg.* **73:**233–237.

Godfrey, K., 1985, Comparing the means of several groups, *N. Engl. J. Med.* **313:**1450–1456.

Goodlad, R. A., Lenton, W., Ghatei, M. A., Adrian, T. E., Bloom, S. R., and Wright, N. A., 1987, Effects of an elemental diet, inert bulk and different types of dietary fibre on the response of the intestinal epithelium to refeeding in the rat and relationship to plasma gastrin, enteroglucagon, and PYY concentrations, *Gut* **28:**171–180.

Jacobs, L. R., 1983, Enhancement of rat colon carcinogenesis by wheat bran consumption during the stage of 1,2-dimethylhydrazine administration, *Cancer Res.* **43:**4057–4061.

Jacobs, L. R., and Lupton, J. R., 1982, Dietary wheat bran lowers colonic pH in rats, *J. Nutr.* **112:**592–594.

Jacobs, L. R., and Lupton, J. R., 1986, Relationship between colonic luminal pH, cell proliferation, and colon carcinogenesis in 1,2-dimethylhydrazine treated rats fed high fiber diets, *Cancer Res.* **46:**1727–1734.

Klurfeld, D. M., 1988, The role of dietary fiber in gastrointestinal disease, *J. Am. Diet. Assoc.* **87:**1172–1177.

Klurfeld, D. M., and Kritchevsky, D., 1986, Update on dietary fat and cancer, *Proc. Soc. Exp. Biol. Med.* **183:**287–292.

Klurfeld, D. M., Weber, M. M., Buck, C. L., and Kritchevsky, D., 1986, Dose-response of colonic carcinogenesis to different amounts and types of cellulose, *Fed. Proc.* **45:**1076.

Klurfeld, D. M., Weber, M. M., and Kritchevsky, D., 1987, Inhibition of chemically induced mammary and colon tumor promotion by caloric restriction in rats fed increased dietary fat, *Cancer Res.* **47:**2759–2762.

Kritchevsky, D., Weber, M. M., and Klurfeld, D. M., 1987, Weight gain in dimethylhydrazine-treated rats: Effect of commercial or semipurified diet, *Nutr. Rep. Int.* **35:**871–876.

LaMont, J. T., and O'Gorman, T. A., 1978, Experimental colon cancer, *Gastroenterology* **75:**1157–1169.

Nigro, N. D., Bull, A. W., Kloper, B. A., Pak, M. S., and Campbell, R. L., 1979, Effect of dietary fiber on azoxymethane-induced intestinal carcinogenesis in rats, *J. Natl. Cancer Inst.* **62:**1097–1102.

Pozharisski, K. M., 1973, Tumours of the intestines, in: *Pathology of Tumours in Laboratory Animals,*

Vol. 1, Part 1 (V. S. Turusov, ed.), International Agency for Research on Cancer, Lyons, pp. 87–100.

Prizont, R., 1987, Absence of large bowel tumors in rats injected with 1,2-dimethylhydrazine and fed high dietary cellulose, *Dig. Dis. Sci.* **32:**1418–1421.

Reddy, B. S., and Mori, H., 1981, Effect of dietary wheat bran and dehydrated citrus fiber on 3,2′-dimethyl-4-aminobiphenyl-induced intestinal carcinogenesis in F344 rats, *Carcinogenesis* **2:**21–25.

Reddy, B. S., Mori, H., and Nicolais, M., 1981, Effect of dietary wheat bran and dehydrated citrus fiber on azoxymethane-induced intestinal carcinogenesis in Fischer 344 rats, *J. Natl. Cancer Inst.* **66:**553–557.

Reddy, B. S., Maeura, Y., and Wayman, M., 1983, Effect on dietary corn bran and autohydrolyzed lignin on 3,2′-dimethyl-4-aminobiphenyl-induced intestinal carcinogenesis in male F344 rats, *J. Natl. Cancer Inst.* **71:**419–423.

Roberts-Andersen, J., Mehta, T., and Wilson, R. B., 1987, Reduction of DMH-induced colon tumors in rats fed psyllium husk or cellulose, *Nutr. Cancer* **10:**1129–136.

Trudel, J. I., Senterman, M. K., and Brown, R. A., 1983, The fat/fiber antagonism in experimental colon carcinogenesis, *Surgery* **94:**691–695.

Ward, J. M., Yamamoto, R. S., and Weisburger, J. H., 1973, Cellulose dietary bulk and azoxymethane-induced intestinal cancer, *J. Natl. Cancer Inst.* **51:**713–715.

Watanabe, K., Reddy, B. S., Weisburger, J. H., and Kritchevsky, D., 1979, Effect of dietary alfalfa, pectin, and wheat bran on azoxymethane- or methylnitrosourea-induced colon carcinogenesis in F344 rats, *J. Natl. Cancer Inst.* **63:**141–145.

Wilpart, M., and Roberfroid, M., 1987, Intestinal carcinogenesis and dietary fibers: The influence of cellulose or Fybogel chronically given after exposure to DMH, *Nutr. Cancer* **10:**39–51.

Wilson, R. B., Hutcheson, D. P., and Widerman, L., 1977, Dimethylhydrazine-induced colon tumors in rats fed diets containing beef fat or corn oil without wheat bran, *Am. J. Clin. Nutr.* **30:**176–181.

31

Bacterial Metabolism, Fiber, and Colorectal Cancer

MICHAEL J. HILL and FRESIA FERNANDEZ

1. INTRODUCTION

A role for dietary fiber in colorectal carcinogenesis was proposed by Burkitt (1969, 1970, 1971). He proposed that the mechanism by which this occurred was that dietary fiber caused:

1. Stool bulking and consequent dilution of the colonic contents, including the carcinogens and tumor promoters present in the colonic lumen.
2. A faster rate of transit through the colon, decreasing the time available for *in situ* production of carcinogens, promoters, etc. by bacterial action.
3. A change in the bacterial flora of the colon to one that would be less likely to produce carcinogens, promoters, etc.
4. A change in the physicochemical environment to one less favorable to bacterial production of carcinogens, etc.

This is a good hypothesis and is attractive because it is plausible and because it is readily testable. Plausibility is not enough; many hypotheses that appear to state the obvious eventually prove to be erroneous in the light of later knowledge (e.g., the flat earth theory, the geocentric universe, medical theory based on the four humors), and so rigorous examination of all hypotheses is essential. The above hypothesized mechanism has been examined extensively during the last 18 years, and this is a good time to reassess its value. In this chapter I discuss the evidence for the proposed role of dietary fiber in colorectal carcinogenesis and then summarize the current state of the art.

MICHAEL J. HILL and FRESIA FERNANDEZ • PHLS-CAMR, Porton Down, Salisbury, Wiltshire SP4 OJG, United Kingdom.

2. FIBER AND COLORECTAL CANCER

2.1. Stool Bulking and Dilution

The effect of fiber on stool bulk and on the concentration of metabolites in feces was studied in depth by many groups; the results are summarized in Table I and have been reviewed extensively by Hill (1980) and by Kay (1982).

Most of the discussion of dietary fiber in colorectal cancer has centered on the role of wheat bran, historically the major source of fiber in the European, North American, and Australasian diet. When consumed in amounts likely to be acceptable to the general population, wheat bran does not increase the daily loss of bile acids, neutral steroids, or fats; since it causes a large increase in stool bulk, this results in a dilution of steroids and fatty acids in feces, as hypothesized by Burkitt. Other fiber sources are less effective in this respect; oat bran actually causes a large increase in fecal bile acid loss, and, despite the stool-bulking effect, there is a net large increase in the fecal bile acid concentration. Similarly, both pectin and guar gum cause a net increase in steroid concentration despite the modest increase in stool bulk. However, these effects are less important because eating large amounts of fruit has not been suggested as a route to the prevention of colorectal cancer.

When wheat bran in very large amounts was fed to volunteers, it caused an increase in fecal loss of steroids that matched the extra stool bulking achieved. Thus, the effect of wheat bran is as hypothesized, but unfortunately there appears to be a limit to the dilution achievable.

2.2. Fiber, Transit Time, and Bacterial Metabolism

It is well established that a major effect of dietary wheat bran is to speed the rate of transit of the gut contents through the large bowel. This should decrease the time

TABLE I. The Effect of Various Forms of Dietary Fiber on Stool Bulk and the Fecal Concentration of Steroids (Expressed as a Percentage of the Control Value)

Dietary fiber (g/day)	Fecal weight	Acid steroids		Neutral steroids	
		Per day	Concn.	Per day	Concn.
Bagasse (10)	167	165	99	100	60
Lignin (10)	110	168	153	—	—
Cellulose (36)	167	100	60	100	60
Wheat bran (16)	167	100	60	100	60
(39)	170	104	63	108	64
(54)	165	91	55	97	58
(100)	300	171	57	168	56
Oat bran (100)	230	290	125	—	—
Pectin (15)	130	140	108	117	89
(36)	125	146	117	141	113
Guar (36)	162	177	109	168	104

TABLE II. The Relationship between the Extent of
Metabolism of Cholesterol and the Intestinal
Transit Time (ITT) in Persons on a Normal Diet
and on One Supplemented with Baggase[a]

	Normal diet		Test diet	
Subject	ITT	Percentage degradation	ITT	Percentage degradation
1	31	38	46	71
2	28	92	26	81
3	43	65	37	79
4	61	72	45	71
5	32	73	31	78
6	81	83	38	80
7	38	90	24	87
8	28	65	34	73
9	57	70	61	88
10	62	85	35	86

[a]Data from Hill (1982).

available for bacterial metabolism and cause a decrease in the concentration of metabolites.

To test this, we assayed the fecal steroids in ten volunteers on a control diet and after consumption of hagasse (10 g dietary fiber). In most of the volunteers there was an increase in the rate of transit manifest as a decreased mouth-to-anus transit time (Table II), but there was no effect on the extent of cholesterol metabolism. A similar lack of effect was noted when wheat bran was used to speed transit; the lack of effect was noted for fecal bile acids as well as for fecal cholesterol metabolites. An explanation for this anomaly is illustrated in Fig. 1. When the transit time is less than T_1, the contents move through the bowel so rapidly that there is insufficient time for any bacterial metabolism. When the transit time is between T_1 and T_2, bacterial metabolism proceeds to an extent determined by the available time and therefore proportional to the transit time. When the transit time is longer than T_2, time is no longer the factor controlling the extent of metabolism, and other factors assume prominence. On this model, presumably wheat bran and other fiber sources failed to decrease the transit time below T_2.

Thus, although it appears to be reasonable to expect transit time to be a major factor determining bacterial metabolism in the gut, it seems not to be so in practice.

2.3. Fiber and the Bacterial Flora

It is well established from studies *in vitro* that the relative proportions of organisms in a mixed population are greatly affected by the nature of the energy and carbon sources. In consequence, it appears to be reasonable to hypothesize that dietary fiber, a ready source of nutrients and energy to certain genera of bacteria and not to others, should have a profound effect on the bacterial flora of the intestine. Very many groups in

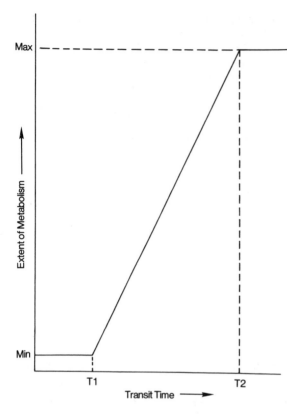

FIGURE 1. A stylized illustration of the relationship between the extent of degradation of a substrate in the colon and the intestinal transit time. Up to a transit time of T_1 transit is too rapid for degradation to occur. Between T_1 and T_2 the extent of degradation is determined mainly by the time available, but above T_2 the transit is low enough to permit maximum degradation and factors other than time limit further degradation.

many centers have attempted to demonstrate this, usually with disappointingly negative results (Table III).

This observation requires an explanation, and this is to be found in the choice of analytical material. All of the early studies were of the fecal flora because feces are readily available in large amounts and because we are familiar with the problems of sampling, transport, cultivation, etc. However, a number of studies have shown that

TABLE III. The Effect of Supplements of Dietary Fiber on the Bacterial Flora of Feces (Data Collected from Various Sources)

Dietary fiber supplement	Amount (g/day)	Effect on the fecal bacterial flora
Bagasse	10	None
Cellulose	16	None
Wheat bran	39	None
Pectin	36	None

TABLE IV. Evidence That the Proximal Colon Is the Major Site of Bacterial Metabolism in the Colon

Type of study	Observation
Stoma patients	Patients with a hemicolectomy have the same extent of bacterial metabolism as in control persons. Total colectomy virtually abolishes such metabolism.
Postmortem subjects	Analysis of samples taken from the ascending, transverse, descending, and rectosigmoid colon showed that steroid metabolism was virtually complete by the transverse colon.
Cannulated pigs	Analyses of samples from the ileum, cecum, midcolon, and feces show that steroid metabolism was virtually completed in the cecum.

very little metabolizable dietary fiber reaches the anus intact; in consequence, dietary fiber has little effect on the energy and carbon sources available to the bacteria in feces. Dietary fiber has a profound effect on the environment in the proximal colon, however, and it has been demonstrated that the proximal colon is the major site of bacterial metabolism in stoma patients (Hill, 1982), in postmortem patients (Boyer *et al.*, 1984), and in the cannulated pig (Fadden *et al.*, 1984) as summarised in Table IV.

The proximal colon is a uniquely inaccessible site in persons with a complete colon, but patients treated by colectomy for colitis were fed diets either rich in fiber or containing little fiber (Berghouse *et al.*, 1984); in such patients fiber was shown to have the expected profound effect on the composition of the ileostomy bacterial flora (Table V). However, the effect was to cause a nonspecific increase in the numbers of all components of the flora, and this would presumably include those organisms producing the carcinogens, promoters, etc.—the opposite of the effect hypothesized. This was disappointing but presumably implies that those organisms able to degrade the fiber (mainly wheat bran) produced extracellular enzymes that left the products available to the whole flora.

TABLE V. The Effect of a High-Fiber Diet on the Bacterial Flora of Ileostomy Effluent[a]

Organisms	Low fiber	High fiber	Significance
Total bacteria	5.45 ± 1.68	6.85 ± 1.15	
Total anaerobes	4.18 ± 1.91	5.08 ± 1.44	
Bacteroides spp.	2.72 ± 0.88	3.65 ± 1.39	
Clostridium spp.	3.65 ± 2.03	4.23 ± 1.60	
Veillonella spp.	3.08 ± 0.83	4.60 ± 1.75	$P < 0.05$
Total facultative organisms	5.34 ± 1.61	6.82 ± 1.14	
Streptococcus spp.	5.14 ± 1.70	6.45 ± 0.91	
Enterobacteria	3.75 ± 1.55	5.30 ± 1.84	
Lactobacillus spp.	3.13 ± 1.22	4.02 ± 1.19	$P < 0.01$

[a]Data from Berghouse *et al.* (1984).

2.4. Fiber and Colon Physicochemistry

The effect of fiber on the physicochemical environment in the gut has received scant attention apart from the effect on the colonic pH. It has been established that disaccharides or sugar alcohols that are resistant to the action of small intestinal brush border saccharidases reach that colon, where they are rapidly fermented by the intestinal bacterial flora. Where the generation of short-chain fatty acid (SCFA) end products of fermentation is faster than the rate of their absorption from the colon, colonic acidification occurs, and cecal pH values averaging 4.5 have been reported in volunteers consuming lactulose (Bown et al., 1974).

It is known that dietary fiber components are metabolized by the gut bacterial flora at rates depending on the composition of the flora and the composition and structure of the fiber matrix. There is little reliable information on the effect of fiber on cecal pH in humans. Pye (1988) studied the effect of ispaghula husk (fybogel) supplements on the pH of the intestine using a radiotelemetry capsule to measure the pH at various subsites. The ispaghula husk had no effect on small bowel pH, which remained at 7.3 in the terminal ileum (the control value) even after 4 weeks of diet supplementation. In contrast, the pH in the proximal colon, which was 6.5 during the control period, was decreased to 5.7 after 4 weeks of supplementation and returned to 6.3 after 2 weeks post-supplementation.

In contrast to the situation in humans, there is a wealth of information on the response of animals to fiber supplements. This has been reviewed by Rowland et al. (1985). In rodents the cecum of animals fed a low-fiber diet was less acidic than that of animals fed diets rich in cereal or vegetable fiber or various purified fiber supplements. In the cannulated pigs studied by Fadden et al. (1984), the cecal pH was more acidic in animals fed a standard diet supplemented with lactulose (pH less than 6) and pectin than in those fed the standard diet alone or one supplemented with wheat bran. The rodent studies gave similar results, with those fiber fractions that are rapidly metabolized (e.g., pectin, guar) having a much greater effect on pH than the slowly metabolized fractions (e.g., cellulose, bagasse) and wheat bran.

There are two further sources of information on the effect of dietary fiber on cecal pH in particular and intestinal pH in general. There have been frequent reports of a lower fecal pH in Africans eating a diet rich in fiber and cereals than in Europeans eating a low-fiber diet. Hill (1971) reported that whereas the pH of feces in English persons was usually 6.5 to 7.0, that in Ugandans was 5.0 to 5.5, and that in Indians was 6.0 to 6.5. However, this is more likely to have been the result of the metabolism of resistant starch than dietary fiber.

2.5. Conclusions

Table VI summarizes the conclusions on the possible mechanism by which dietary fiber might prevent colorectal carcinogenesis. There is no evidence from humans that dietary fiber causes any changes that result in a modified extent of metabolism of various substrates. However, there is good evidence that dietary cellulose or wheat bran causes a dilution of the colonic contents, including putative carcinogens or tumor promoters.

This will be of relevance to the discussion of colorectal cancer prevention only

TABLE VI. An Assessment of the Postulated Mechanisms by Which Dietary Fiber
Might Prevent Colorectal Carcinogenesis

Postulated mechanism	Experimental evidence in humans
1. Fiber causes stool bulking, thereby diluting the fecal contents	Support for this comes from studies of wheat bran and cellulose but not necessarily for studies of lignin or viscous fiber types
2. Fiber speeds transit through the colon, providing less time for bacterial metabolism	Fiber speeds transit through the colon, but there is little evidence that this results in less bacterial metabolism
3. Dietary fiber changes the bacterial flora to one less likely to produce carcinogens	Fiber has no effect on the fecal bacterial flora; it causes a nonspecific increase in all organisms in the proximal colon (which, if anything, would *increase* bacterial metabolism)
4. Dietary fiber modifies the physicochemical conditions in the colon toward those that are less favorable to bacterial metabolism of steroids	There is no evidence of this, and no evidence that dietary fiber supplements give rise to less bacterial metabolism in the colon

when there is good evidence that there is a relationship between dietary fiber intake and the risk of colorectal cancer. This is still eagerly awaited; only when the association has been demonstrated will the hypotheses about mechanisms become important, since they will provide the scientific support for proposed strategies for cancer prevention.

3. CURRENT HYPOTHESIS OF COLORECTAL CARCINOGENESIS

There is general acceptance that colorectal carcinomas arise in areas of dysplasia, the latter usually manifest as a macroscopic adenoma. The mechanism of the adenoma–carcinoma sequence as proposed by Hill *et al.* (1978) is illustrated in Fig. 2. A 10-year follow-up of a cohort of adenoma patients (Hill, 1985) showed that fecal steroids were not related to the rate of adenoma formation, but the fecal bile acid (FBA) concentration was correlated with the size of the largest adenoma seen during the observation period. This suggested that the bile acids are implicated in causing adenoma growth. Most adenomas remain small; the size of the largest adenoma was used since this gives the best marker of the potential for adenoma growth, the small adenomas being a mixture of old ones that are growing very slowly and of ones that *will* grow to a large size but have been observed in the very early stages of their development. In later studies from this laboratory, when the FBA concentration was related to the size of the adenoma at the time of fecal collection, there were no large adenomas (larger than 2 cm diameter) and no correlation (Owen *et al.*, 1987). It should be noted that patients from a follow-up clinic are not suitable for a study of adenoma growth rate unless the size of historical polyps is used; the usual policy of removing all adenomas as soon as they are recognized means that all adenomas seen at follow-up are new ones, with no means of determining their potential rate of growth or increase in epithelial dysplasia. In a study in which a

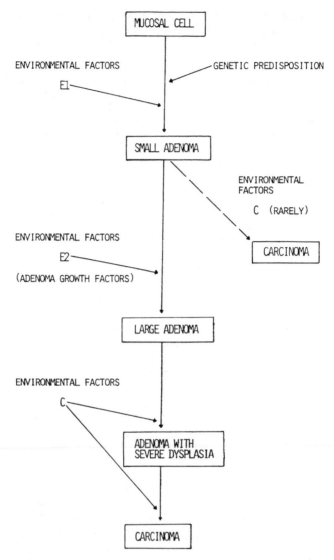

FIGURE 2. The hypothesized mechanism of colorectal carcinogenesis based on that proposed by Hill *et al.* (1978).

cohort of 102 patients with pancolitis of more than 10 years' duration (Hill *et al.*, 1987) was followed for 10 years, there was a clear correlation between the FBA concentration and the severity of epithelial dysplasia developing during the follow-up period. Further, there was a relationship between the FBA concentration and the rate of development of severe dysplasia.

In the early 1980s we changed our method of FBA assay to one that allowed us to

TABLE VII. Summary of Results Implicating the Ratio of Lithocholic to Deoxycholic Acid (LA/DC) as a Risk Factor in Colorectal Carcinogenesis

Study	Observation
Case-control study	Ratio is significantly higher in cases than in controls
Adenoma patients	Ratio is correlated with the size of the largest adenoma
Case-control study	Ratio of the precursor bile acids (chenodeoxycholic to cholic acids) higher in cases than in controls
In vitro mutagenicity	LA and DC are both comutagenic but antagonize each other; comutagenicity of mixtures depends mainly on the relative excess of one over the other

assay the fecal bile acid profile (Owen *et al.*, 1984). Using this method we demonstrated that when the concentrations of the two principal bile acid metabolites in feces (deoxycholic, DC, and lithocholic, LA, both of which are known to be potent promoters of colon carcinogenesis) were studied as markers of cancer risk, neither LA nor DC nor LA plus DC was better than the total FBA concentration already in use. However, the ratio of LA/DC was a very much better discriminant, and for the last 5 years our efforts have been concentrated on the study of this ratio as a marker of cancer risk. Our results have been summarized recently (Owen *et al.*, 1987) together with a rationale for the use of the ratio based on the *in vitro* studies of Wilpart *et al.* (1983a,b); the results are summarized in Table VII.

In summary, our current hypothesis is that fecal bile acids cause small adenomas (with a low malignant potential) to grow to a large size (and with a correspondingly high malignant potential) and to become more severely dysplastic. In this respect, although the total FBA concentration is important, the concentrations of the two principal FBAs, LA and DC, are of particular relevance. Since the two bile acids are mutually antagonistic with respect to tumor promotion, it is the ratio LA/DC that is most important, not the sum of the concentration LA + DC as might be expected.

4. COLORECTAL CANCER PREVENTION

Based on the hypothesis that bile acids are important in colorectal carcinogenesis, strategies to prevent the disease should include steps to decrease the colonic concentration of the bile acid metabolites. These could include:

1. Change to decrease the concentration of bile acids in the colonic contents.
2. Modification of the colonic milieu to one that is less favorable to bacterial metabolic activity.
3. Change in the bacterial flora of the proximal colon to one with less capability of bile acid metabolism.

4.1. Change in the Fecal Bile Acid Concentration

Table I describes the effect of dietary fiber on the fecal bile acid concentration, and it is clear that wheat bran, in large amounts, should produce a change in the desired direction. Similarly, it is known that the fecal bile acid concentration is correlated with the intake of dietary fat (Antonis and Bersohn, 1962; Hill, 1971; Cummings *et al.*, 1978; Turjman *et al.*, 1984; Reddy *et al.*, 1973), although some studies in artificial model situations suggest that this relationship is not a simple one. More recently, studies from Bruce and his co-workers have shown that when bile acids are precipitated as their calcium salts, this ameliorates the toxic effect of bile acids on the colonic mucosa and decreases the severity of epithelial dysplasia in model systems and in humans (War-govich *et al.*, 1983; Bird *et al.*, 1986; Lipkin and Newmark, 1985; Buset *et al.*, 1986; Rafter *et al.*, 1986).

In response to these latter observations, we have been studying the effect of an increased calcium intake on the fecal calcium concentration and on the concentration of "free" and "bound" bile acids. Our first experiment was to measure the effect of a supplement of 2 g calcium as calcium carbonate (500 mg four times per day) on fecal calcium concentrations; we were able to confirm that such an increased intake resulted in a greatly increased fecal calcium concentration, the magnitude of which varied among persons.

We next studied the ability of various fecal bile acid assays to recover calcium salts of bile acids by assaying fecal samples (1) unspiked and (2) spiked with calcium deoxycholate (all of the standard assays give almost quantitative recovery of added free bile acids when carried out by experienced staff). The method of Grundy *et al.* (1965) gave very good recovery of the calcium deoxycholate and therefore measures total fecal bile acids (free, bound, and precipitated). Our current method (Owen *et al.*, 1984) assays the free bile acids but does not recover calcium salts; this was to be expected since calcium deoxycholate is extremely insoluble in chloroform and in petroleum ether. The method of Evrard and Janssen (1968) gave a recovery of calcium deoxycholate that was slow and barely complete in the standard extraction time. Assay of fecal samples using the method of Owen *et al.* (1984) followed by further extraction after saponifica-tion in alkaline alcohol (as in the method of Grundy *et al.*, 1965) showed that in most of the English persons studied, there was very little calcium salts of bile acids present in feces but that in some persons the proportion of calcium salts could be 5–10% of the total and, on occasions, was much larger. Our conclusion from these preliminary studies is that since our current methods give better correlations with bowel cancer risk (and also tend to give lower values for fecal bile acid concentration) than the previous methods, this supports the hypothesis of Bruce *et al.* that the free bile acids are important in colorectal carcinogenesis and the bound bile acids are not. A further conclusion is that we should continue to use our current assay procedure unamended.

A further set of experiments has been the assay of fecal calcium in various patient groups after first developing the methods to assay fecal calcium fractions. Calcium in feces is present in solution in the aqueous phase or "bound" to the insoluble fiber and bacterial fractions or as calcium soaps of bile acids or fatty acids. The total fecal calcium is simply assayed, following lyophilization of the fecal sample, by atomic absorption spectroscopy. The soluble calcium fraction is also simply assayed; the fecal sample is emulsified in an equal weight of distilled water, centrifuged at 20,000 g for 20 min, and calcium assayed in the supernatant fraction. Calcium soaps cannot be assayed by any simple or reliable method; a measure of the soaps can be obtained after soxhlet ex-traction with water, but at present we assay the calcium soaps together with the bound

TABLE VIII. Results to Date on the Relationship between Fecal Calcium Concentration (FCC) and the Risk of Colorectal Carcinogenesis

Study	Observation	Agreement with hypothesis
Comparison of Finnish and Danish populations	FCC higher in Finns than in Danes	+
Comparisons within Finland and Denmark	In both countries, FCC was higher in urban than in rural populations	−
Case-control study	FCC higher in controls than in cases	+
Comparison of Swedish diet groups	No relationship between FCC and LBC risk in the populations	−

calcium in the insoluble fraction. In practice, the soluble calcium fraction is small compared with the insoluble fraction, and so little extra information is gained by assaying fractions. In consequence, we are currently assaying only total fecal calcium. Table VIII summarizes our results to date, which, in general, offer support for the hypothesis that a low calcium intake is associated with a low fecal calcium concentration, a high proportion of free bile acids, and a high risk of colorectal cancer.

4.2. Modification of the Colonic Milieu

Based on the pH optima of the bacterial enzymes that degrade bile acids, particularly the 7-dehydroxylase, it was hypothesized in 1974 that acidification of the colonic contents should decrease the risk of colorectal cancer (Hill, 1974). Support for this came from studies of the fecal pH in colorectal cancer cases compared with control persons (MacDonald et al., 1978), and a detailed case for considering an acidified proximal colon to be protective was proposed by Thornton (1981).

There is evidence that the proximal colon can be acidified by taking dietary supplements of lactulose (Florent et al., 1985; Fadden et al., 1984) and of ispaghula fiber (Pye, 1988).

4.3. Modification of the Cecal Flora

It has still to be established that the bacterial flora of the human proximal or distal colon can be modified long-term either using oral dosage of specific organisms or by dietary change.

5. CONCLUSIONS

Considerable headway has been made in our understanding of the causation of colorectal cancer and of the methods by which the risk might be decreased. It is likely that the next 5 years will yield the first firm results on the prevention of this very important cancer.

ACKNOWLEDGMENTS. The work of this laboratory is financially supported by the Cancer Research Campaign.

REFERENCES

Antonis, A., and Bersohn, I., 1962, The influence of diet on faecal lipids in South African white and bantic prisoners, *Am. J. Clin. Nutr.* **11**:142–55.

Berghouse, L., Hori, S., Hill, M., Hudson, M., Lennard-Jones, J., and Rogers, E., 1984, Comparison between the bacterial and oligosaccharide content of ileostomy effluent in subjects taking diets rich in refined or unrefined carbohydrate, *Gut* **25**:1071–1077.

Bird, R. P., Schneider, R., Stamp, D., and Bruce, W. R., 1986, Effect of dietary calcium and cholic acid on the proliferative indices of muruine colonic epithelium, *Carcinogenesis.* **7**:1656–1661.

Bown, R. L., Gibson, J. A., Sladen, G. E., Hicks, B., and Dawson, A. M., 1974, Effects of lactulose and other laxatives on ileal and colonic pH as measured by a radiotelemetry device, *Gut* **15**:999–1004.

Boyer, J., Day, D. W., and Hill, M. J., 1984, Site of cholesterol degradation in the human gut, *Trans. Biochem. Soc.* **12**:1104–1105.

Burkitt, D. P., 1969, Related disease—related cause? *Lancet* **2**:1229–1231.

Burkitt, D. P., 1970, Relationship as a clue to causation, *Lancet* **2**:1237.

Burkitt, D. P., 1971, Some neglected leads to cancer causation, *J. Natl. Cancer Inst.* **47**:913.

Buset, M., Lipkin, M., Winawer, S., Swaroop, S., and Friedman, E., 1986, Inhibition of human colonic epithelial cell proliferation *in vivo* and *in vitro* by calcium, *Cancer Res.* **46**:5426–5430.

Cummings, J., Wiggins, H., Jenkins, D., Houston, H., Drasar, B., Hill, M., and Jivraj, T., 1978, The influence of different levels of dietary fat intake on faecal composition, microflora and gastrointestinal transit time, *J. Clin. Invest.* **61**:953–961.

Evrard, E., and Janssen, G., 1968, Gas–liquid chromatographic determination of human fecal bile acids, *J. Lipid Res.* **9**:226–236.

Fadden, K., Owen, R., Hill, M., Latymer, E., Low, A. G., and Mason, A. N., 1984, Steroid degradation along the gastrointestinal tract: The use of the cannulated pig as a model system, *Trans. Biochem. Soc.* **12**:1105–1106.

Florent, C., Flourie, B., Leblond, A., Rautureau, M., Bernier, J., and Rambaud, J., 1985, Influence of chronic lactulose ingestion on the colonic metabolism of lactulose in man (an *in vivo* study), *J. Clin. Invest.* **75**:608–613.

Grundy, S. M., Ahrens, E. H., and Miettinen, T. A., 1965, Quantitative isolation and gas–liquid chromatographic analysis of total fecal bile acids, *J. Lipid Res.* **6**:397–410.

Hill, M. J., 1971, The effect of some factors on the faecal concentration of acid steroids, neutral steroids and urobilins, *J. Pathol.* **104**:239–245.

Hill, M. J., 1974, Steroid nuclear dehydrogenation and colon cancer, *Am. J. Clin. Nutr.* **27**:1475–1480.

Hill, M. J., 1980, Conservation of bile acids, in: *Drugs Affecting Lipid Metabolism* (R. Fumagalli, D. Kritchevsky, and R. Paoletti, eds.), Elsevier, Amsterdam, pp. 89–96.

Hill, M. J., 1982, Colonic bacterial activity: effect of fibre on substrate concentration and on enzyme action, in: *Dietary Fiber in Health and Disease* (G. Vahouny and D. Kritchesky, eds.), Plenum Press, New York, pp. 35–44.

Hill, M. J., 1985, Bacteria and colorectal adenomas, *Top. Gastroenterol.* **13**:237–252.

Hill, M. J., Morson, B. C., and Bussey, H. J. R., 1978, Etiology of adenoma–carcinoma sequence in large bowel, *Lancet* **1**:245–247.

Hill, M. J., Melville, D., Lennard-Jones, J., Neale, K., and Ritchie, J. K., 1987, Faecal bile acids dysplasia and carcinoma in ulcerative colitis, *Lancet* **2**:185–186.

Kay, R. M., 1982, Dietary fiber, *J. Lipid Res.* **23**:221–242.

Lipkin, M., and Newmark, H., 1985, Effect of added dietary calcium on colonic epithelial-cell proliferation in subjects at high risk for familial colon cancer, *N. Engl. J. Med.* **313**:1413–1414.

MacDonald, I. A., Webb, G. R., and Mahoney, D. E., 1978, Fecal hydroxysteroid dehydrogenase activity in vegetarian 7th Day Adventists control subjects and bowel cancer patients, *Am. J. Clin. Nutr.* **31**:233–238.

Owen, R., Thompson, M., and Hill, M. J., 1984, Analysis of metabolic profiles of steroids in faeces of healthy subjects undergoing chenodeoxycholic acid treatment by lipid-gel chromatography and gas–lipid chromatography mass spectrometry, *J. Steroid Biochem.* **21**:593–600.

Owen, R. W., Thompson, M. H., Hill, M. J., Wilpart, M., Mainguet, P., and Roberfroid, M., 1987, The importance of the ratio of lithocholic to deoxycholic acid in large bowel carcinogenesis, *Nutr. Cancer* **9**:67–71.

Pye, G., 1988, Gastrointestinal pH and colorectal neoplasia, Ph.D. Thesis, University of Nottingham, Nottingham.

Rafter, J. J., Eng, V. W., Furrer, R., Medline, A., and Bruce, W. R., 1986, Effect of dietary calcium and pH on the mucosal damage produced by deoxycholic acid on the rat colon, *Gut* **27**:1320–1329.

Reddy, B. S., and Wynder, E. L., 1973, Large bowel carcinogenesis: Fecal constituents of populations with diverse incidence rates of colon cancer, *J. Natl. Cancer Inst.* **50**:1437–1442.

Rowland, I. R., Mallett, A. K., and Wise, A., 1985, The effect of diet on the mammalian gut flora and its metabolic activities, *Crit. Rev. Toxicol.* **16**:31–103.

Thornton, J. R., 1981, High colonic pH promote colorectal cancer, *Lancet* **1**:1081–1082.

Turjman, N., Goodman, G. T., Jaeger, B., and Nair, P. P., 1984, Metabolism of bile acids, *Am. J. Clin. Nutr.* **40**:937–941.

Wargovich, M. J., Eng, V. W. S., Newmark, H. L., and Bruce, W. R., 1983, Calcium ameliorates the toxic effect of deoxycholic acid on colonic epithelium, *Carcinogenesis* **4**:1205–1207.

Wilpart, M., Mainguet, P., Maskens, A., and Roberfroid, M., 1983a, Structure–activity relationship amongst biliary acids showing co-mutagenicity towards 1,2-dimethylhydrazine, *Carcinogenesis* **4**:1239–1241.

Wilpart, M., Mainguet, P., Maskens, A., and Roberfroid, M., 1983b, Mutagenicity of 1,2-dimethylhydrazine towards *S. typhimurium*. Co-mutagenic effects of bile acids, *Carcinogenesis* **4**:45–48.

32

The Epidemiology of Fiber and Colorectal Cancer
Why Don't the Analytical Epidemiologic Data Make Better Sense?

JOHN D. POTTER

1. INTRODUCTION

David Kritchevsky has noted on a number of occasions that the first known dietary experiment is recorded in the book of *Daniel* (Chapter 1). In fact, the first dietary study was an apple-feeding experiment on the population of the whole world (*Genesis*). Unfortunately, the scientist involved was so preoccupied with rushing into print on its significance for moral fiber, he never did write up the consequences of ingestion of the dietary fiber.

This chapter addresses the relationship between dietary fiber and large bowel cancer. It discusses the evidence regarding the association in the light of some theoretical arguments, the ecological epidemiologic data, and the analytical epidemiologic data. The animal evidence is discussed elsewhere in this volume (Jacobs, Chapter 29). It then poses the question: "Why are the analytic data apparently at odds with some of the other sources of information?" A connected sequence of explanations is presented. Finally, an interactive model of colon cancer and diet is outlined.

2. SOME THEORETICAL ARGUMENTS

The majority of the temporal existence of the human species (now variously estimated at up to 2 million years) has been spent in gatherer–hunter communities. For most

JOHN D. POTTER • Division of Epidemiology, School of Public Health, University of Minnesota, Minneapolis, Minnesota 55455.

groups, gathering has provided the large majority of caloric intake; this is still the case for extant gatherer–hunter groups, the Tarzanist fantasies of certain pop anthropologists notwithstanding.

As both gatherer–hunters and peasant agriculturalists, humans have largely processed food only to the extent necessary for preparation and palatability. The majority of humanity has consumed, at least until recently, diets relatively high in complex carbohydrate and in vegetable and cereal fiber. When the matter is put to the test, humans appear to be well-adapted to such diets.

There is also a growing body of data regarding the efficacy of diets high in complex carbohydrate and fiber in the treatment of some specific disorders—diabetes mellitus, diverticular disease, irritable colon, and other nonmalignant colonic diseases. There is not yet, however, convincing evidence for the role of such diets (or rather their lack) in the genesis of such diseases.

Finally, there are now coherent physiological data to show that fiber behaves in a variety of ways consistent with a capacity to reduce risk of carcinogenesis in the large bowel. These fiber behaviors include bile acid binding, water holding and thus direct stool bulking (and perhaps reduction of transit time), and the tendency to be fermented by bacteria in the large bowel, thus producing volatile fatty acids (e.g., butyrate). Volatile fatty acids may both act as anticarcinogens and lower colonic pH, which potentially reduces secondary bile acid production. The availability of fuel for fermentation increases bacterial mass and thus, indirectly, stool bulk.

Although it is possible to argue from this historic/evolutionary, therapeutic, and physiological evidence for the importance of fiber to health in general, it is more problematic to argue for a preventive role in relation to a specific disease such as bowel cancer. It may be that the role of fiber is merely neutral. What is clear (and the argument above is merely sketched) is that in order to conclude that fiber is specifically deleterious, some strong and consistent evidence is necessary.

It is argued in this chapter that such evidence against fiber does not exist but that data for a preventive action of fiber against bowel cancer are also wanting.

3. ECOLOGICAL EPIDEMIOLOGIC DATA

In 1971, Burkitt pointed out a pattern of gastrointestinal diseases, including colorectal cancer, that, while common in Western countries, did not occur in Africa. He suggested a protective role for diets high in fiber.

A variety of ecological studies describing the correlation between apparent national fiber consumption and risk of colorectal cancer across a number of countries followed. Although there is a negative relationship described in a number of studies (see Table I), these cannot be regarded as independent tests of the hypothesis, as they have all used essentially the same data. It is rather like reading the same story in several copies of the same newspaper to see if a given event actually occurred. Furthermore, in each of the studies where other nutrient intakes, particularly those related to fat or meat, were controlled, the strength of the correlation diminished or disappeared.

TABLE I. Ecological (Correlation) Studies of Fiber
and Colon Cancer Mortality

Author	Fiber variable	Correlation coefficient
Armstrong and Doll (1975)	Cereals	−0.1 to 0.2[a]
Howell (1975)	Cereals	−0.84
	Pulses	−0.76
Meyer (1977)	Fruits	−0.83
	Potatoes	+0.68
Shrauzer (1976)	Cereals	−0.70
Liu et al. (1979)	Fiber	−0.71
	Fiber	+0.03[b]
Yanai et al. (1979)	Cereals	−0.50
McKeown-Eyssen and Bright-See (1984)	Cereals	−0.73
	Cereal fiber	−0.3 to 0.4[a]
	Pulses	−0.70
	Pulse fiber	−0.2 to 0.3[a]

[a] Controlling for meat, animal protein, or fat.
[b] Controlling for cholesterol.

4. ANALYTICAL EPIDEMIOLOGIC STUDIES

Ecological studies are based on both international variation in apparent consumption and mortality or incidence rates; analytical studies relate individual exposure to individual outcome either in cohort or case-control fashion. The analytical epidemiologic studies are summarized for a variety of cereal, vegetable, fruit, and fiber-related exposures in Table II. What is clear from this summary is that the most consistent association is between a reduced risk of colorectal cancer and vegetable consumption. Indeed, it can be argued that this association is the strongest and most consistent for any relationship in the nutritional epidemiology of cancer (Potter, 1988).

The relationship between fiber and cereals and risk of bowel cancer is much more equivocal. For instance, the studies of Dales et al. (1979) and Modan et al. (1975) using an index of high-fiber foods, found a lower risk of colon cancer among high versus low consumers; a variety of studies have noted little association with risk (e.g., Miller et al., 1983); and some studies have reported higher risk of colorectal cancer in the presence of higher consumption of fiber (e.g., Martinez et al., 1979).

This raises an important question: Given the plausibility of the association, both biologically and historically, why are the analytical data on fiber and cereals so inconsistent? The obvious answer is that there may be no relationship between colon cancer and high fiber consumption and that the real relationship is with vegetables and/or fruit and the specific micronutrients and anticarcinogenic substances they contain. However, lest the kernel be tossed out with the husk, I would like to explore a related sequence of potential explanations for the failure to establish the relationship and suggest what further studies need to be undertaken before this particular hypothesis is rejected.

TABLE II. Analytical Epidemiologic Studies of Fiber
and Colorectal Cancer[a]

Factor	Relationship to risk	Number of studies showing relationship	Number of studies done
Case-control studies			
Vegetables			
Vegetables	−	9	12
String beans	+	1	2
Fiber			
Fiber	−	4	9
Fiber	+	2	9
Cereals			
Rice	−	1	3
Rice	+	2	3
Pasta and rice	+	2	2
Cereals	+	2	6
Fruit			
Fruit	−	2	9
Fruit	+	1	9
Vitamin C	−	3	4
Vitamin A	−	1	4
Cohort studies			
Cereals			
Rice and wheat	−	1	1
Fruit			
Vitamin C	−	1	1

[a]From Berta et al. (1985); Bjelke (1973); Bristol et al. (1985); Dales et al. (1979); Graham et al. (1978); Haenszel et al. (1973, 1980); Higginson (1966); Jain et al. (1980); LaVecchia et al. (1988); Lyon et al. (1987); Macquart-Moulin et al. (1986); Manousos et al. (1983); Martinez et al. (1979); Miller et al. (1983); Modan et al. (1975); Pernu (1960); Phillips (1975); Potter and McMichael (1986); Stocks (1957); Wynder et al. (1969); Wynder and Shigematsu (1967).

5. REASONS FOR THE INCONSISTENCY OF THE ANALYTICAL DATA

Most studies of diet and cancer rely on the collection of usual dietary habits of healthy individuals at the beginning of a study (cohort study) or of both diseased and healthy individuals after the former have developed the condition under study (case-control study). Both study designs have advantages and disadvantages. In both designs, the dietary data (usually in the form of a food-frequency questionnaire) are converted to nutrients using a nutrient data base derived from food analysis and actual or imputed serving sizes. One of the most important sources of problems in this analytical sequence is the dietary data base. In many such data bases, the specific nutrients are not available either because the food sources are highly variable over time and place (e.g., selenium) or the required analysis has not yet been done (e.g., specific fiber types and subtypes).

The problems of fiber classification and measurement are two issues that make epidemiologic studies of cancer and fiber hard to interpret.

5.1. Classification of Fiber

The notion of "crude fiber" with all its unphysiological implications has not been used much in the epidemiologic literature, although at least one relatively recent study did describe results for that exposure (Miller *et al.*, 1983). The more usual description in epidemiologic studies derives from the notion of dietary fiber as described by Trowell in 1972, but as is noted below, there are a variety of ways of operationalizing that definition.

A number of investigators have used specific fiber indices or consumption of certain high-fiber foods (see, for example, Modan *et al.*, 1975; Dales *et al.*, 1979). There have also been attempts to subdivide fiber into that from vegetable and cereal sources (Potter and McMichael, 1986).

There are several problems with this inconsistent classification of fiber. First, the physiological effects pertinent to cancer are not necessarily taken into account; second, even when indices are developed, the criteria for inclusion of specific foods are not consistent across studies; third (see below), the underlying analytical systems from which the dietary data bases are derived are not consistent. Thus, it is not possible to judge which studies, if any, may have happened on the "right" combination and can therefore be said to be an accurate representation of the relationship between exposure and risk in humans.

More recently, a variety of ideas regarding fiber classification schemes have emerged, including specific subtypes such as cellulose, hemicellulose, pectic substances, etc. (Selvendran, 1983); nonstarch polysaccharide (Englyst and Cummings, 1984); resistant starch (Englyst *et al.*, 1982); and soluble starch and insoluble noncellulose polysaccharide (Anderson and Bridges, 1988). These represent a greater problem for epidemiologists because there are no dietary data bases yet that fully incorporate any of these newer fiber classifications (some of which are mutually exclusive). Further, there is as yet no general agreement on the physiological role for some of these fiber subtypes.

One advantage, however, is that a variety of existing studies can be reanalyzed once new classification schemes have been agreed on and operationalized into dietary data bases. In general, the classification issue remains one that makes the interpretation of the fiber–colon cancer relationship extremely difficult.

5.2. Measurement of Fiber Intake

Different studies, even within the same countries, use dietary data bases compiled in somewhat different ways. Because of the lack of comparison studies on dietary fiber data in these data bases, it is rather difficult to show that they are inconsistent. It is possible, however, to show that data bases reportedly derived from the same analytical data (usually USDA data) provide quite different values for daily intake of other nutrients when the same food frequency questionnaires are analyzed, reduced to usual consumption patterns, and compared.

Both Hoover (1983) and Jacobs *et al.* (1985) have shown that there is a major variability in fat, protein, carbohydrates, alcohol, and total calories across nutrient data bases, producing inconsistent analyses when the same food intake data are run against them. This difference may be such as to exaggerate the differences between cases and controls in one study using one data base while producing the reverse effect in another.

Bingham *et al.* (1979) showed a very high correlation ($r > 0.9$) between regional variation in pentosan intake (largely from cereals) in Great Britain and risk of colon cancer. The same authors (Bingham *et al.*, 1985) reanalyzed these original data using what they regarded as a superior data base for fiber and showed that the previously high correlation essentially disappeared ($r < 0.2$). These two analyses demonstrated the extreme sensitivity of the interpretation of relationships to choice of data base and ultimately to analytical methods.

There is an even more problematic issue in the interpretation of fiber data, namely, the fact that because chemical analyses of fiber in food are relatively incomplete, a number of data bases have included data from a variety of sources that use different food analysis methods. This ensures that there may not be consistency even within any given data base.

Thus, the choice of dietary data base crucially determines not only what nutrients may be examined in relation to risk but also the nature and strength of the relationships described. At present, there is no way of predicting what effect the choice of a particular data base will have on the interpretation of the relationship between colorectal cancer and fiber, but some degree of misclassification ensures that the relative risks will be underestimated.

5.3. Controlling for Calorie and Nutrient Intake

A number of authors have noted the strong correlations between intake of specific nutrients and overall energy intake (Lyon *et al.*, 1983; Willett and Stampfer, 1986; Potter and McMichael, 1986). Correlation coefficients among fat, saturated fat, protein, and energy are typically greater than 0.7. Those between dietary fiber intake and total energy are somewhat lower, but it is possible to show that there is significant confounding between energy and fiber intakes (Lyon *et al.*, 1983; Potter and McMichael, 1986).

The issue of how to control for this confounding in relation to fat, especially, has received considerable attention because there is not only the statistical problem of multicollinearity but the biological interpretation of the data (Willett and Stampfer, 1986; Howe *et al.*, 1986). One important question, for instance, is whether it is the calories that fat contributes to the diet that is the significant factor in colon carcinogenesis or whether it is some direct chemical/toxicological effect of lipid itself. The way the epidemiologic data are handled crucially limits the way in which this question can be answered.

The issue with fiber is somewhat more straightforward in that fiber does not contribute calories directly (although some have argued that fiber may contribute a small proportion of energy needs via fermentation). Nonetheless, the appropriate choice of variable for inclusion in an analysis of the relationship between diet and cancer is predetermined by the proposed biological action of fiber in carcinogenesis. For instance, absolute fiber intake, fiber per 1000 kcal, and a fat–fiber ratio all suggest potentially different modes of action for fiber. All have been used, usually without specifying the implications or the mode of fiber action being tested.

Table III shows data from studies in which different ways of expressing fiber are consistent with different interpretations of the relationship between fiber consumption and large bowel cancer. In the study by Lyon *et al.* (1987), the crude estimates of relative risk for higher levels of fiber consumption are in general greater than 1.0. After adjusting for caloric intake, these are mostly less than 1.0. In the Potter and McMichael (1986) study, the crude estimates of relative risk for higher fiber consumption are statistically significantly greater than 1.0 but are reduced considerably and cease to be significantly different from 1.0 by adjustment for caloric intake.

Future epidemiologic studies should state clearly their hypothesis for the action of fiber and operationalize the exposure data in ways consistent with that hypothesis, including especially the issue of the relationships between both fat and calories on the one hand and fiber on the other. (This clearly is dependent on the resolution of the issue of fiber in dietary data bases outlined above.) The meaning of specific relationships between fiber and other nutrients may be considered in relation to the model of colon carcinogenesis presented in the last part of this chapter.

5.4. Metabolic Effects of Fiber

The previous problem for the interpretation of fiber-cancer relationship is directly related to the issue of which metabolic effects of fiber are directly related to the process of carcinogenesis. Elsewhere in this volume, Selvandran and Verne (Chapter 1) describe specific properties of fiber types, but the epidemiologic data have not yet been examined to determine their relationship to cancer risk at a population level.

At least three properties have been proposed for dietary fiber in relation to carcinogenesis: bile acid binding, fermentability, and stool bulking. Unfortunately, even the

TABLE III. Effects of Controlling for Calorie Intake on Estimates of Association between Fiber and Relative Risk (RR) of Colorectal Cancer

	RR for crude fiber (g/day)			
	<3.2	3.2–4.5	4.6–6.0	>6.0
Lyon *et al.* (1987)				
Male (crude)	1.0	0.9	1.4	1.2
(adjusted)	1.0	0.7	1.2	0.8
Female (crude)	1.0	0.8	1.7	1.4
(adjusted)	1.0	0.5	0.9	0.6

	RR for dietary fiber quintiles			
	1	2	3	4	5
Potter and McMichael (1986)				
Female (crude)	1.0	1.8	2.2	2.9	4.1
(adjusted)	1.0	0.8	2.5	1.8	2.0

animal literature is not very helpful in relation to bile acid binding, with at least some of the bile-acid-binding fibers and associated agents being associated in some studies with increased carcinogenesis (e.g., pectin) and others (e.g., saponins) being as yet largely unexplored.

Stool bulking, which is caused directly by water-holding fibers and indirectly by fermentable fibers, is a likely candidate for explaining some of the international variability in colon cancer risk (IARC Microecology Group, 1977; Jensen et al., 1982), but neither the animal evidence nor the epidemiology indicates that wheat bran significantly reduces risk.

The next step for the nutritional epidemiology of cancer in this area might be to specify in greater depth the likely roles of each fiber type and to measure these using data bases with the appropriate degree of precision. This, again, may mean waiting until the data bases are improved.

5.5. Metabolic Interactions with Other Nutrients

The fifth reason the analytic epidemiologic data are so unclear is that studies have not often been designed in ways that allow the exploration of interaction between nutrients. There are hints in the literature, however, of important interactions between protein and fiber and between meat and vegetables.

In the study of Potter and McMichael (1986), there was a strong association between protein consumption and risk of colon cancer, particularly in women. When this was examined taking into account fiber consumption, the majority of this effect could be seen to be accounted for by the group consuming the lowest amount of dietary fiber.

Hirayama (1981) showed that among those who eat meat less than daily, the frequency of consumption of vegetables is largely irrelevant to the risk of colorectal cancer. Among those who consume meat daily, however, frequency of vegetable consumption is a major determinant of risk.

The identification of such interactions is crucial to our understanding of the dietary causes of colorectal cancer, but this research question requires more tightly specified hypotheses and larger studies than have been undertaken to date.

5.6. Metabolic Interactions with Endogenous Processes

One of the important issues that has been ignored in the epidemiologic literature investigating the relationship between fiber and colon cancer is the evidence for interactions with endogenous processes.

Evidence for the importance of endogenous processes includes the observation that the subsite distribution of colonic cancers differs between the sexes (McMichael and Potter, 1980). Specifically, women have higher rates of colon cancer at young ages and men at older ages. There is a higher rate of proximal colon cancer in women throughout life. Thus, women have a tendency to develop cancer more proximally and somewhat younger than men, who are more prone to distal cancers at an older age.

Further, there are data from studies of human physiology to show that these differences can be related to differences in gut biology between the sexes (McMichael and

Potter, 1983). Specifically, women have slower transit times, higher gut pH, lower stool bulk (even on comparable diets), and higher levels of secondary bile acids than men—all differences consistent with a greater proneness to colonic cancer in women than men (McMichael and Potter, 1983) and therefore consistent with its onset at an earlier age. There are also data to show that reproductive status is associated with colon cancer risk in women (Weiss *et al.*, 1981; Potter and McMichael, 1983; Howe *et al.*, 1985).

Finally, there are some data that suggest that the effects of nutrients may vary by age and sex (McMichael and Potter, 1985) in ways consistent with the descriptive data. McMichael and Potter reported that the strength of the association between dietary variables and risk was greater in younger women than older women, and the reverse for men.

The implications for epidemiologic studies of colonic cancer of examining this kind of interaction are, first, a need for greater precision in hypothesis specification; second, a need for larger studies (sample sizes for studies that provide estimates of relative risk for interactive effects between two variables may be increased fourfold over those needed to estimate main effects only); and third, the wider application of metabolic epidemiologic techniques.

5.7. Nutrients and Factors Associated with Fiber

Most food analyses and data bases were originally developed to fill the needs of studies of growth and development. They thus focus on macronutrients and on vitamins and trace elements. (Indeed, the absence of good data on fiber is itself a manifestation of the same focus.) There are many food-borne substances that may be related particularly to anticarcinogenesis (Wattenberg, 1985), the intake of which are not able to be estimated using nutrient data bases. These include indoles, thiols, flavonoids, phenols, and lignans. It may be that the presence of these substances in many fruit, vegetables, cereals, or their production by microbes from plant precursors is as important to the association between reduced risk of colon cancer and vegetable consumption as is the presence of fiber.

The inclusion of appropriate analyses in food data bases and relevant methods for assessing exposure (especially biological measures) are crucial to the next generation of epidemiologic studies. Their absence is a major hindrance to the testing of alternative hypotheses to the proposed fiber–colon cancer link in the epidemiologic setting.

5.8. Missing Epidemiologic Data

There are a great many questions the answers to which would help clarify the fiber–colon cancer relationship. These include the following:

1. Given the evidence in favor of both the fermentability of fiber and the production of volatile fatty acids (VFA) being important in colon carcinogenesis, are there any ecological differences in colon VFA production that may be examined for consistency with the international variation in risk of bowel cancer?
2. Are there differences in colonic VFA production in populations with known differences in susceptibility, e.g., Japanese versus Americans, vegetarians versus omnivores?

3. Is there any association between VFA production and the growing list of biological markers of colonic cancer such as the K-*ras* or *myc* oncogenes or the development of specific chromosome deletions?
4. A similar set of questions may be asked about colonic pH.
5. Further studies could be undertaken, in this area of biological epidemiology, on groups migrating from areas of low risk to areas of high risk.

6. MULTIFACTORIAL THINKING: DEVELOPING A BIOLOGICAL MODEL FOR EPIDEMIOLOGY

Finally, perhaps the most significant block to the understanding of the analytical epidemiologic data on colon cancer and fiber is the failure to think in multifactorial terms. Despite evidence to the contrary, much epidemiologic research and most public health messages are couched as though fiber was a single substance and acted in just one way: both scientists and policy makers often appear to think only of wheat bran and only of stool bulking.

We have elsewhere suggested that a broader view of the interactions between the elements that make up the colonic environment might be to regard digesta as culture medium (McMichael and Potter, 1986). By this we mean a medium not only in which bacteria flourish and die out and are selected for and against but also in which the colonic cells themselves are nourished or damaged and experience specific survival pressures and opportunities. If such a biological model is explored a little further in relation to fiber, its components, and its related substances, a variety of important and complex relationships emerge.

First, consider the model depicted in Fig. 1 as representing the components of the gut environmental milieu: endogenous (bile acids, mucopolysaccharides, water), exogenous (carbohydrate, fiber), and bacterial. These three components are influenced by ingested food and biological processes and in turn interact with both gut musculature (specifically the amount of work done and the gut transit time) and the epithelium (changing cell replication rates and perhaps total epithelial surface area by altering fuel availability and affecting cell damage and cell repair via the effects of potentially toxic agents such as bile acids and potentially beneficial agents such as butyrate).

In looking at any one part of this model, it is possible to describe a potential cascade of events; e.g., ingested vegetables increase the fiber content of the large bowel, which may bind bile acids and be fermented by bacteria increasing bacterial mass and VFA production. This, in turn, produces effects on cell replication and cell damage and repair and inhibits secondary bile acid production by reducing gut pH. Meanwhile, the fiber and the increased bacterial mass increase stool bulk and influence the work load of gut musculature, transit time, and cell turnover rates. Also present are potential anticarcinogens, which are widespread in vegetables, and nitrites, which may be capable of producing toxic or carcinogenic compounds.

Ingested meat, on the other hand, will increase hepatic bile acid production. Other factors, including bacterial species present, enzyme production, and transit time, will alter the effect of conversion to secondary bile acids, thereby altering the likelihood of cell damage and repair and possible changes in gut surface area. The increase in cell workload will increase cell replication rates.

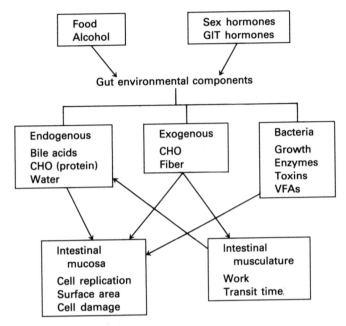

FIGURE 1. An interactive model of colon carcinogenesis.

Table IV details the likely effect that each of the components in these foods will have on the gut milieu. It also shows the likely effect on carcinogenesis that each of these "cascade steps" represents. It can be seen that the net effect of the vegetable exposure on carcinogenesis (if all of these effects are given equal weight) is to reduce its likelihood. The reverse is true for meat.

This model may be helpful on several counts. First, it makes more detailed our description of digesta as culture medium (McMichael and Potter, 1986). Second, it makes explicit the multifactorial nature of colon carcinogenesis without even entering into the established differences in familial/genetic susceptibility for the disease. Third, it defines a variety of testable hypotheses that may be pursued as further techniques for metabolic epidemiology become available. Fourth, it provides the setting for understanding the importance of the extensive epidemiologic data that are, at present, still missing on the relationship between fiber and colorectal cancer. Finally, it suggests that even when a specific dietary exposure is associated with reduced risk of colon cancer, not every component nutrient or factor operates in the same direction, and similarly for exposures that are, overall, likely to increase the risk of carcinogenesis.

7. SUMMARY

This chapter is an attempt to examine the relationship between fiber consumption and risk of colorectal cancer. It has been argued that there are a variety of reasons why the analytical epidemiologic data cannot, at this stage of our knowledge, be expected to

TABLE IV. Relationships between Dietary Exposures and Elements in the Model

Constituents	Relationship with other parts of system	Likely effect on carcinogenesis
VEGETABLES		
Carbohydrate	Binding	−
Fiber	Bile acids	
	Other steroids	
	?Carcinogens	
	Fermentation	
	VFAs	−
	Increase in stool bulk	−
	General workload	
	Increase in cell replication	+
Specific anticarcinogenic substances	Binding, blocking, etc.	−
Nitrites	Endogenous formation of nitrosamines	+
Vitamins	Integrity of biological systems	−
Trace elements	Anticarcinogenic effects	−
Calories	General workload	
	Increase in cell replication	+
	Fuel (tumor cells)	+
	Fuel (normal cells)	−
MEAT		
Protein	Immunogenic load	+
	Integrity of cell systems	−
Fat	Stimulation of bile acids	+
	Integrity of cell membranes	?
Calories	General workload	
	Increase in cell replication	+
	Fuel (tumor cells)	+
	Fuel (normal cells)	−
Vitamins	Integrity of biological systems	−
Trace elements	Anticarcinogenic effects	−

show a clear relationship between an exposure as complex as dietary fiber and bowel cancer even if such a relationship exists. The explanations for this include the inconsistency of both the classification and measurement of fiber, the problem of controlling for total caloric intake, the failure to account for specific metabolic effects of different fiber sources, and potential interactions with both other nutrients and endogenous processes. There are as yet only incomplete data both on the potential anticarcinogenic factors that are highly correlated with fiber (but not measured in epidemiologic studies) and on a variety of aspects of the metabolic epidemiology of the disease. An interactive model of colon carcinogenesis has been sketched out that attempts to acknowledge the complexity of the colonic milieu and the way in which even individual food items can be expected, in sum, to have different effects on different parts of the system. This makes prediction of the overall effect of patterns of exposure extremely complex. Even Daniel and the apple-feeding scientist could have expected this puzzle to take a while to solve.

REFERENCES

Anderson, J. W., and Bridges, S. R., 1988, Dietary fiber content of selected foods, *Am. J. Clin. Nutr.* **47:**440–447.

Armstrong, B., and Doll, R., 1975, Environmental factors and cancer incidence and mortality in different countries, with special reference to dietary practices, *Int. J. Cancer* **15:**617–631.

Berta, J. L., Coste, T., Rautureau, J., Guilloud-Bataille, M., and Pequignot, G., 1985, Alimentation et cancers rectocoliques. Resultats d'une étude "cas-témoin," *Gastroenterol. Clin. Biol.* **9:**348–353.

Bingham, S. A., Williams, D. D. R., Cole, T. J., and James, W. P. T., 1979, Dietary fibre consumption and regional large bowel mortality in Britain, *Br. J. Cancer* **40:**456–463.

Bingham, S. A., Williams, D. D. R., and Cummings, J. H., 1985, Dietary fibre consumption in Britain; new estimates and their relation to large bowel cancer mortality, *Br. J. Cancer* **52:**399–402.

Bjelke, E., 1973, *Epidemiological Studies of Cancer of the Stomach, Colon and Rectum: vol. III, Case-control study of gastrointestinal cancer in Norway; Vol. IV, Case-Control Study of Digestive Tract Cancers in Minnesota,* University Microfilms, Ann Arbor.

Bristol, J. B., Emmett, P. M., Heaton, K. W., and Williamson, R. C. N., 1985, Sugar, fat, and the risk of colorectal cancer, *Br. Med. J.* **291:**1467–1470.

Burkitt, D. P., 1971, Epidemiology of cancer of the colon and rectum, *Cancer* **28:**3–13.

Dales, L. G., Friedman, G. D., Ury, H. K., Grossman, S., and Williams, S. R., 1979, A case-control study of the relationships of diets and other traits to colorectal cancer in American blacks, *Am. J. Epidemiol.* **109:**132–144.

Englyst, H. N., and Cummings, J. H., 1984, Simplified method for the measurement of total non-starch polysaccharides by gas–liquid chromatography of constituent sugars as alditol acetates, *Analyst* **109:**937

J. H., 1982, Determination of the nonstarch polysaccharides in plant foods by gas–liquid chromatography of constituent sugars as alditol acetates, *Analyst* **107:**307–318.

Graham, S., Dayal, H., Swanson, M., Mittelman, A., and Wilkinson, G., 1978, Diet in the epidemiology of cancer of the colon and rectum, *J. Natl. Cancer Inst.* **61:**709–714.

Haenszel, W., Berg, J. W., Segi, M., Kurihara, M., and Locke, F. B., 1973, Large bowel cancer in Hawaiian Japanese, *J. Natl. Cancer Inst.* **51:**1765–1799.

Haenszel, W., Locke, F. B., and Segi, M., 1980, A case-control study of large bowel cancer in Japan, *J. Natl. Cancer Inst.* **64:**17–22.

Higginson, J., 1966, Etiological factors in gastrointestinal cancer in man, *J. Natl. Cancer Inst.* **37:**527–545.

Hirayama, T. A., 1981, A large-scale cohort study on the relationship between diet and selected cancers of the digestive organs, in: *Gastrointestinal Cancer: Endogenous Factors* (W. R. Bruce, P. Correa, M. Lipkin, S. R. Tannenbaum, and T. D. Wilkins, eds.), Banbury Report, No. 7, Cold Spring Harbor Laboratory, Cold Spring Harbor, NY, pp. 409–429.

Hoover, L. W., 1983, Computerized nutrient data bases I. Comparison of nutrient analysis systems, *J. Am. Diet. Assoc.* **82:**501–505.

Howe, G. R., Craib, K. J. P., and Miller, A. B., 1985, Age at first pregnancy and risk of colorectal cancer: A case-control study, *J. Natl. Cancer Inst.* **74:**1155–1159.

Howe, G. R., Miller, A. B., and Jain, M., 1986, Re: "Total energy intake: Implications for epidemiologic analyses," *Am. J. Epidemiol.* **124:**157–159.

Howell, M. A., 1975, Diet as an etiological factor in the development of cancers of the colon and rectum, *J. Chron. Dis.* **28:**67–80.

IARC Intestinal Microcology Group, 1977, Dietary fibre, transit time, faecal bacteria, steroids and colon cancer in two Scandinavian populations, *Lancet* **2:**207–210.

Jacobs, D. R., Elmer, P. G., Gorder, D., Hall, Y., and Moss, D., 1985, Comparison of nutrient calculation systems, *Am. J. Epidemiol.* **121**:580–592.

Jain, M., Cook, G. M., Davis, F. G., Grace, M. G., and Miller, A. B., 1980, A case-control study of diet and colorectal cancer, *Int. J. Cancer* **26**:757–768.

Jensen, O. M., MacLennan, R., and Wahrendorf, J., 1982, Diet, bowel function, faecal characteristics and large bowel cancer in Denmark and Finland, *Nutr. Cancer.* **4**:5–19.

LaVecchia, C., Negri, E., Decarli, A., D'Avanzo, B., Galliotti, L., Gentile, A., and Franceschi, S., 1988, A case-control study of diet and colorectal cancer in Northern Italy, *Int. J. Cancer* **41**:492–498.

Liu, K., Stamler, J., Moss, D., Garside, D., Persky, V., and Soltero, I., 1979, Dietary cholesterol, fat and fiber, and colon cancer, *Lancet* **2**:782–785.

Lyon, J. L., Gardner, J. W., West, D. W., and Mahony, A. M., 1983, Methodologic issues in epidemiologic studies of diet and cancer, *Cancer Res.* **43**:2392s–2396s.

Lyon, J. L., Mahoney, A. W., West, D. W., Gardner, J. W., Smith, K. R., Sorenson, A. W., and Stanish, W., 1987, Energy intake: Its relationship to colon cancer risk, *J. Natl. Cancer Inst.* **78**: 853–861.

Macquart-Moulin, G., Riboli, E., Cornée, J., Charnay, B., Berthèzene, P., and Day, N., 1986, Case-control study on colorectal cancer and diet in Marseilles, *Int. J. Cancer.* **38**:183–191.

Manousos, O., Day, N. E., Trichopoulos, D., Gerovassilis, F., Tzonou, A., and Polychronopoulou, A., 1983, Diet and colorectal cancer: A case-control study in Greece, *Int. J. Cancer.* **32**:1–5.

Martinez, I., Torres, R., Frias, Z., Colon, J. R., and Fernandez, N., 1979, Factors associated with adrenocarcinomas of the large bowel in Puerto Rico, **3**:45–52.

McKeown-Eyssen, G. E., and Bright-See, E., 1984, Dietary factors in colon cancer: International relationships, *Nutr. Cancer* **6**:160–170.

McMichael, A. J., and Potter, J. D., 1980, Reproduction, endogenous and exogenous sex hormones, and colon cancer: A review and hypothesis, *J. Natl. Cancer Inst.* **65**:1201–1207.

McMichael, A. J., and Potter, J. D., 1983, Do intrinsic sex differences in lower alimentary tract physiology influence the sex-specific risks of bowel cancer and other biliary and intestinal diseases? *Am. J. Epidemiol.* **118**:620–627.

McMichael, A. J., and Potter, J. D., 1985, Diet and colon cancer: Integration of the descriptive, analytic and metabolic epidemiology, *Natl. Cancer Inst. Monogr.* **69**:223–228.

McMichael, A. J., and Potter, J. D., 1986, Dietary influences upon colon carcinogenesis, in: *Diet, Nutrition and Cancer* (Y. Hayashi, M. Nagao, T. Sugimura, S. Takayama, L. Tomatis, L. W. Wattenberg, and G. N. Wogan, eds.), Japan Science Society Press, Tokyo, pp. 275–290.

Meyer, F., 1977, Relations alimentation—cancer en France, *Gastroenterol. Clin. Biol.* **1**:971–982.

Miller, A. B., Howe, G. R., Jain, M., Craib, K. J. P., and Harrison, L., 1983, Food items and food groups as risk factors in a case-control study of diet and colorectal cancer, *Int. J. Cancer.* **32**:155–161.

Modan, B., Barell, V., Lubin, F., Modan, M., Greenberg, R. A., and Graham, S., 1975, Low-fiber intake as an etiologic factor in cancer of the colon, *J. Natl. Cancer Inst.* **55**:15–18.

Pernu, J., 1960, An epidemiological study on cancer of the digestive organs and respiratory system. A study based on 7078 cases, *Ann. Med. Int. Fenn.* **49**(Suppl. 3):1–117.

Phillips, R., 1975, Role of lifestyle and dietary habits in risk of cancer among Seventh-Day Adventists, *Cancer Res.* **35**:3513–3522.

Potter, J. D., 1988, Dietary fiber, vegetables, and cancer, *J., Nutr.* **118**:1591–1592.

Potter, J. D., and McMichael, A. J., 1983, Large bowel cancer in women in relation to reproductive and hormonal factors: A case-control study, *J. Natl. Cancer Inst.* **71**:703–709.

Potter, J. D., and McMichael, A. J., 1986, Diet and cancer of the colon and rectum. A case-control study, *J. Natl. Cancer Inst.* **76**:557–569.

Selvendran, R. R., 1983, The chemistry of plant cell walls, in: *Dietary Fiber* (G. G. Birch, K. J. Parker, eds.), Applied Scientific Publishers, London, New York, pp. 95–147.

Shrauzer, G. N., 1976, Cancer mortality correlation studies II. Regional associations of mortalities with the consumption of food and other commodities, *Med. Hypoth.* **2**:39–49.

Stocks, P., 1957, Cancer incidence in North Wales and Liverpool region in relation to habits and environments, in: *British Empire Cancer Campaign 35th Annual Report*, Suppl. to part 2, Her Majesty's Stationery Office, London, pp. 1–127.

Trowell, H., 1972, Dietary fibre and coronary heart disease, *Rev. Eur. Etud. Clin. Biol.* **17**:345–349.

Wattenberg, L. W., 1985, Chemoprevention of cancer, *Cancer Res.* **45**:1–8.

Weiss, N. S., Daling, J. R., and Chow, W. H., 1981, Incidence of cancer of the large bowel in women in relation to reproductive and hormonal factors, *J. Natl. Cancer Inst.* **67**:57–60.

Willett, W., and Stampfer, M. J., 1986, Total energy intake: Implications for epidemiologic analyses, *Am. J. Epidemiol.* **124**:17–27.

Wynder, E. L., and Shigematsu, T., 1967, Environmental factors of cancer of the colon and rectum, *Cancer* **20**:1520–1561.

Wynder, E. L., Kajitani, T., Ishikana, S., Dodo, H., and Tako, A., 1969, Environmental factors of cancer of the colon and rectum II. Japanese epidemiologic data, *Cancer* **23**:1210–1220.

Yanai, H., Inaba, Y., Takagi, H., and Yamamoto, S., 1979, Multivariate analysis of cancer mortalities for selected sites in 24 countries, *Environ. Health Perspect.* **32**:83–101.

33

Starch, Nonstarch Polysaccharides, and the Large Gut
Epidemiologic Aspects

SHEILA A. BINGHAM

1. INTRODUCTION

Since the inception of the dietary fiber hypothesis, much progress has been made in our understanding of the chemistry and definition of dietary fiber. Recent work has established an accurate method for the determination of dietary fiber based on the chemical definition of nonstarch polysaccharides (Cummings *et al.*, 1981). Nonstarch polysaccharides (NSP) include all the carbohydrate fractions and types of dietary fiber: soluble and insoluble, pectins, gums, hemicelluloses, cellulose, β-glucans, and noncellulosic polysaccharides. The NSP are chemically distinct (Southgate and Englyst, 1985), measurable (Englyst and Cummings, 1984), and, regardless of the physical form of food or cooking and processing, are not hydrolyzed by digestive enzymes in the small gut (Chacko and Cummings, 1988).

Comparisons of the values for dietary fiber included in the food tables (Paul and Southgate, 1978) used in the majority of epidemiologic studies to date have however shown large discrepancies when compared with NSP values (Englyst *et al.*, 1988). Figure 1 shows the values for NSP contents of nonstarchy vegetables compared with their food table value of dietary fiber. Although there is no significant difference between the two sets of data [mean for NSP 2.7 ± 1.3 (S.D.) g, mean food tables 2.7 ± 1.6 g ($t = 1.17; P = 0.25; n = 39$)], and the data are significantly correlated ($r = 0.54$), as the graph shows there is considerable variation in the agreement among the pairs of data. Starchy vegetables (Fig. 2) show a consistent difference between the two methods,

SHEILA A. BINGHAM • University of Cambridge and Medical Research Council, Dunn Clinical Nutrition Centre, Cambridge CB2 1QL, United Kingdom.

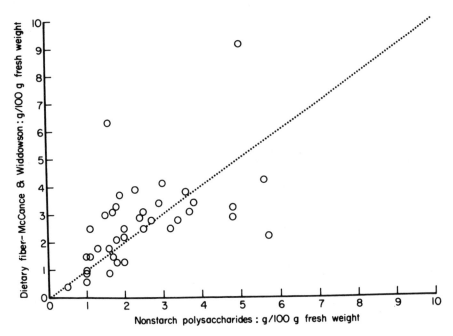

FIGURE 1. Dietary fiber in 39 vegetables measured as nonstarch polysaccharides or as given in *McCance and Widdowson's Composition of Foods,* 4th ed. (Paul and Southgate, 1978); $r = 0.54$. The line of complete agreement is shown.

the food table values being significantly higher for most but not all foods [mean for NSP 5.2 ± 4.5 g; food tables 8.5 ± 6.7 g ($t = -5.68$; $P = 2.7 \times 10^{-5}$; $n = 34$)].

Figure 3 shows the relationship between NSP values for fruit and those found in the food tables. As with starchy vegetables, there is a consistent and significant difference between the two sets of values. Dietary fiber measured in the food tables is 4.6 ± 5.2 g and as NSP is 2.4 ± 1.8 g ($t = -4.13$; $P = 3.5 \times 10^{-4}$; $n = 44$). The two sets of results were significantly correlated ($r = 0.9$; $P < 0.01$), with dietary fiber (food tables) $= 2.5$ NSP $- 1.4$. As Fig. 4 shows, the NSP values for nuts also are consistently and significantly lower than those found in the food tables [dietary fiber 10.2 ± 5.5; NSP 6.5 ± 2.8 g ($t = -3.83$; $P = 4.3 \times 10^{-3}$; $n = 10$)].

Values for the NSP content of 28 cereals also are significantly different from those for the "dietary fiber" content, largely because of the contamination of the analytical method used in the food table values with starch, as is the case with the starchy vegetables and nuts. Figure 5 shows that the NSP values are highly correlated ($r = 0.95$; $P < 0.01$), largely because of the close agreement of the two methods for bran and Allbran. The values are, however, significantly different: those for "dietary fiber" are 1.7 ± 2.7 g per 100 g ($P < 0.01$). There are important differences: values for white rice are 0.5 g per 100 g as NSP and 2.4 g per 100 g as "dietary fiber," and those for white bread are 1.6 g and 2.7 g, respectively. These differences will lead to substantial overestimates of "dietary fiber" in average diets.

The reason for the poor agreement in the case of fruit (Fig. 3) is not clear. Certainly,

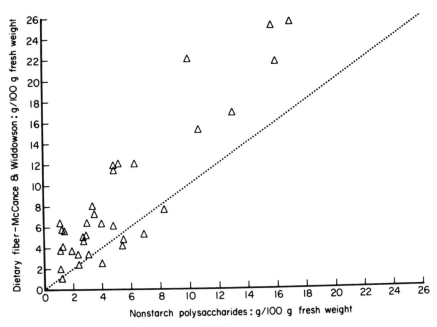

FIGURE 2. Dietary fiber in 34 starchy vegetables measured as nonstarch polysaccharides or as given in *McCance and Widdowson's Composition of Foods*, 4th ed. (Paul and Southgate, 1978); *r* = 0.90. The line of complete agreement is shown.

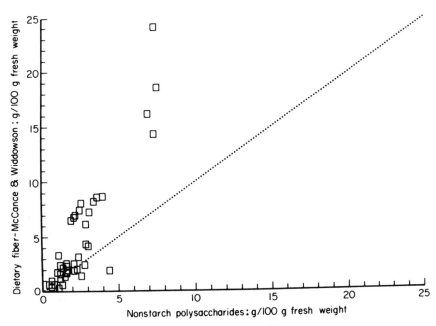

FIGURE 3. Dietary fiber in 44 fruits measured as nonstarch polysaccharides or as given in *McCance and Widdowson's Composition of Foods*, 4th ed. (Paul and Southgate, 1978); *r* = 0.90. The line of complete agreement is shown.

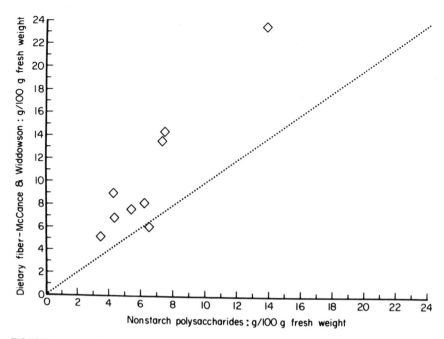

FIGURE 4. Dietary fiber in ten types of nuts measured as nonstarch polysaccharides or as given in *McCance and Widdowson's Composition of Foods,* 4th ed. (Paul and Southgate, 1978); $r = 0.92$. The line of complete agreement is shown.

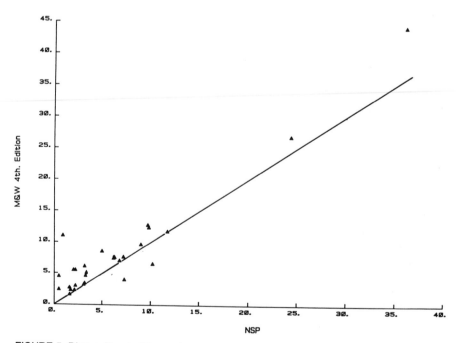

FIGURE 5. Dietary fiber in 28 cereals measured as nonstarch polysaccharides or as given in *McCance and Widdowson's Composition of Foods,* 4th ed. (Paul and Southgate, 1978); $r = 0.95$. The line of complete agreement is shown.

however, all epidemiologic studies that have used *McCance and Widdowson's Composition of Foods* (Paul and Southgate, 1978) as a data base for the determination of dietary fiber in average diets will have found an overestimation of the contribution of cereal and starchy vegetable products to total intakes of dietary fiber when compared with NSP values.

Given these problems with existing data bases, it is necessary to reconsider the epidemiologic evidence concerning dietary fiber and protection from large-gut disorders. What are the data that can be used, and what is needed in future studies?

2. FIBER INTAKE IN DIFFERENT POPULATIONS

The nonstarch polysaccharide (fiber) intakes of the six populations for which data are available are shown in Table I together with age-standardized cancer incidence rates (Waterhouse *et al.*, 1982). The dietary intakes are derived from duplicate collections of food (IARC Large Bowel Cancer Group, 1982; Bingham *et al.*, 1985; Kuratsune *et al.*, 1986) and are not strictly comparable, since those for the United Kingdom and Japan were obtained from household survey estimates, and those for Scandinavia from men aged 50–60 years.

In Britain regional household intakes of foodstuffs have been documented every year since 1940 by the British National Food Survey under the auspices of the Ministry of Agriculture, Fisheries and Food (1953–1983). The housewife is asked to keep a record of her food purchases for 1 week, and from this the average quantity of food eaten per person per day in the household is calculated. Between 300 and 2400 households in nine standard regions of the United Kingdom take part every year. To determine regional NSP intake, compositive diets were made up of all fiber-containing foods for the years 1969–1973 and analyzed by the method of Englyst (Englyst *et al.*, 1982; Englyst and Cummings, 1984). When intakes were compared with age- and sex-standardized truncated colon and rectal cancer rates for 1969–1973, a significant ($P < 0.05$) and negative association ($r = -0.74$) between NSP intake and colon cancer emerged (Bingham *et al.*, 1985). There was no significant association with NSP and colorectal cancer. The strong inverse relationship between the pentose fraction of dietary fiber and colon cancer (Bingham *et al.*, 1979) was not confirmed, although the association with vegetable consumption ($r = -0.940$) remained.

TABLE I. Nonstarch Polysaccharide Intakes
in Different Populations

Location	NSP intake (g/day)
Rural Finland	18.4 ± 7/8
Rural Denmark	18.0 ± 6.4
Helsinki	14.5 ± 5.4
Copenhagen	13.2 ± 4.0
U.K.	12.4
Japan	10.9

In the four Scandinavian populations, average intakes of fat and animal protein were high, fat intake ranging from 102 to 146 g per day and meat consumption from 148 to 214 g per day. There were no significant associations between fat, cholesterol, or meat consumption and large bowel cancer incidence. The simple correlation coefficient between NSP consumption and large bowel cancer incidence was -0.78, and more sophisticated statistical techniques demonstrated that this relationship was significant (IARC Large Bowel Cancer Group, 1982).

It is clear from Table I that although bowel cancer risk is low in Japan, fiber intakes are also low. The nonstarch polysaccharide (NSP) content of rice is low (Englyst et al., 1983), and there have been no major changes in the Japanese fiber intake over the past 50 years (Minowa et al., 1983). Despite increasing Westernization, however, the Japanese diet remains comparatively low in meat and fat, containing approximately 70 g of meat compared with 150 g in Britain and only half the fat (Minowa et al., 1983; Ministry of Agriculture, Fisheries and Food, 1953–1983). This suggests that in the presence of reduced levels of possible risk factors, such as meat and fat, a protective role for dietary fiber might be unnecessary.

However, in all populations so far studied remarkably little dietary fiber is consumed. Table I shows that the greatest amount found in average diets is 18 g per day, which is substantially less than the other major polysaccharide in food, starch.

3. STARCH AND THE LARGE INTESTINE

One of the most important findings in recent years with regard to large bowel function is in relation to starch digestion in man, since it is now clear that substantial starch escapes digestion in the small intestine. The first indication of the importance of starch was the finding of "resistant starch" contaminating methods of dietary fiber analysis in foods (Englyst et al., 1982). Resistant starch consists of retrograded amylose, but there are at least three other types of starch that are poorly digested, such as whole or partly milled grains and seeds and ungelatinized potato and banana starch granules. Retrograded amylopectin may also be present in cooked and processed foods, although the retrogradation is reversible during reheating (Englyst and Cummings, 1987). Taken together, these different types of starch reaching the large intestine could be quantitatively at least as great as NSP and in some countries greater.

Although human studies have yet to be completed, there is evidence to suggest that starch reaching the large intestine may have a similar fate to the NSP, namely, fermentation by the bacterial flora with the production of short-chain fatty acids, gas, and an increase in microbial mass. The pattern of fermentation differs, with a greater proportion of butyrate being produced from starch compared with other NSP (Englyst and Mac-farlane, 1986), and an increase in microbial mass will reduce transit time. Both of these factors could be important in cancer prevention, since a reduced transit time may reduce contact of putative carcinogens with the large gut mucosa. Butyrate production is likely to be important because it is the preferred energy substrate for the lower gut (Roediger, 1980) and because it is a well-known antiproliferative agent and differentiating agent (Kruh, 1982; Smith, 1986).

Hence, the realization that starch in addition to NSP escapes digestion in the small gut means that there are now two classes of polysaccharides available for bacterial fermentation in the large gut. However, in order for the physiological and epidemiologic roles of starch and NSP to be investigated, it is essential that methods of chemical analysis clearly distinguish between them.

4. FUTURE STUDIES

The new findings in relation to starch digestion have major implications for the bowel cancer story. If starch undergoes a similar fermentation in the large bowel, it is therefore capable of fitting into the protective mechanism against cancer proposed for NSP. Moreover, much more starch than NSP may get into the large bowel. In fact, if the role of butyrate specifically as a differentiating agent for the colon epithelium is important, and the potential for starch to produce more butyrate on fermentation than NSP is confirmed *in vivo*, then starch may be the crucial protective dietary component.

An essential piece of this jigsaw that is lacking is information on the epidemiologic relationship between starch intake and large bowel cancer. However, it is not just total starch that needs to be looked at but the amount of starch in the diet that passes into the large intestine.

A new analytical method that is capable of providing this information is required, since identification of retrograded amylose only will underestimate the amount of starch that is not digested in the small gut. For this detailed information on the processing, ripeness, duration of storage, and cooking–cooling cycles during preparation before eating of food will be needed.

An appropriate method is not yet available, but when it is, future epidemiologic studies, whether cross country or prospective, should provide new insights into the etiology of large bowel cancer.

REFERENCES

Bingham, S. A., Williams, D. R. R., Cole, T. J., and James, W. P. T., 1979, Dietary fibre consumption and regional large bowel cancer mortality in Britain, *Br. J. Cancer* **40**:456–463.

Bingham, S., Williams, D. R. R., and Cummings, J. H., 1985, Dietary fibre consumption in Britain; new estimates and their relation to large bowel cancer mortality, *Br. J. Cancer* **52**:399–402.

Chacko, A., and Cummings, J. H., 1988, Nitrogen losses from the human small bowel; obligatory losses and the effect of physical form of food, *Gut* **29**:808–815.

Cummings, J. H., Stephen, A. M., and Branch, W. J., 1981, Implication of dietary fiber breakdown in the human colon, in: *Banbury Report 7, Gastrointestinal Cancer; Endogenous Factors* (R. Bruce, P. Correa, M. Lipkin, S. Tannenbaum, and T. Wilkins, eds.), Cold Spring Harbor Press, Cold Spring Harbor, NY, pp. 71–81.

Englyst, H. N., and Cummings, J. H., 1984, Simplified method for the measurement of total NSP by GLC of constituent sugars as alditol acetates, *Analyst* **109**:937–942.

Englyst, H. N., and Cummings, J. H., 1987, Resistant starch, a 'new' food component: A classification of starch for nutritional purposes, in: *Cereals in a European Context* (I. D. Morton, ed.), Ellis Horwood, Chichester, pp. 221–233.

Englyst, H. N., and Macfarlane, G. T., 1986, Breakdown of resistant and readily digestible starch by human gut bacteria, *J. Sci. Food Agr.* **37**:699–706.

Englyst, H., Wiggins, H. S., and Cummings, J. H., 1982, Determination of the non-starch polysaccharides in plant foods by gas–liquid chromatography of constituent sugars as alditol acetates, *Analyst* **107**:307–318.

Englyst, H. N., Anderson, V., and Cummings, J. H., 1983, Starch and NSP in some cereal foods, *J. Sci. Food Agr.* **34**:1434–1440.

Englyst, H. N., Bingham, S. A., Collinson, E., Runswick, S., and Cummings, J. H., 1988, Dietary fibre in fruit, vegetables and nuts, *J. Hum. Nutr. Dietet.* **1**:233–272.

IARC Large Bowel Cancer Group, 1982, Second IARC international collaborative study on diet and large bowel cancer in Denmark and Finland, *Nutr. Cancer* **4**:3–79.

Kruh, J., 1982, Effects of sodium butyrate, a new pharmacological agent, on cells in culture, *Mol. Cell. Biochem.* **42**:65–82.

Kuratsune, M., Honda, T., Englyst, H. N., and Cummings, J. H., 1986, Dietary fibre in the Japanese diet, *Jpn. J. Cancer Res.* **77**:736–738.

Ministry of Agriculture, Fisheries and Food, 1953–1983, *Household Food Consumption and Expenditure, 1950–1981,* Annual Reports of the National Food Survey Committee, Her Majesty's Stationery Office, London.

Minowa, M., Bingham, S., and Cummings, J. H., 1983, Dietary fibre intake in Japan, *Hum. Nutr. Clin. Nutr.* **37A**:113–119.

Paul, A. A., and Southgate, D. A. T., 1978, *McCance and Widdowson's The Composition of Foods,* 4th ed., Her Majesty's Stationery Office, London.

Roediger, W. E. W., 1980, Role of anerobic bacteria in the metabolic welfare of the colonic mucosa in man, *Gut* **21**:793–798.

Smith, P. J., 1986, *n*-Butyrate alters chromatin accessibility to DNA repair enzymes, *Carcinogenesis* **7**:423–429.

Southgate, D. A. T., and Englyst, H. N., 1985, Dietary fibre: Chemistry, physical properties and analysis, in: *Dietary Fibre, Fibre-Depleted Foods and Disease* (H. Trowell, D. Burkitt, and K. Heaton, eds.), Academic Press, London, pp. 31–55.

Waterhouse, J., Muir, C., Shanmugaratnam, K., and Powell, J., 1982, *Cancer Incidence in Five Continents,* Vol. IV, IARC Scientific Publications No. 42, International Agency for Research on Cancer, Lyon.

34

The Epidemiology of Cancer and Its Risk Factors in South African Populations

A. R. P. WALKER and B. F. WALKER

1. INTRODUCTION

South Africa, a seventh of the area of the United States, has a population of about 29 million, including 21 million blacks, 4½ million whites, 3 million coloreds (Eur-African-Malays), and ¾ million Indians. The country is self-supporting regarding food production and needs.

Information on the epidemiology of cancer in South African populations is based largely on (1) mortality rates (Bradshaw and Harington, 1982; Wyndham, 1985), (2) the pattern of patients admitted to hospital with cancer (Isaacson *et al.*, 1978; Kloppers *et al.*, 1983; Gilpin *et al.*, 1989), (3) limited regional studies on cancer incidence (Higginson and Oettlé, 1960; Rose and Fellingham, 1981), and (4) the newly inaugurated Cancer Registry of South Africa, which went into operation in 1986.

The present contribution provides an update of information on the epidemiology of cancer and its risk factors in the South African populations.

2. GENERAL PATTERN OF DIET-RELATED DISEASES IN SOUTH AFRICAN POPULATIONS

The cancer pattern and frequency situation can only be fairly appreciated and judged when juxtaposed against the general patterns of nutritionally related diseases, those of prosperity and those of deficiency and underprivilege, as they prevail in the four populations (Walker, 1987a).

A. R. P. WALKER and B. F. WALKER • Department of Tropical Pathology, School of Pathology, University of the Witwatersrand and the South African Institute for Medical Research, Johannesburg, South Africa 2000.

In the white population there are high representations of diseases of prosperity. In contrast, at the extreme, there are low representations of these diseases among the rural black population (Table I).

Within an hour's car drive from Johannesburg, three quarters of rural black children of 12 years are caries-free (Walker *et al.*, 1988). Yet with urbanization, the deterioration of teeth is so marked that nowadays their total mean DMFT (decayed–missing–filled–total) scores may actually exceed those of contemporary white children. This is the case in Cape Town (Steyn and Albertse, 1987). White children's caries scores, as is well known, have decreased almost dramatically in the last decade for reasons not wholly understood (Naylor, 1985).

In the near past rural blacks experienced very little rise in weight or in blood pressure with age (Walker, 1964). However, with urbanization and associated changes, there have been huge increases in the frequency of obesity, especially affecting women, as well as in blood pressure levels (Walker, 1964; Seftel *et al.*, 1980). In the latter regard, the increase has been such that in big cities, hypertension frequency (WHO criteria) is extraordinarily high, reaching 28% in adults (Seedat *et al.*, 1982).

Within a 2-hr car drive from Johannesburg, there are rural hospitals where no case of coronary heart disease (CHD) in blacks has yet been detected (Walker and Walker, 1985b), although there are huge numbers of middle-aged and elderly. Among Indians, the high frequency of CHD is notable; indeed, wherever Indians have migrated, they have tended to have higher frequencies of the disease than people in the host countries (Leading Article, 1986; Walker and Walker, 1986). Interestingly, the first description of anginal pain was written by Indian physicians about 700–600 B.C. (Mukerjee, 1975), 200 years before the time of Hippocrates. These early Indian physicians also knew of the sweetness of the urine of diabetic patients. Indians are particularly prone to both of these diseases.

The situation regarding cancer is discussed in Section 3, but to illustrate the extreme, again within a 2-hr drive from Johannesburg, in most rural hospitals no case of colorectal cancer in blacks has yet been encountered (Walker and Segal, 1988). In the Indian population, the epidemiologic situation regarding cancer is roughly intermediate between that of the black and white populations. In the case of the colored population, the occurrence of total cancer is lower than in the white population.

As to diet-related diseases of underprivilege (Table II), in the white population they

TABLE I. Nutritionally Related Diseases of Prosperity in South African Populations

	Whites	Indians	Coloreds	Urban blacks	Rural blacks
Caries	++++	+++	+++	++	+
Obesity	++++	+++	+++	+++	+
Hypertension	+++	+++	+++	++++	+
Diabetes	+++	+++	++++	++	+
CHD	++++	+++++	+++	+	−
Stroke	++	+++	+++	++	+
Cancer (colon, breast)	++++	++	++	+	+

TABLE II. Nutritionally Related Diseases of Deficiency and Imbalance

	Whites	Indians	Coloreds	Urban blacks	Rural blacks
PEM	–	–	+	+	+ +
Rickets	–	–	–	+	+
Pellagra	–	+	+	+	+ +
Beri-beri	–	–	–	+	–
Iron deficiency anemia	+	+ + +	+	+	+

are near absent, but they are still of frequent occurrence among the black population. In the latter, all the diseases listed, together with others usual to poor developing populations such as gastroenteritis, pneumonia, tuberculosis, have greatly decreased in frequency compared with situations in the past (Retief, 1987a,b). Yet they remain tremendous public health burdens, as evidenced by their high representations in admissions of rural blacks to hospital (Reeve and Falkner, 1986; Gilpin et al., 1989).

In brief, in South African populations there are huge extremes in the occurrence of the diseases of prosperity. All such diseases are increasing in occurrence in all populations. Although the diseases of poverty and underprivilege are decreasing, they still constitute huge ill-health problems.

3. DEATH RATES FROM CANCER

Death rates are subject to the usual errors of under- and overreporting. The data on the blacks are less reliable than those on the three other populations, whose data are deemed reasonably accurate.

In Table III, mortality rates for cancer, calculated for "world" population (Doll et al., 1970), are given for the four populations for the year 1984 as derived from reports from the Central Statistical Services, Pretoria. The figures given relate to total ethnic populations save in the case of blacks, whose data refer to that moiety of the population, about 40%, who live in urban areas. Briefly, the comparative situations are as follows.

The rates for whites are relatively low, appreciably lower than those reported for populations in the United States and the United Kingdom (National Cancer Institute,

TABLE III. Cancer Mortality Rates
of Populations in South Africa, 1984,
per 100,000 World Population

	Males	Females
Whites	171	113
Blacks	138	87
Coloreds	229	124
Indians	101	90

1982; World Health Organization, 1982). Total rates for blacks are low. Among these people there are high rates for cancers of the esophagus, lung, liver, and cervix. There are, however, low rates for stomach, breast, and colorectal cancers (Bradshaw and Harington, 1982; Wyndham, 1985). The total rates for coloreds are higher than those of the white population. Coloreds have especially high rates for lung, stomach, esophagus, and cervix cancers. The total rates for Indians are low. Their highest rates are for stomach, lung, and breast cancers. Their data are similar in profile to the cancer incidence rates reported for Indian populations in Bombay (Jussawalla *et al.*, 1985a,b).

4. HOSPITAL ADMISSIONS FOR CANCER

4.1. Baragwanath Hospital, Soweto, Johannesburg

This hospital, which serves the black population, has 2800 beds. Annually, about 125,000 patients are admitted, and there are 1,600,000 outpatients drawn from an urban population of 1½ to 2 million. A major study of admissions for cancer in the period 1971–1977 was published by Isaacson *et al.* (1978). The present study, carried out by the authors, concerns admissions for the year 1986 (Table IV).

The results obtained differ in some respects from those in the earlier investigation mentioned. The present general picture is as follows.

In males, the number of patients with esophageal cancer, by far the most common cancer, remains high. However, taking into account changes in the size and composition of the population at risk, the number of patients appears to be decreasing. Liver cancer is still common, although here again the number of patients is decreasing. In contrast, the number with prostate cancer is certainly rising and undoubtedly will continue to do so as it has done in other African populations (Parkin, 1986). There are still very few patients admitted with stomach and colorectal cancers.

Among females, patients with cervix cancer predominate by far, although the number appears to be steady if not decreasing. Esophageal cancer is also decreasing in occurrence. In contrast, the number of patients with breast cancer is rising, albeit slowly. As with males, patients with stomach and colorectal cancers remain few; indeed, hardly any increases seem to have occurred.

TABLE IV. Urban: Baragwanath Hospital
(2800 beds), 1985, Admissions of Blacks for Cancer

Males		Females	
Site	Cases	Site	Cases
Esophagus	142	Cervix	189
Prostate	94	Breast	75
Liver	67	Esophagus	54
Larynx	36	Colorectal	8
Colorectal	21	Skin, etc.	27
Other	183	Other	263
Total	543	Total	616

4.2. Murchison Hospital, South Natal

This rural hospital has 320 beds, and admits about 8000 patients annually.

The pattern of admissions for cancer (Table V) is similar to that observed at Baragwanath Hospital, with the same trend of changes as far as can be gathered from the data of previous years. Admissions for esophagus, cervix, and liver cancers are high. Admissions for breast cancer, although very few, are rising. This is also the case, only more so, with prostate cancer. In contrast, there was only one patient with stomach cancer and one with colorectal cancer (Gilpin *et al.*, 1989).

Unfortunately, no studies have been published on Indians and coloreds regarding the numbers of cancer patients admitted to hospital.

5. STUDIES ON CANCER INCIDENCE

Early investigations on cancer incidence were made by Higginson and Oettlé (1960) on urban black and colored populations and by Robertson *et al.* (1971) on a rural black population. Rose and Fellingham (1981) reported an investigation on the rural population of Transkei, the results of which are given in Table VI.

This investigation was a field and hospital-based inquiry. Highlights included in males high esophagus and in females high cervix cancer rates. As will be apparent, there were low rates for stomach, colorectal, and prostate cancers. It is of interest that for total cancers, the rates are similar to those reported for African blacks in Dakar (Waterhouse *et al.*, 1982). However, in that city, liver cancer by far predominates, and, surprisingly, esophageal cancer appears near absent.

5.1. Contrasts in Cancer Occurrence in Adjacent Areas

In the Transkei, in two districts only 15 miles apart, inquiries indicated that there was a severalfold difference in esophageal cancer incidence rate (Rose, 1978). Remarkably, in Linxian, China, also concerning esophageal cancer, Yang (1980) reported a 100-fold difference in frequency in two areas 100 or so miles apart. Reasons for such enormous disparities remain obscure. But it is noteworthy that the same type of phe-

TABLE V. Murchison Hospital (320 beds),
Port Shepstone, Natal: Total Admissions of Patients,
8000 Cancer Admissions, Black Patients

Males		Females	
Site	Cases	Site	Cases
Esophagus	33	Esophagus	16
Lung	15	Cervix	16
Liver	9	Liver	3
Other	15	Other	12
Total	72	Total	47

TABLE VI. Cancer Incidence Rates
of Rural Blacks in Transkei
per 100,000 World Population

	Males	Females
Esophagus	41	19
Stomach	3	2
Colon–rectum	2	2
Liver	13	3
Lung	3	1
Breast	1	5
Cervix	—	16
Prostate	3	—
Other	24	15
All sites	90	63

nomenon has been reported to occur in developed populations. Thus, in the *Atlas of Cancer in Scotland* by Kemp *et al.* (1985), the maps given showed two adjacent districts, Badanoch and Annandale, that had a twofold difference in colon cancer incidence rate. Moreover, in two other adjacent districts, Moray and Gordon, there was a fourfold difference in the rate for cancer of the pancreas. For the latter cancer, the comment was made that "there is as yet no explanation for the regional differences in Scotland." As to unexplained excessive cancer proneness in other particular regions, recently there was a contribution entitled "Does New Jersey cause cancer?" (Greenberg, 1986).

5.2. The South African Cancer Registry

In 1986, the South African Cancer Registry included 32,000 diagnoses of cancer. The incidence rates of leading cancers in the four populations are given in Tables VII and VIII. Since the Registry is currently only pathology based, cancers diagnosed from clinical, radiological, and other features will tend to have reduced representations. In this respect the cancer by far most concerned is that of the lung. Broadly, most rates for whites, as previously concluded from other data, are slightly lower than those published on whites in most Western countries (Waterhouse *et al.*, 1982). The very high rates for "other skin cancers" are conspicuous. Rates for the mainly urban blacks are similar to those reported for blacks in other parts of Africa (Waterhouse *et al.*, 1982; Parkin, 1986), and far lower than those for whites. For the Indian population, total rates for males and females are much lower than those for whites and only slightly higher than those published for Indians in Bombay (Jussawalla *et al.*, 1985a,b). Local Indian men have higher incidence rates for stomach, prostate, and bladder cancers but have lower rates for cancers of the esophagus and pancreas. Local Indian women have higher incidence rates for colorectal, breast, and uterine cancers but lower rates for esophagus and cervix cancers. Data on South African Indians are similar to those published on Indians in Birmingham, U.K. (Potter *et al.*, 1984).

For coloreds, total rates for both sexes are lower than those of whites. Colored

TABLE VII. Cancer Incidence Rates of South African White, Black, Indian, and Colored Males in 1986[a]

Site		White	Black	Indian	Colored
150	Esophagus	4.6	28.4	6.3	18.0
151	Stomach	9.8	3.6	15.1	13.3
153	Colon	12.7	1.0	3.3	3.0
154	Rectum	7.9	1.2	6.6	2.8
155	Liver	2.5	6.0	3.3	3.4
162	Bronchus trachea	16.5	6.6	2.7	15.4
172	Skin, melanoma	9.8	0.7	0.7	0.5
173	Skin, basal cell	168.5	0.5	7.6	2.9
	Skin, squamous	63.7	3.3	5.6	3.3
174	Breast	1.4	0.7	0.7	0.8
180	Cervix uteri				
185	Prostate	30.6	10.7	15.9	19.9
	Other and unspecified	141.6	53.4	63.6	86.4
140–209	All sites	469.6	116.1	131.4	170.7
	All sites but 173	237.4	112.4	118.2	164.5

[a]Rates per 100,000 World population.

males have higher rates of esophagus and stomach but lower rates for colorectal, melanoma, and bladder cancers. Females have higher rates for esophagus, stomach, and cervical cancers, but lower rates for colorectal and breast cancers, compared with respective data on white populations. The low rates for skin cancer in the black, Indian, and colored populations are especially noteworthy. To reiterate, these data, of doubtless

TABLE VIII. Cancer Incidence Rates of South African White, Black, Indian, and Colored Females in 1986[a]

Site		White	Black	Indian	Colored
150	Esophagus	1.3	7.8	5.6	3.6
151	Stomach	5.0	1.3	7.4	5.7
153	Colon	10.7	0.6	7.2	2.4
154	Rectum	5.0	1.1	6.9	1.7
155	Liver	1.2	1.6	1.5	1.3
162	Bronchus trachea	5.4	0.7	0.6	3.1
172	Skin, melanoma	10.5	0.8	1.0	0.7
173	Skin, basal cell	86.4	0.3	3.6	4.1
	Skin, squamous	24.5	1.5	0.6	1.8
174	Breast	47.7	9.7	43.7	25.7
180	Cervix uteri	9.7	30.8	15.5	30.9
185	Prostate				
	Other and unspecified	100.8	29.9	74.0	50.3
140–209	All sites	308.2	86.1	167.6	131.6
	All sites but 173	197.9	84.3	163.4	125.7

[a]Rates per 100,000 World population.

greater accuracy, largely confirm the impressions reached from death certificates, from admissions to hospital, and from other sources of data. It is necessary to keep in mind that blacks, coloreds, and Indians, in diet and lifestyle, are in transition to a variable extent, respecting proneness not only to cancer but additionally to other degenerative diseases associated with rise in socioeconomic circumstances.

6. EPIDEMIOLOGY OF CANCER RISK FACTORS IN SOUTH AFRICAN POPULATIONS

In a recent monograph by the National Cancer Institute, Greenwald and Sondik (1986) considered that responsibilities of contributing factors could be roughly apportioned: diet 35%, smoking 30%, alcohol 3%, reproductive behavior 7%, and pollution and industrial products 3%. Other influencing factors include genetic influence and obesity.

6.1. Genetic Influence

As to breast cancer, familiality, i.e., cancer in a patient's mother or sister, can more than triple proneness to the disease (Dupont and Page, 1987). A genetic component has been reported to prevail in about 7–14% of white breast cancer patients (De Waard and Wang, 1988). The component was found present in 7% of black patients studied in the United States (Schatzkin et al., 1987). In South Africa, in an investigation on a series of 55 black breast cancer patients in Soweto, a familial factor appeared present in about 5% (Walker et al., 1989). Concerning colorectal cancer, a familial factor may be present in 10–20% of patients (Jacobs, 1987). It tends to increase proneness considerably, by four to five times (Rosen et al., 1987). In black patients in Soweto, probably because of small case numbers, no familial factor has yet been detected. It is of interest that in urban blacks, polyps, previously rare, are beginning to appear, although mildly (Segal et al., 1981). Local workers consider that colon cancer in blacks arises only rarely from polyps (Bremer and Ackerman, 1970; Segal et al., 1981).

6.2. Dietary Patterns

6.2.1. Whites

Their pattern of diet is similar to that of most prosperous populations. Fat supplies about 40% of energy, and fiber intake is about 15 g daily (Walker and Segal, 1988), slightly more than that in the United States (Block and Lanza, 1987).

6.2.2. Blacks

Previously, in rural areas, fat supplied about 10–15% of energy, and fiber intake was 30 to 40 g or more daily. Nowadays, fat intake has increased to supply about 15–20% and 25–30% of energy in rural and urban populations, respectively. Yet fiber intake has fallen to 15–25 g and 10–15 g, respectively (Walker and Segal, 1988). There are four reasons for the fall in fiber intake. (1) At present almost all of the staple maize meal eaten is produced by a few huge milling concerns. The meal is of low extraction rate, about 70%, and hence has a low concentration of fiber. (2) Consumption of maize meal is falling, but that of bread is increasing. Although the price of brown bread is nationally subsidized and cheaper by 20%, many blacks still prefer to eat white bread. (3) The consumption of various beans, good sources of fiber, previously high, has nearly halved. (4) Vegetables and especially fruit are becoming increasingly expensive. Hence, for the reasons given, urban blacks nowadays consume relatively little fiber. Although the intake of rural blacks is higher, the amount ingested is far less than formerly. As to other relevant dietary components, consumption of vitamins and mineral salts, in general, tend to be low in both rural and urban blacks (Langenhoven et al., 1988).

6.2.3. Indians

Their diet includes about 40% fat as energy, and their daily fiber intake is about 10–15 g. These data are similar to those reported for Indians studied in Birmingham, U.K. (Eaton et al., 1984).

6.2.4. Coloreds

Their diet, respecting fat and fiber intakes, resembles that of Indians.

6.3. Reproduction

In black girls, mean onset of menarche is 1–2 years later than that in white girls. By 18 years, irregularity—more than 3 days early or late—has been noted to prevail in over a third of a series of black girls studied, who also had missed periods occasionally. This behavior was present in fewer than 10% of the white girls studied. Late onset and irregularity of cycles are reported to be consistent with lesser proneness to breast cancer (Olsson et al., 1983; Apter and Vihko, 1983; Henderson et al., 1985; La Vecchia et al., 1985).

In black adolescent girls, a half to three quarters have their first child before 20 years of age, whereas the proportion with white girls is stated to be 10–15% or less (Ferguson, 1987). Early reproduction is a protective factor against breast cancer (Dupont and Page, 1987; Rose et al., 1987).

Until recently, many black women had half a dozen children, with 9–12 months of lactation and its associated amenorrhea. This implies that menses are absent for a quarter or so of their reproductive life. This phenomenon, with the associated protective effect of reduction in circulating serum prolactin levels, is believed to inhibit the development of breast cancer (Wang et al., 1987).

Nowadays, in big cities, the situation is changing. Family size is rapidly decreasing, so much so that in Soweto reproduction is near or even below replacement rate (Medical Officer of Health, 1985). This phenomenon, in gross contrast to the excessively high parity in all countries in Africa, should it become widespread, would tremendously lessen the current almost insoluble socioeconomic burdens in these countries. The change, however, is likely to increase proneness to breast cancer.

In colored and Indian populations, reproduction characteristics have been insufficiently studied to permit comment.

6.4. Smoking Practice

In South African populations, among white males, smoking practice is decreasing. But it is high or is rising among men in the other ethnic groups. Among them, the proportion who smoke is about 70% (Van der Burgh, 1979; Yach, 1982). Among white females, smoking practice is decreasing only slowly, as in the United States (Fielding, 1987). However, that in black women, previously low, is rising. Among Indian females, smoking, frowned on by their culture, remains low in frequency (Crofton, 1987). It is high, however, in colored females.

In Africa, unfortunately, the consumption of tobacco is rapidly increasing. One report indicates a doubling of consumption in the last 10 years (Notes and News, 1987). Accordingly, all tobacco-related cancers almost certainly will increase in all save the white population.

6.5 Obesity

Obesity has been associated with increased risk of cancers of the breast, colon–rectum, prostate, endometrium, kidney, cervix, ovary, thyroid, and bladder (Willett *et al.,* 1987; Albanes, 1987; Osler, 1987). Overweight and obesity are very common in emerging populations, especially among those in urban areas (Walker, 1964; Trowell, 1981). Locally, in limited studies, using the criteria of Bray (1985) of body mass index of 30 and over, obesity was found to be present in 6–12% in the male groups and 7–33% in the female groups. Black, colored, and Indian women were most affected. The highest prevalences of obesity were found in the poorer black populations, as also noted in the United States (Garn, 1986).

6.6. Alcohol Consumption

Consumption is rising in all Third World and in many First World populations (Walsh and Grant, 1985; Walker, 1987a). In South African populations, although no precise data are available, the greatest increases in consumption are occurring in the colored and black populations, especially among the males. Alcohol consumption has a promotive effect on cancers of the liver, esophagus, lung, pancreas, and rectum (Pollack *et al.,* 1984; Kune *et al.,* 1987).

6.7. Toxic Environmental Hazards

In South Africa, atmospheric pollution is relatively low. In certain industries, however, there are marked dangers. This is especially so in the case of workers in areas of asbestos mining and use, among whom mesothelioma is a serious threat (Davies, 1988).

7. STUDIES ON THE OCCURRENCE AND RISK FACTORS OF PARTICULAR CANCERS

7.1. Breast Cancer

7.1.1. The Validity of Its Low Occurrence in Black Women

A very important question is: could there be gross underreporting of the occurrence of breast cancer (and possibly of other cancers) among black women in rural areas? This is a highly plausible question. The possibility is unlikely, however, for the following reason. In all rural complexes of villages there are informal groupings of older women who know of everything that takes place in each family, regarding family problems, sickness, employment, crime, and so forth. Such women usually include the local chief's wife, church leaders, school teachers, and clinic sisters. On a number of occasions they have been closely questioned, and it was found that their knowledge regarding those sick was in agreement with information gathered from attendance at local clinics and hospitals. In brief, we are assured that black women with breast disease, although they will neglect it for a time, and although they will usually seek help first from local traditional healers or *sangomas*, ultimately will all go to hospital. Hence, it is believed that the very low frequency of breast cancer admissions to rural hospitals, which is the rule throughout the country, is valid. The low frequency described is in agreement with the low incidence of breast cancer in black women noted in the cancer registry in Dakar, Senegal (Waterhouse *et al.*, 1982), in the registries and hospital admissions in other African populations (Parkin, 1986), and in the South African Cancer Registry.

7.1.2. Cycle Length

Endeavors are being made to acquire more accurate information on cycle onset, regularity, and length in adolescents in the different populations, who differ in their subsequent proneness to breast cancer. One current investigation concerns seeking to learn whether local Indian girls have cycles a mean of 2 days longer, as was noted in a recent report on Asian girls in California (Bernstein *et al.*, 1987). This does not appear to be the case with local Indian girls. An attempt is also being made to learn whether rural black girls who are very physically active, i.e., those who walk long distances attending school, have longer cycles. This investigation is being carried out because a mean of an extra 2.4 days was reported for physically active white girls in the report just cited.

7.1.3. Serum Prolactin Levels

A beginning has been made to learn to what extent low, medium, and extremely high parities in black women depress serum prolactin levels (Wang *et al.*, 1987).

7.1.4. Receptor Status

In a 5-year follow-up study by Pegoraro *et al.* (1986) in Durban on white, black, and Indian breast cancer patients, cytoplasmic estrogen receptor-positive tumors were present in 67%, 49%, and 41% of patients, respectively. Receptor status had no influence on prognosis with white patients, but it did promote longer disease-free periods in patients in the other ethnic groups. Another local study, done in Johannesburg, concerned clinical features, hormone receptor status, and response to therapy in male breast cancer patients (Bezwoda et al., 1987). The variables in pattern were the same as those prevailing with female patients.

7.2. Colon Cancer

7.2.1. Bowel Behavior

Because of the already mentioned fall in fiber intake in blacks, it was decided to learn whether this change has caused increases in transit time and decreases in defecation frequency in interethnic series of preschool children. Previous studies on school children had indicated markedly slower transit times and less frequent defecation in white compared with black pupils (Walker, 1975). In the series of preschool children investigated, it transpired that, with sweet corn as a crude marker, first appearance in white children's feces averaged 28 hr, whereas that in the children in the other ethnic groups—the blacks, coloreds, and Indians—averaged far less, 6–8 hr. The daily defecation frequencies in the white and in the other groups (combined) were 1.5 and 1.9, respectively (Walker and Walker, 1985a). There is no obvious explanation for these very marked differences. A start has been made to determine whether elderly urban and rural blacks, accustomed for decades to high fiber intake but now on much reduced fiber intake, still have shorter transit times than elderly whites on their habitually low fiber intakes. Initial studies indicate that blacks still have significantly shorter transit times.

7.2.2. Fecal pH

Mean values for fecal pH in black preschool and school children and adults were found to be significantly lower than mean values for corresponding groups of white subjects; values for Indian and colored subjects were intermediate (Walker et al., 1986a). Studies showed that only major changes in diet can alter fecal pH value (Walker et al., 1979). In investigations on immigrant groups, e.g., Japanese in California and Indians in Birmingham, U.K., proneness to colorectal cancer has increased little in the first generation despite alterations in respective traditional diets. Accordingly, some aspects of bowel milieu would seem to be constitutional characteristics, possibly relatively fixed. Thus, conceivably, a low fecal pH in youth could give life-long protection against colorectal cancer. No studies have yet been reported on serial values for fecal pH in individuals. Accordingly, investigations were initiated in the first instance on 30 rural blacks at Hekpoort, collections being made every eighth day for five consecutive occasions. Variations were slight, with individual standard deviations ranging from 0.02 to 0.56 (average 0.21). Analogous studies on Indian and colored schoolchildren are now in progress.

7.2.3. Breath Methane

Investigations on breath methane excretion in different ethnic groups have been made in association with Dr. J. H. Cummings, Dunn Clinical Nutritional Centre,

Cambridge. Briefly, about half of white subjects but about three quarters of blacks were producers (Segal *et al.*, 1988). These observations are in agreement with reports of other workers (Pitt *et al.*, 1980; McKay *et al.*, 1985). However, it was found that only a third of Indians studied were producers. There appears to be no obvious explanation for these diverse responses.

7.2.4. Breath Hydrogen

Series of urban and rural blacks as well as white subjects are now being investigated in association with Professor I. Segal in regard to their breath hydrogen responses to maize, wheat, rice, and other staple carbohydrate-containing foodstuffs. It is hoped to learn, *inter alia,* from serial studies, whether the very marked inter- and intraindividual variability noted by Rumessen *et al.* (1987) in white subjects also occurs with black subjects.

7.3. Cervix Cancer

In view of reports of the fourfold higher incidence of cervix cancer in white women in social class V compared with those in class I (Notes and News, 1988), an attempt is being made to learn to what extent socioeconomic differentiation prevails among urban black cervix cancer patients.

8. STUDIES ON SURVIVAL OF URBAN BLACK CANCER PATIENTS

All studies on series of urban black patients with different cancers in Soweto (Walker *et al.*, 1984a,b, 1985, 1986b,c) have shown the time of 50% mortality to be half or less than the corresponding times reported for white patients in the United Kingdom (Cancer Statistics Groups, 1982) (Table IX). In the United States, shorter survival has been reported for indigent white, black, Mexican Indian, and other similarly placed groups of patients (Dayel *et al.*, 1982; Samet *et al.*, 1987). South African urban black patients, as with the series of patients just mentioned, characteristically present late.

TABLE IX. Studies on Survival of Urban Black Patients from Cancer: Time of 50% Mortality

	Blacks	Whites (U.K.)[a]
Esophagus	3½ months	8 months
Stomach	3 months	8 months
Colon	4 months	10 months
Prostate	1½ years	2–3 years
Breast	1½ years	4–5 years
Cervix	1½ years	4–5 years
Liver	6 weeks	7 months

[a]Cancer Statistics Group (1982).

However, it is doubtful whether this factor and the lesser availability or utilization of medical services are the full explanations for shorter survival. Locally, in an attempt to learn the reasons for late presentation, the only large number of patients available for study were urban cervix cancer patients. Attempts are therefore being made to learn more precisely when they first became aware of the disease, the reasons for their delay in seeking help, what treatments were sought before hospitalization, and subsequent treatments thereafter. Additionally, inquiries are being made into the mean survival times of groups of black cervix cancer patients at clinical stages from I to IV.

9. OUTLOOK FOR CANCER

9.1. Prosperous Populations

The general picture is that death rates from cancer are steady or are only very slightly decreasing (Bailar and Chelimsky, 1987; Davis and Schwartz, 1988). As to incidence rates, those of many cancers are still increasing; moreover, improvements in survival times have been disappointingly small (Cancer Statistics Group, 1982; Cancer Patient Survival, 1987). Significant changes in diet, at least of the magnitude deemed necessary, would seem very unlikely to occur (Anonymous, 1987; Willett et al., 1987; Walker, 1987b), as are also changes in reproduction behavior. Accordingly, apart from stopping smoking, little meaningful avoiding action appears practicable. Hence, as repeatedly stressed, for amelioration of the cancer position, principal emphasis is on earlier detection and treatment.

9.2. Developing Populations

In Third-World developing populations, no avoiding action against cancer in respect to restrictions relating to diet, smoking, and alcohol consumption can even be contemplated. Locally, there is, of course, considerable interest in the likely future situation regarding cancer among the relatively rapidly westernizing urban black population. In Los Angeles it is noteworthy that the black population now has a higher total cancer incidence than the white population; this highly adverse situation obtains especially for cancers of the esophagus, stomach, lung, and prostate (Waterhouse et al., 1982). It is hoped, of course, that in Africa, ultimately, this will not happen to the same degree. But, as stated, it is virtually impossible to restrain changing factors that militate against cancer control—changes that inevitably accompany rise in socioeconomic state.

10. CONCLUSION

In South Africa, cancer in the white population presents a smaller problem than it does in many of such populations elsewhere. In the black, colored, and Indian populations, the present high frequencies of esophageal, liver, and cervix cancers will undoubtedly fall. Simultaneously, although with varying speed, there will be rises in cancers of the prostate, breast, colon–rectum, and lung. Examination and discussion of the risk factors for cancer, as they prevail in the four populations, indicate that apart from stopping smoking, minimal if any avoiding action can be entertained. Accordingly, to lessen morbidity and possibly mortality in patients, the development of means to facilitate early recognition of the disease, together with encouragement to seek help as early as possible, is mandatory.

ACKNOWLEDGMENTS. For financial support for the research studies described, grateful thanks are due to the National Cancer Association of South Africa, South African Medical Research Council, South African Sugar Association, and De Beers Anglo-American Corporation Chairman's Educational Trust. Thanks are due to Miss F. A. Cassim for typing the manuscript.

REFERENCES

Albanes, D., 1987, Caloric intake, body weight, and cancer: A review, *Nutr. Cancer* **9**:199–218.

Anonymous, 1987, Nutrition and cancer, facts, fallacies, and ACS activities, *Cancer News* (Summer):18–19.

Apter, D., and Vihko, R., 1983, Early menarche, a risk factor for breast cancer, indicates early onset of ovulatory cycles, *J. Clin. Endocrinol. Metab.* **57**:82–86.

Bailar, J. C., and Chelimsky, E., 1987, Cancer control, *Science* **236**:1049–1050.

Bernstein, L., Ross, R. K., Labo, R. A., Hanisch, R., Krailo, M. D., and Henderson, B. E., 1987, The effects of moderate physical activity on menstrual cycle patterns in adolescents: Implications for breast cancer prevention, *Br. J. Cancer* **55**:681–685.

Bezwoda, W. R., Hesdorffer, C., Dansey, R., de Moor, N., Demran, D. P., Browde, S., and Lange, M., 1987, Breast cancer in men, *Cancer* **60**:1337–1340.

Block, G., and Lanza, E., 1987, Dietary fiber sources in the United States by demographic group, *J. Natl. Cancer Inst.* **79**:83–91.

Bradshaw, E., and Harington, J. S., 1982, A comparison of the cancer mortality rates in South Africa with those in other countries, *S. Afr. Med. J.* **61**:943–946.

Bray, G. A., 1985, Obesity: Definition, diagnosis and disadvantages, *Med. J. Aust.* **142**:S2–S8.

Bremner, C. G., and Ackerman, L. V., 1970, Polyps and carcinoma of the large bowel in the South African blacks, *Cancer* **26**:991–999.

Cancer Patient Survival: What Progress Has Been Made? 1987, FAO Report PEMD-87-13, General Accounting Office, Washington, DC.

Cancer Statistics Group, 1982, *Trends in Cancer Survival in Great Britain*, Cancer Research Campaign, London, pp. 35–41.

Crofton, E., 1987, Women and smoking, *World Health* (Dec.):28–30.

Davies, J. C. A., 1988, Mesothelioma is a fibre-specific tumour, *S. Afr. Med. J.* **73**:327–328.

Davis, D. L., and Schwartz, J., 1988, Trends in cancer mortality: U.S. white males and females, 1968–83, *Lancet* **1**:633–635.

Dayel, H. H., Power, R. N., and Chiu, C., 1982, Rate and socio-economic status: Survival from breast cancer, *J. Chron. Dis.* **35**:675–683.

De Waard, F., and Wang, D. Y., 1988, Epidemiology and prevention: Workshop report, *Eur. J. Cancer Clin. Oncol.* **24**:45–48.

Doll, R., Muir, C., and Waterhouse, J., 1970, *Cancer Incidence in Five Continents*, Vol. II, International Agency for Research on Cancer, Lyons.

Dupont, W. D., and Page, D. L., 1987, Breast cancer risk associated with proliferative disease, age at first birth, and a family history of breast cancer. *Am. J. Epidemiol.* **125**:769–779.

Eaton, P. M., Wharton, P. A., and Wharton, B. A., 1984, Nutrient intake of pregnant Asian women at Sorrento Maternity Hospital, Birmingham, *Br. J. Nutr.* **52:**457–468.

Ferguson, J., 1987, Reproductive health of adolescent girls, *World Health Stat. Q.* **40:**211–213.

Fielding, J. E., 1987, Smoking and women, *N. Engl. J. Med.* **317:**1343–1345.

Garn, S. M., 1986, Family-line and socioeconomic factors in fatness and obesity, *Nutr. Rev.* **44:**381–386.

Gilpin, T. P., Walker, A. R. P., Walker, B. F., and Evans, J., 1989, Causes of admission of rural black patients in Murchison Hospital, Kwazulu, Natal, *S. Afr. J. Food Sci.* (in press).

Greenberg, M. R., 1986, Does New Jersey cause cancer? *Sciences* (Jan./Feb.):40–46.

Greenwald, P., and Sondik, E. J. (eds.), 1986, Cancer control objectives, *NCI Monogr.* **2:**3–11.

Henderson, B. E., Ross, R. K., and Judd, H. F., 1985, Do regular ovulatory cycles increase breast cancer risk? *Cancer* **56:**1206–1208.

Higginson, J., and Oettlé, A. G., 1960, Cancer incidence in the Bantu and "Cape Coloured" races of South Africa: Report of a cancer survey in Transvaal (1953–55), *J. Natl. Cancer Inst.* **24:**589–671.

Isaacson, C., Selzer, G., Kaye, V., Greenberg, M., Woodruff, D., Davies, J., Ninin, D., Vetten, D., and Andrew, M., 1978, Cancer in the urban blacks of South Africa, *S. Afr. Cancer Bull.* **22:**49–84.

Jacobs, L. R., 1987, Dietary fiber and cancer, *J. Nutr.* **117:**1319–1321.

Jussawalla, D. J., Yoele, B. B., and Natekar, M. V., 1985a, Cancer incidence in Indian Christians, *Br. J. Cancer* **51:**883–891.

Jussawalla, D. J., Yeole, B. B., and Natekar, M. V., 1985b, Cancer in Indian Moslems, *Cancer* **55:** 1149–1158.

Kemp, I., Boyle, P., Smans, M., and Muir, C., 1985, *Atlas of Cancer in Scotland 1975–1980: Incidence and Epidemiological Perspective,* International Agency for Research on Cancer and the Cancer Registries of Scotland, IARC Scientific Publication 72, IARC, Lyon.

Kloppers, P. J., Van Staden, D. A., Fehrsen, J. P., and Van der Walt, E., 1983, Die voorkomspatroon van sekere karsinome in die RSA, *S. Afr. Med. J.* **64:**1062–1063.

Kune, S., Kune, G. A., and Watson, L. F., 1987, Case-control study of alcoholic beverages as etiological factors: The Melbourne colorectal cancer study, *Nutr. Cancer* **9:**43–56.

Langenhoven, M. L., Wolmarans, P., Groenewald, G., Richter, M. J. C., and Eck, M., 1988, Nutrient intakes and food and meal patterns in three South African population groups, in: *Progress in Diet and Nutrition* (C. Horwitz and P. Rozen, eds.), S. Karger, Basel, pp. 41–48.

Lanza, E., Jones, D. Y., Block, G., and Kessler, L., 1987, Dietary fiber intake in the US population, *Am. J. Clin. Nutr.* **46:**790–797.

La Vecchia, C., Decarli, A., and Di Pietro, S., 1985, Menstrual pattern and the risk of breast disease, *Eur. J. Cancer Cin. Oncol.* **21:**417–422.

Leading Article, 1986, Coronary heart disease in Indians overseas, *Lancet* **1:**1307–1308.

McKay, L. F., Eastwood, M. A., and Brydon, W. G., 1985, Methane excretion in man—a study of breath, flatus and faeces, *Gut* **26:**69–74.

Medical Officer of Health, 1985, *Annual Report,* City Health Department, Johannesburg.

Mukerjee, A. B., 1975, Heart diseases in India, *J. Ind. Med. Assoc.* **65:**156–158.

National Cancer Institute, 1982, *Cancer Mortality in the United States, 1950–1977,* NCI Monograph 59, Bethesda.

Naylor, M. N., 1985, Possible factors underlying the decline in caries prevalence, *J. R. Soc. Med.* **78:** 23–24.

Notes and News, 1987, Tobacco smoking in the Third World, *Lancet* **1:**1275.

Notes and News, 1988, Incidence of cancer and social class, *Lancet* **1:**602.

Olsson, H., Landin-Olsson, M., and Gulberg, B., 1983, Retrospective assessment of menstrual cycle length in patients with breast cancer, in patients with benign breast disease, and in women without breast disease, *J. Natl. Cancer Inst.* **70:**17–20.

Osler, M., 1987, Obesity and cancer, *Dan. Med. Bull.* **34:**267–274.

Parkin, D. M., 1986, *Cancer Occurrence in Developing Populations,* International Agency for Research on Cancer, IARC Scientific Publication No. 75, Lyons.

Pegoraro, R. J., Nirmul, D., Reinach, S. G., Jordaan, J. P., and Joubert, S. M., 1986, Breast cancer prognosis in three different racial groups in relation to steroid hormone receptor status, *Breast Cancer Res. Treat.* **7**:111–118.

Pitt, P., de Bruijn, K. M., Beeching, M. F., Goldberg, E., and Blendis, L. M., 1980, Studies on breath methane: The effect of ethnic origins and lactulose, *Gut* **21**:951–959.

Pollack, E. S., Nomura, A. M. Y., Heilbruin, L. K., Stemmermann, G. N., and Green, S. B., 1984, Prospective study of alcohol consumption and cancer, *N. Engl. J. Med.* **310**:617–621.

Potter, J. F., Pandha, H. S., Dawkins, D. M., and Beevers, D. G., 1984, Cancer in blacks, whites and Asians in a British Hospital, *J. R. Coll. Physicians* **18**:231–235.

Reeve, P. A., and Falkner, M. J., 1986, Disease patterns in a rural black population, *S. Afr. Med. J.* **69**:551–552.

Retief, F. P., 1987a, Boycotts and medicine, *Lancet* **2**:278.

Retief, F. P., 1987b, Improvements in health, *S. Afr. Med. J.* **71**:402.

Robertson, M. A., Harington, J. S., and Bradshaw, E., 1971, The cancer pattern in Africans of the Transvaal Lowveld, *Br. J. Cancer* **25**:377–384.

Rose, D. P., Laakso, K., Wynder, E. L., and Kettunen, K., 1987, Breast cancer risk factors in Finland and the United States, *Eur. J. Cancer Clin. Oncol.* **23**:1794.

Rose, E. F., 1978, Patterns of occurrence of oesophageal cancer with particular reference to the Transkei, in: *Carcinoma of the Oesophagus* (W. Silber, ed.), Balkema, Rotterdam, pp. 36–42.

Rose, E. F., and Fellingham, S. A., 1981, Cancer patterns in the Transkei, *S. Afr. J. Sci.* **77**:555–561.

Rosen, P., Fireman, Z., Figer, A., Legum, C., Ron, E., and Lynch, H. T., 1987, Family history of colorectal cancer as a marker of potential malignancy within a screening program, *Cancer* **60**:248–254.

Rumessen, J. J., Kokholm, G., and Gudmand-Høyer, E., 1987, Methodological aspects of breath hydrogen (H_2) analysis. Evaluation of a H_2 test, *Scand. J. Clin. Lab. Invest.* **47**:555–560.

Samet, J. M., Key, C. R., Hunt, W. C., and Goodwin, J. S., 1987, Survival of American Indian and Hispanic cancer patients in New Mexico and Arizona, 1969–82, *J. Natl. Cancer Inst.* **79**:457–463.

Schatzkin, A., Palmer, J. R., Rosenberg, L., Helmrich, S. P., Miller, D. R., Kaufman, D. W., Lesko, S. M., and Shapiro, S., 1987, Risk factors for breast cancer in black women, *J. Natl. Cancer Inst.* **78**:213–217.

Seedat, Y. K., Seedat, M. A., and Hackland, D. B. T., 1982, Biosocial factors and hypertension in the urban and rural Zulu, *S. Afr. Med. J.* **61**:999–1002.

Seftel, H. C., Johnston, S., and Muller, E. A., 1980, Distribution and biosocial correlations of blood pressure levels in Johannesburg blacks, *S. Afr. Med. J.* **57**:313–320.

Segal, I., Cooke, S. A. R., Hamilton, D. G., and Ou Tim, L., 1981, Polyps and colorectal cancer in South African Blacks, *Gut* **22**:653–657.

Segal, I., Walker, A. R. P., Lord, S., and Cummings, J. H., 1988, Breath methane and bowel disease: Risk in contrasting populations, *Gut* **29**:608–613.

South African Cancer Registry, 1986, Johannesburg: South African Institute for Medical Research.

Steyn, N. P., and Albertse, E. M., 1987, Sucrose consumption and dental caries in twelve-year-old children residing in Cape Town, *J. Dent. Assoc. S. Afr.* **42**:43–49.

Trowell, H., 1981, Hypertension obesity, diabetes mellitus and coronary heart disease, in: *Western Diseases: Their Emergence and Prevention* (H. C. Trowell and D. P. Burkitt, eds.), Edward Arnold, London, pp. 3–32.

Van der Burgh, C., 1979, Smoking behaviour of white, black, coloured and Indian South Africans: Some statistical data on a major public health hazard, *S. Afr. Med. J.* **55**:975–978.

Walker, A. R. P., 1964, Overweight and hypertension in emerging populations, *Am. Heart J.* **68**:581–585.

Walker, A. R. P., 1975, Effect of high crude fiber intake on the transit time and the absorption of nutrients in South African Negro school-children, *Am. J. Clin. Nutr.* **28**:1161–1169.

Walker, A. R. P., 1987a, Towards an alcoholic holocaust? *S. Afr. Med. J.* **71**:679–680.

Walker, A. R. P., 1987b, Nutrition–disease changes in western populations—Quo Vadis? *J. Dietet. Home Econ.* **5**:55–60.

Walker, A. R. P., and Segal, I., 1988, Colorectal cancer: Some aspects of epidemiology, risk factors, treatment, screening and survival, *S. Afr. Med. J.* **73:**653–657.

Walker, A. R. P., and Walker, B. F., 1985a, The bowel behaviour in young black and white children, *Arch. Dis. Child.* **60:**967–970.

Walker, A. R. P., and Walker, B. F., 1985b, Coronary heart disease in blacks in underdeveloped populations, *Am. Heart J.* **109:**1410.

Walker, A. R. P., and Walker, B. F., 1986, Coronary heart disease in Indians overseas, *Lancet* **2:**158.

Walker, A. R. P., Walker, B. F., and Segal, I., 1979, Faecal pH value and its modification by dietary means in South African black and white schoolchildren, *S. Afr. Med. J.* **55:**495–498.

Walker, A. R. P., Walker, B. F., Isaacson, C., Segal, I., and Pryor, S., 1984a, Low survival of blacks with oesophageal cancer in Johannesburg, South Africa, *S. Afr. Med. J.* **66:**877–878.

Walker, A. R. P., Walker, B. F., Tshabalala, E. N., Isaacson, C., and Segal, I., 1984b, Low survival of South African urban black women with breast cancer, *Br. J. Cancer* **49:**241–245.

Walker, A. R. P., Walker, B. F., Siwedi, D., Isaacson, C., Van Gelderen, C. J., Andronikou, A., and Segal, I., 1985, Low survival of South African urban black women with cervical cancer, *Br. J. Obstet. Gynaecol.* **92:**1272–1278.

Walker, A. R. P., Walker, B. F., and Walker, A. J., 1986a, Faecal pH, dietary fibre intake, and proneness to colon cancer in four South African populations, *Br. J. Cancer* **53:**489–495.

Walker, A. R. P., Walker, B. F., Serobe, W., Paterson, A., Isaacson, C., and Segal, I., 1986b, Survival of blacks with liver cancer in Soweto, Johannesburg, South Africa, *Trop. Gastroenterol.* **7:**169–172.

Walker, A. R. P., Walker, B. F., Isaacson, C., Doodha, M. I., and Segal, I., 1986c, Survival of black men with prostatic cancer in Soweto, Johannesburg, South Africa, *J. Urol.* **135:**58–59.

Walker, A. R. P., Walker, B. F., Dison, E., and Walker, C., 1988, Dental caries and malnutrition in rural South African black ten to twelve year olds, *J. Dent. Assoc. S. Afr.* **43:**581–583.

Walsh, B., and Grant, M., 1985, The alcohol trade and its effects on public health, *World Health Forum* **6:**195–203.

Wang, D. Y., Stavola, B. L., Bulbrook, R. D., Allen, D. S., and Kwa, H. G., 1987, The relationship between blood prolactin levels and risk of breast cancer in premenopausal women, *Eur. J. Cancer Clin. Oncol.* **23:**1541–1548.

Waterhouse, J., Shanmugaratnam, K., and Muir, C., 1982, *Cancer Incidence in Five Continents,* Vol. IV, IARC Scientific Publication No. 42, International Agency for Research on Cancer, Lyons.

Willett, W., Stampfer, M. J., Colditz, G. A., Rosner, B., Hennekens, C., and Speizer, F. E., 1987, Dietary fat and the risk of breast cancer, *N. Engl. J. Med.* **317:**165–166.

World Health Organization, 1982, Vital statistics and causes of death, in: *World Health Annual Statistics,* World Health Organization, Geneva.

Wyndham, C. H., 1985, Leading causes of cancer mortality in various populations in the RSA, *S. Afr. Med. J.* **67:**584–587.

Yach, D., 1982, Economic aspects of smoking in South Africa, *S. Afr. Med. J.* **62:**167–170.

Yang, C. S., 1980, Research on esophageal cancer in China: A review, *Cancer Res.* **40:**2633–2644.

35

Dietary Fiber Intake and Colon Cancer Mortality in The People's Republic of China

T. COLIN CAMPBELL, WANG GUANGYA,
CHEN JUNSHI, JAMES ROBERTSON,
CHAO ZHONGLIN, and BANOO PARPIA

There are many unresolved questions concerning the relationship between dietary fiber and cancer of the large bowel. These include (1) inadequate knowledge of the physiological activities of the various dietary fiber constituents, (2) uncertainties about the food composition of these constituents, (3) variable constituent content for the same food grown in different geographic regions and at various stages of harvest, and (4) uncertainties about earlier intake levels of various fiber-containing foods during the time when cancers were forming. It would therefore appear to be inappropriate to put too much emphasis on the effects of individual fiber fractions not only because of these uncertainties of intake but also because the intakes of individual fiber fractions are strongly correlated with each other. Moreover, fiber intakes also correlate with many other dietary constituents that may be associated with the prevalence of large bowel cancer.

An alternative strategy that may be useful in determining the relevance of dietary fiber for the promotion of health and avoidance of diseases such as large bowel cancer is to examine various dietary fiber intakes within the context of a broad array of dietary constituents, disease rates, and lifestyle patterns. This was a major objective of a 1983 study by Chen *et al.* (1989), who surveyed a sample of mostly rural counties within the People's Republic of China for a large number of dietary and lifestyle characteristics and various disease mortality rates. The mortality rates used in this study were those re-

T. COLIN CAMPBELL and BANOO PARPIA • Division of Nutritional Sciences, Cornell University, Ithaca, New York 14853. JAMES ROBERTSON • Department of Animal Sciences, Cornell University, Ithaca, New York 14853. WANG GUANGYA, CHEN JUNSHI, and CHAO ZHONGLIN • Institute of Nutrition and Food Hygiene, Chinese Academy of Preventive Medicine, Beijing, China.

corded for 1973–1975 by the Chinese Academy of Medical Sciences (Li *et al.*, 1981). These rates were available as sex-specific, site-specific, and age-standardized data for more than 2300 counties (Table I). The ranges of mortality rates observed for different parts of China are far greater than those observed within Western industrialized countries. Moreover, the geographic patterns of distribution are unique for each cancer, demonstrating intense disease localization. Even for large bowel cancer, where the national average is considerably below that of Western countries, a generous range is observed.

The survey by Chen *et al.* (1989) was undertaken in the fall of 1983 (September– December) in 65 mostly rural counties, which were selected to represent the full range of mortality rates for seven cancers (nasopharynx, esophagus, stomach, liver, colorectal, lung, and leukemia). In addition, the rates for a few additional cancers and about three dozen noncancer diseases were also available for this study.

The locations of the 65 survey counties within China are shown in Fig. 1, and the study design is shown in Table II. Five types of information and samples were collected in each county. The 3-day dietary survey included direct measurements of food intakes in 30 households in each county. Based on these food records and on Chinese food composition tables, intakes of in-season foods and 14 nutrients were obtained. In each commune, questionnaires on selected lifestyle characteristics and annual food frequency patterns were completed by survey interviewers for 50 individuals aged 35–64 years (25 of each sex) in whom cancers were theoretically forming and for whom diagnoses were most accurate. In addition, each subject in this age group donated a 10-ml sample of blood. Next, in one of the communes in each county, these same individuals also provided either a 4-hr or overnight urine sample. And finally, approximately 600 food samples representing all important foods being consumed in the survey households were collected.

Blood samples were prepared as shown in Fig. 2. Within 2 hr of blood draw

TABLE I. Ranges of Cancer Mortality Rates of Survey Population[a] (Fold Ranges in Parentheses, Zeros Terated as Ones)

Cancer site	Males	Females
All cancers	35–721 (21)	35–491 (14)
Nasopharynx[b]	0–75 (75)	0–26 (26)
Esophagus[b]	1–435 (435)	0–286 (286)
Stomach[b]	6–386 (64)	2–141 (70)
Liver[b]	7–248 (35)	3–67 (22)
Colorectal[b,c]	2–67 (34)	2–61 (30)
Lung[b]	3–59 (20)	0–26 (26)
Breast	—	0–20 (20)
Cervix	—	4–97 (24)
Leukemia[b]	0–9 (9)	0–7 (7)

[a] Annual cases per 100,000, truncated for ages 35–64, for 65 survey counties.
[b] Cancers used to select survey counties.
[c] Separate rates for colon and rectum are available in 49 of the 65 survey counties.

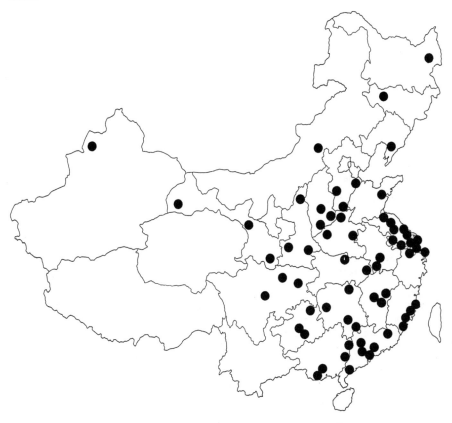

FIGURE 1. People's Republic of China survey counties.

(identified by the dashed line), samples were separated into RBC and plasma fractions, saved in appropriate buffers, and then frozen at −20°C. Within 6 weeks all samples were transported in the frozen state to Dr. Chen Junshi's central laboratory in Beijing, where aliquots of individual samples were taken to prepare age- and sex-specific pools for each of the 130 communes.

Except for viral factors and individual hemoglobin determinations, subsequent analyses were undertaken on pooled blood samples, pooled urine samples, and county

TABLE II. Nature of Data Collected in Survey

	Commune I production brigade		Commune II production brigade	
	Production team IA	Production team IB	Production team IIA	Production team IIB
3-Day food intakes	+	+		
Questionnaire	+	+	+	+
Blood samples	+	+	+	+
Urine samples[a]	+	+	+	+
Food samples	+	+	+	+

[a]Only collected in one of the two communes.

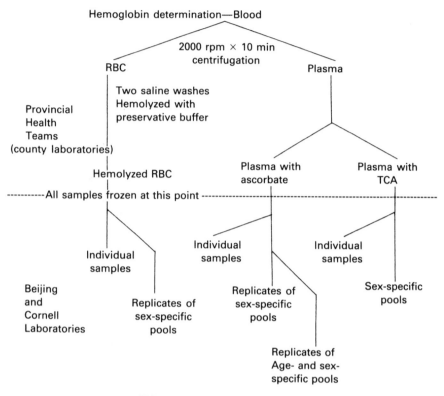

FIGURE 2. Blood-pooling scheme.

composites of food samples. Most of these assays were carried out between 1984 and 1987 to give a final total of 366 items of information (including various disease mortality rates) for each of these 65 counties. A summary of the types of assays undertaken in this study is shown in Table III.

Univariate correlations between various dietary fiber and complex carbohydrate intakes and cancers of the large bowel are shown in Table IV. Weak inverse associations were observed for most intakes. Although most all of these associations of large bowel cancer and fiber intakes were inverse, almost none were statistically significant. The lack of strong inverse associations, as generally hypothesized for dietary fiber and large bowel cancer, may not be surprising because, by Western standards, Chinese fiber intake is already very high, and colorectal cancer rates are very low. That is, a protective effect of dietary fiber may already have become almost fully expressed, as illustrated by a comparison of United States and Chinese data (Table V). The correlations shown in Table IV also show that no particular fiber or complex carbohydrate intake is uniquely associated with disease mortality rates, although a disease association with the intake of the rhamnose-containing constituent was statistically significant ($P < 0.05$). Because the weak association for this particular constituent could well be a chance occurrence,

TABLE III. Summary of Studies Undertaken

Disease mortality rates: cancers (15), noncancers (31)
Nutritional status
  Lipid (varied blood cholesterol fractions, triglycerides, 12 fatty acids; lipid intakes)
  Protein (blood indicators and intakes)
  Carbohydrate (blood glucose, 15 fiber/complex carbohydrate intakes)
  Vitamins (blood indicators, A, E, C; intakes of B_1, B_2, niacin, C)
  Minerals (blood activities, excretion levels, intakes of Se, Cu, Fe, K, Mg, P, Zn, Cl, Na, Ca, Mn, I)
  Foods (all major food intakes)
Antioxidant status (blood)
  Antioxidant nutrients
  SOD
  Lipid peroxide
  Catalase
Viral status (antibody/antigen)
  Hepatitis B
  Epstein–Barr
  Candida
  Herpes simplex
Hormonal status (plasma)
  Testosterone
  Estrogen
  SHBG
  Prolactin
Reproductive characteristics
  Menstrual/menopausal onset
  Number of children
  Child survival
  Reproductive hormone profiles
Contaminants
  Heavy metal intakes (As, Cd, Hg, Pb)
  Pesticide intakes (from plant foods)
  Nitrate/nitrite/nitrosamine excretion
  Thioether excretion
  Aflatoxin excretion
Smoking habits
  Duration, onset, type of tobacco use
  Blood cotinine
Drinking water source
Anthropometry
  Height
  Weight
Geographic characteristics

and because intake calculations were based on analyses of composite food samples in each county, no particular biological relevance is assigned to the rhamose-containing constituent. Analyses of individual foods comprising these composites are currently being undertaken in order to determine which food contributes which type of fiber.

It was earlier noted that variation in fiber composition of individual foods according

TABLE IV. Correlations of Colon and Rectal Cancer Mortality Rates
with Various Fiber/Complex Carbohydrate Fractions

Fibrous constituent[a]	Colon[b]	Rectum[b]	Colorectum[c]
Total fiber	−0.01	−0.18	−0.03
Neutral detergent fiber	−0.14	−0.23	−0.13
Hemicellulose	−0.09	−0.21	−0.10
Cellulose	−0.18	−0.19	−0.13
Lignin	−0.05	−0.11	−0.09
Cutin	−0.14	−0.14	−0.14
Starch	−0.13	−0.04	0
Pectin	0	0.03	−0.03
Rhamnose	−0.33[d]	−0.20	−0.26[d]
Fucose	0.05	−0.04	−0.02
Arabinose	−0.27	−0.24	−0.18
Xylose	−0.22	−0.16	−0.15
Mannose	−0.13	−0.17	−0.13
Galactose	−0.27	−0.22	−0.24

[a] All intakes were based on assays of food composite representing average daily intake of all foods. Total dietary fiber, neutral detergent fiber, hemicellulose, cellulose, lignin, and cutin determined by Robertson and VanSoest as described by Chen et al. (1989). Starch, pectin, and remaining nonstarch monosaccharide residue constituents were determined by Lewis as described by Chen et al. (1989).
[b] Forty-nine counties only.
[c] All 65 counties.
[d] $P < 0.05$.

to geographical location is likely to be an important source of error that limits the ability to detect fiber–cancer associations. Analyses of 158 samples of 13 foods collected in different parts of China (Table IV) show surprisingly large geographic coefficients of variation in the composition of five fiber fractions. When variation of this magnitude is coupled with variation in the estimation of individual food intakes, it would indeed be difficult to reliably isolate cancer protective effects of individual fiber fractions.

Embodied within virtually all dietary guidelines of Western nations is the recommendation to increase the intake of dietary fiber (see review of O'Connor and Campbell, 1987). Most of these guidelines have implicitly recommended a higher fiber intake through increased consumption of high-fiber foods, but some have explicitly recom-

TABLE V. Comparison of Mean Colon Cancer Mortality Rates with Dietary Fiber
Intake in China and the United States

	Dietary fiber (g/day)	Male colon cancer rate[a]	Female colon cancer rate[a]
China	33.3	6.6	4.3
United States[b]	11.1	16.9	13.1

[a] Average annual age-standardized mortality rates per 100,000 population; truncated for ages 35–64.
[b] NCI (1987).

TABLE VI. Variation in Levels of Selected Fiber Fractions in 158 Individual Foods Collected from Different Parts of China

Food	Number of samples	Coefficient of variation (%)[a]				
		NDR	ADR	HC	CE	LIG
Chinese cabbage	20	11.1	14.5	12.5	14.1	34.9
Common cabbage	10	11.3	13.5	16.0	11.1	45.7
Carrot	7	19.4	17.0	29.7	15.1	35.1
Chinese celery	7	10.9	13.1	15.6	13.3	23.3
Chinese leak	10	8.7	8.4	20.6	11.5	23.7
Mustard	11	11.2	13.5	9.6	13.1	34.7
Noodles	5	9.2	13.2	8.7	14.5	17.7
Radish	27	20.0	20.5	23.6	17.4	55.1
Radish leaf	6	15.0	10.8	35.0	11.7	33.3
Rape	8	9.7	7.5	22.8	9.3	34.0
Eggplant	18	13.6	14.4	16.2	10.8	38.2
Gourd sponge	8	28.1	30.7	21.9	24.4	52.0
Pumpkin	21	17.9	17.8	30.6	12.1	60.9
Mean		14.3	15.0	20.2	13.7	37.6

[a]Coefficient of variation (mean divided by standard deviation) for within-same-sample runs is approximately 1–2%, based on analyses of several thousand samples. Abbreviations are NDR (neutral detergent fiber), ADR (acid detergent residue), HC (hemicellulose), CE (cellulose), and LIG (lignin).

mended specific amounts of dietary fiber. The Life Sciences Research Office of the Federation of American Societies for Experimental Biology (1987), for example, recommended an intake of 20–35 g/day (in the form of foods), which represents a substantial increase over current levels. However, some concern has been expressed (Food and Chemical News, 1987) that fiber intakes as high as these may compromise mineral status because of the finding that dietary fiber binds minerals. The data from China (Chen et al., 1989) suggest that, at least for iron status among adults, this may not be of concern (Beard et al., 1988). For example, mean hemoglobin levels are positively, not negatively correlated ($r = +0.27$, $P < 0.05$) with dietary fiber intake.

In summary, the evidence from this Chinese survey is consistent with the dietary fiber–colon cancer hypothesis; i.e., colon cancer rates are relatively low and dietary fiber intakes are relatively high when compared with the United States. However, within China, even though consistent inverse associations between colorectal cancer and various fiber–complex carbohydrate intakes were observed, they were weak and generally insignificant. This may be the case because the cancer protective effect of dietary fiber may already be fully expressed. If the low prevalence of large bowel cancer in China is accounted for, in part, by the high intake of dietary fiber, there is no evidence that any one fiber fraction is better than another. No evidence of an adverse effect on mineral status was observed.

ACKNOWLEDGMENTS. Supported in part by NIH Grant 5RO1CA33638-05, China Academy of Preventive Medicine, United Kingdom ICRF, United States FDA, American Association for Cancer Research, and several American industry groups.

REFERENCES

Beard, J. L., Campbell, T. C., and Chen, T., 1988, Iron nutriture in the Cornell–China Diet Cancer Survey, *Am. J. Clin. Nutr.* **47:**771.

Chen, J., Campbell, T. C., Li, J.-Y., and Peto, R., 1989, *A Preliminary Study of Dietary, Lifestyle and Mortality Characteristics of 65 Rural Populations in the Peoples Republic of China,* Oxford University Press, Oxford, Cornell University Press, Ithaca, NY, and Peoples Medical Publishing House, Beijing.

Federation of American Societies for Experimental Biology, 1987, *Physiological Effects and Health Consequences of Dietary Fiber,* Center for Food Safety and Applied Nutrition, Food and Drug Administration, Department of Health and Human Services, Washington, DC.

Food and Chemical News, 1987, Vanderveen says high fiber levels will decrease nutrient absorption, Food and Chemical News, Inc., Washington, DC, July 13, pp. 52–53.

Li, J.-Y., Liu, B.-Q., Li, G.-Y., Chen, Z.-J., Sun, X.-D., and Rong, S.-D., 1981, Atlas of cancer mortality in the People's Republic of China. An aid for cancer control and research, *Int. J. Epidemiol.* **10:**127–133.

National Cancer Institute, Division of Cancer Prevention and Control, 1987, *1987 Annual Cancer Statistics Review,* NIH Pub. No. 88-2789, National Institutes of Health, Washington, D.C.

O'Connor, T. P., and Campbell, T. C., 1987, The influence of nutrition on carcinogenesis, *Nutrition* **3:** 155–162.

36

Rationale for Intervention Trials of Dietary Fiber and Adenomatous Polyps

ROBERT MacLENNAN

1. INTRODUCTION AND SUMMARY

The high rates of colorectal cancer in many affluent countries and the rapid increase that occurs in migrants to these countries from countries where rates are low suggest the importance in causation of environmental factors, especially dietary. The relevant aspects of diet are not clear despite correlation, migrant, case-control, and cohort studies, and it appears unlikely that simply repeating these same types of studies will elucidate specific dietary risk factors and provide a rational basis for the primary prevention of colorectal cancer. Prevention trials that test putative factors based on epidemiologic and laboratory evidence could help resolve etiological issues as well as test the efficacy of interventions. Such trials are feasible with precursor colorectal adenoma as the outcome variable rather than adenocarcinoma. Diets or supplements that are shown to prevent adenomas would also be expected to prevent adenocarcinomas, but this would require further testing. If possible, prevention trials should be replicated with a variety of dietary interventions. This would be facilitated internationally, where the range of available and traditional food patterns for inclusion in trials is greater than within any single country. The goal of such research is the primary prevention of colorectal cancer.

2. GREENWALD–CULLEN MODEL FOR PREVENTION RESEARCH

Greenwald and Cullen (1985) have classified all cancer control research efforts according to an orderly sequence of five stepwise phases of research process. In applying this model to the primary prevention of colorectal cancer, the results of basic

ROBERT MacLENNAN • Queensland Institute of Medical Research, Brisbane, Australia.

research and epidemiology would first be reviewed (phase I) to formulate testable hypotheses about the results of applying an intervention, e.g., reduction in fat and increased fiber intake. Methods development (phase II) would include finding culturally appropriate means of dietary change and valid and feasible measures of compliance and would be followed by controlled clinical trials (phase III) to test the efficacy of the intervention in limited and well-defined, but not necessarily representative populations.

3. EPIDEMIOLOGIC EVIDENCE FOR FAT, FIBER, AND COLORECTAL CANCER

The increases in colorectal cancer among migrants from China, Japan, Norway, and Poland to the United States have been known for more than 20 years. Similar increases in colon cancer have been found among migrants to Australia from Southern Europe (McMichael et al., 1980). Meats and fats are highly valued foods, and their low intake in countries of birth may be determined not primarily by choice but by lack of availability. Hopkins et al. (1980) found that 68% of Italian migrants to Australia increased their meat consumption at a time when Australians decreased their intake, but the migrants also tended to increase their fat consumption and decrease sugar and starch, vegetables, fruit, salt, and dairy products. The specific aspects of diet associated with risk of colorectal cancer have been sought in numerous epidemiologic studies.

3.1. Fat

Although there is a large literature suggesting that dietary fat might be causally related to colorectal cancer, the evidence for a causal association is inconsistent, and much is of low validity (MacLennan, 1985). Many of the positive associations with colorectal cancer have been found in international correlation studies in which total national food disappearance data are correlated with national cancer incidence or mortality. Case-control studies have yielded inconsistent findings.

Several sources of error in case-control studies, such as poor recall of past diet and disease-produced changes in diet, are avoided in cohort studies in which the diet of well individuals is first documented and their subsequent cancer incidence is determined. In Hawaii after 15 years of follow-up, Stemmermann et al. (1984) found a statistically significant negative association between colon cancer and fat intake. Kinlen (1982) found that colorectal cancer mortality was not significantly lower in nuns who ate little or no meat than in the general population and concluded that there are etiologically important factors in diet other than fat or meat intake.

There is experimental evidence that fat may act as a promoter (Bull et al., 1979). The effect of fat is attributed to increased secretion of bile leading to greater concentrations of bile acids as substrates for bacterial degradation to promoters in the large bowel. Newmark et al. (1984) have proposed that fatty acids from cellular and microbial debris in the colon and free bile acids in the nonionized form are highly irritating, with cell loss, increased cellular proliferation, and thus promotion.

3.2. Fiber

Populations with low colorectal cancer risk generally have high intakes of dietary fiber. Despite widely publicized claims for a protective effect, there have been few specific studies of the relation of dietary fiber to colorectal cancer in man. The concept of dietary fiber (in contrast to crude fiber) is recent, methods for measurement are still under development, and values for specific fractions of total dietary fiber, which have different biological effects, are not yet published in food tables. The limited data available from population studies suggest that unrefined cereal foods are associated with lower risk of colorectal cancer in Denmark and Finland (MacLennan et al., 1978; Jensen et al., 1982). An inverse correlation in England and Wales of colon cancer mortality with the pentose fraction of wheat bran fiber (Bingham et al., 1979) was not found on reanalysis using a more accurate method of fiber analysis (Bingham et al., 1985).

The possibility that the risk of a low-fiber diet is conditional on a high fat intake, i.e., that fiber intake is relevant to colorectal carcinogenesis only with high fat intake, is consistent with the Nordic data and with the apparently low colorectal cancer rates in some Asian populations with little fiber intake on a polished rice staple diet.

3.3. Why Is the Epidemiologic Evidence Weak?

Nonexperimental epidemiologic studies are inherently difficult and subject to error. They may gain strength collectively and taken as a whole. The successes in relation to smoking and lung cancer have not yet been repeated in relation to diet and colorectal cancer.

3.3.1. Dose–Response Curve

As an explanation for the inconsistent findings in epidemiologic studies concerning dietary fat as a risk factor, McMichael and Potter (1985) proposed a dose–response model for fat intake and colorectal cancer with a steep linear rise in incidence up to around 20% of total energy as fat, with the slope then reducing so that large increases in intake would lead to only small increases in incidence. They argue that an association with diet could be missed with a low dose–response slope unless there were a large dietary heterogeneity, e.g., by socioeconomic status as in Hong Kong (Hill et al., 1979). If the dose–response model applies to fat and colorectal cancer, the implications for prevention may be that very large reductions in fat intake would be required to prevent colorectal cancer in those genetically susceptible. This caveat may not apply to postulated protective factors such as dietary fiber, which may reduce colorectal cancer risk despite a high fat intake. Data from IARC studies in the Nordic countries (MacLennan et al., 1978; Jensen et al., 1982) are consistent with this possibility, since despite similar high fat intakes, colon cancer incidence varies threefold in urban and rural populations.

3.3.2. Dietary Measurement Error

Case-control studies may have yielded weak, inconsistent associations because dietary methodology is not precise and associations are likely to be detected only where there are large contrasts in diet, as in groups in transition such as migrants to Hawaii or rural to urban migrants in Greece or where there are large risks from diet.

3.3.3. Relative Dietary and Genetic Variation

Another explanation for the lack of clear dietary relationships with colorectal cancer may be the effect of study design on what may be attributable to genetic and dietary factors. In international comparisons of populations, if diet is considerably more variable between populations than is their average genetic susceptibility, it is to be expected that variation in colorectal cancer will be attributed mainly to diet. In studies (e.g., case-control) of individuals within a population, diet is likely to be more homogeneous than internationally, whereas genetic variabity in individual susceptibility is likely to be great. This may explain the inconsistent findings and small relative risks of diet in case-control studies.

Data on colon cancer in spouses give some support to this concept. Since colorectal cancer mortality increases with length of residence among migrants who came to Australia as adults, it is inferred that diet during adult life can increase the risk of colorectal cancer. Jensen *et al.* (1980) have shown in Sweden that the spouses of persons who died from colorectal cancer are not at increased risk of colorectal cancer. Although we know relatively little about the diet of spouses, there is evidence that it is similar qualitatively (Lee and Kolonel, 1982).

4. POTENTIAL FOR DIETARY PREVENTION OF CANCER

Predictions by Doll and Peto (1981) that future research may show between 10% and 70% of cancer to be attributable to diet are well known. Less well publicized is that Peto (1985) does not include diet among reliably established means of preventing deaths from cancer. The potential for future discoveries in cancer prevention may thus be greatest for diet. Wahrendorf (1987a) has estimated, mainly on the basis of case-control studies with published dose–response estimates, how much colorectal cancer might be prevented through dietary changes in fat, fiber, and cabbage. If 50% of persons who currently have dietary habits associated with an increased risk of colorectal cancer were to change their diet to the next category of intake associated in general with a lower risk, the preventable proportion is estimated to be around 10% or 20%. Wahrendorf's main concern was with a statistical methodology and not with an etiological evaluation of the dietary factors that remain uncertain.

5. RANDOMIZED INTERVENTION TRIALS

Given the complexity of diet, the problems of accurate measurement, and the inconsistency of nonexperimental epidemiologic evidence, it appears unlikely that more of the latter will resolve etiological questions. An experimental approach is needed in human studies. Randomized phase III trials have a powerful potential not only for prevention research but at the same time for etiological research. The prospects for intervention trials for the prevention of colorectal cancer have been reviewed by Wahren-dorf (1987b).

Randomized trials are difficult and expensive, and appropriate strategies are needed to maximize their efficiency, including the use of factorial designs, appropriate interventions and end points, and suitable geographical areas and populations. Factorial designs would allow dietary fiber and other factors to be tested in the same trial. They have great efficiency since all of the data are used for estimating the effect of each intervention (Byar, 1984). Interactions can also be explored, but much larger sample sizes are needed for high statistical power. The types of dietary fiber that could feasibly be tested in intervention trials might be greater internationally than within a single country, since potential food sources should be readily available and the trial's food plans compatible with the food culture of subjects.

5.1. End Points

The choice of end points reflects, to some degree, a compromise between validity and relevance and feasibility.

5.1.1. Colorectal Adenocarcinoma

Despite the ultimate in relevance, it is not practical to do randomized intervention trials with adenocarcinoma as the end point because of the long time scale, the relatively low incidence of adenocarcinoma, and the consequent very large sample sizes required. Although sample size would be reduced by restriction to high-risk individuals, this would be unlikely to increase feasibility because of the greater difficulty in recruiting high-risk subjects.

5.1.2. Adenomatous Polyps

It has been proposed and is commonly accepted that most large bowel cancer arises in adenomatous polyps, especially those that are large, have a villous growth pattern, or show severe dysplasia (Morson, 1974). Persons with a history of adenomas have been shown to be at higher risk for developing future adenomas than those without such a history (Brahme et al., 1974). A three-stage model proposed by Hill et al. (1978) postulates different dietary factors in the progression from normal cell to small adenoma from those in further progression to large adenoma and to carcinoma.

Adenomas have a relatively high prevalence in the population and are now routinely removed from the entire large bowel through flexible fiberoptic colonoscopy. It is feasible to use adenomatous polyps as end points because clinical surveillance of patients includes routine colonoscopy. Such surveillance colonoscopy has shown a cumulative incidence of approximately 40% in the first 24 months. This incidence has been observed in patients with confirmed adenomas initially and where the colonoscopist was confident that the colon was clean of visible polyps at the conclusion of the colonoscopy. In addition to the number and size of polyps at surveillance colonoscopy, the degree of dysplasia could also be used as an end point.

Incident lesions are considered to be the result of growth of already initiated cells or previously very small and colonoscopically not detectable adenomas. They may thus be a measure of promotion in the colorectal microenvironment.

5.1.3. Intestinal Cell Proliferation

Early short-term end points such as measures of intestinal cell proliferation (Lipkin and Newmark, 1985) could be used to select the particular type of intervention to be used, or they could be interim end points. However, their relationship to the ultimate risk of adenocarcinoma may be less relevant than that of adenomatous polyps. It could be argued that increased proliferation is a marker of initiation and possibly of the transition from normal cells to small adenomas.

5.2. Purified Fibers or Source Foods

Fiber might reduce the putative promoting effects of fat and associated bile acids by several mechanisms summarized by Kritchevsky (1986), including dilution of colonic contents with increased fecal bulk, shorter transit time adsorption of bile acids, lowering of colonic pH from production of short-chain fatty acids, inhibition of dehydroxylation of bile acids, and regulation of energy intake. Different types of fiber have different physiological effects and would be expected to have different effects on colorectal cancer risk. Intervention trials of candidate fiber species, especially with factorial designs, might be done using intestinal cell proliferation as an end point. An alternative would be to include foods that are major sources. If as suggested by Kritchevsky (1982) dietary patterns may be ultimately more important than dietary components, then the practicality of including such dietary patterns in clinical trials will depend on the food ethnology of populations including availability of foods, cuisine, and perceptions of what is an acceptable diet.

5.3. The Australian Polyp Prevention Trial

The Queensland Institute of Medical Research is coordinating a collaborative multicenter clinical trial in leading clinical units in Brisbane, Sydney, and Melbourne to prevent colorectal adenomatous polyps. The effects of reduced fat intake, increased dietary fiber, and a supplement of β-carotene on the incidence and size of adenomas after 2 years of follow-up are being measured. A $2 \times 2 \times 2$ randomized factorial design is used with eight groups—half the subjects receive each treatment: reduction in fat to at least 30% of total energy and to 25% if feasible; 25 g additional wheat bran daily; and 20 mg β-carotene (or placebo capsule) daily. Only major colonoscopy centers with experienced colonoscopists in Brisbane, Sydney, and Melbourne are participating. This has limited the recruitment rate to the project but has maintained quality of colonoscopy. Eligibility of subjects is strictly defined. Counseling by study dietitians is used to initiate and continue dietary change. Persons with precancerous lesions are highly motivated, and the food plans do not require major changes in lifestyle.

Recruitment of suitable patients has been limited by both precolonoscopy eligibility criteria and postcolonoscopy criteria, which include confirmed histological diagnosis. Overall, 2542 polyp patients were registered in project clinics during the period October

1985 to December 1987; 1208 were potentially eligible for entry at the time of colonoscopy, and 1334 were ineligible (138 not literate in English, 308 above the age limit, 440 not living in the corresponding metropolitan area, 77 already on a special diet, 155 with cancer in the previous 5 years, 133 with other bowel conditions, 70 with other medical conditions, and 13 refusals). Of the 1208 potentially eligible, 537 were definitely eligible on the basis of histological confirmation of at least one adenoma and confidence by the colonoscopist of a "clean" colon (where there was no fecal residue, all areas were clearly seen, and all polyps were totally removed): 409 (76%) have been recruited (Brisbane 117, Melbourne 154, Sydney 138) with 128 refusals. Of the remaining 671 who were potentially eligible, the reasons for nonrecruitment were: 75 with no specimen taken, 261 where an adenoma was not confirmed at histology, 82 with other bowel disease, 26 with another medical condition precluding participation, 81 with cancer or malignant polyps, and 146 for other miscellaneous reasons. The above numbers illustrate the large amount of documentation and screening needed before patients are recruited into a trial.

Objective measures of compliance are important in such trials. Serum cholesterol is being measured, but its utility as a measure of compliance with a reduced fat intake is uncertain. Fat intake in the diet is monitored by the counseling dietitians, but because of the potential for biased reporting 4-day food records are collected every 6 months by research nurses who act independently of the dietitians.

There are several other such trials now in progress in North America. Our experience is that patients are very highly motivated, and there have been very few who discontinue after entry.

REFERENCES

Bingham, S. A., Williams, D. R. R., Cole, T. J., and James, W. P. T., 1979, Dietary fibre and regional large bowel cancer mortality in Britain, *Br. J. Cancer* **40**:456–463.

Bingham, S. A., Williams, D. R. R., and Cummings, J. H., 1985, Dietary fibre consumption in Britain: New estimates and their relation to large bowel cancer mortality, *Br. J. Cancer* **52**:399–402.

Brahme, F., Ekelund, G. R., Norden, J. G., and Wenekert, A., 1974, Metachronous colorectal polyps comparison of development of colorectal polyps in persons with and without histories of polyps, *Dis. Colon Rectum* **17**:166–171.

Bull, A. W., Soullier, P. S., Wilson, P. S., Hayden, M. T., and Nigro, N. D., 1979, Promotion of azoxymethane induced intestinal cancer in high-fat diet in rats, *Cancer Res.* **39**:4956–4959.

Byar, D. P., 1984, Sample size considerations for prevention studies, in: *Chemoprevention Clinical Trials. Problems and Solutions* (M. A. Sestili, ed.), United States Department of Health and Human Services, Washington, DC, pp. 14–20.

Doll, R., and Peto, R., 1981, The causes of cancer, *J. Natl. Cancer Inst.* **66**:1191–1308.

Greenwald, P., and Cullen, J. W., 1985, The new emphasis in cancer control, *J. Natl. Cancer Inst.* **74**: 543–551.

Hill, M. J., Morson, B. C., and Bussey, H. J. R., 1978, Aetiology of adenoma–carcinoma sequence in the large bowel, *Lancet* **1**:245–247.

Hill, M., MacLennan, R., and Newcombe, K., 1979, Diet and large-bowel cancer in three socioeconomic groups in Hong Kong, *Lancet* **1**:436.

Hopkins, S., Margetts, B. M., Cohen, J., and Armstrong, B. K., 1980, Dietary change among Italians and Australians in Perth, *Commun. Health Studies* **4**:67–75.

Jensen, O. M., Bolander, A. M., Sigtryggsson, P., Vercelli, M., Nguyen-Dinh, H., and MacLennan,

R., 1980, Large-bowel cancer in married couples in Sweden. A follow-up study, *Lancet* **1**:1161–1163.

Jensen, O. M., MacLennan, R., and Wahrendorf, J., 1982, Diet, bowel function, fecal characteristics and large bowel cancer in Denmark and Finland, *Nutr. Cancer* **4**:5–19.

Kinlen, L. J., 1982, Meat and fat consumption and cancer mortality: A study of strict religious orders in Britain, *Lancet* **1**:946–949.

Kritchevsky, D., 1982, Can dietary change prevent disease? in: *Colon and Nutrition, Proceedings of the 32nd Falk Symposium* (H. Kasper and H. Goebell, eds.), MTP Press, Boston, pp. 269–278.

Kritchevsky, D., 1986, Diet, nutrition, and cancer. The role of fiber, *Cancer* **58**:1830–1836.

Lee, J., and Kolonel, L., 1982, Nutrient intakes of husbands and wives: Implications for epidemiologic research, *Am. J. Epidemiol.* **115**:515–525.

Lipkin, M., and Newmark, H., 1985, Effect of added dietary calcium on colonic epithelial-cell proliferation in subjects at high risk for familial colonic cancer, *N. Engl. J. Med.* **313**:1381–1384.

MacLennan, R., 1985, Fat intake and cancer of the gastrointestinal tract and prostate, *Med. Oncol. Tumor Pharmacother.* **2**:137–142.

MacLennan, R., Jensen, O. M., Mosbech, J., and Vuori, H., 1978, Diet, transit time, stool weight, and colon cancer in two Scandinavian populations, *Am. J. Clin. Nutr.* **31**:S239–242.

McMichael, A. J., and Potter, J., 1985, Diet and colon cancer: Integration of the descriptive, analytic, and metabolic epidemiology, *Natl. Cancer Inst. Monogr.* **69**:223–228.

McMichael, A. J., McCall, M. G., Hartshorne, J. M., and Woodings, T. L., 1980, Patterns of gastrointestinal cancer in European migrants to Australia. The role of dietary change, *Int. J. Cancer* **25**:431–437.

Morson, B. C., 1974, The polyp–cancer sequence in the large bowel, *Proc. R. Soc. Med.* **67**:451–457.

Newmark, H. L., Wargovich, M. J., and Bruce, W. R., 1984, Colon cancer and dietary fat, phosphate and calcium: A hypothesis, *J. Natl. Cancer Inst.* **72**:1323–1325.

Peto, R., 1985, The preventability of cancer, in: *Cancer Risks and Prevention* (M. P. Vessey and M. Gray, eds.), Oxford University Press, Oxford, pp. 1–14.

Stemmermann, G. N., Nomura, A. M. Y., and Heilbrun, L. K., 1984, Dietary fat and the risk of colorectal cancer, *Cancer Res.* **44**:4633–4637.

Wahrendorf, J., 1987a, An estimate of the proportion of colo-rectal and stomach cancers which might be prevented by certain changes in dietary habits, *Int. J. Cancer* **40**:625–628.

Wahrendorf, J., 1987b, Prevention of colorectal cancer: Prospects of intervention trials, in: *Causation and Prevention of Colorectal Cancer*, (J. Faivre and M. J. Hill, eds.), Excerpta Medica, Amsterdam, pp. 155–160.

37

Future Research Directions, Including Clinical Trials

PETER GREENWALD and ELAINE LANZA

1. RISK FACTORS FOR CANCER

Approximately 850,000 new cases of cancer are diagnosed in the United States each year, with 450,000 annual deaths attributed to this disease. For several decades, it has become increasingly apparent that most cancer is caused or promoted by lifestyle and environmental factors and that only a fraction of cancer is completely genetic in origin. Some genetic or susceptibility factors must be involved because most people similarly exposed to environmental factors do not develop cancer. Today, however, research data support the estimate that lifestyle and environmental factors are related to the development of roughly 80% of cancer incidence and, therefore, support the conclusion that most cancer is theoretically preventable. The prevention component of cancer control refers to the lowering of cancer incidence and, thus, cancer mortality by changes in those lifestyle and environmental factors that influence the occurrence and progression of cancer.

2. TYPES OF RISK FACTORS

Three types of factors, individually or in combination, increase an individual's risk of developing cancer, i.e., lifestyle, environmental, and genetic factors. For convenience, lifestyle and environment are treated as separate factors, although lifestyle may be considered a subset of environment. Lifestyle factors are behaviors over which the individual has some control, especially tobacco use, diet, alcohol, excessive exposure to sunlight, sexual behavior patterns, and general personal hygiene. Environmental factors include both occupational exposure to carcinogens and exposure to carcinogens and radiation in medical procedures as well as factors that are naturally occurring or man-

PETER GREENWALD • Division of Cancer Prevention and Control, National Cancer Institute, National Institutes of Health, Bethesda, Maryland 20892.   ELAINE LANZA • Cancer Prevention Research Program, Division of Cancer Prevention and Control, National Cancer Institute, National Institutes of Health, Bethesda, Maryland 20892.

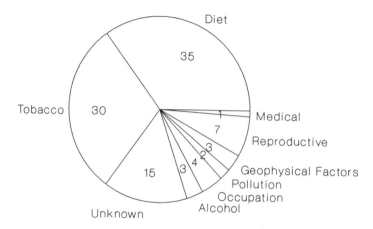

FIGURE 1. Cancer risk factors. Estimates from Doll and Peto (1981).

made causes of cancer that contaminate water, air, and earth. These factors are largely beyond an individual's control and thus require broad social actions or system changes to achieve effective control. Genetic factors are conditions inherited at conception. Control of these factors is at present largely not feasible for cancer except through genetic counseling activities.

Research shows that the largest cancer risk factors are the ones we can control (Fig. 1). Dietary factors are significant in cancer occurrence, possibly as significant as tobacco use. The consensus is that 35% or more of cancer mortality could be related to dietary factors (Doll and Peto, 1981). This estimate is based on a large number of studies, although uncertainty surrounds the exact magnitude of the association and the biological mechanisms involved.

3. RESEARCH FUNDING

The NIH as a whole spends roughly $200 million on nutrition research with the three lead institutes—the National Cancer Institute, the National Heart, Lung and Blood Institute, and the National Institute of Diabetes, Digestive and Kidney Disease–accounting for approximately three quarters of the total. Right now, NCI spends about $52 million per year on research on diet and cancer, which is a huge increase over the past several years. The reason that the NCI has increased spending in this area is that evidence in the aggregate demonstrates that diets have a very important influence on cancer rates.

Of all the research on nutrition, only a small part is devoted to dietary fiber research (Table I). The NCI still takes the lead in funding dietary fiber research. To date the funding emphasis is on developing analytical methods for dietary fiber and fiber component analyses and compiling data for a fiber data base and on etiological studies involving animal models. Fewer projects on dietary fiber involving human subjects are currently being funded. These latter studies are examining the differences between soluble and insoluble fibers and cereal and vegetable fibers.

TABLE I. NIH-Funded Dietary Fiber
Projects: FY 87 Awards

Institute[a]	Dollars awarded	Number of projects
NCI	$4,407,474	18
NHLBI	1,860,973	5
NIDDK	625,775	6
DRR	325,110	26
NIAID	85,804	1
Total	$7,295,136	56

[a] NCI, National Cancer Institute; NHLBI, National Heart Lung
and Blood Institute; NIDDK, National Institute of Diabetes,
Digestive and Kidney Diseases; DRR, Division of Research
Resources; NIAID, National Institute of Allergy and Infectious
Diseases.

4. CANCER PREVENTION

At the Division of Cancer Prevention and Control the major objective is the reduction in cancer incidence, mortality, and morbidity and increases in cancer survival rates. The aim is to reduce these rates through thoroughly validated intervention methods. In order to give cancer control its required objectivity and to help research proceed more effectively, the NCI has developed a strategic decision-making model for all cancer control research. Research is thus classified according to an orderly sequence of five stepwise phases of the overall research process (Fig. 2): (1) hypothesis development, (2) method development, (3) controlled intervention trials, (4) deferred population studies, and (5) demonstration and implementation trials. It should be made very clear that the phase I studies, hypothesis development, are not hypotheses for understanding etiology but rather on a hypothesis for prevention. In prevention the hypothesis is that you intervene or change a variable in a way that lowers cancer incidence. Today there are many research findings from laboratory and epidemiologic studies on dietary fiber and colon cancer, and the evidence now appears compelling enough to move into cancer control phases, yet this is not occurring. What are the barriers that need to be overcome to move into cancer control research, especially phase III, randomized controlled clinical trials.

There appear to be a continual number of studies that try to define hypotheses. For example, the results from an etiological study result in a further etiological study rather than following through into human research. Scientists see one interesting avenue, then another, and another and never quite get to a point where they're ready to do human intervention research. Studies are thus needed that will facilitate human intervention research. Methods development studies in a couple of different categories are needed. One need is for markers. Markers of dietary intake are needed for human clinical trials or in population studies to really know what people are eating. These markers of dietary compliance will greatly increase the validity of human studies. A second set of markers might be markers of cancer risk or markers that correlate so strongly with cancer that one can use them as end points in clinical trials. These markers are often referred to as

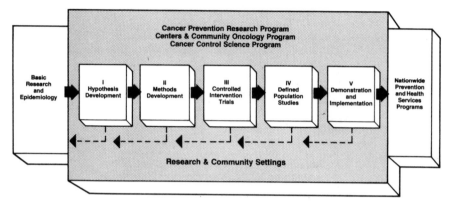

FIGURE 2. Cancer control phases.

intermediate end points. The choice of a precursor or precancerous lesion as the end point for a clinical intervention study should be based on the strength of its association with a specific cancer, the ease with which the end point can be quantitatively evaluated, and the prevalence of this lesion in a study population.

Dietary intervention studies in colon cancer have successfully employed protocols using a number of different hypothesized precursor lesions as the clinical end point of interest. For example, the occurrence of rectal polyps in patients with familial polyposis was reduced by the administration of vitamin C (DeCosse et al., 1975). Over a period of 3 to 13 months, rectal polyps were found to disappear or regress in five of the eight treated patients. A later randomized trial of ascorbic acid intervention in 49 patients confirmed these results (Bussey et al., 1982). Also evaluated in a study of colon cancer has been fecal mutagenicity, possibly associated with colon cancer. Fecal extracts contain mutagenic substances, as assessed by the Ames Salmonella test. The occurrence and prevalence of such compounds may be influenced by diet, and, therefore, the exposure of colonic epithelium to mutagenic substances may be reduced. Supplemental ascorbic acid and α-tocopherol (vitamin E) were used by Dion and colleagues to successfully reduce mutagenic substances in 20 healthy subjects (Dion et al., 1982).

Other clinical work related to colon cancer concerns the observation that Seventh-Day Adventist vegetarians have a significantly lower incidence of this cancer than the general population (Phillips, 1975). Research by Lipkin and colleagues (1985) indicates that the proliferation of colonic mucosal epithelial cells is markedly less in these vegetarians compared with a control group. Further, Lipkin and Newmark (1985) found that calcium carbonate supplementation for 2 to 3 months reduced epithelial cell proliferation in persons at high risk for familial colon cancer. Although the association described is of great interest, equally important is the development of a sensitive assay usable in large populations to measure the proliferation of colonic epithelial cells.

Although there are many possibilities for markers, few of them have really been developed—in fact, none to the point where they're validated as true predictors of cancer. It should be emphasized that the use of precursor lesions as predictors of future cancer incidence has not been experimentally validated in large-scale clinical trials. This important step must be taken before we can be fully confident of any marker end point in

TABLE II. Research Needs for Dietary Fiber and Colon Cancer

1. What are the normal mechanisms in the development, cell differentiation, and hormonal regulation of the colon?
2. Can animal carcinogenesis models be made more relevant to human cancer? Can animal models be used to study dietary interaction or modulation of potential intermediate end points for colon cancer?
3. What types of dietary fiber (cereal, vegetable, cellulose, pectin, hemicellulose, lignin) are most important in reducing colon cancer risk? What is the optimal relationship between fiber and other dietary constituents in reducing risk in humans?
4. What is the level of fiber intake in the United States? Is it changing? Do the changes correlate with changes in colon cancer rates?
5. Are there intermediate end points that can be used to predict colon cancer risk? Are these end points modulated by dietary fiber? By diet?
6. Is colonic polyp recurrence related to diet? Does dietary fiber prevent the recurrence of colonic polyps?

a clinical trial. If these markers existed, however, then studies on the modulation of these markers (and hence cancer) could be more easily carried out.

Another type of methods development may well be logistical. Colon cancer, although one of the most common cancers, is still rare in terms of the probability of an individual getting this cancer during the next year or the next 5 years. In the United States the current rate for colon cancer is 37.1/100,000 for males and females combined, all races. This rate increases to 52.1 if rectal cancer is included. So the question is, how do you deal with these statistics and develop a controlled clinical trial?

5. FUTURE RESEARCH DIRECTIONS

Table II summarizes five areas of research that are important to dietary fiber and colon cancer. Future research directions must emphasize studies that make the animal carcinogenesis model more relevant to human cancer. Studies should be designed that take into account overall features of diet as well as specific or chemically defined single fibers. The interaction between dietary fiber and other dietary components such as fat and calories should be examined. The question is, how can animal studies help bridge the gap to initiate more human studies on dietary fiber and colon cancer? Animal studies must not only test hypotheses but also develop methodology that makes the human studies more valid and rationale. Research on dietary fiber and colon cancer must advance beyond etiological studies into human interventions.

6. DIETARY RECOMMENDATIONS

At the NCI we think that while research is continuing to explore the relationship between colon cancer and dietary fiber, dietary recommendations to the general public are prudent to make. Table III lists some of the dietary fiber recommendations in North America since 1977 by a number of organizations. These recommendations have generally been made in terms of what is best for an individual to do in the interest of good health rather than disease prevention.

TABLE III. Dietary Fiber Recommendations

Dietary Goals, 1977	Increase consumption of vegetables, fruits, and whole grains
National Academy of Sciences, 1980	Consume each day legumes, fruits, vegetables, cereals, breads
National Academy of Sciences, 1982	Include fruits, vegetables, and whole-grain cereals daily
American Cancer Society, 1984	Eat more high-fiber foods—whole-grain cereals, fruits, vegetables
Canada, 1985	Adult Canadian population should double dietary fiber intake
USDA/HHS, 1985	Eat foods with adequate starch and fiber
American Heart Association, 1985	Include at least three daily servings of fruits and vegetables
National Cancer Institute, 1986	Eat a variety of foods that provide 20 to 30 g/day of fiber
FASEB, 1987	Healthy adults eat whole grains, fruits, vegetables—20 to 35 g/day fiber

A step forward was taken by the Canadian Government Expert Advisory Committee on Dietary Fiber (Expert Advisory Committee, 1985) when they quantitated the fiber recommendation (double the intake of dietary fiber). The reason this is a step forward is that people in communications say that we need a message that the public can get a handle on and interpret. Scientists, on the other hand, want to qualify every message in a way that makes it extremely difficult for the our 13 million illiterate in this country, or 26 million semiilliterate, to understand. So we're always caught between a clear and simple message and one that satisfies the pure scientist completely. The NCI recommendation is to eat a variety of foods that provide 20 to 30 g of fiber a day, roughly doubling our current dietary fiber intake.

REFERENCES

Bussey, H. J. R., DeCosse, J. J., Deschner, E. E., Eyers, A. A., Lesser, M. L., Morson, B. C., Ritchie, S. M., Thomson, J. P. S., and Wadsworth, J., 1982, A randomized trial of ascorbic acid in polyposis coli, Cancer 50:1434–1439.

DeCosse, J. J., Adams, M. B., Kumza, J. F., and Codon, R. E., 1975, Effect of ascorbic acid on rectal polyps of patients with familial polyposis, Surgery 78:608–612.

Dion, P. W., Bright-See, E. B., Smith, C. C., and Bruce, W. R., 1982, The effect of dietary ascorbic acid and alpha-tocopherol on fecal mutagenicity, Mutat. Res. 102:27–37.

Doll, R., and Peto, R., 1981, The causes of cancer: Quantitative estimates of avoidable risk of cancer in the United States today, J. Natl. Cancer Inst. 66:1191–1308.

Expert Advisory Committee on Dietary Fiber, 1985, Report, Minister of Health and Welfare Canada, Ottawa, pp. 1–30.

Lipkin, M., and Newmark, H., 1985, Effect of added dietary calcium on colonic epithelial cell proliferation in subjects at high risk for familial colon cancer, N. Engl. J. Med. 313:1381–1384.

Lipkin, M., Uehara, K., Winawer, S., Sanchez, A., Bauer, C., Phillips, R., Lynch, H. T., Blattner, W. A., and Fraumeni, J. F., 1985, Seventh-Day Adventist vegetarians have a quiescent proliferative activity in colonic mucosa, Cancer Lett. 26:139–144.

Phillips, R., 1975, Role of life-style and dietary habits in risk of cancer among Seventh-Day Adventists, Cancer Res. 35:3513–3522.

Index